Current Topics in Microbiology and Immunology

Volume 358

Current Topics in Microbiology and Immunology

Previously published volumes

Further volumes can be found at www.springer.com

Vol. 328: **van Etten, (Ed.)**:
Lesser Known Large dsDNA Viruses. 2008.
ISBN 978-3-540-68617-0

Vol. 329: **Griffin, Diane E.; Oldstone, Michael B. A. (Eds.)**: Measles. 2009.
ISBN ISBN 978-3-540-70616-8

Vol. 330: **Griffin, Diane E.; Oldstone, Michael B. A. (Eds.)**:
Measles. 2009.
ISBN 978-3-540-70616-8

Vol. 331: **Villiers, E. M. de (Ed.)**:
TT Viruses. 2009.
ISBN 978-3-540-70917-8

Vol. 332: **Karasev A. (Ed.)**:
Plant produced Microbial Vaccines. 2009.
ISBN 978-3-540-70857-5

Vol. 333: **Compans, Richard W.; Orenstein, Walter A. (Eds.)**:
Vaccines for Pandemic Influenza. 2009.
ISBN 978-3-540-92164-6

Vol. 334: **McGavern, Dorian; Dustin, Micheal (Eds.)**:
Visualizing Immunity. 2009.
ISBN 978-3-540-93862-0

Vol. 335: **Levine, Beth; Yoshimori, Tamotsu; Deretic, Vojo (Eds.)**:
Autophagy in Infection and Immunity. 2009.
ISBN 978-3-642-00301-1

Vol. 336: **Kielian, Tammy (Ed.)**:
Toll-like Receptors: Roles in Infection and Neuropathology. 2009.
ISBN 978-3-642-00548-0

Vol. 337: **Sasakawa, Chihiro (Ed.)**:
Molecular Mechanisms of Bacterial Infection via the Gut. 2009.
ISBN 978-3-642-01845-9

Vol. 338: **Rothman, Alan L. (Ed.)**:
Dengue Virus. 2009.
ISBN 978-3-642-02214-2

Vol. 339: **Spearman, Paul; Freed, Eric O. (Eds.)**:
HIV Interactions with Host Cell Proteins. 2009.
ISBN 978-3-642-02174-9

Vol. 340: **Saito, Takashi; Batista, Facundo D. (Eds.)**:
Immunological Synapse. 2010.
ISBN 978-3-642-03857-0

Vol. 341: **Bruserud, Øystein (Ed.)**:
The Chemokine System in Clinical and Experimental Hematology. 2010.
ISBN 978-3-642-12638-3

Vol. 342: **Arvin, Ann M. (Ed.)**:
Varicella-zoster Virus. 2010.
ISBN 978-3-642-12727-4

Vol. 343: **Johnson, John E. (Ed.)**:
Cell Entry by Non-Enveloped Viruses. 2010.
ISBN 978-3-642-13331-2

Vol. 344: **Dranoff, Glenn (Ed.)**:
Cancer Immunology and Immunotherapy. 2011.
ISBN 978-3-642-14135-5

Vol. 345: **Simon, M. Celeste (Ed.)**:
Diverse Effects of Hypoxia on Tumor Progression. 2010.
ISBN 978-3-642-13328-2

Vol. 346: **Christian Rommel; Bart Vanhaesebroeck; Peter K. Vogt (Ed.)**:
Phosphoinositide 3-kinase in Health and Disease. 2010.
ISBN 978-3-642-13662-7

Vol. 347: **Christian Rommel; Bart Vanhaesebroeck; Peter K. Vogt (Ed.)**:
Phosphoinositide 3-kinase in Health and Disease. 2010.
ISBN 978-3-642-14815-6

Vol. 348: **Lyubomir Vassilev; David Fry (Eds.)**:
Small-Molecule Inhibitors of Protein-Protein Interactions. 2011.
ISBN 978-3-642-17082-9

Vol. 349: **Michael Kartin (Eds.)**:
NF-kB in Health and Disease. 2011.
ISBN 978-3-642-16017-2

Vol. 350: **Rafi Ahmed; Tasuku Honjo (Eds.)**:
Negative Co-receptors and Ligands. 2011.
ISBN 978-3-642-19545-7

Vol. 351: **Marcel, B. M. Teunissen (Ed.)**:
Intradermal Immunization. 2011.
ISBN 978-3-642-23689-1

Vol. 352: **Rudolf Valenta; Robert L. Coffman (Eds.)**
Vaccines against Allergies. 2011.
ISBN 978-3-642-20054-0

Vol. 353: **Charles E. Samuel (Ed.)**:
Adenosine Deaminases Acting on RNA (ADARs) and A-to-I Editing. 2011.
ISBN 978-3-642-22800-1

Vol. 354: **Pamela A. Kozlowski (Ed.)**:
Mucosal Vaccines. 2012.
ISBN 978-3-642-23692-1

Vol. 355: **Ingo K. Mellinghoff; Charles L. Sawyers (Eds.)**:
Therapeutic Kinase Inhibitors. 2012.
ISBN 978-3-642-28295-9

Vol. 356: **Cornelis Murre (Ed.)**:
Epigenetic Regulation of Lymphocyte Development. 2012.
ISBN 978-3-642-24102-4

Vol. 357: **Nicholas J. Mantis (Ed.)**:
Ricin and Shiga Toxins. 2012.
ISBN 978-3-642-27469-5

Ulrich Dobrindt · Jörg H. Hacker
Catharina Svanborg
Editors

Between Pathogenicity and Commensalism

Responsible Series Editor: Fritz Melchers

Springer

Editors
Ulrich Dobrindt
Institut für Hygiene
Universitätsklinikum Münster
Münster
Germany

Jörg H. Hacker
Nationale Akademie der Wissenschaften
Deutsche Akademie der Naturforscher
Leopoldina
Halle
Germany

Catharina Svanborg
Section of Microbiology, Immunology
and Glycobiology
Lund University
Lund
Sweden

ISSN 0070-217X
ISBN 978-3-642-36559-1 ISBN 978-3-642-36560-7 (eBook)
DOI 10.1007/978-3-642-36560-7
Springer Heidelberg New York Dordrecht London

Library of Congress Control Number: 2013933115

Printed on acid-free paper

Springer is part of Springer Science+Business Media (www.springer.com)

Preface

Microbes colonize nearly every biotic and abiotic niche on our planet. This includes also our human body which is densely populated with microbes, the majority of which interacts with us in a commensal, sometimes even mutualistic, relationship. Only a minority of our microbiota represents pathogenic organisms with the ability to cause infection.

Traditionally, the microbiota colonizing a niche or the entire body is usually distinguished into non-pathogenic and pathogenic organisms based on their potential to cause disease. Based on Jakob Henle's assumption that microorganisms are the causative agents of infectious diseases (Henle 1840), his student Robert Koch and Friedrich Loeffler formulated already in 1884 four criteria, to establish a causal relationship between a causative microbe and a disease (Koch 1884; Loeffler 1884). According to the so-called Henle–Koch postulates, bacteria isolated from disease cases have been compared with isolates from healthy individuals with regard to the presence of virulence-associated traits that contribute to the establishment of an infection in healthy hosts and that distinguish pathogenic from non-pathogenic variants.

The term commensalism has been introduced in the second half of the nineteenth century by Pierre-Joseph van Beneden, defining a relation between individuals of two species where one species obtains benefits from the other without harming or benefiting the latter. The commensal, which benefits from the association, may obtain nutrients, shelter, or locomotion from the host species, which is substantially unaffected (Boucher 1985).

Whereas the Henle–Koch postulates and the definition of commensalism allow(ed) to categorize obligate pathogens, this is often difficult for opportunistic pathogens which may belong to the normal microbiota of healthy individuals. Successful infection by such facultative pathogens does often not only rely on virulence-associated bacterial traits, but also requires susceptible hosts with an impaired mucosal barrier and/or immune response defects. Identifying individual host susceptibility factors as well as the characterization of compositional changes of the microbiota, their gene expression and metabolic profiles in patient subsets will thus be essential for an improved treatment of infectious diseases in the future.

The vast majority of microbes colonizing a healthy host, can probably be considered commensals, although for most of them we still lack information regarding the nature of their relationship with the host (whether they hurt or help…). Estimated 500–1000 bacterial species belong to the resident human intestinal flora and the same holds true for the microbial consortium colonizing the skin (Grice et al. 2009; Sears 2005). The microorganisms living inside or on "us" outnumber "our" somatic and germ cells by a factor of ten (Turnbaugh et al. 2007), but so far only a minor fraction of this microbiota has been characterized and identified (Marcy et al. 2007). The human microbiome project will help us to further increase our knowledge on the microbial diversity associated with our healthy body and the contribution of microbiota to disease and infection. Although we already know that, e.g., the individual composition of the microbial flora differs from person to person, and that different diseases are associated with deviations in the composition of the microbial community in the diseased niche, we are far from understanding the commensals' exact role for human development, physiology, protection against opportunistic pathogens, immunity, and nutrition.

This book covers various aspects of the interplay between commensal or pathogenic bacteria with their hosts. The chapters summarize the recent knowledge on geno- and phenotypic traits of opportunistic bacterial pathogens, such as *Escherichia coli*, staphylococci or *Pseudomonas aeruginosa*, as well as the impact of commensal and probiotic bacteria on intestinal physiology and health. The differential interaction of pathogenic, commensal, or probiotic bacteria with their host is reviewed from the bacterial and from the host perspective to complete this compilation of articles on differences and similarities of pathogenic and commensal microorganisms.

Ulrich Dobrindt
Jörg H. Hacker
Catharina Svanborg

References

Boucher DH (1985) The idea of mutualism, past and future. In: Boucher DH (ed) The biology of mutualism: ecology and evolution, Oxford University Press, New York, pp. 1–28

Grice EA et al (2009) Topographical and temporal diversity of the human skin microbiome. Science 324:1190–1192

Henle J (1840) Von den Contagien und Miasmen und den contagiös-miasmatischen Krankheiten. Berlin

Koch R (1884) Die Aetiologie der Tuberkulose. Mittheilungen aus dem kaiserlichen Gesundheitsamte 2:1–81

Loeffler F (1884) Untersuchung über die Bedeutung der Mikroorganismen für die Entstehung der Diphtherie beim Menschen, bei der Taube und beim Kalbe. Mittheilungen aus dem kaiserlichen Gesundheitsamte 2:421–499

Marcy Y et al (2007). Dissecting biological "dark matter" with single-cell genetic analysis of rare and uncultivated TM7 microbes from the human mouth. Proc Natl Acad Sci USA 104:11889–11894

Sears CL (2005) A dynamic partnership: celebrating our gut flora. Anaerobe 11:247–251

Turnbaugh PJ, Ley RE, Hamady M, Fraser-Liggett CM, Knight R, Gordon JI (2007) The human microbiome project. Nature 449:804–810

Contents

Part II Host–Microbe Interaction in the Intestinal Tract

Contributors

Ingo B. Autenrieth Institute for Medical Microbiology and Hygiene, University Hospital of Tübingen, Elfriede-Aulhorn-Str. 6, 72076 Tübingen, Germany, e-mail: ingo.autenrieth@med.uni-tuebingen.de

Peter van Baarlen Host Microbe Interactomics Group, Wageningen University, P.O. box 338, 6700 AH Wageningen, The Netherlands, e-mail: peter.vanbaarlen@wur.nl

Michael Blaut Department of Gastrointestinal Microbiology, German Institute of Human Nutrition Potsdam-Rehbruecke (DIfE), Arthur-Scheunert-Allee 114-116, 14558 Nuthetal, Germany, e-mail: blaut@dife.de

Jing Cheng Department of Veterinary Biosciences, University of Helsinki, P.O. Box 66, 00014 Helsinki, Finland, e-mail: jing.cheng@helsinki.fi

Willem M. de Vos Department of Veterinary Biosciences and Department of Bacteriology and Immunology, University of Helsinki, Helsinki, Finland; Laboratory of Microbiology, Wageningen University, Dreijenplein 10, 6703 HB Wageningen, The Netherlands, e-mail: willem.devos@wur.nl

Ulrich Dobrindt Institute of Hygiene, University Hospital of Münster, Robert-Koch-Straße 41, 48149 Münster, Germany, e-mail: dobrindt@uni-muenster.de

Julia-Stefanie Frick Institute for Medical Microbiology and Hygiene, University Hospital of Tübingen, Elfriede-Aulhorn-Str. 6, 72076 Tübingen, Germany, e-mail: julia-stefanie.frick@med.uni-tuebingen.de

Friedrich Götz Microbial Genetics, Institute of Microbiology and Infection Medicine Tübingen, University of Tübingen, Waldhäuser Str. 70/8, 72076 Tübingen, Germany, e-mail: friedrich.goetz@uni-tuebingen.de

Jörg Hacker National German Academy of Sciences Leopoldina Emil-Abderhalden-Straße 37, 06108 Halle/Saale, Germany, e-mail: joerg.hacker@leopoldina. de

Jürgen Heesemann Max von Pettenkofer-Institut für Hygiene and Medical Microbiology, Ludwig-Maximilians-University Munich, Pettenkoferstraße 9a, 80336 Munich, Germany, e-mail: heesemann@mvp.uni-muenchen.de

Brian Henderson Department of Microbial Diseases, UCL-Eastman Dental Institute, University College London, 256 Gray's Inn Road, London WC1X 8LD, UK, e-mail: Brian.Henderson@ucl.ac.uk

Michael Hogardt Department of Infectiology, Bavarian Health and Food Safety Authority, Veterinärstraße 2, 85764 Oberschleissheim, Germany, e-mail: Michael.Hogardt@lgl.bayern.de

Amparo Latorre Institut Cavanilles de Biodiversitat i Biologia Evolutiva, Universitat de València, Apartado Postal 22085, 46071 Valencia, Spain, e-mail: amparo.latorre@uv.es

Andreas Leimbach Institute of Hygiene, University Hospital of Münster, Robert-Koch-Straße 41, 48149 Münster, Germany; Göttingen Genomics Laboratory, University of Göttingen, Grisebachstraße 8, 37077 Göttingen, Germany, e-mail: andreas.leimbach@ukmuenster.de

Andrew C. R. Martin Institute of Structural and Molecular Biology, Division of Biosciences, University College London, Darwin Building, Gower Street, WC1E 6BT London, UK, e-mail: andrew.martin@ucl.ac.uk

Andrés Moya Área de Genómica y Salud, Centro Superior de Investigación en Salud Pública (CSISP), Avenida de Cataluña 21, 46020 Valencia, Spain; Institut Cavanilles de Biodiversitat i Biologia Evolutiva, Universitat de València, Apartado Postal 22085, 46071 Valencia, Spain, e-mail: andres.moya@uv.es

John O'Callaghan Department of Microbiology and Alimentary Pharmabiotic Centre, University College Cork, Cork, Ireland, e-mail: j.ocallaghan@ucc.ie

Paul W. O'Toole Department of Microbiology and Alimentary Pharmabiotic Centre, University College Cork, Cork, Ireland, e-mail: pwotoole@ucc.ie

Airi M. Palva Department of Veterinary Biosciences, University of Helsinki, P.O. Box 66, 00014 Helsinki, Finland, e-mail: airi.palva@helsinki.fi

Vicente Pérez-Brocal Área de Genómica y Salud, Centro Superior de Investigación en Salud Pública (CSISP), Avenida de Cataluña 21, 46020 Valencia, Spain, e-mail: perez_vicbro@gva.es

Ralf Rosenstein Microbial Genetics, Institute of Microbiology and Infection Medicine Tübingen, University of Tübingen, Waldhäuser Str. 70/8, 72076 Tübingen, Germany, e-mail: ralf.rosenstein@uni-tuebingen.de

Oriana Rossi Host Microbe Interactomics Group, Wageningen University, P.O. box 338, 6700 AH Wageningen, The Netherlands, e-mail: oriana.rossi@wur.nl

Reetta Satokari Department of Veterinary Biosciences, University of Helsinki, P.O. Box 66, 00014 Helsinki, Finland, e-mail: reetta.satokari@helsinki.fi

Jerry M. Wells Host Microbe Interactomics Group, Wageningen University, P.O. box 338, 6700 AH Wageningen, The Netherlands, e-mail: jerry.wells@wur.nl

Part I
Between Commensalism and Pathogenicity: Bacterial and Host Aspects

E. coli as an All-Rounder: The Thin Line Between Commensalism and Pathogenicity

Andreas Leimbach, Jörg Hacker and Ulrich Dobrindt

Abstract *Escherichia coli* is a paradigm for a versatile bacterial species which comprises harmless commensal as well as different pathogenic variants with the ability to either cause intestinal or extraintestinal diseases in humans and many animal hosts. Because of this broad spectrum of lifestyles and phenotypes, *E. coli* is a well-suited model organism to study bacterial evolution and adaptation to different growth conditions and niches. The geno- and phenotypic diversity, however, also hampers risk assessment and strain typing. A marked genome plasticity is the key to the great variability seen in this species. Acquisition of genetic information by horizontal gene transfer, gene loss as well as other genomic modifications, like DNA rearrangements and point mutations, can constantly alter the genome content and thus the fitness and competitiveness of individual variants in certain niches. Specific gene subsets and traits have been correlated with an increased potential of *E. coli* strains to cause intestinal or extraintestinal disease. Intestinal pathogenic *E. coli* strains can be reliably discriminated from non-pathogenic, commensal, or from extraintestinal *E. coli* pathogens based on genome content and phenotypic traits. An unambiguous distinction of extraintestinal pathogenic *E. coli* and commensals is, nevertheless, not so easy, as strains with the ability to cause extraintestinal infection are facultative pathogens and belong to the

A. Leimbach · U. Dobrindt (✉)
Institute of Hygiene, University of Münster, Münster, Germany
e-mail: dobrindt@uni-muenster.de

A. Leimbach
e-mail: andreas.leimbach@ukmuenster.de

A. Leimbach
Göttingen Genomics Laboratory, University of Göttingen, Göttingen, Germany

J. Hacker
German National Academy of Sciences Leopoldina, Halle/Saale, Germany
e-mail: joerg.hacker@leopoldina.de

Current Topics in Microbiology and Immunology (2013) 358: 3–32
DOI: 10.1007/82_2012_303

Published Online: 23 January 2013

normal flora of many healthy individuals. Here, we compare insights into phylogeny, geno-, and phenotypic traits of commensal and pathogenic *E. coli*. We demonstrate that the borderline between extraintestinal virulence and intestinal fitness can be blurred as improved adaptability and competitiveness may promote intestinal colonization as well as extraintestinal infection by *E. coli*.

Contents

1 *E. coli*: A Versatile Species

The bacterial species *Escherichia coli* (*E. coli*) is a member of the family *Enterobacteriaceae*, located taxonomically within the gamma subdivision of the phylum *Proteobacteria*. *E. coli* is best known as a ubiquitous member of the normal intestinal bacterial microflora in humans, other warm-blooded animals, and reptiles (Kaper et al. 2004; Lukjancenko et al. 2010).

Normally, *E. coli* persists as a harmless commensal in the mucous layer of the cecum and colon. The Gram-negative, motile bacterium has adapted its metabolism very successfully to this nutritional ecological niche, holding its ground against more than 500 other bacterial species (Tenaillon et al. 2010). *E. coli* colonizes the infant gut within hours of birth and establishes itself as the most abundant facultative anaerobe of the human intestinal microflora for the remainder of life, equipped with the abilities to grow in the ever-changing environment in the gut and cope with the mammalian host interaction. Nevertheless, *E. coli* can survive in many different ecological habitats, including abiotic environments, and is considered a highly versatile species. Population expansion paired with a differential niche adaptation in the last 5 million years led to disparate lifestyles of *E. coli* strains, while adapting to a multitude of environments under specific selective pressures. The astonishing metabolic and regulatory capabilities of *E. coli* facilitate the colonization of different ecological niches, as well as survival under long periods of non-growth. Known habitats of *E. coli* include soil, water, sediment, and food. Some strains of *E. coli* have evolved and adapted to a pathogenic lifestyle and can cause different disease pathologies (Kaper et al. 2004; Crossman et al. 2010; Diaz et al. 2001; Hendrickson 2009; Wirth et al. 2006).

Pathogenic *E. coli* strains can be divided into intestinal pathogenic *E. coli* (IPEC) and extraintestinal pathogenic *E. coli* (ExPEC), depending on the site of infection. Both are further subcategorized into distinct pathotypes, defined as a group of strains of a single species with certain pathogenic traits. Pathotype classification is based on the clinical manifestation of disease, the virulence factors (VFs) involved, and the phylogenetic background. The most prominent IPEC pathotypes are enteroaggregative *E. coli* (EAEC), enterohaemorrhagic *E. coli* (EHEC), enteroinvasive *E. coli* (EIEC), enteropathogenic *E. coli* (EPEC), enterotoxigenic *E. coli* (ETEC), diffusely adherent *E. coli* (DAEC), and adherent invasive *E. coli* (AIEC). Uropathogenic *E. coli* (UPEC), meningitis-associated *E. coli* (MNEC), septicemia-associated *E. coli* (SEPEC), and avian pathogenic *E. coli* (APEC) are the most common ExPEC pathotypes (Kaper et al. 2004; Crossman et al. 2010; Croxen and Finlay 2010). The different lifestyles make *E. coli* a good candidate to study the interplay between host and bacterium, and the relationship between mutualism, commensalism, and pathogenicity.

2 Population Genetics of *E. coli*

The ECOR (*E. coli* reference) strain collection was established by Ochman and Selander based on multi-locus enzyme electrophoresis (MLEE) results (Ochman and Selander 1984). This collection comprises 72 isolates from human and 16 other mammalian hosts and was chosen to represent the genetic diversity of the species *E. coli*. Surprisingly, even today in the age of genomics this holds true in most cases. The collection is classified into five major phylogenetic lineages, A, B1, B2, D, and E (Fig. 1). Group A, including mostly commensal *E. coli*, and B1 are sister taxa and the youngest lineages in *E. coli* phylogeny. Phylogroup B1 is constituted of an assortment of different pathotypes and commensals, including non-O157 EHEC. Phylogroups B2 and D diverged simultaneously early in the history of *E. coli* evolution. B2 comprises many of the ExPEC strains and shows the highest diversity in gene content and on the nucleotide level, consistent with the early emergence of the group in the phylogenetic *E. coli* tree. Group D is polyphyletic and split into two clades by the root of the phylogenetic tree (with *Escherichia fergusonii* as outgroup, a close relative to *E. coli*). Group D1, composed of UPEC and EAEC isolates, clusters close to A, B1, and E, whereas group D2, containing ExPEC and environmental strains, clusters with phylogroup B2.

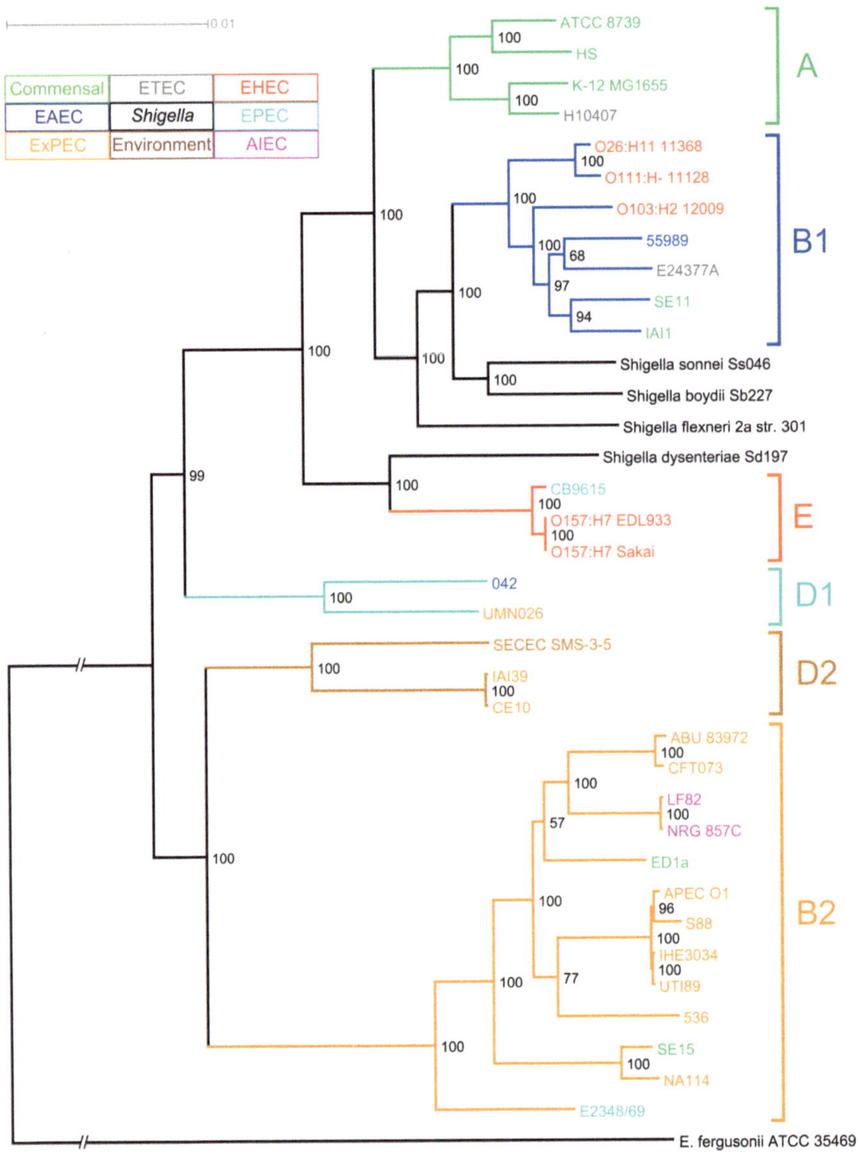

Fig. 1 Phylogeny of a selection of complete *E. coli* genome sequences based on a whole genome alignment. The alignment was calculated with Mugsy (Angiuoli et al. 2011) and only alignment regions present in all analyzed *E. coli* were extracted. These regions were concatenated and positions with gaps removed (Sahl et al. 2012). The resulting core alignment (2.45 Mb) was used to infer a maximum likelihood tree with RAxML (version 7.3.2) and its rapid bootstrapping algorithm (Stamatakis 2006; Stamatakis and Ott 2008). The GTRGAMMA model for nucleotide substitution and rate heterogeneity was utilized, bootstrap support values of 1000 replicates are shown at the nodes. The tree was visualized with Dendroscope (version 3.2.2) (Huson and Scornavacca 2012). The ECOR phylogroups are indicated and the pathotype of each *E. coli* strain is given in the legend. *Escherichia fergusonii* was used as outgroup

AIEC, which are commonly found in ileal lesions of Crohn's Disease patients, also cluster in phylogroup B2, with a close relationship to ExPEC strains. Finally, group E which forms a separate clade of O157:H7 EHEC and O55:H7 EPEC strains, lies in the middle of these *E. coli* histories. The "*E. coli* pathotype" *Shigella*, retained as a genus for historical reasons, is phylogenetically close to groups A, B1, and E. Although contradictory results were obtained in the past, the overall topology described above was confirmed by several methods, including MLST (multi-locus sequence typing), feature frequency profiles, and whole genome phylogeny of the core genome of several *E. coli* strains (Fig. 1) (Wirth et al. 2006; Chaudhuri and Henderson 2012; Chaudhuri et al. 2010; Escobar-Paramo et al. 2004a; Ogura et al. 2009; Sims and Kim 2011; Touchon et al. 2009). The phylogenetic neighborhood of geographically remote *E. coli* isolates supports the notion of a rapid worldwide spread of an evolutionary common ancestor (maybe with the advent of mammals) and selection in specific habitats (Chaudhuri and Henderson 2012).

An MLST analysis by Escobar-Páramo et al. suggested that a certain phylogenetic core genome is necessary to support expression, regulation, and maintenance of VFs (Escobar-Paramo et al. 2004b, 2006). However, a closer look at the ECOR collection in connection with a large-scale analysis of diverse *E. coli* isolates (462 isolates) by the group of Mark Achtman came later to another conclusion. Their application of a different MLST scheme showed that much more homologous recombination takes place in the species *E. coli* than initially thought. Thus, the proposed predominant clonal evolution of *E. coli* was second-guessed, as recombination obscures any phylogenetic association with pathovar or habitat (Wirth et al. 2006; Leopold et al. 2011). However, MLST schemes only analyze a very small portion of the genome by observing a small number of genes. This is a problem in MLST, as one of these genes might even be subject of lateral transfer between strains, leading to somewhat incongruent phylogenetic trees that do not correlate with the genome content of a bacterium (Ochman and Selander 1984; Chaudhuri and Henderson 2012; Sims and Kim 2011).

But despite the dynamic nature of the *E. coli* genome, the overall chromosome structure is stable and the core genome largely co-linear between genomes; only few rearrangements are detected. Most of the variation takes place by insertion or deletion events in chromosomal hotspots, hotspots, like tRNA-neighboring regions (Tenaillon et al. 2010). Homologous recombination can obscure the phylogenetic signal in the *E. coli* core genome and can lead both to divergence as well as convergence. Touchon et al. hypothesized that a single nucleotide was 100 times more prone to be involved in genetic transfer than mutation (Touchon et al. 2009). Nevertheless, an estimation of recombination within the core genome of *E. coli* estimated that only about 10 % of the core genome is affected and identified recombination events were in most cases small (<2 kb) (Mau et al. 2006). Thus, recombination in *E. coli* has not disrupted the phylogenetic signal of the core genome as long as the analyzed sequence is long enough. Hence, the aforementioned problem with MLST techniques based on a few selected chromosomal loci is genuine.

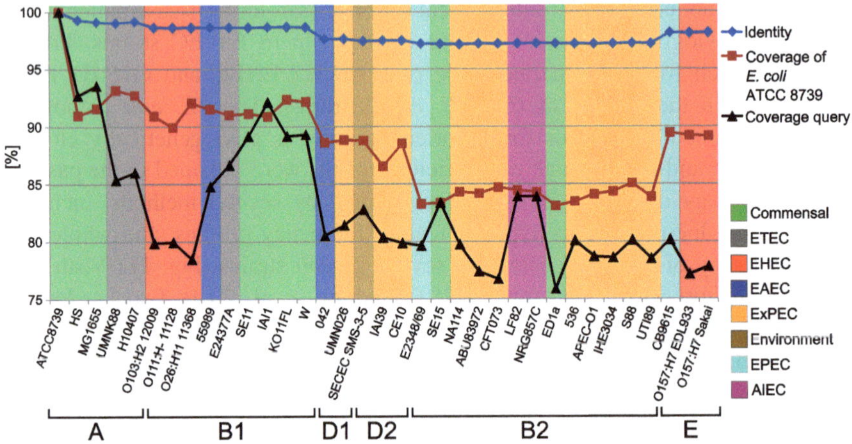

Fig. 2 Genome content comparison of completely sequenced *E. coli* strains. The commensal *E. coli* isolate ATCC 8739 was chosen as a reference. All the other *E. coli* genomes were aligned against the reference with the dnadiff script of the MUMmer package (version 3.22) (Kurtz et al. 2004). Coverage of *E. coli* ATCC 8739 is indicated in *red*, coverage of the respective query in *black*, and identity of the aligned sequences in *blue*. No correlation between pathotype and genome coverage of the reference or the query can be seen, but rather correlation between phylogroups

As solution emerged the new "gold standard" of phylogenetic analysis, whole genome phylogeny. The method leads to clear phylogenetic signals in *E. coli*, as exemplified by robust tree phylogenies calculated with different methods. As a result, the predominantly clonal population structure of *E. coli* can be used to delineate the major phylogenetic groups described above (Fig. 1) (Tenaillon et al. 2010; Chaudhuri and Henderson 2012; Sims and Kim 2011; Leopold et al. 2011). With this information, several genomic features can be related to the major phylogenetic groups outlined by the ECOR collection. For example, Rhs elements are arranged according to the phylogroups (Hill et al. 1995), but also VFs like the *Yersisina* 'high pathogenicity island' (HPI) (Clermont et al. 2001) and the putative type III secretion system ETT2 (Ren et al. 2004). Additionally, a connection can also be found in the distribution of extraintestinal pathotypes and phylogenetic ancestry. While strains from ECOR phylogroups A and B1 usually do not exhibit ExPEC phenotypes and lack ExPEC VFs, ECOR B2 and D cluster the majority of ExPEC strains (Boyd and Hartl 1998). The five major phylogenetic groups might even represent diverse ecological niches, as they have a different distribution in humans, domesticated animals, and wild animals (Tenaillon et al. 2010). There is, however, no direct correlation between the pathotype and phylogenetic lineage. Comparison of genome coverage and nucleotide identity of selected IPEC, ExPEC and commensal *E. coli* relative to non-pathogenic *E. coli* ATCC8739 could not reveal marked overlaps. No correlation among pathotype and genome coverage of the reference or the query can be seen, but rather correlation between phylogroups (Fig. 2).

Each pathotype forms multiple phylogenetic clades and has arisen polyphyletically several times via parallel evolution. Phylogenetic trees based on the complete genome sequences of *E. coli* strains support these observations. Thus, convergent evolution of *E. coli* strains resulted in nowadays pathotypes (Fig. 1). This supports the notion of extensive horizontal gene transfer (HGT) in *E. coli* and the transmission of the genetic source of whole pathotypes in a single step via mobile elements, like PAIs, plasmids, and phages (Reid et al. 2000; Whittam et al. 1993). Moreover, a comparison of whole genome phylogeny to metabolic distance estimation also showed that pathotypes cannot be grouped together. Differences in metabolic reactions and networks between strains rather evolve in a phylogenetic manner and follow the ECOR phylogroups. As an exception ECOR A and B1 exhibit a major intersection in their metabolic capabilities, consistent with their most recent differentiation. Only the "pathotype" *Shigella* has converged phenotypically to a distinct metabolic profile relative to the other *E. coli*. This is in contrast to the close phylogenetic clustering of *Shigella* and *E. coli* strains based on genomic data (Chaudhuri and Henderson 2012; Vieira et al. 2011).

Whole genome sequencing shed light on the parallel evolution of several *E. coli* IPEC pathotypes, like EPEC (Iguchi et al. 2009; Rasko et al. 2008) EAEC (Chaudhuri et al. 2010; Touchon et al. 2009; Rasko et al. 2008), and ETEC (Crossman et al. 2010; Rasko et al. 2008; Sahl et al. 2011; Shepard et al. 2012). All of these pathotypes are phylogenetically diverse, occur in different phylogenetic *E. coli* lineages, and have only a few pathotype-specific genes, except for the most common virulence markers. In the case of ETEC, no pathotype-specific genes could be detected at all. ETEC, however, do share a genomic core with each other in comparison to other pathotypes (Crossman et al. 2010; Sahl et al. 2011). Genome sequence analysis of several distantly related AIEC isolates of different serotypes came to the same result (Clarke et al. 2011; Krause et al. 2011; Miquel et al. 2010; Nash et al. 2010). Early genomic studies on EHEC O157:H7 confined phylogenetic analysis in the context of the diversity of the pathotype and the species. O157:H7 strains share the same phylogenetic history and therefore have similar genome content (Hayashi et al. 2001; Perna et al. 2001). EHEC O157:H7 strains are hypothesized to have arisen from an O55:H7 EPEC precursor by the acquisition of additional VFs, like the phage-encoded Shiga toxin. This is reinforced by their close clustering in phylogenetic analyses, as well as the additional whole genome sequencing of O55:H7 EPEC isolates and draft genome sequencing of intermediates (Rump et al. 2011; Zhou et al. 2010). However, also the EHEC pathotype evolved on several occasions, as exemplified by non-O157:H7 EHEC, which are phylogenetically ranked in ECOR phylogroup B1 (Fig. 1) (Ogura et al. 2009). These studies highlight the need of choosing phylogenetically diverse *E. coli* isolates for sequencing, without the bias of selecting the most clinically relevant strains. Only the whole genome analysis of sufficient isolates can establish a significant phylogenetic ancestry and the parallel emergence of distinct pathotypes.

3 Genome Plasticity: The Key to Diversity

Underlying the amazing metabolic and phenotypic diversity of *E. coli* is a very dynamic genome structure. A genome of a species can be classified in two categories. On one the hand, the core genome is defined as the genes that are present in all strains of one bacterial species. It includes mostly essential housekeeping genes involved in replication, transcription, and translation. The core genome makes up the genomic backbone of a bacterial species, defining the basic metabolic functions. On the other hand, the flexible/dispensable genome comprises genes that are only present in a few strains or unique to single isolates, so-called singletons. These genes are responsible for diverse phenotypes and adaptations to specific environmental conditions in a population or species (Medini et al. 2005). They often show high rates of nucleotide sequence variability. Examples for the flexible gene pool are mobile elements, like plasmids, phages, and genomic islands (GEIs), summarized as the "mobilome". In the context of pathogenicity, the flexible gene pool encodes for fitness and VFs, which give the pathogen the potential to colonize the host and cause disease. The combination of the core and the flexible genome makes up the pangenome, i.e., the total gene repertoire of a species (Tettelin et al. 2005). The pangenome of a species is many times larger than the genome of a single bacterium (Medini et al. 2005, 2008; Tettelin et al. 2005, 2008; Hacker and Dobrindt 2006).

3.1 Mobile Elements and Their Role in E. coli Evolution

Gene acquisition by HGT, together with homologous recombination as well as genome reduction events, account for a large fraction of genetic flexibility in a bacterial species, between species, and also higher taxa (Dobrindt et al. 2004; Hacker et al. 2003). Therefore, a pangenome might not only be constricted to the species taxonomic level.

Intra-strain factors for genome diversity are the occurrence of point mutations, genome reduction by deletion events, and the function of accessory elements, like insertion sequence (IS)-elements, transposons, and integrons, that can jump into different sites on replicons. These accessory elements further enhance homologous recombination, which can lead to large-scale genomic rearrangements. In addition, accessory elements are also shuttled by mobile elements between strains (except for the special case of conjugative transposons, which can transmit themselves) (Dobrindt et al. 2004; Ambur et al. 2009; Jackson et al. 2011; Schubert et al. 2009). The resulting genomic diversity can then propagate vertically inside a population by clonal proliferation.

Mobile elements are the driving force of HGT, as well as the major origin of the flexible gene pool. Thus, vectors for inter-strain transfer are plasmids, phages, GEIs, or chromosomal DNA by the mechanisms of conjugation, transduction, or natural transformation, respectively (Dobrindt et al. 2004; Juhas et al. 2009; Wiedenbeck and Cohan 2011). Although *E. coli* was traditionally not considered to be naturally competent, recent reports indicate otherwise and show a yet uncharacterized transformation mechanism under certain environmental conditions (Etchuuya et al. 2011). Because *E. coli* thrives in contact to the gut microbiome with a diverse bacterial community, a manifold flexible gene pool is available for HGT (Tenaillon et al. 2010).

Bacterial chromosomes are highly organized in relation to their interaction with cellular processes like replication, segregation, transcription, and translation, as well as regulatory elements and operons. Thus, insertion or deletion of DNA regions can disrupt this organization or regulation structure. Selection should therefore allow insertion/deletion only at certain positions in the genome, which are not restricted by organizational constraints. Touchon et al. suggested that, once a rare, large integration event disrupts the chromosome order in a permissive region, this less perfectly adapted region opens the way to future recombination events through a "founder effect" for additional HGT, leading to an integration hotspot (Touchon et al. 2009). GEIs and their subgroup pathogenicity islands (PAIs) most likely originate by genome integration and loss of mobile function events of former lysogenic bacteriophages and plasmids. Afterwards, demobilized elements are passed on in a vertical fashion within different *E. coli* phylogenetic lineages. These islands cluster novel genes because they are often genetically unstable, serve as integration hotspots and undergo further evolution with the help of mobility genes (Hacker et al. 2003). Integrases, transposases, and their associated elements, integrons, IS elements and transposons, are key in these processes. Successive integrations of (foreign) genes or deletions in islands result in their typical mosaic-like genetic structure. Also bacteriophages show a very high composite structure and positional diversity, therefore contributing extensively to genome diversity and rearrangement. As a consequence, novel genes in a bacterium occur in higher proportions on mobile or formerly mobile elements, which is in accordance with the concept of the flexible gene pool and the mobilome. The above-mentioned mobility genes are also required for chromosomal integration and excision of GEIs and phages, with a possible subsequent transfer to other recipients (Dobrindt et al. 2004; Juhas et al. 2009; Ho Sui et al. 2009; Schneider et al. 2011). Although island contents differ, many of the same chromosomal regions serve as insertion sites via site-specific recombination. Especially, tRNA-encoding genes are hotspots of bacteriophage and island insertions, as different non-related mobile elements can be found associated with them.

GEIs and PAIs can store a large pool of novel genes accessible for adaptation and innovation. Bacterial strains can thus draw foreign genes from the environment for short-term adaptation and survival strategies. The large size of such islands makes it possible to transfer new phenotypes depending on several genes or operons in a single step (Dobrindt et al. 2004; Ho Sui et al. 2009). Both, gene acquisition via HGT and genome reduction, are reflected in the variable sizes of *E. coli* genomes, which range from $\sim$4.6 to $\sim$5.7 Mb. More than 1 Mb of DNA can be absent between one *E. coli* and another! As with novel genes, also VFs are over-represented and clustered on mobile, or formerly mobile genetic entities, especially PAIs. Accordingly, HGT plays an important role in propagating virulence determinants between different bacterial strains and species. This is especially disturbing with respect to antibiotic resistances, often encoded by integrons and resistance islands. As PAIs serve as integration sites for other accessory genetic elements encoding for VFs, concentrating virulence genes in specific genomic regions, distinct pathogenic and resistance phenotypes can be rapidly and simultaneously acquired. This ensures successful uptake and integration into existing regulatory networks in the recipient. GEIs or PAIs can, however, also be deleted in a single step (Tenaillon et al. 2010; Medini et al. 2008; Ho Sui et al. 2009; Dobrindt 2005).

Gene acquisition by HGT and gene loss is extensive in *E. coli* resulting in the above-described pathotypes with distinct pathogenic capabilities, independent of phylogenetic lineages. The *E. coli* genomic backbone is composed of clonally evolving DNA segments, disrupted by dispensable DNA fragments introduced via homologous recombination and insertion of horizontally acquired DNA. Early on, a model of clonal frames of different ages was suggested, which proposes clonal propagation of chromosomes with advantageous mutations. The clonal frames are punctuated by region of differences introduced by mobile elements, resulting in a mosaic genome structure (Dobrindt 2005; Milkman and Bridges 1990). In the light of the previous observations, this model holds still true today.

Because of these interesting discoveries, Goldenfeld and Woese challenged the traditional bacterial taxonomy and species concept. They hypothesize, that the flexible gene pool is a possibility for bacteria to absorb and discard genes as dictated by selective pressures. In their view, single genomes do not exist, but a continuum of genomic possibilities, discarding the microbial species concept. Especially phages act as a repository and memory of genetic information, i.e. the flexible gene pool, and contribute to the genetic dynamics and stability of bacterial communities. Their assumptions are in accordance with the concept of the pangenome and a mosaic-like genome structure, albeit different terminology (Goldenfeld and Woese 2007). It has also been suggested, that the diversity and overlap in the gene content of *Enterobacteriaceae*, like *Shigella* and *Escherichia*, reflects a continuum rather than sharp species borders (Lukjancenko et al. 2010).

3.2 Genome Content and Phenotypic Variation

The recent drastic accumulation of genomic data revealed some surprising results, supporting the concept of core, dispensable, and pangenome. On the one hand, the core genome within the species *E. coli* is largely co-linear between genomes. Conserved syntenic DNA regions compared between any two *E. coli* strains show only up to 3 % nucleotide divergence. On the other hand, sequencing projects discovered a surprisingly high intra-species diversity in *E. coli*, in an order of magnitude never dreamed of in the pre-genomics era. It is estimated that only

~40 % of the combined *E. coli* proteins are conserved among all strains, a set of ~2,200 genes with high homology constituting the core genome (Tenaillon et al. 2010; Chaudhuri and Henderson 2012; Touchon et al. 2009). The residual 60 % make up the dispensable gene pool of paralogs, alleles and singletons often colocalized on mobile elements. The unexpected low number of genes, which make up the core genome, exemplifies the high plasticity of the *E. coli* genome, which results in the diverse adaptation strategies of different strains (Tenaillon et al. 2010). The divergence of *E. coli* lifestyles is based on a high versatility and adaptability to manifold environments, which in turn promotes HGT and results in an open pangenome structure. Sequencing of new *E. coli* genomes leads to the discovery of novel singletons within the species, extending the size of the pangenome and characterizing an open pangenome. Recent calculations of the *E. coli* pangenome resulted in more than 18,000 genes, while a typical *E. coli* genome has around 5,000 genes (Chaudhuri and Henderson 2012; Touchon et al. 2009; Rasko et al. 2008; Tettelin et al. 2008; Halachev et al. 2011).

Because a bacterial genome size is finite, non-essential adaptations have to rely on the flexible genome via mobile elements and a tradeoff between gene loss and acquisition. The described *E. coli* genome plasticity illustrates the diversity of phenotypic adaptations present in the species. Accordingly, genes responsible for a certain phenotype, e.g. packaged on islands, should be only found in strains, in which these genes contribute to adaptation to a specific environment. Also, different alleles or alternative combination of genes can promote adaptation to a given environment (Tenaillon et al. 2010). This is also the case for VFs and their associated *E. coli* pathotypes. Although in vitro studies are somewhat artificial, it was shown that only a small amount of genes can support the life of a bacterial cell. Databases like the Online GEne Essentiality database (OGEE) (Chen et al. 2012) and the PEC database (Profiling of *E. coli* Chromosome) (Hashimoto et al. 2005; Kato and Hashimoto 2007) report on about 300 genes in *E. coli* K-12 (both MG1655 and W3110), which are essential for robust aerobic growth in rich media (of a total of ~4,500 genes). This is reinforced by targeted mutagenesis studies, like the Keio collection of single *E. coli* mutants, which resulted in the detection of ~300 essential genes (Baba et al. 2006). The residual genes, which make up the core genome of the species *E. coli*, are most likely genes important for in vivo colonization and growth in the mammalian intestine. Hence, the small amount of genes of the *E. coli* core genome, described above, seems quite possible.

In contrast to the low number of core genes that were detected in *E. coli*, the core reactions of the *E. coli* metabolome have a broader scale. Of 1,545 metabolic reactions forming the *E. coli* panmetabolome, 57 % are core reactions common to all strains analyzed. Anabolic reactions are the majority in the core metabolome, whereas catabolic reactions are over-represented in the dispensable metabolome. This can be a result of specific niche-adapted catabolic processes. In contrast to the open pangenome structure in *E. coli*, the panmetabolome already reached a plateau with the analysis of 29 *E. coli* strains. Hence, metabolic functions are less diverse than overall gene functions, a possible result of the conservation of genes encoding for enzymes. Additionally, phenotypic comparisons between *E. coli* strains show even less diversity than predicted by in silico metabolic constructions. This might be an indication of redundant uncharacterized pathways and regulation mechanisms of novel metabolic pathways (Chaudhuri and Henderson 2012; Vieira et al. 2011).

Genome sequencing of environmental *E. coli* isolates widened the horizon on the genomic capabilities of the highly adaptive species. Although *E. coli* is traditionally considered a commensal of the mammal intestinal systems and used as an indicator of fecal contaminations, *E. coli* strains can also adapt to abiotic environments. A saprophytic lifestyle in sediment and water, depending on nutrient availability and temperature, has been proposed (Tenaillon et al. 2010; Berger et al. 2010; Holden et al. 2009). Although isolates from the gastrointestinal tract have dominated the sequencing facilities, environmental isolates have also been sequenced. They might even contribute to the spread of antibiotic resistances between *E. coli* strains, as survival for longer periods outside of animals is feasible. But, because of their geographic isolation, HGT most likely is limited (Fricke et al. 2008; Luo et al. 2011). Environmental strains illustrate that the genomic diversity represented in the ECOR collection does not cover the whole diversity of the species *E. coli*. The application of MLST analyses brought the idea of isolates lying outside the ECOR diversity (Wirth et al. 2006). More detailed analysis with extended MLST and whole genome sequencing, demonstrated phenotypically undistinguishable, but genotypically divergent *E. coli* isolates. These were classified into five *Escherichia* clades, C–I to C–V. *Escherichia* isolates from clades C–II to C–V are more prominent in the environment than being enteric, with exception of C–I, which is closest related to ECOR strains. Thus, these strains might be better adapted for an abiotic lifestyle, shown in the absence of certain nutrient transporters/utilization systems abundant in the gastrointestinal tract. The final nomenclature of these new clades, as new species in the genus *Escherichia* or as divergent *E. coli* species, is still under debate (Luo et al. 2011; Walk et al. 2009).

3.3 Genome Plasticity and Evolution of Pathogenic E. coli

Bacteria have to face changes in their environment. This is especially true for commensal or pathogenic bacteria as they have to deal with extensive and dynamic variations in their co-evolving hosts (Medini et al. 2008). Nevertheless, the versatile pathogen *E. coli* kills about two million humans per year, both through intestinal and extraintestinal diseases (Tenaillon et al. 2010). Genome structure and size reflects bacterial lifestyle and seem to be driven by evolutionary forces. Strictly host-dependent bacteria, like intracellular ones, have reduced genomes via deletion mechanisms, because they rely on the host metabolism for the functions they have lost. This is a specific case of niche adaptation. On the contrary, gene acquisition via HGT is a common trait among extracellular bacteria, including facultative pathogens, symbionts, and environmental bacteria. Here, mobile elements increase adaptability to ever-changing environments and need a larger gene pool to address different metabolic needs (Medini et al. 2008; Dobrindt et al. 2004). Moreover, commensal *E. coli* strains mostly have smaller genome sizes than pathogenic strains. This might be an indication of reductive convergent evolution, but probably just reflects the shedding of unnecessary virulence-associated genes (Chaudhuri and Henderson 2012; Sims and Kim 2011).

IPEC and ExPEC strains differ in their genetic makeup as well as their phylogenetic past. Various IPEC pathotypes were traditionally considered to be clonal, characterized by common serotypes, which have evolved under adaptation to the respective niches as distinct genetic types. Early HGT events played a vital role in the emergence and subsequent divergence of these clones. However, recombination keeps evolution in progress, resulting in very dynamic and diverse genome structures (Kaper et al. 2004; Hacker and Dobrindt 2006; Castillo et al. 2005; Didelot et al. 2012; Laing et al. 2009). Additionally, mobile elements mediate the ordered gain and loss of genetic elements in various *E. coli* pathotypes and enable the parallel evolution of separate clones with a polyphyletic phylogenetic root that undergo convergent evolution to specific pathogenic capabilities. Thus, with the availability of genomes from several strains from one pathotype the present-day view of IPEC pathotype emergence took shape (Fig. 1) (Chaudhuri and Henderson 2012; Reid et al. 2000). The potential to interact with one another, e.g. in the intestinal environment, makes the acquisition of complex pathogenic phenotypes possible, as described above (Ren et al. 2004). Novel combinations of VFs increase the bacterium's capacity to adapt to new niches and allow these *E. coli* clones to cause a broad spectrum of diseases. Only the most successful sets of VFs develop into pathotypes of *E. coli*, capable of causing disease in healthy individuals (Kaper et al. 2004; Hacker and Dobrindt 2006).

ExPEC differ from IPEC, because these facultative pathogens were traditionally already regarded as derived from different phylogenetic groups, illustrated for instance by their diversity of serotypes. Additionally, they do not host an unambiguous distinctive repertoire of VFs characteristic for a specific type of disease (Dobrindt 2005; Köhler and Dobrindt 2011). Various combinations of VFs can lead to the same extraintestinal disease outcome, which solely defines an ExPEC pathotype. Genome sequencing projects revealed extensive genome diversity among ExPEC, but also identified some pathotype-specific genes including toxins, iron acquisition systems, adhesins, lipopolysaccharides (LPS), polysaccharide capsules, proteases, and invasins. Again, these factors are frequently encoded on mobile elements (Dobrindt 2005; Köhler and Dobrindt 2011; Brzuszkiewicz et al. 2006; Chen et al. 2006; Johnson et al. 2007; Lu et al. 2011; Moriel et al. 2010; Welch et al. 2002).

4 Genomic Differences Between ExPEC and Commensal *E. coli*

Many ExPEC virulence-associated features are also present in commensal *E. coli*. Whereas the role of *E. coli* as an extraintestinal pathogen has been intensely studied for decades, much less is known about specific traits of commensal variants and how they may be adapted to the mammalian gut. Studies on the diversity of the *E. coli* fecal flora from individual human hosts indicated that intra-host diversity is variable: usually one predominant strain exists at a given time-point which is accompanied by other strains which are less frequent. The predominant strain often colonizes for longer time periods, i.e. months or even years, whereas the less frequent strains are transient, colonizing only for days or weeks (Escobar-Paramo et al. 2004a; Caugant et al. 1981; Sears and Brownlee 1952; Sears et al. 1950). Commensal *E. coli* isolated from the porcine intestine were shown to be genetically quite diverse. A large fraction of these commensals carried at least one bacteriocin gene which is frequently plasmid-encoded. The prevalence and type of colicin determinants varied among the isolates with respect to the gut region from which they have been isolated. Similarly, these isolates exhibited a non-random distribution of several plasmid replicon types. In conclusion, a broad variety of commensal *E. coli* exists in the porcine intestine with different characteristics depending on the intestinal region from which they have been isolated (Abraham et al. 2012).

Different *E. coli* phylogroups have been associated with different gut niches before (Dixit et al. 2004). When plasmid relatedness and diversity of colicin determinants were compared between different APEC, UPEC, and *E. coli* from avian or human fecal samples, a great overall plasmid variability was observed as well. Interestingly, IncFIB plasmids occurred significantly more frequent in APEC relative to UPEC and avian or human fecal *E. coli*. APEC also carried more frequently colicin genes than UPEC, or fecal isolates from birds or humans. As a result, some commensals might be distinguished from extraintestinal pathogenic variants because of their plasmid content. The ability to acquire and propagate certain plasmid types can differ between commensal and pathogenic *E. coli* subgroups (Johnson et al. 2007; Smajs et al. 2010).

A survey of phylogenetic groups and PAI markers in commensal *E. coli* from Chinese individuals indicated that phylogroup A strains were the most common. In addition, almost 50 % of all randomly selected fecal strains carried known PAIs (Li et al. 2010). Other screenings of ExPEC and fecal *E. coli* indicate that although the mean number of PAIs per isolate was higher among UPEC than in commensals, statistical differences among group B2 UPEC or commensals could not be observed, suggesting that the intestinal flora may act as a reservoir for bacteria that can cause urinary tract infection (Tenaillon et al. 2010; Grasselli et al. 2008; Sabaté et al. 2006). On the other hand, several ExPEC virulence genes, such as *hlyA* (α-hemolysin), *fyuA* (yersiniabactin receptor), *traT* (serum resistance-associated outer membrane protein), and *iutA* (aerobactin receptor) were found to be independent predictors for pathogenicity. Especially two of them, *iutA* and *traT*, were significantly more common in *E. coli* isolates carrying certain antibiotic resistance genes as well (Lee et al. 2010). The observed differences in the prevalence of certain phylogroups and gene contents are assumed to depend on host characteristics, such as diet or the physical complexity of the hindgut (Gordon and Cowling 2003), as well as on the environment in which a given animal or human population lives (Escobar-Paramo et al. 2006).

4.1 ExPEC Virulence or Fitness Traits: A Matter of Perspective, Niche or Strain Background

From the fine line that distinguishes commensal *E. coli* from ExPEC two questions arise: What is an ExPEC virulence factor and can we exclude that these factors solely promote ExPEC pathogenesis? Several bacterial traits and so-called VFs have been described to contribute to extraintestinal infection (Table 1). Although their role in ExPEC pathogenesis and their prevalence in ExPEC isolates has been demonstrated, several of them can be found in commensal *E. coli* strains as well, thus questioning their exclusive role during ExPEC pathogenesis and our understanding of the evolution and adaptation of ExPEC. So-called ExPEC virulence-associated genes are often located on PAIs and plasmids. Several of these PAIs and plasmids are found in ExPEC, but their role in commensal bowel colonization and persistence is unknown. Interestingly, commensal *E. coli* capable of long-term intestinal colonization often belong to phylogroup B2 and D, and frequently express adhesins (P fimbriae and type 1 fimbriae), capsular antigens (K1 and K5), the toxin α-hemolysin, as well as the siderophore system aerobactin. With regard to the prevalence of these virulence—or fitness-associated genes and phylogroup allocation, these strains resemble typical ExPEC isolates. The accumulation of these PAI markers in commensal *E. coli* correlated positively with their time of persistence in the colon. In addition, ECOR group B2 and D strains which usually carry many of the above-mentioned genes were shown to have superior capacity to persist in the infantile colonic microbiota. Accordingly, certain ExPEC virulence traits improve the colonizing capacity of phylogroup B2 strains and thus intestinal persistence. They have probably evolved primarily because they increase the fitness of *E. coli* in its natural niche and thus enhance their survival in the intestine (Diard et al. 2010; Le Gall et al. 2007; Nowrouzian et al. 2001, 2003, 2005, 2009; Ostblom et al. 2011; Schierack et al. 2008; Wold et al. 1992). Comparative genomics of commensal *E. coli* strain SE15 revealed that this strain carries fewer known ExPEC virulence genes than other commensal strains of phylogroup B2, e.g. ED1a and EcN. Consequently, analysis of traits present in strain SE15, but absent from ED1a and EcN, may help to identify traditional ExPEC virulence-related genes which may be necessary for commensal *E. coli* to colonize the human gut (Toh et al. 2010). Genome sequence analysis of commensal isolate

Table 1 Fitness and virulence traits of extraintestinal pathogenic *E. coli*

Trait	Example	Role during infection	Role during commensalism or in secondary habitat	Reference
Adhesins	Type 1 fimbriae	Adhesion, niche tropism, biofilm formation	Adhesion, niche tropism, biofilm formation	Bouckaert et al. (2006), Hung et al. (2002), Stahlhut et al. (2009)
Siderophore receptors	Yersiniabactin receptor Salmochelin receptor IrgA homolog adhesin (Iha)	Iron acquisition, adhesion, invasion, biofilm formation	Iron acquisition, adhesion, biofilm formation	Bielaszewska et al. (2011), Feldmann et al. (2007), Hancock et al. (2008), Léveillé et al. (2006)
Extracellular polysaccharides, cellulose, capsule, LPS	Capsule, cellulose, LPS	Serum resistance, protection against immune response; interaction with eukaryotic cells	Protection against predation, desiccation, intestinal colonization	Diard et al. (2010), Monteiro et al. (2009), Wang et al. (2006), Hafez et al. (2009), Mordhorst et al. (2009)
Toxins	α-Hemolysin	Cell/tissue destruction, release of nutrients	Signaling	Söderblom et al. (2002), Uhlén et al. (2000)
Flagella		Motility/chemotaxis	Motility/chemotaxis	Adler et al. (1973), Lane et al. (2007), (2005), Mesibov and Adler (1972), Schwan (2008)
Metabolic traits	Utilization of D-serine, fructooligosaccharides	Growth advantage, niche colonization	Growth advantage, niche colonization	Fabich et al. (2008), Bernier-Febreau et al. (2004), Le Bouguénec and Schouler (2011), Porcheron et al. (2012), Schouler et al. (2009), Rouquet et al. (2009)

SE11 also identified large horizontally acquired regions in the chromosome or in plasmids, which frequently comprise fimbrial and autotransporter determinants. This finding led to the hypothesis that these cell surface-associated factors may contribute to the adherence of *E. coli* SE11 to host cells or to conjugation. Accordingly, *E. coli* SE11 probably accumulated functions which promote stable colonization of intestinal cells. These data support the idea that adhesion-associated functions are important for the commensality of *E. coli* in the human gut (Oshima et al. 2008). Most likely, these factors may, however, also promote bacterial adhesion in niches outside of the intestine.

The *pks* genomic island present in *E. coli* strains of phylogroup B2 encodes colibactin, a hybrid polyketide/non-ribosomal peptide that causes DNA damage and cell cycle arrest of eukaryotes (Nougayrède et al. 2006). The colibactin-encoding determinant has been detected primarily in extraintestinal pathogenic isolates of *E. coli*, *Klebsiella pneumonia*, *Enterobacter aerogenes* and *Citrobacter koseri*, but also in commensal *E. coli*. The presence of the *pks* island in mainly extraintestinal pathogens may indicate that colibactin contributes to fitness or virulence during extraintestinal infection (Johnson et al. 2008; Krieger et al. 2011; Putze et al. 2009). The frequent detection of the *pks* island and other ExPEC PAIs in *E. coli* isolates from biopsy material of patients suffering from colon cancer again raises the question whether traits encoded on ExPEC PAIs, including colibactin, may contribute to long-term intestinal colonization or pathogenicity of *E. coli* strains, here associated with colon cancer (Bronowski et al. 2008). Screening of the gut microbiota of Swedish infants from birth to 18 months of age revealed that *E. coli* with the capacity to persist in the microbiota carried significantly more often the *pks* island than either intermediate-term colonizers or transient strains. This finding suggests that the *pks* island contributes to the gut-colonizing capacity of group B2 strains (Nowrouzian and Oswald 2012). The recent observation that the probiotic effects of strain EcN to ameliorate colitis severity and modulate cytokine expression cannot be separated from the strain's ability to express functional colibactin (Olier et al. 2012) also demonstrates that, depending on the niche or context, colibactin can be considered a virulence and/or a probiotic factor.

The commensal *E. coli* strain A0 34/86 (O83:K24:H31) has proven for several decades to be clinically safe and efficient in the prophylaxis and treatment of nosocomial infections and diarrhea of preterm and newborn infants. Interestingly, many PAI-associated genes were detected in this strain, including those coding for the important ExPEC toxins α-hemolysin and cytotoxic necrotizing factor 1 (CNF-1). The search for genomic regions specific for *E. coli* A0 34/86 identified some genes to be implicated in the colonization capacity of the strain, enabling it to outcompete pathogens. A genomic fragment coding for gluconate and mannonate metabolism, adhesion (*fim*), invasion (*ibe*), and restriction/modification functions reproducibly enhanced persistence in the intestine of newborn piglets on laboratory strain DH10B (Hejnova et al. 2005). The presence of many ExPEC virulence-associated genes in the genome sequence of this efficient colonizer strain underlines the thin line between ExPEC virulence and bacterial fitness in the intestine. Similar results have been obtained upon comparative genomic and phenotypic analysis of different collections of ExPEC, IPEC, and fecal *E. coli* isolates (Dobrindt et al. 2003; Salvador et al. 2012).

Horizontally acquired determinants which support fitness and competitiveness of *E. coli* pathogens also code for metabolic traits. As an intestinal bacterium, *E. coli* is adapted to utilize energy sources in the mammalian intestine and live and multiply at this site. Analyses of the metabolic versatility of pathogenic and non-pathogenic *E. coli* variants indicate, however, that *E. coli* pathogens can use sugars or other carbon sources that are not used by commensal *E. coli* to colonize the mouse intestine. This strategy enables the pathogen to gain advantage by simultaneously consuming several C-sources that may be available because they are not consumed by the commensal intestinal microbiota (Anfora et al. 2007; Anfora and Welch 2006; Fabich et al. 2008; Roesch et al. 2003). Similarly, studies using various animal models of intestinal colonization showed that the metabolism of short-chain fructooligosaccharides and deoxyribose help avian and human pathogenic *E. coli* to outcompete the normal flora and colonize the intestine. Furthermore, phosphotransferase system (PTS) and non-PTS sugar transporters can expand metabolic capabilities and modulate ExPEC virulence (Bernier-Febreau et al. 2004; Le Bouguénec and Schouler 2011; Porcheron et al. 2012; Schouler et al. 2009). It will, however, be interesting to see, how widespread such metabolic capabilities are among commensals. In conclusion, there is a thin line between the definition of virulence and fitness factors in ExPEC and commensals. In consequence, a clear distinction between ExPEC and commensal *E. coli* strains can be difficult (Tenaillon et al. 2010; Dobrindt 2005; Köhler and Dobrindt 2011; Diard et al. 2010). Nevertheless, commensal fitness determinants required for efficient intestinal colonization and competitiveness serve as a reservoir for virulent strains, in respect to the concept of the pangenome. The virulence genes probably evolved by adaptation to the intestinal growth environment and were selected for a commensal lifestyle. As a consequence, many of these features can be rather considered fitness traits (e.g. iron uptake systems, bacteriocins, toxins, proteases, flagella, adhesins, extracellular polysaccharides), that contribute to the overall ability to colonize the host. They also serve as fitness factors to occupy a niche in a secondary habitat as some ExPEC VFs might even protect against predation by protozoa or nematodes (Alsam et al. 2006; Diard et al. 2007; Steinberg and Levin 2007) (Table 1). This suggests, that ExPEC virulence might just be a by-product of the “main” non-pathogenic or commensal lifestyle (Tenaillon et al. 2010; Diard et al. 2010; Le Gall et al. 2007).

Despite the presence or absence of specific determinants promoting virulence or fitness, the pathogenic potential of *E. coli* can be markedly affected by the individual strain background and different gene regulation patterns. For example, production of the extracellular polysaccharide cellulose in EcN is required for its efficient adhesion to gastrointestinal epithelial cells in vitro as well as to mouse epithelium in vivo, and for enhanced cytokine production by immune cell lines. Accordingly, cellulose expression has been assumed to potentially contribute to the long-term colonization capability of EcN in vivo. However, this contribution of cellulose to bacterial adhesion on epithelial cells seems to depend on the strain background: In contrast to EcN, adherence of commensal strain TOB1 to HT-29 cells was enhanced upon the loss of cellulose production (Monteiro et al. 2009; Wang et al. 2006). *E. coli* K-12 is well established as a harmless laboratory strain. Recent evidences, however, indicate that the typical non-invasive nature of this *E. coli* strain can be reversed under specific circumstances even in the absence of any major genomic flux. Introduction of a mutated histone-like protein HU into *E. coli* K-12 resulted in significant changes in nucleoid organization and global transcription. These changes transformed the mutant *E. coli* into an almost obligate intracellular bacterium. This result demonstrates that even without gross changes in its genome content, changes of the cellular transcription program can lead to widely divergent lifestyles of *E. coli* K-12 in relation to mammalian host cells (Koli et al. 2011).

4.2 Impact of Genome Plasticity on Pathogenicity and Fitness of E. coli B2 Strains: Three Closely Related Strains with Markedly Divergent Phenotypes

The comparison of three closely related *E. coli* sequence type ST73 isolates CFT073, 83972, and Nissle 1917 (EcN) exemplifies the difficulty to distinguish certain ExPEC and non-pathogenic *E. coli* variants. Strain CFT073 is a highly virulent archetypal uropathogenic isolate, whereas *E. coli* strains 83972 and Nissle 1917 are non-pathogenic strains derived from the urinary tract and the fecal flora, respectively. The three strains belong to the same clonal group (ST73) and are thus phylogenetically very closely related, despite their different environmental origins and disease-causing ability (Grozdanov et al. 2004; Zdziarski et al. 2008).

UPEC strain CFT073 has been isolated from the blood and urine of a woman with acute pyelonephritis and is widely used a model organism to study UPEC pathogenicity. CFT073 expresses a multitude of virulence genes which promote successful colonization and infection of the urinary tract, including several determinants coding for e.g. adhesins, toxins, iron uptake systems, proteases, flagella, and capsule (Welch et al. 2002; Gunther et al. 2002; Lloyd et al. 2009). The probiotic *E. coli* strain Nissle 1917 shows large overall genome content similarity with *E. coli* CFT073. Nevertheless, EcN lacks functional important virulence gene clusters, such as P-fimbrial and α-hemolysin determinants. The absence of a long-chain O-antigen due to a point mutation in the O-antigen polymerase gene *wzy* renders EcN serum-sensitive (Grozdanov et al. 2002). Among others, these traits are held responsible for the non-pathogenic character of this isolate. Beyond that, EcN has been used as a safe and efficient probiotic strain against a variety of intestinal disorders in humans and animals (Kruis et al. 2012; Schultz 2008; von Buenau et al. 2005). EcN was reported to protect gnotobiotic piglets from infection with invasive bacterial pathogens. Furthermore, Nissle 1917 is a good biofilm former and its efficient adhesion to epithelial cells interferes in vitro with the invasion of several bacterial pathogens. Besides its bactericidal activity against many bacterial pathogens (Storm et al. 2011), EcN has also been demonstrated to negatively affect adhesion of bacterial pathogens through secretory components. This seems to be a common mechanism of *E. coli* strains with strong adhesive capacity (Storm et al. 2011; Altenhoefer et al. 2004; Huebner et al. 2011; Schierack et al. 2011). Immunomodulatory, anti-inflammatory properties have been described for EcN as well (Adam et al. 2010; Güttsches et al. 2012; Ukena et al. 2005). The inducible antimicrobial peptide human β-defensin 2 (hBD-2) is synthesized by the epithelium to counteract bacterial adherence and invasion. Flagellin expressed by EcN induces hBD-2 expression and can thus promote host defenses against bacterial infection (Schlee et al. 2007). In addition, EcN is able to restore disrupted epithelial barriers and to increase its resistance to microbial pathogens (Stetinova et al. 2010; Ukena et al. 2007).

E. coli 83972 is an asymptomatic bacteriuria (ABU) isolate with the ability to colonize the human urinary bladder without inducing an immune response. Similar to EcN, mutations in the *E. coli* 83972 genes encoding type 1-, F1C- and P fimbriae as well as α-hemolysin result in the loss of its ability to express these important virulence-associated genes as a result of host-driven adaptation. This strain also exhibits a semi-rough phenotype indicative of the absence of a long-chain O-antigen. *E. coli* 83972 has beneficial features as it outcompetes UPEC isolates for growth in urine and thus has a selective advantage over UPEC. This can be exploited for preventative and/or therapeutic approaches based on bacterial interference (Sundén et al. 2010). Strain 83972 has been established as an example of bacterial adaptation from pathogenicity to commensalism through virulence factor loss. It is assumed that prolonged asymptomatic bladder colonization selects for such attenuated variants where VFs have been inactivated, e.g. by point mutations and small deletions (Salvador et al. 2012; Zdziarski et al. 2008, 2010; Sundén et al. 2010; Klemm et al. 2006; Roos et al. 2006). Comparative genomic analyses indicated relatively few variations in genome content between these three isolates, thus suggesting that genetic variations (e.g. mutations, rearrangements, deletions) and expression differences, rather than a markedly different genome content, contribute to the divergent phenotypes of these strains. Notably, the two *E. coli* strains Nissle 1917 and 83972 with beneficial traits are deconstructed, attenuated pathogens (Grozdanov et al. 2002, 2004; Zdziarski et al. 2008, 2010; Hancock 2010a, b; Vejborg et al. 2010). To date, it is unknown whether strain CFT073 is also able to express beneficial traits, which could generally contribute to fitness and competitiveness, and whether they are just superimposed by the functional VFs expressed. Alternatively, specific genomic features of EcN and ABU isolate 83972, which are absent in UPEC CFT073, may account for their specific interaction with human epithelial cells or other bacteria. It will be an interesting and important future task to further characterize beneficial traits of strains EcN and 83972, to identify the underlying molecular mechanisms, and correlate them with genomic and phenotypic differences between UPEC CFT073, probiotic EcN, and ABU isolate 83972.

5 Outlook on Future *E. coli* Genomic Studies, Challenges, and What Can Be Expected

Due to genome plasticity, new virulence gene combinations and thus *E. coli* pathotypes with altered features can quickly arise. The large EHEC outbreak in May and June 2011 in central Europe that was caused by an *E. coli* O104:H4 strain combining characteristics of EHEC and EAEC demonstrated how new combinations of virulence genes can result in dangerous pathogenic variants (Brzuszkiewicz et al. 2011; Mellmann et al. 2011; Rasko et al. 2011; Rohde et al. 2011). Parallel evolution and the generation of new virulence gene combinations due to horizontal transfer of large mobile genetic elements will constantly result in the emergence of new variants of already existing *E. coli* pathotypes. Thus, an improved and accelerated strain typing and risk assessment of such new variants is required including the determination of the phylogenetic background and pathotype markers (Karch et al. 2012). Fast phylogenetic analyses will have to be combined with rapid whole genome sequencing (next generation sequencing) to quickly assess the complete (virulence- and resistance-associated) gene content of pathogenic isolates.

Because *E. coli* is such a diverse organism, thriving in very different environments, and having a huge genomic diversity, it is an ideal candidate to study adaptation and evolutionary events involved in the diversification and adaptation of pathogenic and commensal variants. In this respect, analysis of the interplay between the host and commensal or pathogenic *E. coli* strains is very promising. This includes e.g. studies on the intra-host evolution of bacterial strains, host factors contributing to susceptibility of infection (interaction of *E. coli* pathogens and commensals with the immune system), differential regulation of conserved genes in commensals and pathogens as well as the interplay of *E. coli* variants with the complex accompanying intestinal microbiota that also characterizes the healthy or diseased intestinal niche of *E. coli*. Recent technological advances in transcriptomics, (meta-)genomics, and metabolomics will be very helpful to further analyze (disease) ecology of niches colonized or infected by *E. coli* as well as the bacterial traits distinguishing commensal and pathogenic variants.

Acknowledgments Our work related to this subject was supported by the German Research Foundation (SFB 479, TP A1; DO 789/3-1; DO 789/4-1). Our studies were carried out within the European Virtual Institute for Functional Genomics of Bacterial Pathogens (CEE LSHB-CT-2005-512061) and the ERA-NET PathoGenoMics I consortium ‘Deciphering the intersection of extraintestinal pathogenic and commensal *Escherichia coli*’ (Federal Ministry of Education and Research (BMBF) grant no. 0313937A).

References

Abraham S, Gordon DM, Chin J et al (2012) Molecular characterization of commensal *Escherichia coli* adapted to different compartments of the Porcine Gastrointestinal tract. Appl Environ Microbiol 78:6799–6803

Adam E, Delbrassine L, Bouillot C et al (2010) Probiotic *Escherichia coli* Nissle 1917 activates DC and prevents house dust mite allergy through a TLR4-dependent pathway. Eur J Immunol 40:1995–2005

Adler J, Hazelbauer GL, Dahl MM (1973) Chemotaxis toward sugars in *Escherichia coli*. J Bacteriol 115:824–847

Alsam S, Jeong SR, Sissons J, Dudley R, Kim KS, Khan NA (2006) *Escherichia coli* interactions with *Acanthamoeba*: a symbiosis with environmental and clinical implications. J Med Microbiol 55:689–694

Altenhoefer A, Oswald S, Sonnenborn U et al (2004) The probiotic *Escherichia coli* strain Nissle 1917 interferes with invasion of human intestinal epithelial cells by different enteroinvasive bacterial pathogens. FEMS Immunol Med Microbiol 40:223–229

Ambur OH, Davidsen T, Frye SA et al (2009) Genome dynamics in major bacterial pathogens. FEMS Microbiol Rev 33:453–470

Anfora AT, Welch RA (2006) DsdX is the second D-serine transporter in uropathogenic *Escherichia coli* clinical isolate CFT073. J Bacteriol 188:6622–6628

Anfora AT, Haugen BJ, Roesch P, Redford P, Welch RA (2007) Roles of serine accumulation and catabolism in the colonization of the murine urinary tract by *Escherichia coli* CFT073. Infect Immun 75:5298–5304

Angiuoli SV, Salzberg SL (2011) Mugsy: fast multiple alignment of closely related whole genomes. Bioinformatics 27(3):334–342

Baba T, Ara T, Hasegawa M et al (2006) Construction of *Escherichia coli* K-12 in-frame, single-gene knockout mutants: the Keio collection. Mol Syst Biol 2(2006):0008

Berger CN, Sodha SV, Shaw RK et al (2010) Fresh fruit and vegetables as vehicles for the transmission of human pathogens. Environ Microbiol 12:2385–2397

Bernier-Febreau C, du Merle L, Turlin E et al (2004) Use of deoxyribose by intestinal and extraintestinal pathogenic *Escherichia coli* strains: a metabolic adaptation involved in competitiveness. Infect Immun 72:6151–6156

Bielaszewska M, Middendorf B, Tarr PI et al (2011) Chromosomal instability in enterohaemorrhagic *Escherichia coli* O157:H7: impact on adherence, tellurite resistance and colony phenotype. Mol Microbiol 79:1024–1044

Bouckaert J, Mackenzie J, de Paz JL et al (2006) The affinity of the FimH fimbrial adhesin is receptor-driven and quasi-independent of *Escherichia coli* pathotypes. Mol Microbiol 61:1556–1568

Boyd EF, Hartl DL (1998) Chromosomal regions specific to pathogenic isolates of *Escherichia coli* have a phylogenetically clustered distribution. J Bacteriol 180:1159–1165

Bronowski C, Smith SL, Yokota K et al (2008) A subset of mucosa-associated *Escherichia coli* isolates from patients with colon cancer, but not Crohn's disease, share pathogenicity islands with urinary pathogenic *E. coli*. Microbiology 154:571–583

Brzuszkiewicz E, Brüggemann H, Liesegang H et al (2006) How to become a uropathogen: comparative genomic analysis of extraintestinal pathogenic *Escherichia coli* strains. Proc Natl Acad Sci USA 103:12879–12884

Brzuszkiewicz E, Thürmer A, Schuldes J et al (2011) Genome sequence analyses of two isolates from the recent *Escherichia coli* outbreak in Germany reveal the emergence of a new pathotype: Entero-Aggregative-Haemorrhagic *Escherichia coli* (EAHEC). Arch Microbiol 193:883–891

Castillo A, Eguiarte LE, Souza V (2005) A genomic population genetics analysis of the pathogenic enterocyte effacement island in *Escherichia coli*: the search for the unit of selection. Proc Natl Acad Sci USA 102:1542–1547

Caugant DA, Levin BR, Selander RK (1981) Genetic diversity and temporal variation in the *E. coli* population of a human host. Genetics 98:467–490
Chaudhuri RR, Henderson IR (2012) The evolution of the *Escherichia coli* phylogeny. Infect Genet Evol 12:214–226
Chaudhuri RR, Sebaihia M, Hobman JL et al (2010) Complete genome sequence and comparative metabolic profiling of the prototypical enteroaggregative *Escherichia coli* strain 042. PLoS One 5:e8801
Chen SL, Hung CS, Xu J et al (2006) Identification of genes subject to positive selection in uropathogenic strains of *Escherichia coli*: a comparative genomics approach. Proc Natl Acad Sci USA 103:5977–5982
Chen WH, Minguez P, Lercher MJ, Bork P (2012) OGEE: an online gene essentiality database. Nucl Acids Res 40:901–906
Clarke DJ, Chaudhuri RR, Martin HM et al (2011) Complete genome sequence of the Crohn's disease-associated adherent-invasive *Escherichia coli* strain HM605. J Bacteriol 193:4540
Clermont O, Bonacorsi S, Bingen E (2001) The *Yersinia* high-pathogenicity island is highly predominant in virulence-associated phylogenetic groups of *Escherichia coli*. FEMS Microbiol Lett 196:153–157
Crossman LC, Chaudhuri RR, Beatson SA et al (2010) A commensal gone bad: complete genome sequence of the prototypical enterotoxigenic *Escherichia coli* strain H10407. J Bacteriol 192:5822–5831
Croxen MA, Finlay BB (2010) Molecular mechanisms of *Escherichia coli* pathogenicity. Nat Rev Microbiol 8:26–38
Diard M, Baeriswyl S, Clermont O et al (2007) *Caenorhabditis elegans* as a simple model to study phenotypic and genetic virulence determinants of extraintestinal pathogenic *Escherichia coli*. Microbes Infect 9:214–223
Diard M, Garry L, Selva M, Mosser T, Denamur E, Matic I (2010) Pathogenicity-associated islands in extraintestinal pathogenic *Escherichia coli* are fitness elements involved in intestinal colonization. J Bacteriol 192:4885–4893
Diaz E, Ferrandez A, Prieto MA, Garcia JL (2001) Biodegradation of aromatic compounds by *Escherichia coli*. Microbiol Mol Bio Rev 65:523–569
Didelot X, Meric G, Falush D, Darling AE (2012) Impact of homologous and non-homologous recombination in the genomic evolution of *Escherichia coli*. BMC Genomics 13:256
Dixit SM, Gordon DM, Wu XY, Chapman T, Kailasapathy K, Chin JJ (2004) Diversity analysis of commensal porcine *Escherichia coli*—associations between genotypes and habitat in the porcine gastrointestinal tract. Microbiology 150:1735–1740
Dobrindt U (2005) (Patho-)Genomics of *Escherichia coli*. Int J Med Microbiol 295:357–371
Dobrindt U, Agerer F, Michaelis K et al (2003) Analysis of genome plasticity in pathogenic and commensal *Escherichia coli* isolates by use of DNA arrays. J Bacteriol 185:1831–1840
Dobrindt U, Hochhut B, Hentschel U, Hacker J (2004) Genomic islands in pathogenic and environmental microorganisms. Nat Rev Microbiol 2:414–424
Escobar-Paramo P, Grenet K, Le Menac'h A et al (2004a) Large-scale population structure of human commensal *Escherichia coli* isolates. Appl Environ Microbiol 70:5698–5700
Escobar-Paramo P, Clermont O, Blanc-Potard AB, Bui H, Le Bouguenec C, Denamur E (2004b) A specific genetic background is required for acquisition and expression of virulence factors in *Escherichia coli*. Mol Biol Evol 21:1085–1094
Escobar-Paramo P, Le Menac'h A, Le Gall T et al (2006) Identification of forces shaping the commensal *Escherichia coli* genetic structure by comparing animal and human isolates. Environ Microbiol 8:1975–1984
Etchuuya R, Ito M, Kitano S, Shigi F, Sobue R, Maeda S (2011) Cell-to-cell transformation in *Escherichia coli:* a novel type of natural transformation involving cell-derived DNA and a putative promoting pheromone. PLoS One 6:e16355
Fabich AJ, Jones SA, Chowdhury FZ et al (2008) Comparison of carbon nutrition for pathogenic and commensal *Escherichia coli* strains in the mouse intestine. Infect Immun 76:1143–1152

Feldmann F, Sorsa LJ, Hildinger K, Schubert S (2007) The salmochelin siderophore receptor IroN contributes to invasion of urothelial cells by extraintestinal pathogenic *Escherichia coli* in vitro. Infect Immun 75:3183–3187
Fricke WF, Wright MS, Lindell AH et al (2008) Insights into the environmental resistance gene pool from the genome sequence of the multidrug-resistant environmental isolate *Escherichia coli* SMS-3-5. J Bacteriol 190:6779–6794
Goldenfeld N, Woese C (2007) Biology's next revolution. Nature 445:369
Gordon DM, Cowling A (2003) The distribution and genetic structure of *Escherichia coli* in Australian vertebrates: host and geographic effects. Microbiology 149:3575–3586
Grasselli E, Francois P, Gutacker M et al (2008) Evidence of horizontal gene transfer between human and animal commensal *Escherichia coli* strains identified by microarray. FEMS Immunol Med Microbiol 53:351–358
Grozdanov L, Zähringer U, Blum-Oehler G et al (2002) A single nucleotide exchange in the *wzy* gene is responsible for the semi-rough O6 lipopolysaccharide phenotype and serum sensitivity of *Escherichia coli* strain Nissle 1917. J Bacteriol 184:5912–5925
Grozdanov L, Raasch C, Schulze J et al (2004) Analysis of the genome structure of the non-pathogenic probiotic *Escherichia coli* strain Nissle 1917. J Bacteriol 186:5432–5441
Gunther NWt, Snyder JA, Lockatell V, Blomfield I, Johnson DE, Mobley HL (2002) Assessment of virulence of uropathogenic *Escherichia coli* type 1 fimbrial mutants in which the invertible element is phase-locked on or off. Infect Immun 70:3344–3354
Güttsches AK, Loseke S, Zähringer U et al (2012) Anti-inflammatory modulation of immune response by probiotic *Escherichia coli* Nissle 1917 in human blood mononuclear cells. Innate Immun 18:204–216
Hacker J, Dobrindt U (eds) (2006) Pathogenomics: Genome analysis of pathogenic microbes. Wiley-VCH, Weinheim
Hacker J, Hentschel U, Dobrindt U (2003) Prokaryotic chromosomes and disease. Science 301:790–793
Hafez M, Hayes K, Goldrick M, Warhurst G, Grencis R, Roberts IS (2009) The K5 capsule of *Escherichia coli* strain Nissle 1917 is important in mediating interactions with intestinal epithelial cells and chemokine induction. Infect Immun 77:2995–3003
Halachev MR, Loman NJ, Pallen MJ (2011) Calculating orthologs in *Bacteria* and *Archaea*: a divide and conquer approach. PLoS One 6:e28388
Hancock V, Ferrieres L, Klemm P (2008) The ferric yersiniabactin uptake receptor FyuA is required for efficient biofilm formation by urinary tract infectious *Escherichia coli* in human urine. Microbiology 154:167–175
Hancock V, Dahl M, Klemm P (2010a) Probiotic *Escherichia coli* strain Nissle 1917 outcompetes intestinal pathogens during biofilm formation. J Med Microbiol 59:392–399
Hancock V, Vejborg RM, Klemm P (2010b) Functional genomics of probiotic *Escherichia coli* Nissle 1917 and 83972, and UPEC strain CFT073: comparison of transcriptomes, growth and biofilm formation. Mol Genet Genomics 284:437–454
Hashimoto M, Ichimura T, Mizoguchi H et al (2005) Cell size and nucleoid organization of engineered *Escherichia coli* cells with a reduced genome. Mol Microbiol 55:137–149
Hayashi T, Makino K, Ohnishi M et al (2001) Complete genome sequence of enterohemorrhagic *Escherichia coli* O157:H7 and genomic comparison with a laboratory strain K-12. DNA Res 8:11–22
Hejnova J, Dobrindt U, Nemcova R et al (2005) Characterization of the flexible genome complement of the commensal *Escherichia coli* strain A0 34/86 (O83:K24:H31). Microbiology 151:385–398
Hendrickson H (2009) Order and disorder during *Escherichia coli* divergence. PLoS Genet 5:e1000335
Hill CW, Feulner G, Brody MS, Zhao S, Sadosky AB, Sandt CH (1995) Correlation of Rhs elements with *Escherichia coli* population structure. Genetics 141:15–24
Ho Sui SJ, Fedynak A, Hsiao WW, Langille MG, Brinkman FS (2009) The association of virulence factors with genomic islands. PLoS One 4:e8094

Holden N, Pritchard L, Toth I (2009) Colonization outwith the colon: plants as an alternative environmental reservoir for human pathogenic enterobacteria. FEMS Microbiol Rev 33:689–703

Huebner C, Ding Y, Petermann I, Knapp C, Ferguson LR (2011) The probiotic *Escherichia coli* Nissle 1917 reduces pathogen invasion and modulates cytokine expression in Caco-2 cells infected with Crohn's disease-associated *E. coli* LF82. Appl Environ Microbiol 77:2541–2544

Hung CS, Bouckaert J, Hung D et al (2002) Structural basis of tropism of *Escherichia coli* to the bladder during urinary tract infection. Mol Microbiol 44:903–915

Huson DH, Scornavacca C (2012) Dendroscope 3: an interactive tool for rooted phylogenetic trees and networks. Syst Biol 61:1061–1067

Iguchi A, Thomson NR, Ogura Y et al (2009) Complete genome sequence and comparative genome analysis of enteropathogenic *Escherichia coli* O127:H6 strain E2348/69. J Bacteriol 191:347–354

Jackson RW, Vinatzer B, Arnold DL, Dorus S, Murillo J (2011) The influence of the accessory genome on bacterial pathogen evolution. Mob Genet Elements 1:55–65

Johnson TJ, Kariyawasam S, Wannemuehler Y et al (2007) The genome sequence of avian pathogenic *Escherichia coli* strain O1:K1:H7 shares strong similarities with human extraintestinal pathogenic *E. coli* genomes. J Bacteriol 189:3228–3236

Johnson JR, Johnston B, Kuskowski MA, Nougayrède JP, Oswald E (2008) Molecular epidemiology and phylogenetic distribution of the *Escherichia coli pks* genomic island. J Clin Microbiol 46:3906–3911

Juhas M, van der Meer JR, Gaillard M, Harding RM, Hood DW, Crook DW (2009) Genomic islands: tools of bacterial horizontal gene transfer and evolution. FEMS Microbiol Rev 33:376–393

Kaper JB, Nataro JP, Mobley HL (2004) Pathogenic *Escherichia coli*. Nat Rev Microbiol 2:123–140

Karch H, Denamur E, Dobrindt U et al (2012) The enemy within us: lessons from the 2011 European *Escherichia coli* O104:H4 outbreak. EMBO Mol Med 4:841–848

Kato J, Hashimoto M (2007) Construction of consecutive deletions of the *Escherichia coli* chromosome. Mol Systems Biol 3:132

Klemm P, Roos V, Ulett GC, Svanborg C, Schembri MA (2006) Molecular characterization of the *Escherichia coli* asymptomatic bacteriuria strain 83972: the taming of a pathogen. Infect Immun 74:781–785

Köhler CD, Dobrindt U (2011) What defines extraintestinal pathogenic *Escherichia coli*? Int J Med Microbiol 301:642–647

Koli P, Sudan S, Fitzgerald D, Adhya S, Kar S (2011) Conversion of commensal *Escherichia coli* K-12 to an invasive form via expression of a mutant histone-like protein. MBio 2(5):e00182–11

Krause DO, Little AC, Dowd SE, Bernstein CN (2011) Complete genome sequence of adherent invasive *Escherichia coli* UM146 isolated from Ileal Crohn's disease biopsy tissue. J Bacteriol 193:583

Krieger JN, Dobrindt U, Riley DE, Oswald E (2011) Acute *Escherichia coli* prostatitis in previously health young men: bacterial virulence factors, antimicrobial resistance, and clinical outcomes. Urology 77:1420–1425

Kruis W, Chrubasik S, Boehm S, Stange C, Schulze J (2012) A double-blind placebo-controlled trial to study therapeutic effects of probiotic *Escherichia coli* Nissle 1917 in subgroups of patients with irritable bowel syndrome. Int J Colorectal Dis 27:467–474

Kurtz S, Phillippy A, Delcher AL et al (2004) Versatile and open software for comparing large genomes. Genome Biol 5:R12

Laing CR, Buchanan C, Taboada EN et al (2009) *In silico* genomic analyses reveal three distinct lineages of *Escherichia coli* O157:H7, one of which is associated with hyper-virulence. BMC Genomics 10:287

Lane MC, Lockatell V, Monterosso G et al (2005) Role of motility in the colonization of uropathogenic *Escherichia coli* in the urinary tract. Infect Immun 73:7644–7656

Lane MC, Alteri CJ, Smith SN, Mobley HL (2007) Expression of flagella is coincident with uropathogenic *Escherichia coli* ascension to the upper urinary tract. Proc Natl Acad Sci U S A 104:16669–16674

Le Bouguénec C, Schouler C (2011) Sugar metabolism, an additional virulence factor in enterobacteria. Int J Med Microbiol 301:1–6

Le Gall T, Clermont O, Gouriou S et al (2007) Extraintestinal virulence is a coincidental by-product of commensalism in B2 phylogenetic group *Escherichia coli* strains. Mol Biol Evol 24:2373–2384

Lee S, Yu JK, Park K, Oh EJ, Kim SY, Park YJ (2010) Phylogenetic groups and virulence factors in pathogenic and commensal strains of *Escherichia coli* and their association with *bla*CTX-M. Ann Clin Lab Sci 40:361–367

Leopold SR, Sawyer SA, Whittam TS, Tarr PI (2011) Obscured phylogeny and possible recombinational dormancy in *Escherichia coli*. BMC Evol Biol 11:183

Léveillé S, Caza M, Johnson JR, Clabots C, Sabri M, Dozois CM (2006) Iha from an *Escherichia coli* urinary tract infection outbreak clonal group A strain is expressed in vivo in the mouse urinary tract and functions as a catecholate siderophore receptor. Infect Immun 74:3427–3436

Li B, Sun JY, Han LZ, Huang XH, Fu Q, Ni YX (2010) Phylogenetic groups and pathogenicity island markers in fecal *Escherichia coli* isolates from asymptomatic humans in China. Appl Environ Microbiol 76:6698–6700

Lloyd AL, Henderson TA, Vigil PD, Mobley HL (2009) Genomic islands of uropathogenic *Escherichia coli* contribute to virulence. J Bacteriol 191:3469–3481

Lu S, Zhang X, Zhu Y, Kim KS, Yang J, Jin Q (2011) Complete genome sequence of the neonatal-meningitis-associated *Escherichia coli* strain CE10. J Bacteriol 193:7005

Lukjancenko O, Wassenaar TM, Ussery DW (2010) Comparison of 61 sequenced *Escherichia coli* genomes. Microb Ecol 60:708–720

Luo C, Walk ST, Gordon DM, Feldgarden M, Tiedje JM, Konstantinidis KT (2011) Genome sequencing of environmental *Escherichia coli* expands understanding of the ecology and speciation of the model bacterial species. Proc Natl Acad Sci U S A 108:7200–7205

Mau B, Glasner JD, Darling AE, Perna NT (2006) Genome-wide detection and analysis of homologous recombination among sequenced strains of *Escherichia coli*. Genome Biol 7:R44

Medini D, Donati C, Tettelin H, Masignani V, Rappuoli R (2005) The microbial pan-genome. Curr Opin Genet Dev 15:589–594

Medini D, Serruto D, Parkhill J et al (2008) Microbiology in the post-genomic era. Nat Rev Microbiol 6:419–430

Mellmann A, Harmsen D, Cummings CA et al (2011) Prospective genomic characterization of the German enterohemorrhagic *Escherichia coli* O104:H4 outbreak by rapid next generation sequencing technology. PLoS One 6:e22751

Mesibov R, Adler J (1972) Chemotaxis toward amino acids in *Escherichia coli*. J Bacteriol 112:315–326

Milkman R, Bridges MM (1990) Molecular evolution of the *Escherichia coli* chromosome. III. Clonal frames. Genetics 126:505–517

Miquel S, Peyretaillade E, Claret L et al. (2010) Complete genome sequence of Crohn's disease-associated adherent-invasive *E. coli* strain LF82. PLoS One 5:e12714

Monteiro C, Saxena I, Wang X et al (2009) Characterization of cellulose production in *Escherichia coli* Nissle 1917 and its biological consequences. Environ Microbiol 11:1105–1116

Mordhorst IL, Claus H, Ewers C et al (2009) O-acetyltransferase gene *neuO* is segregated according to phylogenetic background and contributes to environmental desiccation resistance in *Escherichia coli* K1. Environ Microbiol 11:3154–3165

Moriel DG, Bertoldi I, Spagnuolo A et al (2010) Identification of protective and broadly conserved vaccine antigens from the genome of extraintestinal pathogenic *Escherichia coli*. Proc Natl Acad Sci U S A 107:9072–9077

Nash JH, Villegas A, Kropinski AM et al (2010) Genome sequence of adherent-invasive *Escherichia coli* and comparative genomic analysis with other *E. coli* pathotypes. BMC Genomics 11:667

Nougayrède JP, Homburg S, Taieb F et al (2006) *Escherichia coli* induces DNA double-strand breaks in eukaryotic cells. Science 313:848–851

Nowrouzian FL, Oswald E (2012) *Escherichia coli* strains with the capacity for long-term persistence in the bowel microbiota carry the potentially genotoxic *pks* island. Microb Pathog 53:180–182

Nowrouzian F, Adlerberth I, Wold AE (2001) P fimbriae, capsule and aerobactin characterize colonic resident *Escherichia coli*. Epidemiol Infect 126:11–18

Nowrouzian F, Hesselmar B, Saalman R et al (2003) *Escherichia coli* in infants' intestinal microflora: colonization rate, strain turnover, and virulence gene carriage. Pediatr Res 54:8–14

Nowrouzian FL, Wold AE, Adlerberth I (2005) *Escherichia coli* strains belonging to phylogenetic group B2 have superior capacity to persist in the intestinal microflora of infants. J Infect Dis 191:1078–1083

Nowrouzian FL, Adlerberth I, Wold AE (2006) Enhanced persistence in the colonic microbiota of *Escherichia coli* strains belonging to phylogenetic group B2: role of virulence factors and adherence to colonic cells. Microbes Infect 8:834–840

Nowrouzian FL, Ostblom AE, Wold AE, Adlerberth I (2009) Phylogenetic group B2 *Escherichia coli* strains from the bowel microbiota of Pakistani infants carry few virulence genes and lack the capacity for long-term persistence. Clin Microbiol Infect 15:466–472

Ochman H, Selander RK (1984) Standard reference strains of *Escherichia coli* from natural populations. J Bacteriol 157:690–693

Ogura Y, Ooka T, Iguchi A et al (2009) Comparative genomics reveal the mechanism of the parallel evolution of O157 and non-O157 enterohemorrhagic *Escherichia coli*. Proc Natl Acad Sci USA 106:17939–17944

Olier M, Marcq I, Salvador-Cartier C et al. (2012) Genotoxicity of *Escherichia coli* Nissle 1917 strain cannot be dissociated from its probiotic activity. Gut Microbes 3(6):501–509

Oshima K, Toh H, Ogura Y et al (2008) Complete genome sequence and comparative analysis of the wild-type commensal *Escherichia coli* strain SE11 isolated from a healthy adult. DNA Res 15:375–386

Ostblom A, Adlerberth I, Wold AE, Nowrouzian FL (2011) Pathogenicity island markers, virulence determinants *malX* and *usp*, and the capacity of *Escherichia coli* to persist in infants' commensal microbiotas. Appl Environ Microbiol 77:2303–2308

Perna NT, Plunkett G 3rd, Burland V et al (2001) Genome sequence of enterohaemorrhagic *Escherichia coli* O157:H7. Nature 409:529–533

Porcheron G, Chanteloup NK, Trotereau A, Bree A, Schouler C (2012) Effect of fructooligosaccharide metabolism on chicken colonization by an extra-intestinal pathogenic *Escherichia coli* strain. PLoS One 7:e35475

Putze J, Hennequin C, Nougayrède JP et al (2009) Genetic structure and distribution of the colibactin genomic island among members of the family *Enterobacteriaceae*. Infect Immun 77:4696–4703

Rasko DA, Rosovitz MJ, Myers GS et al (2008) The pangenome structure of *Escherichia coli*: comparative genomic analysis of *E. coli* commensal and pathogenic isolates. J Bacteriol 190:6881–6893

Rasko DA, Webster DR, Sahl JW et al (2011) Origins of the *E. coli* strain causing an outbreak of hemolytic-uremic syndrome in Germany. N Engl J Med 365:709–717

Reid SD, Herbelin CJ, Bumbaugh AC, Selander RK, Whittam TS (2000) Parallel evolution of virulence in pathogenic *Escherichia coli*. Nature 406:64–67

Ren CP, Chaudhuri RR, Fivian A et al (2004) The ETT2 gene cluster, encoding a second type III secretion system from *Escherichia coli*, is present in the majority of strains but has undergone widespread mutational attrition. J Bacteriology 186:3547–3560

Roesch PL, Redford P, Batchelet S et al (2003) Uropathogenic *Escherichia coli* use D-serine deaminase to modulate infection of the murine urinary tract. Mol Microbiol 49:55–67

Rohde H, Qin J, Cui Y et al (2011) Open-source genomic analysis of Shiga-toxin-producing *E. coli* O104:H4. N Engl J Med 365:718–724

Roos V, Schembri MA, Ulett GC, Klemm P (2006) Asymptomatic bacteriuria *Escherichia coli* strain 83972 carries mutations in the foc locus and is unable to express F1C fimbriae. Microbiology 152:1799–1806

Rouquet G, Porcheron G, Barra C et al (2009) A metabolic operon in extraintestinal pathogenic *Escherichia coli* promotes fitness under stressful conditions and invasion of eukaryotic cells. J Bacteriol 191:4427–4440

Rump LV, Strain EA, Cao G et al (2011) Draft genome sequences of six *Escherichia coli* isolates from the stepwise model of emergence of *Escherichia coli* O157:H7. J Bacteriol 193:2058–2059

Sabaté M, Moreno E, Perez T, Andreu A, Prats G (2006) Pathogenicity island markers in commensal and uropathogenic *Escherichia coli* isolates. Clin Microbiol Infect 12:880–886

Sahl JW, Steinsland H, Redman JC et al (2011) A comparative genomic analysis of diverse clonal types of enterotoxigenic *Escherichia coli* reveals pathovar-specific conservation. Infect Immun 79:950–960

Sahl JW, Matalka MN, Rasko DA (2012) Phylomark, a tool to identify conserved phylogenetic markers from whole-genome alignments. Appl Environ Microbiol 78:4884–4892

Salvador E, Wagenlehner F, Köhler CD et al (2012) Comparison of asymptomatic bacteriuria *Escherichia coli* isolates from healthy individuals versus those from hospital patients shows that long-term bladder colonization selects for attenuated virulence phenotypes. Infect Immun 80:668–678

Schierack P, Walk N, Ewers C et al (2008) ExPEC-typical virulence-associated genes correlate with successful colonization by intestinal *E. coli* in a small piglet group. Environ Microbiol 10:1742–1751

Schierack P, Kleta S, Tedin K et al. (2011) *E. coli* Nissle 1917 affects *Salmonella* adhesion to porcine intestinal epithelial cells. PLoS One 6:e14712

Schlee M, Wehkamp J, Altenhoefer A, Oelschlaeger TA, Stange EF, Fellermann K (2007) Induction of human beta-defensin 2 by the probiotic *Escherichia coli* Nissle 1917 is mediated through flagellin. Infect Immun 75:2399–2407

Schneider G, Dobrindt U, Middendorf B et al (2011) Mobilisation and remobilisation of a large archetypal pathogenicity island of uropathogenic *Escherichia coli* in vitro support the role of conjugation for horizontal transfer of genomic islands. BMC Microbiol 11:210

Schouler C, Taki A, Chouikha I, Moulin-Schouleur M, Gilot P (2009) A genomic island of an extraintestinal pathogenic *Escherichia coli* Strain enables the metabolism of fructooligosaccharides, which improves intestinal colonization. J Bacteriol 191:388–393

Schubert S, Darlu P, Clermont O et al (2009) Role of intraspecies recombination in the spread of pathogenicity islands within the *Escherichia coli* species. PLoS Pathog 5:e1000257

Schultz M (2008) Clinical use of *E. coli* Nissle 1917 in inflammatory bowel disease. Inflamm Bowel Dis 14:1012–1018

Schwan WR (2008) Flagella allow uropathogenic *Escherichia coli* ascension into murine kidneys. Int J Med Microbiol 298:441–447

Sears HJ, Brownlee I (1952) Further observations on the persistence of individual strains of *Escherichia coli* in the intestinal tract of man. J Bacteriol 63:47–57

Sears HJ, Brownlee I, Uchiyama JK (1950) Persistence of individual strains of *Escherichia coli* in the intestinal tract of man. J Bacteriol 59:293–301

Shepard SM, Danzeisen JL, Isaacson RE, Seemann T, Achtman M, Johnson TJ (2012) Genome sequences and phylogenetic analysis of K88- and F18-positive porcine enterotoxigenic *Escherichia coli*. J Bacteriol 194:395–405

Sims GE, Kim SH (2011) Whole-genome phylogeny of *Escherichia coli/Shigella* group by feature frequency profiles (FFPs). Proc Natl Acad Sci USA 108:8329–8334

Smajs D, Micenkova L, Smarda J et al (2010) Bacteriocin synthesis in uropathogenic and commensal *Escherichia coli*: colicin E1 is a potential virulence factor. BMC Microbiol 10:288

Söderblom T, Laestadius A, Oxhamre C, Aperia A, Richter-Dahlfors A (2002) Toxin-induced calcium oscillations: a novel strategy to affect gene regulation in target cells. Int J Med Microbiol 291:511–515

Stahlhut SG, Tchesnokova V, Struve C et al (2009) Comparative structure-function analysis of mannose-specific FimH adhesins from *Klebsiella pneumoniae* and *Escherichia coli*. J Bacteriol 191:6592–6601

Stamatakis A (2006) RAxML-VI-HPC: maximum likelihood-based phylogenetic analyses with thousands of taxa and mixed models. Bioinformatics 22:2688–2690

Stamatakis A, Ott M (2008) Efficient computation of the phylogenetic likelihood function on multi-gene alignments and multi-core architectures. Philos Trans R Soc Lond B Biol Sci 363:3977–3984

Steinberg KM, Levin BR (2007) Grazing protozoa and the evolution of the *Escherichia coli* O157:H7 Shiga toxin-encoding prophage. Proc Biol Sci 274:1921–1929

Stetinova V, Smetanova L, Kvetina J, Svoboda Z, Zidek Z, Tlaskalova-Hogenova H (2010) Caco-2 cell monolayer integrity and effect of probiotic *Escherichia coli* Nissle 1917 components. Neuro Endocrinol Lett 31(Suppl 2):51–56

Storm DW, Koff SA, Horvath DJ Jr, Li B, Justice SS (2011) *In vitro* analysis of the bactericidal activity of *Escherichia coli* Nissle 1917 against pediatric uropathogens. J Urol 186:1678–1683

Sundén F, Håkansson L, Ljunggren E, Wullt B (2010) *Escherichia coli* 83972 bacteriuria protects against recurrent lower urinary tract infections in patients with incomplete bladder emptying. J Urol 184:179–185

Tenaillon O, Skurnik D, Picard B, Denamur E (2010) The population genetics of commensal *Escherichia coli*. Nat Rev Microbiol 8:207–217

Tettelin H, Masignani V, Cieslewicz MJ et al (2005) Genome analysis of multiple pathogenic isolates of *Streptococcus agalactiae:* implications for the microbial "pan-genome". Proc Natl Acad Sci USA 102:13950–13955

Tettelin H, Riley D, Cattuto C, Medini D (2008) Comparative genomics: the bacterial pan-genome. Curr Opin Microbiol 11:472–477

Toh H, Oshima K, Toyoda A et al (2010) Complete genome sequence of the wild-type commensal *Escherichia coli* strain SE15, belonging to phylogenetic group B2. J Bacteriol 192:1165–1166

Touchon M, Hoede C, Tenaillon O et al (2009) Organised genome dynamics in the *Escherichia coli* species results in highly diverse adaptive paths. PLoS Genet 5:e1000344

Uhlén P, Laestadius A, Jahnukainen T et al (2000) Alpha-haemolysin of uropathogenic *E. coli* induces Ca2+ oscillations in renal epithelial cells. Nature 405:694–697

Ukena SN, Westendorf AM, Hansen W et al (2005) The host response to the probiotic *Escherichia coli* strain Nissle 1917: specific up-regulation of the proinflammatory chemokine MCP-1. BMC Med Genet 6:43

Ukena SN, Singh A, Dringenberg U et al (2007) Probiotic *Escherichia coli* Nissle 1917 inhibits leaky gut by enhancing mucosal integrity. PLoS One 2:e1308

Vejborg RM, Friis C, Hancock V, Schembri MA, Klemm P (2010) A virulent parent with probiotic progeny: comparative genomics of *Escherichia coli* strains CFT073, Nissle 1917 and ABU 83972. Mol Genet Genomics 283:469–484

Vieira G, Sabarly V, Bourguignon PY et al (2011) Core and panmetabolism in *Escherichia coli*. J Bacteriol 193:1461–1472

von Buenau R, Jaekel L, Schubotz E, Schwarz S, Stroff T, Krueger M (2005) *Escherichia coli* strain Nissle 1917: significant reduction of neonatal calf diarrhea. J Dairy Sci 88:317–323

Walk ST, Alm EW, Gordon DM et al (2009) Cryptic lineages of the genus *Escherichia*. Appl Environ Microbiol 75:6534–6544

Wang X, Rochon M, Lamprokostopoulou A, Lunsdorf H, Nimtz M, Römling U (2006) Impact of biofilm matrix components on interaction of commensal *Escherichia coli* with the gastrointestinal cell line HT-29. Cell Mol Life Sci 63:2352–2363

Welch RA, Burland V, Plunkett G 3rd et al (2002) Extensive mosaic structure revealed by the complete genome sequence of uropathogenic *Escherichia coli*. Proc Natl Acad Sci USA 99:17020–17024

Whittam TS, Wolfe ML, Wachsmuth IK, Ørskov F, Ørskov I, Wilson RA (1993) Clonal relationships among *Escherichia coli* strains that cause hemorrhagic colitis and infantile diarrhea. Infect Immun 61:1619–1629

Wiedenbeck J, Cohan FM (2011) Origins of bacterial diversity through horizontal genetic transfer and adaptation to new ecological niches. FEMS Microbiol Rev 35:957–976

Wirth T, Falush D, Lan R et al (2006) Sex and virulence in *Escherichia coli*: an evolutionary perspective. Mol Microbiol 60:1136–1151

Wold AE, Caugant DA, Lidin-Janson G, de Man P, Svanborg C (1992) Resident colonic *Escherichia coli* strains frequently display uropathogenic characteristics. J Infect Dis 165:46–52

Zdziarski J, Svanborg C, Wullt B, Hacker J, Dobrindt U (2008) Molecular basis of commensalism in the urinary tract: low virulence or virulence attenuation? Infect Immun 76:695–703

Zdziarski J, Brzuszkiewicz E, Wullt B et al (2010) Host imprints on bacterial genomes—rapid, divergent evolution in individual patients. PLoS Pathog 6:e1001078

Zhou Z, Li X, Liu B et al (2010) Derivation of *Escherichia coli* O157:H7 from its O55:H7 precursor. PLoS One 5:e8700

What Distinguishes Highly Pathogenic Staphylococci from Medium- and Non-pathogenic?

Ralf Rosenstein and Friedrich Götz

Abstract Members of the genus *Staphylococcus* are widespread as commensals of humans and animals where they colonize the skin or mucous membranes. While this coexistence remains mostly untroubled, especially for the healthy host, the bacteria may pose a serious threat for the human or animal host when they get access to inner layers of the body through breaches in skin or membranes. Among the members of the genus a wide span exists in the ability to cope with the hostile conditions encountered in the bloodstream of the living host as a scarce supply of certain nutrients, attacks of the immune system, or anti-infective measures undertaken in the clinical field. In this respect, *Staphylococcus aureus* is by far the most versatile species of the genus. Its equipment with a huge repertoire of different virulence factors and additional supportive gene products that increase the capability to survive within the living host makes *S. aureus* the leading pathogen not only within the genus but also one of the most threatening microorganisms regarding hospitally and community-acquired infections. Compared with *S. aureus*, the other virulent species of the genus like *S. epidermidis*, *S. lugdunensis*, *S. saprophyticus*, and *S. haemolyticus* have a more limited arsenal of virulence factors resulting in a specialized spectrum of diseases and a generally lower degree of pathogenicity. Besides the highly and medium-pathogenic staphylococci, the genus comprises also species like *S. carnosus*, *S. xylosus*, and *S. equorum* that are generally inconspicuous regarding clinical occurrences. Some strains of this group

R. Rosenstein
Microbial Genetics, Institute of Microbiology and Infection Medicine Tübingen,
University of Tübingen, Tübingen, Germany

F. Götz (✉)
Microbial Genetics, University of Tübingen, Waldhäuser Str. 70/8
72076 Tübingen, Germany
e-mail: friedrich.goetz@uni-tuebingen.de

Current Topics in Microbiology and Immunology (2013) 358: 33–89
DOI: 10.1007/82_2012_286

Published Online: 8 December 2012

are used in the food industry and can be graded as non-pathogenic. This review aims to work out the differences between the pathogenic properties of highly and medium-pathogenic staphylococcal species and to draw a comparison between the pathogenic species and the food-grade *S. carnosus* TM300.

Contents

1 Introduction

The genus *Staphylococcus* comprises more than 40 species that differ in their potential to endanger human and animal health, ranging from non-pathogenic food-grade members to dangerous pathogens causing severe infections and being resistant to the treatment by most of the commonly applied antibiotics.

Many staphylococci are found on humans, mammals, or birds where they are located on the skin, skin glands, or mucous membranes. They are either found to coexist indigenously as commensals or to be transiently present as colonizers of their hosts. The transitory presence of some staphylococcal cohabitants often hampers the identification of their natural host range (Götz et al. 2006).

The coexistence with living humans or animals accounts for the equipment of many staphylococci with factors which allow the inhabitation of their ecological niche on skin or mucous membranes by dealing with the existing environmental conditions. The capability to "only" colonize may be extended to elaborate survival strategies that become necessary when the bacteria are able to pass through breaches of skin or membranes and to enter the blood vessels of the host organism. The highly pathogenic *Staphylococcus aureus* strains developed an immane variety of mechanisms that enable them to specifically interact with host factors, to escape the hosts defense mechanisms, to enhance their fitness in the competition for rare substrates, to detach nutritions by disintegration of host tissue, and to cope with antibacterial compounds like antibiotics.

The impact of *Staphylococcus* on human life and health is reflected by numerous genome sequencing projects which aim at a better understanding of the molecular basis of staphylococcal pathogenicity. The so far sequenced staphylococcal species representatives are listed in Table 1.

2 Graduation of Pathogenicity in Staphylococcal Species

According to the equipment with virulence factors (or the ability to acquire them by horizontal gene transfer) and the resulting pathogenic capabilities, staphylococcal species may be arranged along a pathogenicity scale ranging from highly- to non-pathogenic.

2.1 Highly Pathogenic Staphylococci

Among the staphylococci, *S. aureus* is the species with the highest pathogenic potential. *S. aureus* is responsible for a variety of nosocomial or community-acquired infections ranging from boils, furuncles, styes, impetigo, and other superficial skin infections to more serious infections particularly in the chronically

Table 1 Completed genome sequences of staphylococcal species

Species/Strain	Characteristics	Reference
Non-S. aureus		
S. epidermidis RP62A	Biofilm former, catheter-associated sepsis	Gill et al. (2005)
S. epidermidis ATCC 12228	Non-biofilm, not infection associated	Zhang et al. (2003)
S. lugdunensis HKU09-01	Clinical isolate from human pus swab	Tse et al. (2010)
S. lugdunensis N920143	Clinical isolate from breast abscess	Heilbronner et al. (2011)
S. haemolyticus JCSC1435	Clinical strain, multidrug-resistant	Takeuchi et al. (2005)
S. saprophyticus ATCC 15305	Isolated from human urine specimen, uropathogenic	Kuroda et al. (2005)
S. pseudintermedius ED99	Canine pyoderma, coagulase-positive	Ben Zakour et al. (2011)
S. pseudintermedius HKU10-03	Canine pyoderma, coagulase-positive	Tse et al. (2011)
S. carnosus TM300	Non-pathogenic food grade organism	Rosenstein et al. (2009)
S. aureus strains		
04-02981	MRSA, For comparative analysis	Nubel et al. (2010)
RF 122	Mastitis in cattle	Herron et al. (2002)
11819-97	Clinical isolate from skin abscess; MRSA, ST80-IV, CA	Stegger et al. (2012)
71193	Clinical isolate, ST398	Uhlemann et al. (2012)
COL	Early MRSA isolate	Gill et al. (2005)
ECT-R 2	Multiresistant MSSA	Lindqvist et al. (2012)
ED 133	Bovine isolate	Guinane et al. (2010)
ED 98	Poultry isolate, bacterial chondronecrosis with osteomyelitis	Lowder et al. (2009)
HO 5096 0412	Neonatal MRSA outbreak	Wellcome trust sanger institute
JH1	Bloodstream isolate, MRSA vancomycin-sensitive	Mwangi et al. (2007)
JH9	Bloodstream isolate, MRSA VISA	Mwangi et al. (2007)
JKD6159	cMRSA ST93-IV, severe skin and invasive infection	Chua et al. (2010)
LGA251	Bulk milk isolate	Garcia-Alvarez et al. (2011)
M013	CA-MRSA, pvl-positive, ST59	Huang et al. (2012)
MN8	Isolate from urogenital tract, reference genome	Human microbiome project
MRSA252	HA-MRSA, EMRSA-16	Holden et al. (2004)
MSHR1132	Early-branched SA lineage, staphyloxanthin-negative	Holt et al. (2011)
MSSA476	CA-MSSA	Holden et al. (2004)
MW2	Highly virulent CA-MRSA	Baba et al. (2002)

(continued)

Table 1 (continued)

Species/Strain	Characteristics	Reference
Mu3	Hetero-VISA	Neoh et al. (2008)
Mu50	MRSA *van*^{R}, pus isolate	Kuroda et al. (2001)
N315	MRSA, pharyngeal smear isolate	Kuroda et al. (2001)
NCTC 8325	Prototype strain for SA molecular genetics	GenBank: CP000253; direct submission
ST398	MRSA, from human endocarditis, livestock-associated	Schijffelen et al. (2010)
T0131	MRSA ST239, clinical isolate	Li et al. (2011)
TCH60	Skin isolate, reference genome	Human microbiome project
TW20	Clinical isolate, MRSA, ST239	Holden et al. (2010)
USA300_FPR3757	CA-MRSA, carries ACME	Diep et al. (2006)
USA300_TCH1516	Sepsis isolate, CA-MRSA	Highlander et al. (2007)
USA300_TCH959	Buttock abscess isolate, CA-MSSA	Highlander et al. (2007)
VC40	Highly vancomycin-resistant, *vanA*-negative	Sass et al. (2012)
JKD6008	Bloodstream isolate, MRSA-VISA	Howden et al. (2010)
ssp Newbould 305	Bovine mastitis isolate	http://www.ncbi.nlm.nih.gov/bioproject/PRJNA162721
ssp Newman	Human infection isolate (1952), often used in pathogenesis studies	Baba et al. (2008)

ill or immunocompromised patient. The latter include pneumonia, deep abscesses, osteomyelitis, endocarditis, phlebitis, mastitis, and meningitis. Its pathogenicity is based on a huge spectrum of virulence factors as well as on a variety of fitness factors that support the survival in the host. The use of anti-infective counter-measures in the treatment of diseases caused by *S. aureus* is hindered by rapidly spreading resistance genes that could give rise to multi-resistant *S. aureus* strains (MRSA).

2.2 Medium-pathogenic Staphylococci

Besides *S. aureus* as the leading staphylococcal pathogen, various other members of the genus, mostly belonging to the coagulase-negative staphylococci (CoNS), play roles as infectious agents for human or animal hosts. In comparison with *S. aureus,* they reveal a more restricted palette of virulence factors. Accordingly, the medium-pathogenic species are more specialized in their infective strategies and/or limited to a narrow spectrum of diseases. The clinical appearance of these infections may be characterized as more subtle with subacute or chronical clinical courses without fulminant signs and rarely being life-threatening (von Eiff et al. 2002).

The most prominent representative in this category is *S. epidermidis* (Kleeman et al. 1993; Weinstein et al. 1998). This organism is the most prevalent and persistent staphylococcal species on human skin. Wounds or surgery may open entry ports to the host's bloodstream. In recent years, *S. epidermidis* has emerged as a common cause of hospital-acquired infections, including catheter-associated infections and septicemia, particularly in immunocompromised patients. Similar to *S. aureus*, *S. epidermidis* strains may become highly resistant to many antibiotics including penicillins and cephalosporins. In contrast to the multifaceted character of *S. aureus* infections, the pathogenic potential of *S. epidermidis* is almost exclusively based on its ability to form biofilms (Heilmann et al. 1996a, b, 1997; Mack et al. 1996) on implanted or indwelling polymeric material (Götz and Peters 2000). There are only few indications of infections caused by *S. epidermidis* without the involvement of foreign bodies (von Eiff et al. 2002).

Besides *S. epidermidis*, a number of other staphylococcal species that are rather occasionally observed as infectious agents of humans and animals, often in patients with a compromised immune system belong to the "medium-pathogenic" category. Strains of the species *S. saprophyticus*, *S. haemolyticus*, *S. lugdunensis*, and *S. pseudintermedius* may be assigned to this group. Staphylococcus *saprophyticus*, normally a commensal on human skin and mucous membranes, is now the second most common cause of acute urinary tract infections after *Escherichia coli* (Kuroda et al. 2005). *S. haemolyticus* was originally isolated from human skin. It is known to be involved in opportunistic infections associated with the implantation of foreign bodies, particularly in those with compromised immune systems (Takeuchi et al. 2005). *S. pseudintermedius* is a major veterinary pathogen and is the most important cause of pyoderma in dogs and other animals (Ben Zakour et al. 2011; Tse et al. 2011). *S. lugdunensis* has been reported as causative pathogen of skin and soft tissue infections, catheter-related bacteremia, native valve endocardititis, and osteomyelitis (Celard et al. 1997; Donvito et al. 1997; Heilbronner et al. 2011; Tse et al. 2010; Vandenesch et al. 1993).

2.3 Non-pathogenic Staphylococci

Some staphylococcal species are used in the food industry and are commonly regarded as non-pathogenic. This group is represented by *S. carnosus* and some strains of *S. xylosus* and *S. equorum* that are used in meat or cheese fermentations. *S. carnosus* (Schleifer and Fischer 1982) has been used since the 1950s as a starter culture in the food-industry and is classified as GRAS (generally recognized as safe) organism (Barriere et al. 2001; Marchesini et al. 1992; Niinivaara and Pohja 1956). Because of its food-grade quality it is used as a cloning host to study the function of particular staphylococcal genes (Götz 1990). Recently, the genome of the cloning strain *S. carnosus* TM300 has been analyzed (Rosenstein et al. 2009; Wagner et al. 1998). *S. xylosus* belongs to the novobiocin-resistant coagulase-negative species group of staphylococci and is commonly isolated from the skin of humans and animals (Devriese et al. 1985; Kloos and Musselwhite 1975). The type strain *S. xylosus* DSM20267 carries an arsenate, arsenite, and antimony III resistance plasmid (pSX267), which has been cured (Götz et al. 1983). One of the cured strains, *S. xylosus* C2a has been molecularly characterized in more detail with respect to urease, sucrose, and catabolite repression via a new regulator, the catabolite control protein CcpA (Brückner et al. 1993; Brückner 1997; Fiegler et al. 1999; Jankovic et al. 2001; Jankovic and Brückner 2002, 2007; Wagner et al. 1993). *S. xylosus* C2a is used as a starter culture in the production of sausage and cheese. It contributes to the development of the red color characteristic of sausages through its nitrate reductase activity and to the orange color on the surface of certain cheeses, since some strains of *S. xylosus* are pigmented. Its genome is currently being sequenced (http://www.genoscope.cns.fr/spip/-Staphylococcus-xylosus-C2a-.html). *S. equorum* has been isolated from smear-ripened cheese (Hoppe-Seyler et al. 2004; Place et al. 2003). Some strains secrete a macrocyclic peptide antibiotic, micrococcin P(1), with anti-listerial activity which is exploited in cheese fermentation to cope with the threat of contaminations by *Listeria moncytogenes* (Carnio et al. 2000, 2001).

3 Virulence Factors of Highly and Medium-pathogenic Staphylococci

Staphylococcus aureus represents by far the most versatile and potent pathogen among the staphylococci due to its capability to cause infectious diseases ranging from rather harmless superficial skin lesions to systemic infections with life-threatening symptoms. As a widespread colonizer of human skin and mucous membranes (Lowy 1998), *S. aureus* exhibits two lifestyles, as a tolerated commensal on one hand and a dangerous pathogen on the other hand. Its abilities as human pathogen are based on a comprehensive collection of various virulence factors and supportive fitness factors that play roles during the various steps of the infectious process, like adhesion to host tissue, forming multilayered and encapsulated biofilms, evasion of the hosts immune system, and coping with limited supply of nutrients like iron compounds. Correspondingly, the staphylococcal virulence factors may be subdivided into adhesins or soluble factors that mediate the attachment to host cells or extracellular matrix proteins, exoenzymes that are involved in the destruction of host tissues, toxins that directly exert detrimental effects to the host and a heterogeneous group comprising iron uptake systems, immune system evasion mechanisms, and other factors that enhance the fitness to survive in the host. The staphylococcal pathogenic potential is completed by a variety of genes that are mediating resistance to antibiotics and other antibacterial agents. Its opulent arsenal of factors involved in the course of infection makes *S. aureus* the outstanding pathogen within the genus and thus represents the benchmark to which the other pathogenic species have to be compared. Therefore, in the following sections the molecular components involved in staphylococcal pathogenicity will be presented based on a comparison with the infection principles detected in *S. aureus*.

3.1 Colonization Factors

The prevalence of *S. aureus* in the ability to cause infections is largely caused by a rich assortment of surface-located proteins that mediate attachment to host tissue by adherence to host plasma proteins or extracellular matrix (ECM) components. Many of these adhesins are covalently linked to the cell wall via a C-terminally located, conserved LPXTG-motif by the action of sortase (*srtA*) (Fischetti et al. 1990). Of the 28 surface proteins identified in *S. aureus*, 21 reveal LPXTG motifs (Houston et al. 2011). Although sortase-deficient mutants of *S. aureus* were hardly affected in growth, they showed a reduced virulence indicating the importance of correctly cell wall-anchored adhesins (Mazmanian et al. 2000). Two forms of sortase are present in *S. aureus*. Sortases of type A are responsible for anchoring the majority of surface proteins via the LPXTG-motif (Bentley et al. 2007; Schneewind et al. 1992) while sortase B is specialized to anchor the heme–iron uptake protein IsdC via its NPQTN-motif to the cell wall (Mazmanian et al. 2003). Due to its role in sorting of infection-related proteins to the cell surface, Sortase A is commonly regarded as a virulence factor. On the other hand, the corresponding gene is found throughout the staphylococcal genomes, also in the non-pathogenic *S. carnosus* TM300 with more than 60 % identity to Sortase A of *S. aureus*. This emphasizes that the role of *srtA* in virulence is not an exclusive one and depends on the contribution of the cognate substrate proteins to the infectious pathway. The more specialized sortase B is present in only a few staphylococcal species and, in addition to *S. aureus*, is found in *S. capitis*, *S. caprae* and *S. lugdunensis*. Since all these species have been described to be involved in invasive infections (Götz et al. 2006), it is conclusive that they carry an uptake system for the supply with heme–iron. Accordingly, the *srtB* gene found in these species is a more unambiguous marker for staphylococcal virulence than *srtA*.

3.1.1 Host Matrix Binding Proteins

The *S. aureus* repertoire of surface proteins allows interactions with virtually every structural component of the hosts extracellular matrix and with many plasma proteins. Correspondingly, *S. aureus* cells are able to adhere to fibril-forming collagens of types I, II, and III, laminin, elastin, fibronectin, vitronectin, fibrinogen, von Willebrand factor, and thrombospondin. The majority of these surface-located proteins are subsumed under the term "Microbial Surface Components Recognizing Adhesive Matrix Molecules (MSCRAMMs) of which most are covalently linked to the cell wall (Foster and Hook 1998).

The surface components described for *S. aureus* and those found in medium-pathogenic staphylococci are listed in Table 2. Of 21 proteins with LPXTG sequences in *S. aureus,* 10 had not been characterized initially and were collectively designated as *Staphylococcus aureus* surface (Sas) proteins (Roche et al. 2003a, b). At present, most of these still have no assigned function, with the exceptions of SasG and SasC for which roles in nasal colonization or in biofilm

Table 2 Staphylococcal adhesins and immune evasion factors

Adhesin	Type	Ligand	Species		Reference
Spa	MSCRAMM	Immunoglobulins, von Willebrand factor	*S. aureus*	–	Hartleib et al. (2000)
Cna	MSCRAMM	Collagen	*S. aureus*	Mobile element	Foster and Hook (1998)
FnBPA	MSCRAMM	Fibronectin, fibrinogen	*S. aureus*	–	Wann et al. (2000)
FnBPB	MSCRAMM	Fibronectin, fibrinogen	*S. aureus*	–	Wann et al. (2000)
ClfA	MSCRAMM	Fibrinogen; complement factor I	*S. aureus*	SD-repeats Binds to plasma clots	Hair et al. (2008)
ClfB	MSCRAMM	Fibrinogen, keratin	*S. aureus*	SD-repeats	Walsh et al. (2004)
EbpS	MSCRAMM	Elastin	*S. aureus*	No LPXTG, membrane protein	Park et al. (1996)
SdrC	MSCRAMM	Adhesion to nasal epithelial cells	*S. aureus*	SD-repeats	Corrigan et al. (2009)
SdrD	–	Adhesion to nasal epithelial cells	*S. aureus*	SD-repeats	Corrigan et al. (2009)
SdrE	–	Complement factor H	*S. aureus*	SD-repeats	Sharp et al. (2012)
SdrG (FbeE)	MSCRAMM	Fibrinogen	*S. epidermidis*	SD-repeats	Hartford et al. (2001)
SdrF	MSCRAMM	Collagen, abiotic surfaces	*S. epidermidis*	SD-repeats	Arrecubieta et al. (2007, 2009)
IsdA	MSCRAMM	Fibrinogen, fibronectin	*S. aureus*	Iron regulated; heme uptake	Clarke et al. (2004)
SasC	MSCRAMM	Adhesion to nasal epithelial cells	*S. aureus*	Role in biofilm formation	Schroeder et al. (2009)
SasG	MSCRAMM	Adhesion to nasal epithelial cells	*S. aureus*	Accumulation in biofilms	Geoghegan et al. (2010)
Pls	–	Prevents adhesion	*S. aureus*	SD-repeats; homologue of Aap	Savolainen et al. (2001)
Bbp	MSCRAMM	Bone sialoprotein, Fibrinogen	*S. aureus*	SD repeats	Tung et al. (2000), Vazquez et al. (2011)
AtlA	Autolysin/ Adhesin	Heat shock protein Hsc70	*S. aureus*	Internalization	Hirschhausen et al. (2010)
Aaa	Autolysin/ Adhesin	Fibrinogen, fibronectin	*S. aureus*	–	Heilmann et al. (2005)
Aae	Autolysin/ Adhesin	Fibrinogen, fibronectin, vitronectin	*S. epidermidis*	–	Heilmann et al. (2003)

(continued)

Table 2 (continued)

Adhesin	Type	Ligand	Species		Reference
Aap	–	Intercellular adhesion	*S. epidermidis*	Biofilm formation	Hussain et al. (1997)
AtlE	Autolysin/ Adhesin	Vitronectin	*S. epidermidis*	Primary attachment Internalization	Heilmann et al. (1997), Hirschhausen et al. (2010)
Embp	–	Fibronectin	*S. epidermidis*	Biofilm formation, primary attachment, intercellular adhesion	Christner et al. (2010), Williams et al. (2002,)
Ebh	–	Fibronectin	*S. aureus*	Homologous to *Streptococcus* adhesin Emb	Clarke et al. (2004)
GehD	Lipase	Collagen	*S. epidermidis*	Primary attachment	Bowden et al. (2002)
Eap map					
P70	SERAM	Fibrinogen, fibronectin, prothrombin, vitronectin, bone sialo protein, osteopontin, prothrombin	*S. aureus*	Internalization?	Chavakis et al. (2005)
Coa	SERAM	Prothrombin, Platelet-binding	*S. aureus*	Staphylocoagulase, fibrin formation, blood clotting	Chavakis et al. (2005)
FbpA	–	Fibrinogen	*S. aureus*		Cheung et al. (1995)
vWbp	SERAM	Von Willebrand factor, prothrombin	*S. aureus*	Coagulase activity	Bjerketorp et al. (2002, 2004)
Efb	SERAM	Fibrinogen, complement factor C3b	*S. aureus*	Inhibition of platelet aggregation, antiphagocytic	Lee et al. (2004), Palma et al. (2001)
Ecb	–	Complement factor C3b	*S. aureus*	Antiphagocytic	Jongerius et al. (2010)
Emp	SERAM	Fibronectin, fibrinogen, collagen, vitronectin	*S. aureus*	–	Hussain et al. (2001)
Aas	Autolysin/ Adhesin	Fibronectin	*S. saprophyticus*	Agglutination of sheep erythrocytes	Hell et al. (1998)
UafA	MSCRAMM	Adhesion to bladder cells	*S. saprophyticus*	Hemagglutination	Kuroda et al. (2005)
SdrI	MSCRAMM	Collagen	*S. saprophyticus*	–	Sakinc et al. (2006)

accumulation have been described (Roche et al. 2003a, b; Schroeder et al. 2009). Some of the surface adhesins are characterized by the presence of repeats with serine-aspartate (SD) rich sequences and correspondingly were grouped into the Sdr family of surface proteins (Josefsson et al. 1998). ClfA from *S. aureus* was the first described member of the Sdr protein family (McDevitt et al. 1994). In *S. epidermidis*, eleven cell-wall-anchored proteins have been identified of which seven had not been characterized previously and were bundled under the term *Staphylococcus epidermidis* surface (Ses) proteins (Bowden et al. 2005). Like in *S. aureus*, also in *S. epidermidis* members of the serine-aspartate-rich (Sdr) proteins were identified, of which SdrF and SdrG carry an LPXTG-motif (McCrea et al. 2000). SdrG (also termed Fbe) is known as *S. epidermidis* fibrinogen-binding surface protein (Hartford et al. 2001) while SdrF has collagen-binding activity and plays a role in the attachment to abiotic surfaces (Arrecubieta et al. 2007, 2009). Fbe shows similarity to the clumping factors ClfA and ClfB of *S. aureus*. In contrast to the tested *S. aureus* strains, the adherence to fibrinogen varies significantly in the analyzed *S. epidermidis* strains.

Compared to *S.aureus* and *S. epidermidis*, the coagulase-negative *S. saprophyticus* has a more reduced pathogenic potential that is essentially limited to urinary tract infections (UTI). The species representatives carry a number of virulence factors. The genome sequence of *S. saprophyticus* ATCC15305 revealed a gene encoding a cell-wall anchored protein (UafA) which has been identified as an adhesin that mediates hemagglutination and adherence to human bladder cells (Kuroda et al. 2005). Another cell wall-anchored protein identified in *S. saprophyticus* is SdrI. It belongs to the family of serine-aspartate-rich proteins and was shown to bind to collagen (Sakinc et al. 2006) as well as to fibronectin (Sakinc et al. 2009a, b). A surface-associated lipase, Ssp, is present in high amounts on the cell surface of clinical isolates (Sakinc et al. 2007). Recently, a plasmid-encoded cell-wall anchored *S. saprophyticus* surface protein F (SssF) has been described (King et al. 2012). The *sssF* gene is highly prevalent in *S. saprophyticus* clinical isolates; it shows similarity to the *S. aureus* surface protein SasF and like SasF it mediates resistance to linoleic acid and seems not to be involved in adhesion (King et al. 2012).

The equipment with the surface-located adhesins in *S. aureus* varies from strain to strain. While clumping factor A (ClfA) and the serin-aspartate repeat protein C (SdrC) are found in almost all clinical *S. aureus* isolates, the collagen-binding protein Cna and the colonization factor SdrD are harbored by only some strains (Bartlett and Hulten 2010). Some *S. aureus* adhesins are involved in nasal colonization of about 30 % of the population without symptoms, which illustrates the ambivalent commensal-pathogen nature of *S. aureus* (Wertheim et al. 2005).

3.1.2 SERAMS and Anchorless Adhesins

Another group of adhesins identified in *S. aureus* is composed of structurally unrelated proteins that are secreted and mediate the binding to host molecules, cells, or tissues. These factors have been grouped as "secretable expanded repertoire adhesive molecules" (SERAM) and comprise various fibronectin- and fibrinogen-binding proteins as well as staphylocoagulase and von Willebrand factor-binding protein (Chavakis et al. 2005). Besides the MSCRAMMs and SERAMs, also proteins without secretion or anchoring signal were found to be involved in the recognition of ECM components. Like for streptococci, where anchorless adhesins and invasins have been categorized as a new class of virulence factors (Chhatwal 2002), a cell surface-located alpha-enolase of *S. aureus* has been detected as a laminin-binding protein (Carneiro et al. 2004). The alpha-enolase of *S. aureus* also functions as plasminogen receptor and probably is involved in enhancing the activity of staphylokinase (Carneiro et al. 2004; Molkanen et al. 2002). Furthermore, a binding to laminin and collagen I has been demonstrated (Antikainen et al. 2007). The alpha-enolase is well conserved among the staphylococci and, at least for *S. epidermidis* O47, a similar role for the enolase as virulence factor was indicated by binding to serum components and immune reactivity (Sellman et al. 2005).

3.1.3 Autolysins/Adhesins

Another class of bifunctional proteins for which adhesive properties have been described comprises the autolysins/adhesins including AtlA and Aaa from *S. aureus* (Heilmann et al. 2005), AtlE and Aae from *S. epidermidis* (Heilmann et al. 1997, 2003), Aas from *S. saprophyticus* (Hell et al. 1998) and AtlC from *S. caprae* (Allignet et al. 2002). Members of this group are distinguished by a bi-functional character as autolysins with amidase/glucosaminidase activity that also have adhesive properties (Heilmann et al. 1997). The major autolysins of *S. aureus* and *S. epidermidis,* AtlA and AtlE, interact with peptidoglycan (Biswas et al. 2006; Zoll et al. 2010). They are targeted to the septum region via their repeat domains by an exclusion strategy mediated by wall teichoic acid (WTA) (Schlag et al. 2010) and the receptor at the septum is most likely lipoteichoic acid (Zoll et al. 2012). Analyses of *S. epidermidis* biofilm mutants have shown that AtlE is involved in the initial attachment to polystyrene. Furthermore, a vitronectin-binding activity has been shown for AtlE (Heilmann et al. 1997) and a role in the accumulation of extracellular DNA in *S. epidermidis* biofilms has been proposed (Qin et al. 2007). These findings indicate that AtlE might play an important role in several steps of biofilm formation. Interestingly, while no or low fibronectin-binding activity has been reported for AtlE (Heilmann et al. 1997), the homologous AtlC of *S. caprae* was shown to exhibit a pronounced binding to fibronectin (Allignet et al. 2002). Recently, another function of AtlE and AtlA in internalization of infecting staphylococcal cells to host cells based on an interaction with the host heat shock cognate protein Hsc70 has been reported (Hirschhausen et al. 2010).

Highly similar orthologues of Atl have been found in all sequenced staphylococcal genomes indicating a ubiquitous presence of the corresponding gene in staphylococci, whereas data on functions of the gene products are scarce. The conservation of the *atl* gene sequence has been exploited to develop an Atl-based typing method that provided a phylogenetic tree comparable to those based upon 16S RNA or on the comparison of whole genome sequences (Albrecht et al. 2012).

3.1.4 Elastin-Binding Proteins

In contrast to the LPXTG-anchoring of most MSCRAMMs, EbpS, an elastin-binding protein of *S. aureus* is integrated into the cytoplasmic membrane by two transmembrane domains which are located in the central part of the primary structure (Downer et al. 2002). Binding to elastin, a major component of the human extracellular matrix, is also mediated by FnBPA and FnBPB that had already earlier been identified as adhesins with fibronectin-binding activity (Greene et al. 1995). But the elastin-binding activity differs between both adhesin types since EbpS interacts with soluble elastin and tropoelastin while FnBPA and FnBPB bind to immobilized elastin.

3.1.5 Fibronectin-Binding Proteins

Binding to soluble and insoluble fibronectin (Fn), a major component of the fibrin-matrix in blood clots and one of the plasma proteins that cover surfaces of implants, is mediated by various adhesins in *S. aureus* (see Table 2). Fibronectin-binding adhesins are also found in *S. epidermidis* (Table 2) but in contrast to *S. aureus*, the *S. epidermidis* Fn-binding proteins interact only with immobilized fibronectin (Valentin-Weigand et al. 1993). This differential adhesion behavior reflects the different occurrences of pathogenicity of both species as *S. epidermidis* exerts its virulence predominantly by adhesion to surfaces that are covered by plasma proteins followed by intercellular adhesion while *S. aureus* also acts as a wound pathogen adhering to soluble fibronectin as being present in wound exudates (Valentin-Weigand et al. 1993).

3.1.6 Factors Involved in Intercellular Adhesion During Biofilm Formation

The adhesin-mediated attachment to host tissue or surfaces of implants is the precondition for the subsequent formation of multi-layered cell communities which eventually become entrapped by exopolysaccharides (EPS) or proteinaceous intercellular material—the biofilms. The ability to adhere to surfaces mediated by adherent compounds is the prerequisite for the second stage, the accumulation phase and intercellular aggregation. This attachment either occurs directly to the abiotic surface or indirectly by adhesion to host proteins that cover the implant. There is a great number of surface adhesins known; any of the above described adhesins might contribute to binding to a specific surface. In comparison with the rich spectrum of *S. aureus* adhesins, the repertoire of *S. epidermidis* is restricted to a limited number of adhesive proteins (see Table 2).

3.2 Biofilms

The formation of biofilms is a feature shared by virulent *S. aureus* and *S. epidermidis* strains. In *S. aureus*, the formation of biofilms is one strategy among others to resist unfavorable conditions in the host as exerted by shearing forces of the bloodstream, the immune response of the host, or anti-infective measures during disease treatment. While the self-protective strategies of *S. aureus* involve additional possibilities like formation of encapsulated microcolonies mediated by the action of coagulating enzymes (Guggenberger et al. 2012), *S. epidermidis'* defensive strategy is mainly based on the formation of biofilms.

The compounds mediating intercellular adhesion and leading to multi-layered bacterial cell aggregates during the later steps of biofilm formation are limited. The most obstinate biofilms are still formed by polysaccharide intercellular adhesin (PIA) (Mack et al. 1996) encoded by the *ica* operon (Heilmann et al. 1996a, b). The accumulated cells are tightly imbedded in this polysaccharide matrix like in a chewing gum (Götz 2002; Heilmann et al. 1996a, b).

Production of PIA is dependent on gene products encoded by the *ica* locus (Cramton et al. 1999; Heilmann et al. 1996a, b) that is generally considered as major genetic component correlated with biofilm formation. This view is supported by the deficiency in biofilm formation found in *ica*-negative strains like *S. epidermidis* ATCC 12228, whereas the *ica*-positive clinical isolate *S. epidermidis* RP62A is a potent biofilm former (Gill et al. 2005; Zhang et al. 2003). PIA-dependent biofilm formation by *S. epidermidis* was shown to be a phase variable process as insertion of IS*256* into the *icaC* gene (a frequent integration site) leads to inactivation of biofilm formation—a process which is reversible by precise excision of IS*256* from *icaC* (Ziebuhr et al. 1999). But reversion to a biofilm-positive phenotype has also been achieved without excising IS*256* indicating that *S. epidermidis* is able to form biofilms in a PIA-independent manner (Rohde et al. 2005).

This variation also seems to be achieved by genome rearrangements as indicated by a genome comparison between the *ica*-positive *S. epidermidis* RP62A and the *ica*- and biofilm-negative *S. epidermidis* ATCC12228. The *ica* operon of *S. epidermidis* RP62A is preceded by a genome segment that is inverted in the genome of strain ATCC12228 (see Fig. 3). Since the *ica* operon is located at the "break point" between the inverted fragments, it is tempting to speculate that this inversion caused the loss of the *ica* genes in ATCC 12228 thus leading to a biofilm-negative phenotype.

For a long time the *ica* operon as the genetic basis for PIA production has been regarded as mandatory for biofilm formation but more recently also an *ica*-independent formation of biofilms has been detected in *ica*-negative mutants of *S. epidermidis* (Rohde et al. 2007). The PIA-independent biofilm was demonstrated to be metaperiodate-resistant indicating a proteinaceous rather than a polysaccharide character (Rohde et al. 2005). This kind of biofilm is less compact and can be easily disrupted by proteases (Marti et al. 2010). A number of surface proteins were described that mediate an *ica*-independent intercellular adhesion thus contributing to biofilm formation.

The proteins involved in intercellular aggregation are Aap, Bap, Bhp, Embp, SasC, and SasG (Hussain et al. 1997; Macintosh et al. 2009; Rohde et al. 2005; Schumacher-Perdreau et al. 1994).

The 2276 amino acids biofilm-associated protein (Bap) from *S. aureus* is encoded on a mobile genetic element, pathogenicity island SaPIbov2, which is only found in bovine-associated *S. aureus* strains (Cucarella et al. 2001; Tormo et al. 2005). Thus, it seems to play no role in human infections by *S. aureus*. For *S. aureus*, a phase-variable expression of Bap has been reported which switches between the "on" and "off" states at a similar frequency and might be involved in the detaching of *S. aureus* cells from a biofilm (Tormo et al. 2007).

In *S. epidermidis* and other CoNS homologs of *S. aureus* Bap have been identified and their role in biofilm formation for *bap*-positive and *ica*-negative staphylococci has been reported (Tormo et al. 2005). The primary structure of Bap is characterized by domains comprising sequence repeats and Bap-homologs in *S. epidermidis*, *S. simulans*, *S. chromogenes*, *S. xylosus* and *S. hyicus* reveal different numbers of repeats in these domains (Tormo et al. 2005). In contrast to *S. aureus*, no indications for a location on a mobile element were found for the *bap*-positive non-*S. aureus* species (Tormo et al. 2005).

The strong *ica*-positive biofilm-former *S. epidermidis* RP62A is a methicillin-resistant clinical isolate from an intravascular catheter-associated sepsis (Gill et al. 2005). In the genome sequence of *S. epidermidis* RP62A another homolog of Bap, named Bhp (Bap homologous protein) is encoded that shows a more distant relationship to *S. aureus* Bap and the Bap homologs in the non-*S. aureus* species. The *bhp* gene is also found in a number of *S. epidermidis* strains of which no complete genome sequences are available at the time of writing this article (*S. epidermidis* M23864, VCU126, VCU128, VCU037, VCU045, VCU125) but it is not present in the *ica*-negative strain *S. epidermidis* ATCC 12228 or in the *ica*-positive biofilm-former *S. epidermidis* O47 (unpublished results, our lab). The Bhp proteins can also contribute to biofilm formation, play a role in bacterial infectious processes, and can occasionally be contained in mobile elements (Lasa and Penades 2006). On the other hand, the expression of Bhp was shown to be downregulated in an Aap-dependent biofilm (Hennig et al. 2007).

As a factor involved in the PIA-independent biofilm formation of *S. epidermidis* the surface protein accumulation-associated protein (Aap) was identified which has to be proteolytically processed in order to mediate intercellular adhesion (Rohde et al. 2005).

Aap is an *S. epidermidis* surface protein that was shown to mediate intercellular adhesion in PIA-negative *S. epidermidis* strains leading to an extracellular biofilm matrix of proteinaceous character (Rohde et al. 2005). Aap is a large cell wall-anchored protein which reveals a signal peptide followed by an A-domain that shares similarity with the *S. aureus* surface protein SasG and a B-domain composed of a varying number of repeats, depending on the Aap-producing *S. epidermidis* strain (Rohde et al. 2007). Aap is anchored to the cell wall via its C-terminal LPXTG signal. The intercellular adhesion activity could be assigned to the B-domain of Aap that becomes active after proteolytic excision of the A domain (Rohde et al. 2005).

While the B-domain is responsible for a later step in biofilm formation, the A-domain may have a function in the initial attachment of *S. epidermidis* cells to host tissue. Macintosh et al. demonstrated that the A-domain mediates adhesion of *S. epidermidis* to human corneocytes (Macintosh et al. 2009). Thus, Aap may act as a bi-functional protein that is involved in the initial phase in biofilm formation via the activity of the A- domain and, after proteolytic processing, in the biofilm stabilization by intercellular adhesion promoted by the B-domain. Some *S. epidermidis* strains were isolated that are genotypically *aap*-positive, whereas Aap proteins are only expressed in a subpopulation of the corresponding cultures. It is postulated that cells that do not express Aap might get detached easier from host cells and could be candidates for cells that enter the hosts' bloodstream via implanted materials (Macintosh et al. 2009).

The LPXTG-anchored proteins SasG and SasC of *S. aureus* show intercellular adhesion activity which is involved in biofilm formation. SasG shares similarity with Aap and SasC (Corrigan et al. 2007; Geoghegan et al. 2010) and contributes to nasal adhesion (Roche et al. 2003a, b). SasC conferred production of huge cell aggregates, increased adherence to polystyrene, and enhanced biofilm formation to *S. carnosus* and *S. aureus* (Schroeder et al. 2009). In *S. aureus*, an additional role for the IgG-binding protein A (Mazmanian et al. 2000) seems to be the involvement in *ica*-independent biofilm formation. The reported data indicate that protein A is not covalently linked to the cell wall when acting in biofilm formation (Merino et al. 2009).

Embp is a giant multifunctional cell-wall protein from *S. epidermidis* that mediates fibronectin-binding, biofilm accumulation, and escape from phagocytosis (Christner et al. 2010; Williams et al. 2002). Its *S. aureus* homolog, Ebh, is a 1.1 megadalton surface protein with fibronectin-binding activity (Clarke et al. 2004).

3.3 Immune Escape and Modulating Factors

The versatility of *S. aureus* as a potent pathogen is *inter alia* based on a variety of strategies to evade the hosts' defensive immune response. The factors involved in undermining the immune defense may be roughly categorized into defensive and active measures.

Defensive strategies are based on protective principles such as biofilm formation (see above) and blood clotting by coagulating factors. In this way, *S. aureus* protects itself by mechanical barriers like the intercellular matrix present in biofilms, a pseudocapsule and/or the microcolony-associated meshwork which is built up by the action of coagulase and von Willebrand factor-binding protein (Guggenberger et al. 2012). In addition, comparative transcriptome and proteome analysis revealed that biofilm-forming staphylococci may adopt a state of reduced metabolic activity due to oxygen and nutrient limitations (Beenken et al. 2004; Resch et al. 2005, 2006). The physiological heterogeneity as well as the reduced to non-growth state may be responsible for the high tolerance in biofilms (Lewis 2007). The biofilm-dependent evasion strategy is common to *S. aureus* and the less aggressive pathogen *S. epidermidis* while the defense by encapsulation into microcolonies by a concerted action of coagulase and von Willebrand factor-binding protein (Guggenberger et al. 2012) has up to now exclusively been reported for *S. aureus*.

Classically, the determination of coagulase (staphylocoagulase, *coa*) activity is used as a distinctive feature to differentiate between *S. aureus* and most of the other staphylococcal species that are grouped as coagulase-negative staphylococci (CoNS). At the beginning of staphylococcal taxonomy, the classification as "coagulase-negative" was virtually synonymous to "non-pathogenic". But the ascent of the coagulase-negative *S. epidermidis* as potent pathogen in the clinical field as well as the isolation of *S. aureus* strains that show no coagulase activity (Akineden et al. 2011) have softened the discriminatory power of coagulase activity as a diagnostic criterion.

Coagulase interacts with prothrombin, leading to a complex called staphylothrombin. Coagulase activates prothrombin by conformational changes (not by cleavage like in the physiological pathway) eventually resulting in the conversion of fibrinogen into fibrin.

Besides staphylocoagulase, another factor with coagulating activity has been identified in *S. aureus*: the von Willebrand factor-binding protein (vWbp) (Bjerketorp et al. 2004). The von Willebrand factor plays a role in platelet adhesion and aggregation at the site of vascular damage. The *S. aureus* vWbp comprises around 500 amino acids and reveals sequence similarity (about 30 % identity / 45 % similarity between *S. aureus* N315 *Coa* and vWfbp) to coagulase in its N-terminal half and interacts also with prothrombin but in a more host-specific manner. The coagulating activity of *S. aureus* vWbp is highest with human or porcine plasma, while the activity in rabbit plasma which is used in the classical staphylocoagulase assay is significantly lower (Bjerketorp et al. 2004). Since both, vWbp and staphylocoagulase, exert coagulating activity it can be assumed that plasma coagulation by *S. aureus* can be ascribed to a concerted function of both proteins. Furthermore, it has been demonstrated that staphylocoagulase and von Willebrand factor-binding protein are necessary for abscess formation and that they have a role in the protection of *S. aureus* microcolonies against neutrophils (Cheng et al. 2010; Guggenberger et al. 2012). Of note, the observation of coagulase activity is not exclusive for *S. aureus* since various other staphylococcal species like *S. delphini*, *S. hyicus*, *S. intermedius*, *S. lutrae*, *S. pseudintermedius*, and *S. schleiferi* have been described as coagulase-positive.

The coagulase activity observed in these species makes it tempting to speculate that they might also be able to form a microcolony-like defensive barrier based on the coagulating activity. But at least in the coagulase-positive species *S. pseudintermedius*, of which two complete genome sequences are available (Ben Zakour et al. 2011; Tse et al. 2011), the observed coagulase activity seems to be solely based on the von Willebrand factor-binding protein (Guggenberger et al. 2012), making the formation of *S. aureus*-like two-layered microcolonies unlikely. It will be of interest to see whether the coagulating activity of the other coagulase-positive non-*S. aureus* species can also be exclusively ascribed to the von Willebrand factor-binding protein.

A cell-wall anchored protein with von Willebrand factor-binding activity has been identified also in *S. lugdunensis* (Nilsson et al. 2004). But this protein deviates significantly from the *S. aureus* and *S. pseudintermedius* vWbp in size (about 500 amino acids versus about 2000 amino acids) and lacks sequence similarities with the latter. Furthermore, no coagulase-activity has been detected in *S. lugdunensis*, suggesting that its von Willebrand factor-binding protein does not exert this activity in contrast to the vWfbps in *S. aureus* and *S. pseudintermedius*.

The active strategies against the hosts' immune system involve factors that specifically point towards certain components of the immune response. *S. aureus* underlines its role as the leading staphylococcal pathogen by numerous factors that act in this field and distinguish *S. aureus* from the medium-pathogenic staphylococci. The IgG-binding protein A (Mazmanian et al. 2000) is an exclusive virulence factor of *S. aureus*. Its classical role in host invasion is the binding of immune globulin G via its Fc part, thus counteracting antibody-mediated opsonization by the host immune system. Besides this activity, Spa has also been described to bind the von Willebrand factor (Hartleib et al. 2000) as well as the receptor for tumor necrosis factor α, TNFR1. But the role of these binding activities is not clear (Smeltzer et al. 2009). Besides Spa, *S. aureus* produces various specific immune modulating proteins like the chemotaxis inhibitory protein (CHIPS, *chp*), the staphylococcal complement inhibitor (SCIN, *scn*), staphylokinase (*sak*), and enterotoxin A (*sea*). CHIPS interacts with complement factor C5 (C5aR) and formylated peptide receptor (FPR) of human neutrophils (Postma et al. 2004). SCIN inhibits the conversion of complement factor C3 into C3b thereby hindering phagocytosis of *S. aureus* by human neutrophils (Rooijakkers et al. 2005a, b). Staphylokinase has been reported to destroy defensins (Jin et al. 2004) and to exert an antiopsonic activity (Rooijakkers et al. 2005a, b). Besides its commonly known role in food-poisoning, the superantigenic enterotoxin A interacts with several chemokine receptors (Rahimpour et al. 1999). The genes encoding these immune modulators, *chp*, *scn*, *sea* and *sak*, are located within an innate immune evasion cluster (Chavakis et al. 2005) which is part of a ß-hemolysin converting prophage (van Wamel et al. 2006). Also, clumping factor A (*clfA*) shows a recently identified complement modulating activity by binding to complement regulatory factor I thus increasing the inactivation of opsonin C3b (Hair et al. 2008).

The reputation of *S. aureus* as a "master of complement evasion" (Sharp et al. 2012) is corroborated by further complement modulating activities: the extracellular fibrinogen-binding protein (EfB) exerts an antiphagocytic function by interacting with the complement factor C3 (Lee et al. 2004); the extracellular complement-binding protein (Ecb) interacts in a similar way as Efb with the complement system but lacks the fibrinogen-binding activity of the former (Jongerius et al. 2010); the superantigen-like protein 7 was shown to inhibit specifically the conversion of factor C5 into C5a thus counteracting the chemotaxis of neutrophils (Bestebroer et al. 2010). Recently, also the *S. aureus* surface protein SdrE has been identified as immune evasion factor by binding to complement regulatory factor H (Sharp et al. 2012). Also, in *S. epidermidis* a protein of the serine-aspartate-rich family has been shown to counteract the immune response:

the fibrinogen-binding protein SdrG (Fbe) prevents thrombin-induced fibrinogen clotting by interfering with the release of fibrinopeptide B; this interaction might interfere with the influx of phagocytic neutrophils (von Eiff et al. 2002).

3.3.1 Cytolytic Toxins Interfering with the Immune Response

The most aggressive component of the *S. aureus* immune evasion strategies involves the action of cytolytic toxins that actively attack cells of the host immune system (Foster 2005). In this context, the phenol soluble modulins (PSM) first described by Mehlin et al. in *S. epidermidis* (Mehlin et al. 1999) gained substantial attention during the past years due to the strong lytic activity of α-type PSMs on human neutrophils (Wang et al. 2007). One of the longest known PSM is the delta-toxin (*hld*) which is part of the *agr* regulatory system (Janzon and Arvidson 1990) and which has recently been associated together with beta-toxin with the escape from phago-endosomes of human epithelial and endothelial cells (Giese et al. 2011). The delta-toxin is produced by many staphylococcal species with an intact *agr* system but its activity is boosted by additional cytotoxins mainly present in *S. aureus*. Compared to *S. aureus*, the coagulase-negative staphylococci are commonly regarded as non-toxigenic and as more passive evaders that cope with the immune system mainly by defensive barriers like biofilms. Accordingly, factors with cytolytic activities against neutrophils had generally been assumed to be absent in the coagulase-negative staphylococci. However, this view is going to be challenged somewhat by the identification of the *S. epidermidis* δ-type PSM as potent leukocyte toxin (Cheung et al. 2010). PSMδ exerts a lytic activity towards neutrophils comparable to that of the most active phenol-soluble modulin of *S. aureus*, PSMα3. But the overall cytolytic activity of *S. epidermidis* culture filtrates was observed to be low, probably due to a reduced production level of PSMδ (Cheung et al. 2010).

3.4 Internalization

S. aureus is generally regarded as an extracellular pathogen but it is also able to evade the attacking immune system by internalization into host cells. This invasive strategy opens the opportunity to persist within the host and consequently enhances the risk of recurrent infections. *S. aureus* is able to persist within a variety of non-professional phagocytic host cells (Hirschhausen et al. 2010). The onset of internalization of infecting *S. aureus* is based on the interaction of certain MSCRAMMs with the phagocytic host cells. The fibronectin-binding proteins FnbpA and FnBpB act in internalization in combination with bound fibronectin that interacts with the host cell integrin $\alpha_5\beta_1$ (Sinha et al. 1999) leading to the activation of a signaling pathway and to the uptake of fibronectin-bound bacteria by host cells (Bartlett and Hulten 2010). FnbpA and FnBpB can also mediate internalization via the heat shock protein Hsp60 (Dziewanowska et al. 2000). Moreover, evidence has been provided that the Atl autolysins/adhesins of *S. aureus* and *S. epidermidis*, AtlA and AtlE, contribute to the internalization into epithelial cells via heat shock protein Hsc70 as a receptor (Hirschhausen et al. 2010). Recent reports also indicate that internalization as a strategy to persist within the host and to evade its immune system seems not to be exclusively exerted by *S. aureus* since invasion of coagulase-negative staphylococci like *S. epidermidis* into bone cells and *S. saprophyticus* into urinary bladder cells has been observed (Khalil et al. 2007; Szabados et al. 2008). As Atl homologs are present in all tested staphylococcal species (Albrecht et al. 2012), we assume that it may be involved in internalization of coagulase-negative staphylococci that do not contain the fibronectin-binding proteins FnbpA and FnBpB. The persistence of *S. aureus* may be further enhanced if they adopt the state of small-colony variants (SCV) which is accompanied by significant growth defects and other phenotypic changes like unusual colony morphology, reduced pigmentation, and hemolysis as well as auxotrophisms due to defects in electron transport pathways (see below).

3.5 Toxins—Aggressive Pathogenicity Factors with Host Cell Damaging Activity

The pathogenic capabilities of *S. aureus* are to a large extent characterized by a multifaceted repertoire of secreted proteins that act as toxins (see Table 3). Several diseases caused by *S. aureus* are associated with the production of specific toxins like the staphylococcal scalded skin syndrome caused by exfoliative toxins, food poisoning by enterotoxins, and toxic shock syndrome by TSST-1 (Smeltzer et al. 2009). The staphylococcal exfoliative disease is correlated with *S. aureus* strains producing one of four types of exofoliative toxins, ETA, ETB, ETC, and ETD. The exfoliative toxins share similarity with serine proteases and the epidermolytic effect is correspondingly caused by their proteolytic activity (Smeltzer et al. 2009). Besides their cell-damaging activity, some of these toxins exhibit superantigenic characteristics, giving rise to massive T-cell proliferation and enhanced cytokine production which can lead to toxic shock syndrome and hypotension due to capillary leakage (Schlievert et al. 2009).

Toxins with superantigenic activity are the toxic shock syndrome toxin 1 (TSST-1), the staphylococcal enterotoxins of serotypes A–D and I, enterotoxin-like toxins, G, H, J–X. All staphylococcal superantigens are encoded on mobile genetic elements like phages, plasmids, or pathogenicity islands with the exception of enterotoxin-like superantigen X that is chromosomally encoded (Brosnahan and Schlievert 2011). Another class of staphylococcal exotoxins is represented by hemolytic or cytolytic toxins. These include the hemolysins α, ß, γ, and δ and the Panton-Valentine-leukocidin (PV-leukocidin). The hemolysins α, ß, and δ act as single-component toxins while hemolysin γ and PV-leukocidin are active as

Table 3 Staphylococcal toxins

Toxin	Effect/activity	Features	Reference
Toxic shock syndrome toxin-1	Toxic shock	Superantigen (formerly SEF)	Dinges et al. (2000)
Enterotoxin A, B, C, D, E, G, I	Emetic effects, food poisoning	Superantigen	Dinges et al. (2000), Thomas et al. (2007)
Enterotoxin-like H, J–V	Lack emetic effects or have not been examined	Superantigen	Lina et al. (2004), Smeltzer et al. (2009)
Exfoliative toxins A, B, C, D	Only ETA and ETB associated with Skalded Skin Syndrome	homologs in *S. hyicus*: ExhA, ExhB, ExhC, ExhD	Ahrens and Andresen (2004), Smeltzer et al. (2009)
α-toxin (hemolysin)	Most active against rabbit erythrocytes; dermonecrotic and neurotoxic	Pore-forming (cylindrical heptamers),	Dinges et al. (2000)
ß-toxin (hemolysin)	Highly hemolytic against sheep erythrocytes; "hot–cold" hemolysin; most prevalent in animal isolates	Sphingomyelinase (phosphorylase c activity): on sphingomyelin and lyso-phosphatidylcholine	Dinges et al. (2000)
Pantone-Valentine leukocidin	Attacks leukocytes, tissue necrosis	Bicomponent leukocidin; made of S- and F-components, made by 2–3 % of *S. aureus* strains, common among CA-MRSA	Lina et al. (1999), Szmigielski et al. (1999)
Leukocidin	Target leukocytes	LukM/F' widely found among ruminant isolates	Rainard (2007), Smeltzer et al. (2009)
γ-toxin (hemolysin γ)	Affects neutrophils, macrophages, erythrocytes (agar inhibits toxic activity)	Bicomponent leukocidin, made by most *S. aureus* strains	Dinges et al. (2000)

(continued)

Table 3 (continued)

Toxin	Effect/activity	Features	Reference
Phenol-soluble modulins α	Neutrophil activation and lysis	ca. 20 amino acids in length; associated with CA-MRSA virulence	Wang et al. (2007)
Phenol-soluble modulins ß	Cytokine release	ca. 40 amino acids in length, homologs in several CoNS	Otto et al. (2004), Wang et al. (2007)
δ-hemolysin	Lyses erythrocytes, cytokine release	Similar to PSMα, synergism with ß-toxin	Dinges et al. (2000), Otto et al. (2004)
SLUSH	Activity similar to δ-hemolysin	Three 43 aa peptides, synergistic activity with ß-toxin	Donvito et al. (1997)

bicomponent toxins. α, γ, and PV-leukocidin act by forming pores in membranes of erythrocytes or neutrophils leading to lysis. ß-hemolysin is produced predominantly by animal isolates of *S. aureus*, is highly hemolytic for sheep erythrocytes and was shown to have sphingomyelinase activity (Dinges et al. 2000). δ-toxin is a small, helical, and amphipathic peptide of 26 amino acids that is encoded within RNA III, the regulatory component of the *agr* system (Novick and Geisinger 2008). It causes membrane damage in a variety of mammalian cells and is secreted without a discernible signal peptide but it has been suggested that the whole toxin itself might have signal peptide-properties (Dinges et al. 2000). The bicomponent toxins hemolysin γ and PV-leukocidin are composed of two protein subunits, S and F, that assemble into oligomers in the host cell membrane (Dinges et al. 2000). The S components of γ-toxin are encoded by *hlgA* and *hlgC*; the F-component is encoded by *hlgB*. The components of the bipartite PV-leukocidin are encoded by *lukS-PV* and *lukF-PV*, but additional genes have been identified that code for components of bi-component leukocidins: *lukE*/*lukD* for S components and *luk M*/ lukF'-PV code for F components (Smeltzer et al. 2009). The *hlg* genes are almost ubiquitously present in *S. aureus* strains while the genes for PV-leukocidin are located on a bacteriophage and found in only 1–2 % of *S. aureus* strains, predominantly in CA-MRSA. It is believed that PV-leukocidin is associated with severe community-acquired necrotizing pneumonia (Labandeira-Rey et al. 2007). However, this was controversial discussed due to conflicting results based on studies with different animal models. But it has recently been shown that the PV-leukocidin has a severe cytotoxic effect which is restricted to human and rabbit cells and could not be reproduced in murine or monkey cells (Löffler et al. 2010). Very recent data show that PV-leukocidin uses the complement factor C5a receptor, which is abundant in humans and rabbits but not in mice (Jos van Strijp, personal communication).

Due to their cytolytic activity on leukocytes, the bicomponent leukotoxins play an important role in the evasion of the host's immune system (see above and (Foster 2005)).

3.5.1 Toxins in Non-*Staphylococcus aureus* Species

For a long time the aggressive pathogenic lifestyle involving the production of cytolytic toxins has been exclusively assigned to *S. aureus*. But recently, upcoming reports on toxigenic factors or corresponding genes detected in S. epidermidis and other CoNS have softened this distinctive criterion (Madhusoodanan et al. 2011; Zell et al. 2008). In some strains of *S. hyicus*, exfoliative toxins have been identified (ExhA, ExhB, ExhC, and ExhD) that probably cause exudative epidermitis in pigs, a skin lesion that has several aspects in common with staphylococcal scalded skin syndrome in humans, and share sequence similarities with ETA, ETB, and ETD of *S. aureus* (Ahrens and Andresen 2004). Moreover, a comprehensive analysis of a collection of CoNS has revealed hemolytic activities and led to the identification of enterotoxins, TSST-1 and exfoliative toxin by immunoblot analysis in a significant proportion of the tested strains (Seitter et al. 2011; Zell et al. 2008). In addition to these reports on toxins in CoNS, another class of small peptides, termed phenol soluble modulins, is presently gaining more attention as virulence factors with cytotoxic activity (see above).

Small peptides with cytolytic activity have also been described for some strains of the coagulase-negative species *S. lugdunensis*. They exert a synergistic hemolytic activity and have been named SLUSH (*S. lugdunensis* synergistic hemolysin). The synergistic activity of SLUSH in combination with ß-hemolytic activity resembles that of δ-toxin of *S. aureus*. The hemolytic activity has been assigned to three peptides of 43 amino acids, which share high sequence identity to each other (Donvito et al. 1997). As already indicated by their size, the SLUSH peptides resemble PSMs and comparably act in the attraction and stimulation of human leukocytes (Rautenberg et al. 2011). The similarity in neutrophil response to PSM-like peptides via the formyl peptide receptor 2 led to the assumption that the production of PSM-like peptides by staphylococci might be a criterion to discriminate between virulent and commensal staphylococcal species (Rautenberg et al. 2011).

4 Fitness Factors Involved in Infection

The pathogenic properties of virulent staphylococcal species would not be complete without the sustaining activity of fitness factors that support survival under hostile conditions in the host. These factors are often not discernible from housekeeping genes that are conserved throughout the genus and thus their distinctive power concerning the categorization as virulence factor is ambiguous.

The fitness factors that play an accessory role in virulence comprise exoenzymes involved in degradation of host tissues and compounds, iron uptake systems, enzymatic functions of more specialized pathogens like the urease of *S. saprophyticus*, and additional physiological pathways such as the arginine deiminase pathway encoded by the ACME element found in *S. epidermidis* ATCC 12228 and *S. aureus* USA 300 that may provide selective advantages in the colonization of the hosts.

4.1 Exoenzymes

Degradative exoenzymes like nucleases, proteases, and lipases generally support the growth of staphylococci by the pulping of polymeric substrates in order to get hold of nutritions and components for biosynthetic purposes. Besides this housekeeping character, some of these exoenzymes may also be involved in infectious processes by sustaining the survival in confrontation with hostile conditions and shortage of certain nutritions.

4.1.1 Nuclease

The thermostable nuclease secreted by *S. aureus* is known for decades and is highly conserved within the species which is exploited in the specific detection of *S. aureus* in blood cultures (Kiedrowski et al. 2011; Lagace-Wiens et al. 2007). Besides its habitual activity concerning the metabolic digestion of nucleic acids it seems to play an additional role in the regulation of biofilm formation (Kiedrowski et al. 2011). Reports emerged on extracellular DNA as a component of the *S. aureus* biofilm matrix (Izano et al. 2008). In this context, it is consistent that the *S. aureus* thermonuclease negatively influences the biofilm formation in *S. aureus* (Kiedrowski et al. 2011) and contributes to escape from neutrophil extracellular traps (Berends et al. 2010).

4.1.2 Protease

Since a protein-based matrix has also been proposed for *S. aureus* biofilms, some of the various extracellular proteases produced by this species could have an influence on switching between biofilm-associated and free living cells. This was shown for two *S. aureus* exoproteases, Aur and SspA (Marti et al. 2010). It can be assumed that also the protein-dependent biofilm formation of *S. epidermidis* is influenced by proteases in a similar way as in *S. aureus*. As mentioned above, proteolytic activity also is involved in the processing of the *S. epidermids* autolysin and adhesin Aap in order to activate its intercellular adhesion activity (Rohde et al. 2005). Proteases may play a further role in the inactivation of host defense mechanisms like antibodies and platelet microbicidal proteins and in the destruction of host proteins (von Eiff et al. 2002). In *S. epidermidis*, an extracellular metalloprotease with elastase activity was described (Teufel and Götz 1993) and a cysteine protease which degrades immunoglobulins, serum albumin, fibrinogen, and fibronectin has been reported (Sloot et al. 1992). Furthermore, a serine protease is involved in the processing of the antibiotic epidermin (Geissler et al. 1996). Antimicrobial peptides like epidermin and related compounds are produced by many staphylococcal species and their antimicrobial activity makes it inviting to speculate that they may be important for competition with other microorganisms during skin or mucous membrane colonization but direct evidence for this hypothesis is missing (Otto 2010a, b).

4.1.3 Lipase

The lipase or glycerol ester hydrolase is a secreted enzyme which is biosynthesized as a pre-pro-enzyme that is stepwise processed by removing signal- and propeptide during and after secretion to finally achieve its mature form (Rosenstein and Götz 2000). Besides the nutritional function by hydrolyzation of exogenous substrates a role for lipases in the colonization of the skin by the release of fatty acids which may promote adherence has been proposed (Gribbon et al. 1993). Apart from GehC, a second lipase, GehD, has been identified in *S. epidermidis* (Longshaw et al. 2000). GehD is well conserved among the *S. epidermidis* strains and homologs are also found in *S. aureus*, *S. haemolyticus*, S. capitis, *S. caprae*, and *S. warneri*. Lipases are regarded as possible virulence factors in the pathogenesis of a number of localized infections such as boils or abscesses but a corresponding role has not been clearly determined (Longshaw et al. 2000). Accordingly, a surface-associated lipase of *S. saprophyticus*, Ssp, turned out to be necessary for persistence in bladder and kidney in a murine infection model but its true role in infection remains to be elucidated (Kline et al. 2010). Recent findings indicate a more concrete contribution of staphylococcal lipase in pathogenesis since *S. epidermidis* GehD has been identified as a collagen binding adhesin. Also, for the other *S. epidermidis* lipase, GehC, a role in collagen binding has been proposed but is hitherto unexplored (Bowden et al. 2002).

4.2 Iron Acquisition

The acquisition of iron is an indispensable need in microbial life and all staphylococcal species are outfitted with a variety of iron uptake systems that either work siderophore-mediated or by direct uptake of iron via surface proteins. On the other hand, the lifestyle of invasive pathogens involves a competition with the host iron uptake systems for the rare mineral compound. The number of iron uptake systems is not much different in *S. aureus* and the medium- or non-pathogenic staphylococci indicating that successful competition for iron is also important in other habitats than in the mammalian host. A genomic comparison of *S. aureus*, *S. epidermidis*, *S. haemolyticus*, *S. saprophyticus*, and the non-pathogenic *S. carnosus* (Rosenstein et al. 2009) yielded only the *isd* system as exclusively present in *S. aureus*. The *isd* (iron-responsive surface determinant) genes encode an iron-regulated ATP-driven uptake machinery for heme–iron, which is a preferred iron source for *S. aureus* in the initial stage of infection, while iron carried by siderophores appears to be the preferred source during later stages of infection (Skaar et al. 2004). Since inactivation of the *isd* locus has no detrimental effect on the consumption of heme–iron as sole iron source it became clear that *S. aureus* does not rely on a single system for supply with heme-bound iron (Mazmanian et al. 2003). Skaar et al. identified the *hts*ABC transport system as being also involved in heme–iron uptake by *S. aureus* (Skaar et al. 2004). But it has been reported that *hts* may not exclusively be involved in uptake of heme and recent results with *isd-hts* double mutants indicate that further heme–iron uptake systems may exist in *S. aureus* (Wright and Nair 2012). Another uptake system, most probably specific for inorganic iron, is encoded by the *fepABC operon* which comprises the genes for an iron-binding lipoprotein, an iron-dependent peroxidase (that is translocated by the flanking TatA/C system) and an iron permease (Biswas et al. 2009a, b). This system is present in *S. aureus*, *S. carnosus*, and *S. haemolyticus* but is missing in a number of other staphylococcal species and was shown to contribute to fitness and virulence during infection (Biswas et al. 2009a, b).

4.3 ACME

The arginine catabolic element (ACME) is a located on a mobile genetic element and comprises the *arc* genes which encode an arginine deiminase pathway. ACME has been identified with a high prevalence in *S. epidermidis* and in some community-acquired *S. aureus* strains like USA300 (Diep et al. 2006). The ACME element was found to be integrated into *orfX* of *S. aureus* USA300 at the same attachment site as SCC*mec* and, at a corresponding attachment locus, in the genome of *S. epidermidis* ATCC12228 (Diep et al. 2006). Arginine deiminase has been described as a virulence factor in *Streptococcus pyogenes* (Degnan et al. 1998) and seems to be involved in bacterial survival at low pH values. Furthermore, depletion of arginine by the enzymatic activity of arginine deiminase inhibits nitric oxide production from L-arginine which might be advantageous for the bacteria since nitric oxide is a molecule used in immune responses against microbial infections (Diep et al. 2006). The *arc* genes are also found on the chromosomes of the ACME-containing strains and it is hypothesized that the duplication of the encoded biosynthetic pathway might enhance the ability to survive within the host (Shore et al. 2011). The high prevalence of ACME in *S. epidermidis* isolates indicates that this element might also play an essential role for these species and it is assumed that it might have been horizontally transferred from *S. epidermidis* to *S. aureus* (Diep et al. 2006).

4.4 Urease of **Staphylococcus saprophyticus**

S. saprophyticus is a coagulase-negative *Staphylococcus* belonging to the medium-pathogenic species and is of clinical relevance as a frequent cause of uncomplicated urinary tract infections (UTI) (Wallmark et al. 1978). Its specialization with regard to disease and infection locus is reflected by a restricted genetic equipment with virulence and fitness factors. Only a single LPXTG-containing protein has been identified in the genome of *S. saprophyticus* ATCC 15305 that was shown to mediate adherence to human bladder cells (Kuroda et al. 2005). The urease of *S. saprophyticus* seems also to be involved in virulence since the hydrolyzation of urea by urease is known to be an important factor in urinary tract infection (Gatermann et al. 1989). Although the urease of *S. saprophyticus* is not exclusively present in this species, it shows a remarkably higher activity than its homologs in *S.aureus* or *S. epidermidis*, indicating a regulatory phenomenon that causes an adaptation to persistence in the urinary tract environment based on activity of urease as a fitness factor (Kuroda et al. 2005).

4.5 *D-Serine Deaminase of* **Staphylococcus saprophyticus**

The urinary tract-pathogenic species *S. saprophyticus* appears to be well adapted to this site of infection as it is resistant to D-serine in contrast to other staphylococcal species. Human urine contains a relatively high concentration of D-serine, which is toxic to several non-uropathogenic bacteria, but can be utilized or detoxified by uropathogenic *Escherichia coli* (UPEC). *S. saprophyticus* contains a D-serine deaminase gene (*dsdA*) which is homologous to the corresponding gene in UPEC. The gene is absent in *S. xylosus* and *S. cohnii*, phylogenetically close relatives of *S. saprophyticus*, and is also not found in isolates of *S. aureus*, *S. epidermidis* and 13 other staphylococcal species (Sakinc et al. 2009a, b). It is proposed that D-serine utilization and detoxification may be a general property of uropathogenic bacteria.

5 Physiological Properties Involved in Virulence and Fitness

The virulence of staphylococci is influenced by certain physiological properties that may impact the interplay with competing pathogens, enable persistence in the infected host, mediate resistance to antimicrobial compounds like lysozyme and reactive oxygen substances, and allow for evasion from neutrophil killing. Examples for these physiological influences are presented in the following sections.

5.1 *Cyanide-sensitive Cytochrome bd Oxidase Prevalent in Pathogenic Species*

Pseudomonas aeruginosa and *S. aureus* are opportunistic pathogens and frequently coinfect the lungs of cystic fibrosis patients. With increasing age of the patients the number of *S. aureus* cells declines slightly and that of *P. aeruginosa* increases. One of the main reasons for the decrease in *S. aureus* is its susceptibility to respiratory inhibitors excreted by *P. aeruginosa*, like pyocyanin, hydrogen cyanide, or quinoline N-oxides that may act against the commensal flora as well as against host cells. S. *aureus* and other pathogenic species contain a pyocyanin- and cyanide-sensitive cytochrome bd quinol oxidase, CydAB, while the non-pathogenic species such as *S. carnosus*, *S. piscifermentans* and *S. gallinarum* have a pyocyanin- and cyanide-resistant cytochrome bd quinol oxidase with the subunit CydB determining the resistance (Voggu et al. 2006). On the other hand it was also shown that a subpopulation of *S. aureus* survives these respiratory toxins adopting the small-colony variant (SCV) phenotype (Biswas et al. 2009a, b).

5.2 Small-Colony Variants as a Survival Strategy Under Certain Selective Pressure

Small-colony variants (SCVs) can be produced most likely by all facultative anaerobic microorganisms. In *S. aureus*, they have been recognized for many years and the small-colony variant phenotype is associated with persistent and recurrent infections. Clinical *S. aureus* SCVs are frequently auxotrophic for compounds involved in the biosynthesis of the electron transport chain, like menadione, thymidine, or hemin (Proctor et al. 2006). A stable *S. aureus hemB* mutant showed typical characteristics of clinical SCVs, such as slow growth, decreased pigment formation, low coagulase activity, reduced hemolytic activity, and resistance to aminoglycosides (von Eiff et al. 1997). Furthermore, the mutant was able to persist within cultured endothelial cells due to decreased alpha-toxin production. In Northern and Western blot analyses a markedly reduced expression of alpha-toxin and protein A at the mRNA and protein level was shown. The SCV phenotype of the *hemB* mutant was reversed by growth with supplemented hemin or by complementation with intact *hemB* (von Eiff et al. 1997). Hence, a defect in the electron transport system allows *S. aureus* SCVs to resist aminoglycosides and to persist intracellularly. The SCV phenotype is the answer of pathogenic staphylococcal species to certain antibiotics, to respiratory toxins as produced in the co-infection with *P. aeruginosa* or during respiratory burst in phagocytic cells.

5.3 Structural Alteration of Peptidoglycan and Resistance to Lysozyme

S. aureus belongs to the few bacterial species that are completely resistant to lysozyme, which greatly contributes to their persistence and success in colonizing the skin and mucosal areas of humans and animals. The reason for the lysozyme resistance is mainly based on the O-acetylation of the peptidoglycan structure at the C6-OH position of the muramic acid (Bera et al. 2005). The peptidoglycan-specific O-acetyltransferase is encoded by the *oatA* gene. Interestingly, this *oatA* gene is only present in pathogenic lysozyme-resistant staphylococci (e.g., *S. aureus*, *S. epidermidis*, *S. lugdunensis*, and others). All non-pathogenic species are lysozyme-sensitive. They can be divided into sensitive (e.g. *S. carnosus*, *S. gallinarum*, and *S. xylosus*) and hypersensitive species (e.g. *S. equorum*, *S. lentus*, and *S. arlettae*). In all lysozyme-sensitive species, the analyzed peptidoglycan was de-O-acetylated (Bera et al. 2006). Besides O-acetylation of peptidoglycan, also the wall teichoic acid and a high degree of murein cross-linking play a role in lysozyme resistance (Bera et al. 2007). It turned out that lysozyme acts not only by its muramidase activity but also as an antimicrobial peptide (Herbert et al. 2007).

5.4 Staphyloxanthin

In former times the pigmentation was a main criterion to distinguish *S. aureus* from *S. epidermidis* (originally called *S. albus*) (Götz et al. 2006). The major pigment produced by *S. aureus* is the deep-yellow carotenoid 4,4'-diaponeurosporene which is, after prolonged cultivation, in part converted into the orange pigment staphyloxanthin (Marshall and Wilmoth 1981; Wieland et al. 1994). The staphyloxanthin biosynthesis genes are organized in an operon, *crtOPQMN* which is controlled by a sigma(B)-dependent promoter (Pelz et al. 2005). The first step in the biosynthesis of staphyloxanthin is the head-to-head condensation of two molecules of farnesyl diphosphate catalyzed by the dehydrosqualene synthase CrtM to form dehydrosqualene. Dehydrogenation of dehydrosqualene by CrtN yields the yellow pigment 4,4'-diaponeurosporene. The subsequent conversion into staphyloxanthin is catalyzed in further steps involving oxidation (CrtP) and esterification reactions (CrtQ and CrtO) (Pelz et al. 2005). The membrane-bound staphyloxanthin plays a role in the resistance to reactive oxygen species (ROS) and mediates the evasion from neutrophil killing (Clauditz et al. 2006).

6 Antibiotic Resistances

Besides the fitness factors that support survival under hostile conditions, the quick adaptation to selective pressure exerted by antibiotic treatment represents an indispensable capability of pathogenic staphylococci. Staphylococcal resistance to antibiotics is mediated by genes that are in most cases located on mobile genetic elements thereby allowing their rapid spreading by lateral gene transfer. The archetypical resistance to the ß-lactam derivative methicillin is encoded by the *mec* gene carried by the staphylococcal cassette chromosome SCC*mec* (see below). Methicillin-resistant *S. aureus* strains (MRSA) soon occurred after the introduction of the semi-synthetic penicillins (Barber 1961). The application of alternative antibiotics in the treatment of staphylococcal infections has been quickly counteracted by the appearance of the corresponding resistances thus leading to the synonymous use of the term MRSA for multiple resistant *S. aureus*. Most of the antibiotic resistance genes are carried by plasmids and transposons (Malachowa and DeLeo 2010) leading to the quick spreading of resistance genes under the pressure of antibiotic treatment. A comprehensive and up-to-date compilation of the numerous resistance determinants of staphylococci is presented in a review by Malachawo and DeLeo (Malachowa and DeLeo 2010). The increasing resistance of staphylococci left only a few antibiotics being effective in the treatment of infections. But also the use of glycopeptides, particularly vancomycin, as so-called last-resort antibiotics, runs the risk of getting ineffective since *S. aureus* strains with reduced susceptibility to glycopeptides are emerging. The versatility of *S. aureus* in coping with the threat of being attacked by these antibiotics is demonstrated by different strategies to counteract the effect of the cell-wall antibiotic vancomycin. An intermediary resistance to Vancomycin is achieved by changes in the cell wall and in metabolic pathways (Appelbaum and Bozdogan 2004), while a high level vancomycin resistance is mediated by acquisition of the resistance determinant *vanA* that seems to originate from an enterococcal source (Rehm and Tice 2010). Other antibiotics like macrolides, lincosamides, aminoglycosides, and quinolones were introduced but soon became also compromised by the emergence of resistances (Cameron et al. 2011) rendering the antibiotic resistances an ever-increasing threat in the treatment of staphylococcal infections.

7 Regulation of Virulence Determinants

The versatility and quick adaptability of staphylococcal pathogens depends on the regulated expression and the concerted action of many exoproteins in response to changes in the hostile environment in the host. Several global regulatory systems have been reported for staphylococci of which some are involved in the regulation of virulence factors like the two-component regulatory systems *agr*, *sae*, *arlRS*, *srrAB* and *lytRS* and the transcription factors *sarA* and its homologs *sarR*, *sarT* and *sarS* (Bronner et al. 2004; Novick 2003). Other more specialized regulatory systems, *vra* and *gra*, are involved in the resistance to antibiotics (Gardete et al. 2006; Meehl et al. 2007) or, like *aps*, in the resistance to antimicrobial peptides (Li et al. 2007). The global regulatory systems are well conserved among staphylococci and most of them are also present in the non-pathogenic *S. carnosus* TM300 (Rosenstein et al. 2009). Their role in the regulation of virulence factors has been confirmed by numerous studies on the attenuation for virulence of mutants in the regulatory systems (for a detailed overview of the regulatory systems, see the excellent reviews by Novick (2003) and Bronner et al. (2004). The *rot* and *sar* systems seem to be specific for *S. aureus* and may exclusively function as virulence regulators (Bronner et al. 2004). The *agr* system, albeit ubiquitously present in staphylococci, reveals a significant amount of genetic diversity resulting in various *agr* types of *S. aureus* which were found to correlate with certain diseases (Feng et al. 2008).

8 Genomic Aspects Concerning Staphylococcal Virulence

Due to their impact in infection, staphylococci are also of great interest in genomic studies. In the course of the flourishing genomic research during the past decade, numerous staphylococcal genome sequences have been determined and due to the highly effective next generation sequencing methods the number of staphylococcal genome projects is ever increasing. According to its leading role as human pathogen, most of the determined staphylococcal genome sequences are derived from various *S. aureus* strains. Currently (April 2012), 34 completed genome sequences of different *S. aureus* strains or isolates are available in the public databases (Table 1) and a multiple of this is listed as ongoing bioprojects. Based on this overwhelming amount of sequence information, important aspects concerning the impact of *S. aureus* as leading pathogen could be analyzed. Thus, the genomic changes leading from a vancomycin-susceptible bloodstream isolate of *S. aureus* to an intermediarily vancomycin-resistant *S. aureus* (VISA) could be monitored based on the corresponding genome sequences (Mwangi et al. 2007). Another study reconstructed the jump of an *S. aureus* strain from human to poultry more than 30 years ago (Lowder et al. 2009). Genomic studies helped also to explain the increasing epidemiological success of community-acquired *S. aureus* strains (Diep et al. 2006; Highlander et al. 2007; Huang et al. 2012). Last but not least, the genome analysis of an early-branched, non-pigmented *S. aureus* strain yielded valuable information about *S. aureus* genome evolution (Holt et al. 2011).

Compared to the wealth of genomic information available for *S. aureus*, the amount of sequence data for non-*S. aureus* genomes is sparse. According to their clinical relevance, two genomes of *S. epidermidis* and *S. lugdunensis* strains have been published (Gill 2009; Heilbronner et al. 2011; Tse et al. 2010; Zhang et al. 2003). The determined genome sequences of *S. haemolyticus* and *S. saprophyticus* offered insights into the genetic basis for a multi-resistant phenotype and the virulence of a causative agent of urogenital tract infections (Kuroda et al. 2005; Takeuchi et al. 2005). Recently, the genomes of two canine *S. pseudintermedius* strains were added to the database providing the first genome data on coagulase-positive non-*S. aureus* strains (Ben Zakour et al. 2011; Tse et al. 2011).

Concerning the non-pathogenic staphylococcal species, still only one sequence represented by the genome of the food-grade organism *S. carnosus* TM300 is available which furnished the basis for differential genome studies of pathogenic and non-pathogenic staphylococci (Rosenstein et al. 2009; Rosenstein and Götz 2010).

8.1 Genome Structure

A comparison of the staphylococcal genome sequences revealed a structure which is common to all genomes (Fig. 1). Each genome reveals a region that comprises about 75 % of the genome size and predominantly carries genes that are conserved throughout the genus and reveal a high degree of synteny. Accordingly, this part is designated as the conserved core region (Gill et al. 2005). More pronounced differences among the staphylococcal core regions are mostly due to mobile elements like prophages, genome, and pathogenicity islands, transposons, IS elements, and integrated plasmids (Baba et al. 2002). A comprehensive comparison of dominant sequence type (ST) lineages of *S. aureus* showed that the core genome is interspersed with regions that are highly variable between the lineages: thus, the core

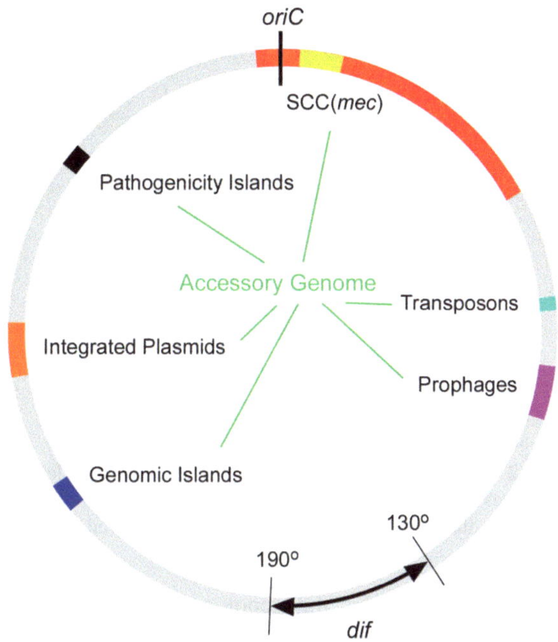

Fig. 1 General structure of staphylococcal genomes. The *light-gray* section represents the conserved core region, the *red* part corresponds to the variable genome region (*oriC* environ). The components of the accessory genome are named accordingly. *oriC* = origin of replication; *dif* = replication termination locus. The *double-headed arrow* indicates the span comprising the replication termination loci in the various staphylococcal genomes (see Fig. 2)

genome has been subdivided in the stable core and the core variable (CV) genome region (Lindsay et al. 2006). Many core variable genes are toxins, superantigens, exoenzymes, and regulatory elements involved in virulence. Furthermore, these genes are distinguished by a higher rate of evolutionary accepted mutations and many repeat regions (Feng et al. 2008). It should be noted that the core variable genes are defined by the comparison of closely related strains of one species. In a genome comparison of different species where the syntenic character of the conserved core has been blurred by the accumulation of evolutionary changes, a core variable region would be much less confined. The remaining 25 % of the genomes correspond to a region that exhibits a high degree of variability even between closely related species. This variable region carries the majority of the species-specific genes like for example the genes encoding protein A (Mazmanian et al. 2000) and staphylocoagulase in *S. aureus* and thus is an important factor in shaping genomic characteristics of species or strains. Since this variable region is located next to the chromosomal replication origin, *oriC*, the term "*oriC* environ" has been coined (Takeuchi et al. 2005).

Whereas the relative composition concerning conserved and variable genome regions is similar in the staphylococcal genomes sequenced so far, a remarkable difference exists regarding the lengths of the replicated genome halves (replichores) as determined by the locations of *oriC* and the replication termination locus which is near the conserved *dif* site (Hendrickson and Lawrence 2007). While all *S. aureus* genomes reveal almost perfectly balanced genomes with replichores of

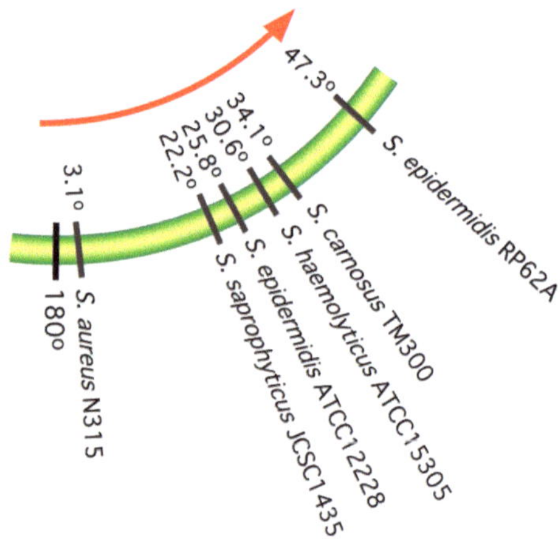

Fig. 2 Positions of the replication termination loci in selected staphylococcal genomes. The 180° indicates the position directly opposite to the origin of replication (at 0°). For each genome sequence the deviation from the 180° position is indicated

about equal lengths, the genomes of non-*S. aureus* species show a significant asymmetry as shown by a location of the *dif* region that remarkably deviates from the 180° coordinate (referring to the location of *oriC* at 0°) (see Fig. 2).

The reason for the differences in genome symmetry is unknown but we hypothesize that it could have arisen from an inversion event during the staphylococcal evolution that comprised the origin of replication in a balanced ancestor genome as can be observed between the genomes of *S. epidermidis* strains RP62A and ATCC12228 (see Fig. 3). If *oriC* were placed at a decentral position in the inverted fragment, the inversion would have led to positioning the replication origin closer to the replication terminator thereby resulting in different replichore lengths in the descendant genome. This might also be the reason for the different replichore lengths of *S. epidermidis*, RP62A and ATCC12228 (Fig. 3).

The *S. aureus* genomes reveal a more stable character with less obvious signs of recombinatorial events than the non-*S. aureus* species. This is assumed to be a consequence of a preservation of the clonal character of the *S. aureus* lineages by restriction-modification systems that control the acquisition of foreign DNA. On the other hand, it has been reported that the reversible inversion of a large chromosomal fragment provokes a switch between small colony and normal colony variants in *S. aureus* Mu50Ω (Cui et al. 2012). Of note, the inverted fragment comprises about one genome half of *S. aureus* Mu50Ω indicating that the symmetry concerning the replichore lengths might not be disturbed.

In contrast to *S. aureus*, the evolution of the *S. epidermidis* lineages has been proposed to be based mainly on recombinational events and lateral transfer of genetic material which is supported by the lack of corresponding restriction-modification sytems (Feng et al. 2008).

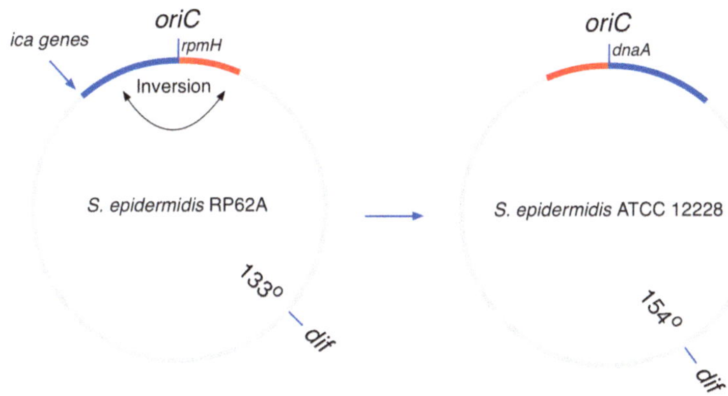

Fig. 3 Inversion of a chromosomal fragment observed between *S. epidermidis* strains RP62A and ATCC 12228. In order to emphasize the decentral position of *oriC* in the inverted fragment, the region downstream of *oriC* (referring to the RP62A genome) is colored *red* while the upstream region is labeled in *dark blue*. The position of the *ica* genes (that are not present in strain ATCC 12228) is indicated. In order to compensate for the inversion, different start genes were chosen by the annotators: *rpmH* in strain RP62A and *dnaA* in strain ATCC 12228. The differing distances between *oriC* and *dif* (given in angular degrees) in both genomes are probably caused by the asymmetric inversion around *oriC*

8.2 Genome Rearrangements

The variable region around *oriC* seems to be a preferred region for chromosomal rearrangements. Genome comparisons indicate inversions comprising the *oriC* region in some staphylococcal genomes which led to an inverted orientation of the conserved genes next to the replication origin: *dnaA* to *gyrA* on one side of *oriC* and *rpmH* to *gidB* on the other side. All *S. aureus* genomes sequenced so far show the same orientation of these conserved genes. In contrast, a pronounced heterogeneity exists among the non-*S. aureus* genomes concerning the orientation of the *ori*C region. Since it is a kind of unwritten rule to annotate *dnaA* as the first gene in newly determined genome sequences, considerable inconvenience may occur in genome alignments as some genomes appear to be inverted with respect to others resulting from inversions around *oriC* and the concomitant insistence on annotating *dnaA* as gene number 1. In order to compensate for this, some staphylococcal genome sequences have been annotated with *rpmH* as the first gene (see Fig. 3). Interestingly, different orientations of *oriC* even occur within one species as revealed by the alignments of *S. epidermidis*, RP62A and ATCC 12228. As mentioned above, this rearrangement is accompanied by the absence of the *ica* genes in strain ATCC12228 and a biofilm-negative phenotype. This example emphasizes the role of chromosomal rearrangements in the evolution of staphylococcal genomes. This is also demonstrated by the different locations of *att* sites with the same core sequences in *S. haemolyticus* and *S. saprophyticus* in comparison with *S. aureus*, which is presumably caused by genome rearrangements (Novick and Subedi 2007).

8.3 *Mobile Genetic Elements in* Staphylococcus

In summary, the comparative studies on staphylococcal genome sequences revealed a high flexibility corresponding to the capability to adapt quickly to the varying challenges exerted by an environment that could drastically change in the course of switching from commensalic to infectious lifestyle due to host immune defenses, anti-infective treatments, or the appearance of competing bacteria. Soon after the first staphyloccal genomes had been sequenced it became clear that the differences in pathogenic potential and the adaptability to environmental changes are to a large extent determined by mobile genetic elements which have been detected in almost all staphylococcal genomes. Their acquisition by horizontal transfer allows the rapid spreading of genetic information in order to cope with selective environmental challenges. Genomic comparisons showed that particular virulence factors are coupled to certain mobile genetic elements and that different degrees of pathogenicity correlated with the presence or absence of these elements. Nearly all toxins that cause specific diseases like pneumonia, toxic shock syndrome, necrotizing fasciitis, and food poisoning are encoded by mobile genetic elements (Novick et al. 2010). The gene coding for the PV-leukocidin is located on prophage ΦSA2pvl that is predominantly found in community-acquired (CA) MRSA (Otto 2010a, b). In *S. aureus*, also the biofilm-associated protein Bap is located on a pathogenicity island, SaPIbov2, that is exclusively present in strains associated with bovine mastitis (Tormo et al. 2005; Ubeda et al. 2003). In contrast, in other staphylococcal species *bap*-homologous genes are not located on mobile genetic elements demonstrating the remarkable genetic variability among the staphylococcal species (Tormo et al. 2005).

In *S. epidermidis*, integrated plasmids play a significant role in providing strain-specific properties. In strain RP62A the plasmid νSe1 and in strain ATCC12228 the plasmid νSe2 carry the genes for a strain-specific sortase and two strain-specific MSCRAMMs (Gill et al. 2005; Zhang et al. 2003). The mobile elements contribute to a large extent to the variable or accessory part of the staphylococcal genomes and comprise bacteriophages, genomic islands, pathogenicity islands, plasmids, transposons, and the staphylococcal cassette chromosome (SCC) (Lindsay and Holden 2006).

8.3.1 Pathogenicity Islands

Virulence factors or resistance determinants are often encoded on prophages or genomic islands. According to their carrying of virulence genes the corresponding genomic islands are referred to as pathogenicity islands. Some uncertainty exists concerning the definition and nomenclature of pathogenicity islands. Therefore, it has been proposed to use the term "pathogenicity island" only for those elements that share a conserved functional and genetic organization and insertion site specificity and to name them "SaPI" in the case of *S. aureus* pathogenicity islands and with a corresponding species-dependant derivation for non-*S. aureus* islands (Novick and Subedi 2007). Correspondingly, for SaPI-type islands detected in *S. haemolyticus* (originally designated νSh2) (Takeuchi et al. 2005) and *S. saprophyticus* (originally designated νSs15305) (Kuroda et al. 2005) the names ShPI2 and SsPI15305 have been proposed, respectively (Novick and Subedi 2007).

The pathogenicity islands are widespread among the *S. aureus* genomes and related elements found also in *S. saprophyticus* and *S. haemolyticus* (Novick and Subedi 2007). The genetic organization of the pathogenicity islands resembles that of temperate phages; their integration occurs integrase-dependent at specific chromosomal sites and their mobilization is mediated by helper phages like 80α or Φ13 (Maiques et al. 2007). As implied by their designation, the pathogenicity islands are a major source of pathogenicity factors encoded by mobile elements in staphylococci.

Staphylococcal superantigens are often encoded by genes located on pathogenicity islands with the toxic shock syndrome toxin being the toxin most frequently encoded by SaPIs (Novick and Subedi 2007). As mentioned above, also the *S. aureus* biofilm associated protein Bap is encoded on a pathogenicity island, SaPI bov2 (Ubeda et al. 2003).

8.3.2 Genomic Islands

Besides the well-defined SaPIs, also other less strictly defined genomic islands are known which comprise clusters of genes that probably have been obtained by horizontal gene transfer. Many of these gene clusters are flanked by direct repeats and could have been mobilized by self-coded transfer functions (Novick et al. 2010). These elements often have obviously become immobilized in the genomes thus representing only remnants of formerly active elements.

Two islands, νSaα und νSaß, are found in nearly all *S. aureus* isolates (Fitzgerald et al. 2003). νSaα carries a cluster of genes encoding superantigen-like proteins, the *set*-cluster (or, according to a new nomenclature, *ssl* cluster) and a lipoprotein (*lpl*) cluster (Feng et al. 2008; Lina et al. 2004). The genomic island νSaß encodes a serine protease cluster (*spl*) and an enterotoxin cluster (Feng et al. 2008).

8.3.3 Prophages

Prophages play an important role in the differential genome compositions of staphylococci and often carry genetic information that contributes significantly to the virulence traits of a certain strain. Prominent virulence factors provided by prophages are enterotoxin A (involved in food poisoning), exfoliative toxin A (localized scalded skin syndrome) and PV-leukocidin which is implicated in the pathogenesis of community-acquired MRSA (Ben Zakour et al. 2008; Loffler et al. 2010). Other prophages provide their host genomes with means to avoid the hosts' immune responses like the chemotactic inhibitory protein (CHIPS), staphylokinase or the staphylococcal complement inhibitor (SCIN) (van Wamel et al. 2006).

8.3.4 Staphylococcal Cassette Chromosome

The staphylococcal cassette chromosome (SCC) represents a unique type of mobile genetic element in *Staphylococcus* that carries the gene for methicillin resistance, *mecA*, and reveals specificity for integration at a specific site downstream to the conserved gene *orfX* that is located near the origin of replication (Katayama et al. 2000). Spreading of the *mec*-containing SCC-elements is the cause for rapidly turning MSSA into MRSA under the corresponding selective pressure.

Besides the *mecA* gene, the SCC element carries genes responsible for mobilization, cassette chromosome recombinase (*ccr*) genes, and interjacent DNA, formerly designated junkyard (J-) DNA whereas the "J" is now standing for "joining" (Turlej et al. 2011). The J-region may function as target for additional mobile elements like plasmids or transposons that could add further resistance genes to the SCC element (Turlej et al. 2011). This region thus represents a hot spot for recombination resulting in several structural variants of SCC*mec*. Thus, various combinations of *mec*, *ccr* and J regions have been described leading to numerous types and subtypes of SCC elements (Turlej et al. 2011). Some SCC types also carry restriction-modification systems but these are only present in a small proportion of *S. aureus* strains (Feng et al. 2008). In addition to *S. aureus*, SCC*mec* has also been described in several coagulase-negative staphylococci (Garza-Gonzalez et al. 2010).

8.4 Repeat Sequences

Besides the presence of various mobile genetic elements, the adaptive flexibility of bacterial genomes is mirrored by repetitive DNA sequences. There are various types of repeat sequences indentified in staphylococcal genomes. They may play a role for genome diversification due to recombinatorial events that might be important for the rapid adaptation to the environment (Aras et al. 2003).

One type of sequence repeats is represented by the CRISPR (Clustered regularly interspaced short palindromic repeats) elements which are present in roughly 50 % of bacterial genomes and in most archae. The CRISPRs are composed of repeats with lengths between 20 and 50 nucleotides which are interspersed by unique spacer sequences (Bhaya et al. 2011; Marraffini and Sontheimer 2008). Recently, it has been recognized that the CRISPR elements act in cooperation with flanking CRISPR-associated (*cas*) genes as an adaptive immune mechanism against foreign DNA like phages and plasmids, which is comparable to eukaryotic RNA interference systems (Wiedenheft et al. 2012). According to the literature, CRISPR elements are seldom found in *S. aureus* strains (Holt et al. 2011; Otto 2012) with only two reported CRISPR-carrying *S. aureus* strains: the livestock-associated sequence-type 398 (Golding et al. 2010) and the early-branched lineage strain MSHR1132 (Holt et al. 2011). In contradiction to this, the database of CRISPR elements, CRISPRdb (Grissa et al. 2007), lists most *S. aureus* genome with at least one predicted CRISPR element. But both sources congruently state that *S. saprophyticus*, *S. haemolyticus*, and *S. carnosus* reveal no CRISPR (Grissa et al. 2007; Holt et al. 2011). A CRISPR element is accordantly reported also for *S. epidermidis* RP62A where it is involved in the maintenance of genetic identity by preventing horizontal gene transfer (Marraffini and Sontheimer 2008). Also, both sequenced *S. lugdunensis* strains (Heilbronner et al. 2011; Tse et al. 2010) and *S. pseudintermdius* ED99 (Ben Zakour et al. 2011) are listed in the CRISPRdb with CRISPR elements.

Another type of repeat element, STAR (= *Staphylococcus Aureus* Repeat), has been detected in *S. aureus* and seems to be present also in some other species of the genus, like in *S. epidermidis*, as indicated by hybridization (Cramton et al. 2000). These repeats are found in intergenic regions as in the region between the *geh* (lipase) gene and *icaC* gene of the *ica*-operon (Cramton et al. 2000). The STAR-comprising intergenic regions differ significantly between various *S. aureus* strains indicating a high genetic variability, probably due to recombinatorial events.

The STAR-containing intergenic region seems to provide also a breakpoint in the rearrangement of chromosomal fragments as indicated by the large inversion observed between the genomes of *S. epidermidis* RP62a and *S. epidermidis* ATCC12228 (see Fig. 3). Since the intergenic region between *icaC* and *geh* flanks the large chromosomal fragment that has been inverted according to the genome comparison of *S. epidermidis* RP62A and *S. epidermidis* ATCC 12228, it can be assumed that this rearrangement caused the concomitant loss of the *ica*-operon in *S. epidermidis* ATCC12228. Since sequence repeats could facilitate genomic diversification by recombinational events (Aras et al. 2003) it is tempting to speculate that the STAR repeats might be prone to recombinatorial events.

8.5 Single Nucleotide Polymorphisms

Also, Single Nucleotide Polymorphisms (SNP) play a role in the variation of virulence among *S. aureus* strains. Of note, this finding is due to the wealth of genomic data available for various *S. aureus* strains or isolates. Since only one or two genome sequences are available for the other non-*S. aureus* species, the role of SNPs still remains uncertain for these. A striking example for the impact of single nucleotide exchanges in *S. aureus* is represented by the study on the decrease of vancomycin susceptibility in an *S. aureus* isolate in the course of the antibiotic treatment of an endocarditis patient. The increasing vancomycin resistance of the infecting strain was found to correlate with the accumulation of 35 point mutations in the isolates taken between the start and end of treatment (Mwangi et al. 2007).

9 The Non-pathogenic *Staphylococcus carnosus* in Comparison with the Other Staphylococcal Species

S. carnosus is a non-pathogenic member of the genus that is applied in starter cultures in the food industry and has been classified as GRAS = generally recognized as safe (Götz 1990; Schleifer and Fischer 1982). Its genome sequence has been determined and provided the means for comparative studies of pathogenic and non-pathogenic staphylococci (Rosenstein et al. 2009; Rosenstein and Götz 2010).

The *S. carnosus* genome is characterized by many features that result from an adaptation to constant environmental conditions as they are present in a defined meat starter culture environment. Phenotypically, *S. carnosus* TM300 displays various metabolic functions that favor its application in meat fermentation, like nitrate/nitrite reduction (Neubauer and Götz 1996; Neubauer et al. 1999), high osmotic stress tolerance, and two catalase genes (Rosenstein et al. 2009). On the other hand, the long-standing habitation of a nutrition-rich environment led to the loss of many functions that are required by staphylococci that live in more challenging surroundings. *S. carnosus* exhibits no noteworthy exoenzyme activities like lipase, protease, and nuclease which are commonly observed in other staphylococci. These signs of a degenerative evolution become also apparent in a comparative analysis of the *S. carnosus* genome data and those of virulent staphylococci. The GC content of nearly 35 % significantly surmounts that of the other staphylococci (Rosenstein et al. 2009). In addition, the relatively small genome size and a significant number of truncated genes including important functions like global regulators (*agr*, *sae*), the signal recognition particle *ffh* and various exoenzyme genes are a clear indication for an adaptation to a nutritionally stable habitat (Rosenstein et al. 2009). Similar signs of degenerative accommodation are also observed in other bacteria that are applied in the food industry (van de Guchte et al. 2006). While the genome data of staphylococcal species that are confronted with quickly changing and challenging habitats present in the living human or animal host are distinguished by a remarkable genome plasticity, *S. carnosus* reveals a rather static genome with few repetitive sequences, few traces of ancient mobile elements like an inactive prophage and a remnant of a genome island. All staphylococcal genomes determined so far carry also an SSC element downstream of the conserved open reading frame *orfX*. In contrast, *S. carnosus* reveals only a *pbp4*-homologous gene as a faint sign of a former SCC in the corresponding region of its genome. Taking these observations together, *S. carnosus* reveals the lowest genome plasticity of all staphylococcal genomes sequenced so far.

On the other hand, also the comparably static *S. carnosus* genome comprises a variable region next *oriC* (*oriC* environ) with an abundance of species-specific genes. A functional categorization of the gene products encoded within the *S. carnosus oriC* environ exhibits a clear prevalence of genes involved in amino acid transport and metabolism, again indicating an adaptation to the specific nutrition-rich environment for a meat fermentation starter bacterium. The remnants of a prophage and a genomic island as well as the faint indication for a former SCC element suggest an ancestor that revealed a higher genome plasticity than *S. carnosus* at the present evolutionary stage. The *S. carnosus* genome data also revealed some genes with weak similarities to virulence factors like hemolysin and exotoxins and to hypothetical proteins found in *S. aureus* pathogenicity islands (Rosenstein et al. 2009).

It is thus tempting to speculate that the non-pathogenic character of *S. carnosus* is the consequence of a degenerative evolution from a pathogenic progenitor as indicated by the genetic relicts in its genome. The original habitat of *S. carnosus* is unknown but the isolation of *S. carnosus* TM300 from fermented meat suggests a former habitation of animals. Since the genetic repertoire necessary for the habitation of humans or animals most likely also involves factors classified as virulence genes it is not surprising that *S. carnosus* reveals signs of a putative pathogenic ancestor.

10 Conclusions

Albeit being implicated by the title of this article, a clearly differentiated characterization of staphylococcal pathogenicity by arbitrary and crude categories like "highly", "medium-" and "non-pathogenic" appears not to be adequate in the face of the tremendous genetic and phenotypic flexibility and variability of staphylococcal virulence. The graduation of staphylococcal pathogenicity based on qualitative and quantitative criteria is hampered by the lack of a universally valid definition of what has to be regarded as a virulence factor. The difficulty to differentiate gene products that are exclusively involved in pathogenicity from those that may have dual roles in supporting bacterial life during infection as well as in "daily life" metabolic functions is evoked by the complex and multivalent nature of pathogenicity itself. The multitude of studies on the multifaceted aspects of staphylococcal pathogenicity of which only the "tip of the iceberg" could be addressed in this summarizing article, create a view of a variegated interplay of specialized virulence factors, ambivalent fitness factors, genetic potential, regulatory phenomenons, and the resulting versatility to cope with quickly changing environmental stimuli. Consequently, if we focus only on obvious virulence genes in genomes, we might overlook the more subtle but powerful factors that lead to chronic and persistent infections, cause colonization of specific sites of body regions including implant materials, contribute to high genetic adaptability to environmental changes, or represent fitness factors with impact on physiological diversity and growth rate.

Examples for the multivalent character of pathogenicity are numerous in staphylococcal pathogens. For instance the janus-headed character of the urease of *S. saprophyticus* as a metabolic enzyme on one hand, and as a virulence factor involved in the urogenital tract infections on the other hand (Kuroda et al. 2005). Another example comes from differential analyses between pathogenic and food-associated staphylococci that reveal numerous univocal virulence factors in the latter by immunoblot analysis (Zell et al. 2008). This at least demonstrates that a simplified approach of "counting the virulence genes" will only partially cover the complex phenomenon of staphylococcal pathogenicity. In addition, the overwhelming amount of clinical data on *S. aureus* have shown that also among the various isolates and sequence types of the leading staphylococcal pathogen the impact on human or animal health varies significantly. Here, the remarkable in vivo study on the development of vancomycin resistance in an *S. aureus* bloodstream isolate has to be mentioned again which has shown that variations in virulence could be related to subtle changes on the level of single nucleotide polymorphism even between different isolates of one strain (Mwangi et al. 2007). Moreover, this study underlines that the staphylococcal genomes are highly dynamic and that genomic studies have to consider that they are based on "genome snapshots" that have been taken under certain conditions with regard to time and environment.

There is no easy answer to discriminate pathogenic from non-pathogenic staphylococci. The role of *S. aureus* as the leading staphylococcal pathogen is by all means undisputed according to its wide and diverse spectrum of pathogenicity factors involved in destruction of host cells and tissues as well as escaping the immune defense. They could be regarded as the benchmark for the other staphylococcal pathogens. Correspondingly, the prevalence of *S. aureus* can be based on an assortment of various factors that are exclusively present in this species like toxins, gene products involved in interaction with the hosts immune defense, adhesins, represented by molecular hallmarks like protein A, thermonuclease, or staphylococagulase. But recent studies on the virulence of non-*S. aureus* pathogens indicate that at least some of these distinctive criteria will be taken over by the other staphylococci. Thus, the hitherto exclusive position of *S. aureus* as toxicogenic pathogen is attacked by the fortifying importance of phenol-soluble modulins in *S. epidermidis* as cytolytic toxins (Cheung et al. 2010) and by recent observations of CoNS as being able to internalize during infection (Hirschhausen et al. 2010).

10.1 Staphylococcus *Between Commensalism and Pathogenicity*

Another feature concerning the potentially harmful impact of staphylococcal pathogens on human and animal health that withstands a clear and distinctive categorization is the dualism between a commensalic and a pathogenic lifestyle that is common to the staphylococcal pathogens. Staphylococci that coexist with their hosts are well equipped with the factors that enable them to colonize their niches in the anterior nares or on the skin of humans or animals. The ambivalent nature of this coexistence is reflected by a possible advantage for the host as the bacterial colonizers might prevent other bacteria from gaining a foothold. While on the one hand this advantage is possibly gained by the risk of getting infected, the benefit of eradication of *S. aureus* from the carriers in order to prevent an infection remains questionable on the other hand (Wertheim et al. 2005). Furthermore, it has been shown that non-*S. aureus* carriers have an increased mortality rate in comparison with nasal carriers (Wertheim et al. 2004). This ambivalent character is also reflected on the molecular level, as some adhesins which might be important in the colonization of the host might also have a role in the adhesion to host cells in the course of an infection (Corrigan et al. 2009). Also, the arginine deiminase pathway encoded by the *arc* operon on the ACME element falls into this category as it might support skin colonization on the one hand and might also be involved in getting rid of nitric oxide compounds during infection on the other hand (Diep et al. 2006).

Last but not least, the manifold occurrences of staphylococcal virulence might be regarded as a consequence of the environment inhabited by the particular species. The adaptation to the conditions of the peculiar habitat clearly correlates with the pathogenic potential. This is demonstrated by comparing the two extremes on the staphylococcal virulence scale. The tremendous versatility and flexibility of *S. aureus* as leading pathogen is surely one consequence of its habituation of the moist squamous epithelium of the anterior nares, an area that is defended by innate and induced immune responses (Foster 2009) and thus may be regarded as a front-line between commensalism and pathogenicity. On the other hand, the non-pathogenic, food-grade *S. carnosus* TM300 reveals striking signs of an adaptation to a static environment as present in sausage meat (Rosenstein et al. 2009) while the gene relicts observable in its genome indicate that it might descend from a pathogenic ancestor.

Also, this converse example of staphylococcal adaptability underlines that the different manifestations of staphylococcal virulence and the versatile switching between commensalic and infectious lifestyles reflect the variably challenging conditions present in the peculiar ecological niches occupied by staphylococcal species on their human or animal hosts and thus supports that staphylococcal pathogenesis "is best viewed as an adaptation to the hostile environment of the host and its formidable antibacterial defences" (Novick 2003).

References

Ahrens P, Andresen LO (2004) Cloning and sequence analysis of genes encoding *Staphylococcus hyicus* exfoliative toxin types A, B, C, and D. J Bacteriol 186:1833–1837

Akineden O, Hassan AA, Schneider E, Usleber E (2011) A coagulase-negative variant of *Staphylococcus aureus* from bovine mastitis milk. J Dairy Res 78:38–42

Albrecht T, Raue S, Rosenstein R, Nieselt K, Götz F (2012) Phylogeny of the staphylococcal major autolysin and its use in genus and species typing. J Bacteriol 194:2630–2636

Allignet J, England P, Old I, El Solh N (2002) Several regions of the repeat domain of the *Staphylococcus caprae* autolysin, AtlC, are involved in fibronectin binding. FEMS Microbiol Lett 213:193–197

Antikainen J, Kuparinen V, Lahteenmaki K, Korhonen TK (2007) Enolases from Gram-positive bacterial pathogens and commensal lactobacilli share functional similarity in virulence-associated traits. FEMS Immunol Med Microbiol 51:526–534

Appelbaum PC, Bozdogan B (2004) Vancomycin resistance in *Staphylococcus aureus*. Clin Lab Med 24:381–402
Aras RA, Kang J, Tschumi AI, Harasaki Y, Blaser MJ (2003) Extensive repetitive DNA facilitates prokaryotic genome plasticity. Proc Natl Acad Sci U S A 100:13579–13584
Arrecubieta C, Lee MH, Macey A, Foster TJ, Lowy FD (2007) SdrF, a *Staphylococcus epidermidis* surface protein, binds type I collagen. J Biol Chem 282:18767–18776
Arrecubieta C, Toba FA, von Bayern M, Akashi H, Deng MC, Naka Y, Lowy FD (2009) SdrF, a *Staphylococcus epidermidis* surface protein, contributes to the initiation of ventricular assist device driveline-related infections. PLoS Path 5:e1000411
Baba T, Takeuchi F, Kuroda M, Yuzawa H, Aoki K, Oguchi A, Nagai Y, Iwama N, Asano K, Naimi T, Kuroda H, Cui L, Yamamoto K, Hiramatsu K (2002) Genome and virulence determinants of high virulence community-acquired MRSA. Lancet 359:1819–1827
Baba T, Bae T, Schneewind O, Takeuchi F, Hiramatsu K (2008) Genome sequence of *Staphylococcus aureus* strain Newman and comparative analysis of staphylococcal genomes: polymorphism and evolution of two major pathogenicity islands. J Bacteriol 190:300–310
Barber M (1961) Methicillin-resistant staphylococci. J Clin Pathol 14:385–393
Barriere C, Leroy-Setrin S, Talon R (2001) Characterization of catalase and superoxide dismutase in *Staphylococcus carnosus* 833 strain. J Appl Microbiol 91:514–519
Bartlett AH, Hulten KG (2010) *Staphylococcus aureus* pathogenesis: secretion systems, adhesins, and invasins. Pediatr Infect Dis J 29:860–861
Beenken KE, Dunman PM, McAleese F, Macapagal D, Murphy E, Projan SJ, Blevins JS, Smeltzer MS (2004) Global gene expression in *Staphylococcus aureus* biofilms. J Bacteriol 186:4665–4684
Ben Zakour NL, Guinane CM, Fitzgerald JR (2008) Pathogenomics of the staphylococci: insights into niche adaptation and the emergence of new virulent strains. FEMS Microbiol Lett 289:1–12
Ben Zakour NL, Bannoehr J, van den Broek AH, Thoday KL, Fitzgerald JR (2011) Complete genome sequence of the canine pathogen *Staphylococcus pseudintermedius*. J Bacteriol 193:2363–2364
Bentley ML, Gaweska H, Kielec JM, McCafferty DG (2007) Engineering the substrate specificity of *Staphylococcus aureus* Sortase A. The beta6/beta7 loop from SrtB confers NPQTN recognition to SrtA. J Biol Chem 282:6571–6581
Bera A, Herbert S, Jakob A, Vollmer W, Götz F (2005) Why are pathogenic staphylococci so lysozyme resistant? The peptidoglycan O-acetyltransferase OatA is the major determinant for lysozyme resistance of *Staphylococcus aureus*. Mol Microbiol 55:778–787
Bera A, Biswas R, Herbert S, Götz F (2006) The presence of peptidoglycan O-acetyltransferase in various staphylococcal species correlates with lysozyme resistance and pathogenicity. Infect Immun 74:4598–4604
Bera A, Biswas R, Herbert S, Kulauzovic E, Weidenmaier C, Peschel A, Götz F (2007) Influence of wall teichoic acid on lysozyme resistance in *Staphylococcus aureus*. J Bacteriol 189:280–283
Berends ET, Horswill AR, Haste NM, Monestier M, Nizet V, von Kockritz-Blickwede M (2010) Nuclease expression by *Staphylococcus aureus* facilitates escape from neutrophil extracellular traps. J Innate Immun 2:576–586
Bestebroer J, Aerts PC, Rooijakkers SH, Pandey MK, Kohl J, van Strijp JA, de Haas CJ (2010) Functional basis for complement evasion by staphylococcal superantigen-like 7. Cell Microbiol 12:1506–1516
Bhaya D, Davison M, Barrangou R (2011) CRISPR-Cas systems in bacteria and archaea: versatile small RNAs for adaptive defense and regulation. Annu Rev Genet 45:273–297
Biswas R, Voggu L, Simon UK, Hentschel P, Thumm G, Götz F (2006) Activity of the major staphylococcal autolysin Atl. FEMS Microbiol Lett 259:260–268
Biswas L, Biswas R, Nerz C, Ohlsen K, Schlag M, Schäfer T, Lamkemeyer T, Ziebandt AK, Hantke K, Rosenstein R, Götz F (2009a) Role of the twin-arginine translocation pathway in *Staphylococcus*. J Bacteriol 191:5921–5929

Biswas L, Biswas R, Schlag M, Bertram R, Götz F (2009b) Small-colony variant selection as a survival strategy for *Staphylococcus aureus* in the presence of *Pseudomonas aeruginosa*. Appl Environ Microbiol 75:6910–6912

Bjerketorp J, Nilsson M, Ljungh A, Flock JI, Jacobsson K, Frykberg L (2002) A novel von Willebrand factor binding protein expressed by *Staphylococcus aureus*. Microbiology 148:2037–2044

Bjerketorp J, Jacobsson K, Frykberg L (2004) The von Willebrand factor-binding protein (vWbp) of *Staphylococcus aureus* is a coagulase. FEMS Microbiol Lett 234:309–314

Bowden MG, Visai L, Longshaw CM, Holland KT, Speziale P, Hook M (2002) Is the GehD lipase from *Staphylococcus epidermidis* a collagen binding adhesin? J Biol Chem 277:43017–43023

Bowden MG, Chen W, Singvall J, Xu Y, Peacock SJ, Valtulina V, Speziale P, Hook M (2005) Identification and preliminary characterization of cell-wall-anchored proteins of *Staphylococcus epidermidis*. Microbiology 151:1453–1464

Bronner S, Monteil H, Prevost G (2004) Regulation of virulence determinants in *Staphylococcus aureus*: complexity and applications. FEMS Microbiol Rev 28:183–200

Brosnahan AJ, Schlievert PM (2011) Gram-positive bacterial superantigen outside-in signaling causes toxic shock syndrome. FEBS J 278:4649–4667

Brückner R (1997) Gene replacement in *Staphylococcus carnosus* and *Staphylococcus xylosus*. FEMS Microbiol Lett 151:1–8

Brückner R, Wagner E, Götz F (1993) Characterization of a sucrase gene from *Staphylococcus xylosus*. J Bacteriol 175:851–857

Cameron DR, Howden BP, Peleg AY (2011) The interface between antibiotic resistance and virulence in *Staphylococcus aureus* and its impact upon clinical outcomes. Clin Infect Dis 53:576–582

Carneiro CR, Postol E, Nomizo R, Reis LF, Brentani RR (2004) Identification of enolase as a laminin-binding protein on the surface of *Staphylococcus aureus*. Microb Infect 6:604–608

Carnio MC, Holtzel A, Rudolf M, Henle T, Jung G, Scherer S (2000) The macrocyclic peptide antibiotic micrococcin P(1) is secreted by the food-borne bacterium *Staphylococcus equorum* WS 2733 and inhibits *Listeria monocytogenes* on soft cheese. Appl Environ Microbiol 66:2378–2384

Carnio MC, Stachelhaus T, Francis KP, Scherer S (2001) Pyridinyl polythiazole class peptide antibiotic micrococcin P1, secreted by foodborne *Staphylococcus equorum* WS2733, is biosynthesized nonribosomally. Eur J Biochem 268:6390–6401

Celard M, Lelievre H, Obadia JF, Chevalier P, Forey F, Vandenesch F, Etienne J (1997) Long-standing bacteremia and endocarditis caused by *Staphylococcus lugdunensis* in a patient with an implantable cardioverter defibrillator. Clin Microbiol Infect 3:387–388

Chavakis T, Wiechmann K, Preissner KT, Herrmann M (2005) *Staphylococcus aureus* interactions with the endothelium: the role of bacterial "secretable expanded repertoire adhesive molecules" (SERAM) in disturbing host defense systems. Thromb Haemost 94:278–285

Cheng AG, McAdow M, Kim HK, Bae T, Missiakas DM, Schneewind O (2010) Contribution of coagulases towards *Staphylococcus aureus* disease and protective immunity. PLoS Path 6:e1001036

Cheung AI, Projan SJ, Edelstein RE, Fischetti VA (1995) Cloning, expression, and nucleotide sequence of a *Staphylococcus aureus* gene (*fbpA*) encoding a fibrinogen-binding protein. Infect Immun 63:1914–1920

Cheung GY, Rigby K, Wang R, Queck SY, Braughton KR, Whitney AR, Teintze M, DeLeo FR, Otto M (2010) *Staphylococcus epidermidis* strategies to avoid killing by human neutrophils. PLoS Path 6:e1001133

Chhatwal GS (2002) Anchorless adhesins and invasins of Gram-positive bacteria: a new class of virulence factors. Trends Microbiol 10:205–208

Christner M, Franke GC, Schommer NN, Wendt U, Wegert K, Pehle P, Kroll G, Schulze C, Buck F, Mack D, Aepfelbacher M, Rohde H (2010) The giant extracellular matrix-binding protein of *Staphylococcus epidermidis* mediates biofilm accumulation and attachment to fibronectin. Mol Microbiol 75:187–207

Chua K, Seemann T, Harrison PF, Davies JK, Coutts SJ, Chen H, Haring V, Moore R, Howden BP, Stinear TP (2010) Complete genome sequence of *Staphylococcus aureus* strain JKD6159, a unique Australian clone of ST93-IV community methicillin-resistant *Staphylococcus aureus*. J Bacteriol 192:5556–5557

Clarke SR, Wiltshire MD, Foster SJ (2004) IsdA of *Staphylococcus aureus* is a broad spectrum, iron-regulated adhesin. Mol Microbiol 51:1509–1519

Clauditz A, Resch A, Wieland KP, Peschel A, Götz F (2006) Staphyloxanthin plays a role in the fitness of *Staphylococcus aureus* and its ability to cope with oxidative stress. Infect Immun 74:4950–4953

Corrigan RM, Rigby D, Handley P, Foster TJ (2007) The role of *Staphylococcus aureus* surface protein SasG in adherence and biofilm formation. Microbiology 153:2435–2446

Corrigan RM, Miajlovic H, Foster TJ (2009) Surface proteins that promote adherence of *Staphylococcus aureus* to human desquamated nasal epithelial cells. BMC Microbiol 9:22

Cramton SE, Gerke C, Schnell NF, Nichols WW, Götz F (1999) The intercellular adhesion (ica) locus is present in *Staphylococcus aureus* and is required for biofilm formation. Infect Immun 67:5427–5433

Cramton SE, Schnell NF, Götz F, Brückner R (2000) Identification of a new repetitive element in *Staphylococcus aureus*. Infect Immun 68:2344–2348

Cucarella C, Solano C, Valle J, Amorena B, Lasa I, Penades JR (2001) Bap, a *Staphylococcus aureus* surface protein involved in biofilm formation. J Bacteriol 183:2888–2896

Cui L, Neoh HM, Iwamoto A, Hiramatsu K (2012) Coordinated phenotype switching with large-scale chromosome flip-flop inversion observed in bacteria. Proc Natl Acad Sci USA 109:E1647–1656

Degnan BA, Palmer JM, Robson T, Jones CE, Fischer M, Glanville M, Mellor GD, Diamond AG, Kehoe MA, Goodacre JA (1998) Inhibition of human peripheral blood mononuclear cell proliferation by *Streptococcus pyogenes* cell extract is associated with arginine deiminase activity. Infect Immun 66:3050–3058

Devriese LA, Schleifer KH, Adegoke GO (1985) Identification of coagulase-negative staphylococci from farm animals. J Appl Bacteriol 58:45–55

Diep BA, Gill SR, Chang RF, Phan TH, Chen JH, Davidson MG, Lin F, Lin J, Carleton HA, Mongodin EF, Sensabaugh GF, Perdreau-Remington F (2006) Complete genome sequence of USA300, an epidemic clone of community-acquired meticillin-resistant *Staphylococcus aureus*. Lancet 367:731–739

Dinges MM, Orwin PM, Schlievert PM (2000) Exotoxins of *Staphylococcus aureus*. Clin Microbiol Rev 13:16–34

Donvito B, Etienne J, Denoroy L, Greenland T, Benito Y, Vandenesch F (1997) Synergistic hemolytic activity of *Staphylococcus lugdunensis* is mediated by three peptides encoded by a non-agr genetic locus. Infect Immun 65:95–100

Downer R, Roche F, Park PW, Mecham RP, Foster TJ (2002) The elastin-binding protein of *Staphylococcus aureus* (EbpS) is expressed at the cell surface as an integral membrane protein and not as a cell wall-associated protein. J Biol Chem 277:243–250

Dziewanowska K, Carson AR, Patti JM, Deobald CF, Bayles KW, Bohach GA (2000) Staphylococcal fibronectin binding protein interacts with heat shock protein 60 and integrins: role in internalization by epithelial cells. Infect Immun 68:6321–6328

Feng Y, Chen CJ, Su LH, Hu S, Yu J, Chiu CH (2008) Evolution and pathogenesis of *Staphylococcus aureus*: lessons learned from genotyping and comparative genomics. FEMS Microbiol Rev 32:23–37

Fiegler H, Bassias J, Jankovic I, Brückner R (1999) Identification of a gene in *Staphylococcus xylosus* encoding a novel glucose uptake protein. J Bacteriol 181:4929–4936

Fischetti VA, Pancholi V, Schneewind O (1990) Conservation of a hexapeptide sequence in the anchor region of surface proteins from gram-positive cocci. Mol Microbiol 4:1603–1605

Fitzgerald JR, Reid SD, Ruotsalainen E, Tripp TJ, Liu M, Cole R, Kuusela P, Schlievert PM, Jarvinen A, Musser JM (2003) Genome diversification in *Staphylococcus aureus*: Molecular evolution of a highly variable chromosomal region encoding the Staphylococcal exotoxin-like family of proteins. Infect Immun 71:2827–2838

Foster TJ (2005) Immune evasion by staphylococci. Nat Rev Microbiol 3:948–958

Foster TJ (2009) Colonization and infection of the human host by staphylococci: adhesion, survival and immune evasion. Vet Dermatol 20:456–470

Foster TJ, Hook M (1998) Surface protein adhesins of *Staphylococcus aureus*. Trends Microbiol 6:484–488

Garcia-Alvarez L, Holden MT, Lindsay H, Webb CR, Brown DF, Curran MD, Walpole E, Brooks K, Pickard DJ, Teale C, Parkhill J, Bentley SD, Edwards GF, Girvan EK, Kearns AM, Pichon B, Hill RL, Larsen AR, Skov RL, Peacock SJ, Maskell DJ, Holmes MA (2011) Meticillin-resistant *Staphylococcus aureus* with a novel mecA homologue in human and bovine populations in the UK and Denmark: a descriptive study. Lancet Infect Dis 11:595–603

Gardete S, Wu SW, Gill S, Tomasz A (2006) Role of VraSR in antibiotic resistance and antibiotic-induced stress response in *Staphylococcus aureus*. Antimicrob Agents Chemother 50:3424–3434

Garza-Gonzalez E, Morfin-Otero R, Llaca-Diaz JM, Rodriguez-Noriega E (2010) Staphylococcal cassette chromosome mec (SCC *mec*) in methicillin-resistant coagulase-negative staphylococci. A review and the experience in a tertiary-care setting. Epidemiol Infect 138:645–654

Gatermann S, John J, Marre R (1989) *Staphylococcus saprophyticus* urease: characterization and contribution to uropathogenicity in unobstructed urinary tract infection of rats. Infect Immun 57:110–116

Geissler S, Götz F, Kupke T (1996) Serine protease EpiP from *Staphylococcus epidermidis* catalyzes the processing of the epidermin precursor peptide. J Bacteriol 178:284–288

Geoghegan JA, Corrigan RM, Gruszka DT, Speziale P, O'Gara JP, Potts JR, Foster TJ (2010) Role of surface protein SasG in biofilm formation by *Staphylococcus aureus*. J Bacteriol 192:5663–5673

Giese B, Glowinski F, Paprotka K, Dittmann S, Steiner T, Sinha B, Fraunholz MJ (2011) Expression of delta-toxin by *Staphylococcus aureus* mediates escape from phago-endosomes of human epithelial and endothelial cells in the presence of beta-toxin. Cell Microbiol 13:316–329

Gill SR (2009) Genomics of the staphylococci. In: Crossley KB et al (eds) Staphylococci in human disease, 2nd edn. Wiley-Blackwell, Chicester

Gill SR, Fouts DE, Archer GL, Mongodin EF, Deboy RT, Ravel J, Paulsen IT, Kolonay JF, Brinkac L, Beanan M, Dodson RJ, Daugherty SC, Madupu R, Angiuoli SV, Durkin AS, Haft DH, Vamathevan J, Khouri H, Utterback T, Lee C, Dimitrov G, Jiang L, Qin H, Weidman J, Tran K, Kang K, Hance IR, Nelson KE, Fraser CM (2005) Insights on evolution of virulence and resistance from the complete genome analysis of an early methicillin-resistant *Staphylococcus aureus* strain and a biofilm-producing methicillin-resistant *Staphylococcus epidermidis* strain. J Bacteriol 187:2426–2438

Golding GR, Bryden L, Levett PN, McDonald RR, Wong A, Wylie J, Graham MR, Tyler S, Van Domselaar G, Simor AE, Gravel D, Mulvey MR (2010) Livestock-associated methicillin-resistant *Staphylococcus aureus* sequence type 398 in humans, Canada. Emerg Infect Dis 16:587–594

Götz F (1990) *Staphylococcus carnosus*: a new host organism for gene cloning and protein production. Soc Appl Bacteriol Symp Ser 19:49–53

Götz F (2002) *Staphylococcus* and biofilms. Mol Microbiol 43:1367–1378

Götz F, Peters G (2000) Colonization of medical devices by coagulase-negative staphylococci. In: Waldvogel FA, Bisno AL (eds) Infections associated with indwelling medical devices, 3rd edn. ASM, Washington, pp 55–58

Götz F, Zabielski J, Philipson L, Lindberg M (1983) DNA homology between the arsenate resistance plasmid pSX267 from *Staphylococcus xylosus* and the penicillinase plasmid pI258 from *Staphylococcus aureus*. Plasmid 9:126–137

Götz F, Bannerman T, Schleifer K-H (2006) The genera *staphylococcus* and *macrococcus*. In: Dworkin M et al (eds) The prokaryotes, 3rd edn. Springer Science-Business Media, LLC, New York

Greene C, McDevitt D, Francois P, Vaudaux PE, Lew DP, Foster TJ (1995) Adhesion properties of mutants of *Staphylococcus aureus* defective in fibronectin-binding proteins and studies on the expression of fnb genes. Mol Microbiol 17:1143–1152

Gribbon EM, Cunliffe WJ, Holland KT (1993) Interaction of *Propionibacterium acnes* with skin lipids in vitro. J Gen Microbiol 139:1745–1751

Grissa I, Vergnaud G, Pourcel C (2007) The CRISPRdb database and tools to display CRISPRs and to generate dictionaries of spacers and repeats. BMC Bioinform 8:172

Guggenberger C, Wolz C, Morrissey JA, Heesemann J (2012) Two distinct coagulase-dependent barriers protect *Staphylococcus aureus* from neutrophils in a three dimensional in vitro infection model. PLoS Path 8:e1002434

Guinane CM, Ben Zakour NL, Tormo-Mas MA, Weinert LA, Lowder BV, Cartwright RA, Smyth DS, Smyth CJ, Lindsay JA, Gould KA, Witney A, Hinds J, Bollback JP, Rambaut A, Penades JR, Fitzgerald JR (2010) Evolutionary genomics of *Staphylococcus aureus* reveals insights into the origin and molecular basis of ruminant host adaptation. Genome Biol Evol 2:454–466

Hair PS, Ward MD, Semmes OJ, Foster TJ, Cunnion KM (2008) *Staphylococcus aureus* clumping factor a binds to complement regulator factor I and increases factor I cleavage of C3b. J Infect Dis 198:125–133

Hartford O, O'Brien L, Schofield K, Wells J, Foster TJ (2001) The fbe (SdrG) protein of *Staphylococcus epidermidis* HB promotes bacterial adherence to fibrinogen. Microbiology 147:2545–2552

Hartleib J, Kohler N, Dickinson RB, Chhatwal GS, Sixma JJ, Hartford OM, Foster TJ, Peters G, Kehrel BE, Herrmann M (2000) Protein A is the von Willebrand factor binding protein on *Staphylococcus aureus*. Blood 96:2149–2156

Heilbronner S, Holden MT, van Tonder A, Geoghegan JA, Foster TJ, Parkhill J, Bentley SD (2011) Genome sequence of *Staphylococcus lugdunensis* N920143 allows identification of putative colonization and virulence factors. FEMS Microbiol Lett 322:60–67

Heilmann C, Gerke C, Perdreau-Remington F, Götz F (1996a) Characterization of Tn917 insertion mutants of *Staphylococcus epidermidis* affected in biofilm formation. Infect Immun 64:277–282

Heilmann C, Schweitzer O, Gerke C, Vanittanakom N, Mack D, Götz F (1996b) Molecular basis of intercellular adhesion in the biofilm-forming *Staphylococcus epidermidis*. Mol Microbiol 20:1083–1091

Heilmann C, Hussain M, Peters G, Götz F (1997) Evidence for autolysin-mediated primary attachment of *Staphylococcus epidermidis* to a polystyrene surface. Mol Microbiol 24:1013–1024

Heilmann C, Thumm G, Chhatwal GS, Hartleib J, Uekotter A, Peters G (2003) Identification and characterization of a novel autolysin (Aae) with adhesive properties from *Staphylococcus epidermidis*. Microbiology 149:2769–2778

Heilmann C, Hartleib J, Hussain MS, Peters G (2005) The multifunctional *Staphylococcus aureus* autolysin aaa mediates adherence to immobilized fibrinogen and fibronectin. Infect Immun 73:4793–4802

Hell W, Meyer HG, Gatermann SG (1998) Cloning of aas, a gene encoding a *Staphylococcus saprophyticus* surface protein with adhesive and autolytic properties. Mol Microbiol 29:871–881

Hendrickson H, Lawrence JG (2007) Mutational bias suggests that replication termination occurs near the *dif* site, not at Ter sites. Mol Microbiol 64:42–56

Hennig S, Nyunt Wai S, Ziebuhr W (2007) Spontaneous switch to PIA-independent biofilm formation in an *ica*-positive *Staphylococcus epidermidis* isolate. Int J Med Microbiol 297:117–122

Herbert S, Bera A, Nerz C, Kraus D, Peschel A, Goerke C, Meehl M, Cheung A, Götz F (2007) Molecular basis of resistance to muramidase and cationic antimicrobial peptide activity of lysozyme in staphylococci. PLoS Pathog 3:e102

Herron LL, Chakravarty R, Dwan C, Fitzgerald JR, Musser JM, Retzel E, Kapur V (2002) Genome sequence survey identifies unique sequences and key virulence genes with unusual rates of amino acid substitution in bovine *Staphylococcus aureus*. Infect Immun 70:3978–3981

Highlander SK, Hulten KG, Qin X, Jiang H, Yerrapragada S, Mason EO Jr, Shang Y, Williams TM, Fortunov RM, Liu Y, Igboeli O, Petrosino J, Tirumalai M, Uzman A, Fox GE, Cardenas AM, Muzny DM, Hemphill L, Ding Y, Dugan S, Blyth PR, Buhay CJ, Dinh HH, Hawes AC, Holder M, Kovar CL, Lee SL, Liu W, Nazareth LV, Wang Q, Zhou J, Kaplan SL, Weinstock GM (2007) Subtle genetic changes enhance virulence of methicillin resistant and sensitive *Staphylococcus aureus*. BMC Microbiol 7:99

Hirschhausen N, Schlesier T, Schmidt MA, Götz F, Peters G, Heilmann C (2010) A novel staphylococcal internalization mechanism involves the major autolysin Atl and heat shock cognate protein Hsc70 as host cell receptor. Cell Microbiol 12:1746–1764

Holden MT, Feil EJ, Lindsay JA, Peacock SJ, Day NP, Enright MC, Foster TJ, Moore CE, Hurst L, Atkin R, Barron A, Bason N, Bentley SD, Chillingworth C, Chillingworth T, Churcher C, Clark L, Corton C, Cronin A, Doggett J, Dowd L, Feltwell T, Hance Z, Harris B, Hauser H, Holroyd S, Jagels K, James KD, Lennard N, Line A, Mayes R, Moule S, Mungall K, Ormond D, Quail MA, Rabbinowitsch E, Rutherford K, Sanders M, Sharp S, Simmonds M, Stevens K, Whitehead S, Barrell BG, Spratt BG, Parkhill J (2004) Complete genomes of two clinical *Staphylococcus aureus* strains: evidence for the rapid evolution of virulence and drug resistance. Proc Natl Acad Sci U S A 101:9786–9791

Holden MT, Lindsay JA, Corton C, Quail MA, Cockfield JD, Pathak S, Batra R, Parkhill J, Bentley SD, Edgeworth JD (2010) Genome sequence of a recently emerged, highly transmissible, multi-antibiotic- and antiseptic-resistant variant of methicillin-resistant *Staphylococcus aureus*, sequence type 239 (TW). J Bacteriol 192:888–892

Holt DC, Holden MT, Tong SY, Castillo-Ramirez S, Clarke L, Quail MA, Currie BJ, Parkhill J, Bentley SD, Feil EJ, Giffard PM (2011) A very early-branching *Staphylococcus aureus* lineage lacking the carotenoid pigment staphyloxanthin. Genome Biol Evolut 3:881–895

Hoppe-Seyler TS, Jaeger B, Bockelmann W, Noordman WH, Geis A, Heller KJ (2004) Molecular identification and differentiation of *Staphylococcus* species and strains of cheese origin. Syst Appl Microbiol 27:211–218

Houston P, Rowe SE, Pozzi C, Waters EM, O'Gara JP (2011) Essential role for the major autolysin in the fibronectin-binding protein-mediated *Staphylococcus aureus* biofilm phenotype. Infect Immun 79:1153–1165

Howden BP, Seemann T, Harrison PF, McEvoy CR, Stanton JA, Rand CJ, Mason CW, Jensen SO, Firth N, Davies JK, Johnson PD, Stinear TP (2010) Complete genome sequence of *Staphylococcus aureus* strain JKD6008, an ST239 clone of methicillin-resistant *Staphylococcus aureus* with intermediate-level vancomycin resistance. J Bacteriol 192:5848–5849

Huang TW, Chen FJ, Miu WC, Liao TL, Lin AC, Huang IW, Wu KM, Tsai SF, Chen YT, Lauderdale TL (2012) Complete genome sequence of *Staphylococcus aureus* M013, a pvl-positive, ST59-SCCmec type V strain isolated in Taiwan. J Bacteriol 194:1256–1257

Hussain M, Herrmann M, von Eiff C, Perdreau-Remington F, Peters G (1997) A 140-kilodalton extracellular protein is essential for the accumulation of *Staphylococcus epidermidis* strains on surfaces. Infect Immun 65:519–524

Hussain M, Becker K, von Eiff C, Schrenzel J, Peters G, Herrmann M (2001) Identification and characterization of a novel 38.5-kilodalton cell surface protein of *Staphylococcus aureus* with extended-spectrum binding activity for extracellular matrix and plasma proteins. J Bacteriol 183:6778–6786

Izano EA, Amarante MA, Kher WB, Kaplan JB (2008) Differential roles of poly-N-acetylglucosamine surface polysaccharide and extracellular DNA in *Staphylococcus aureus* and *Staphylococcus epidermidis* biofilms. Appl Environ Microbiol 74:470–476

Jankovic I, Brückner R (2002) Carbon catabolite repression by the catabolite control protein CcpA in *Staphylococcus xylosus*. J Mol Microbiol Biotechnol 4:309–314
Jankovic I, Brückner R (2007) Carbon catabolite repression of sucrose utilization in *Staphylococcus xylosus*: catabolite control protein CcpA ensures glucose preference and autoregulatory limitation of sucrose utilization. J Mol Microbiol Biotechnol 12:114–120
Jankovic I, Egeter O, Brückner R (2001) Analysis of catabolite control protein A-dependent repression in *Staphylococcus xylosus* by a genomic reporter gene system. J Bacteriol 183:580–586
Janzon L, Arvidson S (1990) The role of the delta-lysin gene (hld) in the regulation of virulence genes by the accessory gene regulator (agr) in *Staphylococcus aureus*. EMBO J 9:1391–1399
Jin T, Bokarewa M, Foster T, Mitchell J, Higgins J, Tarkowski A (2004) *Staphylococcus aureus* resists human defensins by production of staphylokinase, a novel bacterial evasion mechanism. J Immunol 172:1169–1176
Jongerius I, Garcia BL, Geisbrecht BV, van Strijp JA, Rooijakkers SH (2010) Convertase inhibitory properties of Staphylococcal extracellular complement-binding protein. J Biol Chem 285:14973–14979
Josefsson E, McCrea KW, Ni Eidhin D, O'Connell D, Cox J, Hook M, Foster TJ (1998) Three new members of the serine-aspartate repeat protein multigene family of *Staphylococcus aureus*. Microbiology 144(Pt 12):3387–3395
Katayama Y, Ito T, Hiramatsu K (2000) A new class of genetic element, *Staphylococcus* cassette chromosome *mec*, encodes methicillin resistance in *Staphylococcus aureus*. Antimicrob Agents Chemother 44:1549–1555
Khalil H, Williams RJ, Stenbeck G, Henderson B, Meghji S, Nair SP (2007) Invasion of bone cells by *Staphylococcus epidermidis*. Microb Infect 9:460–465
Kiedrowski MR, Kavanaugh JS, Malone CL, Mootz JM, Voyich JM, Smeltzer MS, Bayles KW, Horswill AR (2011) Nuclease modulates biofilm formation in community-associated methicillin-resistant *Staphylococcus aureus*. PloS One 6:e26714
King NP, Sakinc T, Ben Zakour NL, Totsika M, Heras B, Simerska P, Shepherd M, Gatermann SG, Beatson SA, Schembri MA (2012) Characterisation of a cell wall-anchored protein of *Staphylococcus saprophyticus* associated with linoleic acid resistance. BMC Microbiol 12:8
Kleeman KT, Bannerman TL, Kloos WE (1993) Species distribution of coagulase-negative staphylococcal isolates at a community hospital and implications for selection of staphylococcal identification procedures. J Clin Microbiol 31:1318–1321
Kline KA, Ingersoll MA, Nielsen HV, Sakinc T, Henriques-Normark B, Gatermann S, Caparon MG, Hultgren SJ (2010) Characterization of a novel murine model of *Staphylococcus saprophyticus* urinary tract infection reveals roles for Ssp and SdrI in virulence. Infect Immun 78:1943–1951
Kloos WE, Musselwhite MS (1975) Distribution and persistence of *Staphylococcus* and *Micrococcus* species and other aerobic bacteria on human skin. Appl Microbiol 30:381–385
Kuroda M, Ohta T, Uchiyama I, Baba T, Yuzawa H, Kobayashi I, Cui L, Oguchi A, Aoki K, Nagai Y, Lian J, Ito T, Kanamori M, Matsumaru H, Maruyama A, Murakami H, Hosoyama A, Mizutani-Ui Y, Takahashi NK, Sawano T, Inoue R, Kaito C, Sekimizu K, Hirakawa H, Kuhara S, Goto S, Yabuzaki J, Kanehisa M, Yamashita A, Oshima K, Furuya K, Yoshino C, Shiba T, Hattori M, Ogasawara N, Hayashi H, Hiramatsu K (2001) Whole genome sequencing of meticillin-resistant *Staphylococcus aureus*. Lancet 357:1225–1240
Kuroda M, Yamashita A, Hirakawa H, Kumano M, Morikawa K, Higashide M, Maruyama A, Inose Y, Matoba K, Toh H, Kuhara S, Hattori M, Ohta T (2005) Whole genome sequence of *Staphylococcus saprophyticus* reveals the pathogenesis of uncomplicated urinary tract infection. Proc Natl Acad Sci USA 102:13272–13277
Labandeira-Rey M, Couzon F, Boisset S, Brown EL, Bes M, Benito Y, Barbu EM, Vazquez V, Hook M, Etienne J, Vandenesch F, Bowden MG (2007) *Staphylococcus aureus* Panton-Valentine leukocidin causes necrotizing pneumonia. Science 315:1130–1133

Lagace-Wiens PR, Alfa MJ, Manickam K, Karlowsky JA (2007) Thermostable DNase is superior to tube coagulase for direct detection of *Staphylococcus aureus* in positive blood cultures. J Clin Microbiol 45:3478–3479

Lasa I, Penades JR (2006) Bap: a family of surface proteins involved in biofilm formation. Res Microbiol 157:99–107

Lee LY, Liang X, Hook M, Brown EL (2004) Identification and characterization of the C3 binding domain of the *Staphylococcus aureus* extracellular fibrinogen-binding protein (Efb). J Biol Chem 279:50710–50716

Lewis K (2007) Persister cells, dormancy and infectious disease. Nat Rev Microbiol 5:48–56

Li M, Cha DJ, Lai Y, Villaruz AE, Sturdevant DE, Otto M (2007) The antimicrobial peptide-sensing system aps of *Staphylococcus aureus*. Mol Microbiol 66:1136–1147

Li Y, Cao B, Zhang Y, Zhou J, Yang B, Wang L (2011) Complete genome sequence of *Staphylococcus aureus* T0131, an ST239-MRSA-SCCmec type III clone isolated in China. J Bacteriol 193:3411–3412

Lina G, Piemont Y, Godail-Gamot F, Bes M, Peter MO, Gauduchon V, Vandenesch F, Etienne J (1999) Involvement of panton-valentine leukocidin-producing *Staphylococcus aureus* in primary skin infections and pneumonia. Clin Infect Dis 29:1128–1132

Lina G, Bohach GA, Nair SP, Hiramatsu K, Jouvin-Marche E, Mariuzza R (2004) Standard nomenclature for the superantigens expressed by *Staphylococcus*. J Infect Dis 189:2334–2336

Lindqvist M, Isaksson B, Grub C, Jonassen TO, Hallgren A (2012) Detection and characterisation of SCCmec remnants in multiresistant methicillin-susceptible *Staphylococcus aureus* causing a clonal outbreak in a Swedish county. Eur J Clin Microbiol 31:141–147

Lindsay JA, Holden MT (2006) Understanding the rise of the superbug: investigation of the evolution and genomic variation of *Staphylococcus aureus*. Funct Integr Genomics 6:186–201

Lindsay JA, Moore CE, Day NP, Peacock SJ, Witney AA, Stabler RA, Husain SE, Butcher PD, Hinds J (2006) Microarrays reveal that each of the ten dominant lineages of *Staphylococcus aureus* has a unique combination of surface-associated and regulatory genes. J Bacteriol 188:669–676

Loffler B, Hussain M, Grundmeier M, Bruck M, Holzinger D, Varga G, Roth J, Kahl BC, Proctor RA, Peters G (2010) *Staphylococcus aureus* panton-valentine leukocidin is a very potent cytotoxic factor for human neutrophils. PLoS Path 6:e1000715

Longshaw CM, Farrell AM, Wright JD, Holland KT (2000) Identification of a second lipase gene, gehD, in *Staphylococcus epidermidis*: comparison of sequence with those of other staphylococcal lipases. Microbiology 146(Pt 6):1419–1427

Lowder BV, Guinane CM, Ben Zakour NL, Weinert LA, Conway-Morris A, Cartwright RA, Simpson AJ, Rambaut A, Nubel U, Fitzgerald JR (2009) Recent human-to-poultry host jump, adaptation, and pandemic spread of *Staphylococcus aureus*. Proc Natl Acad Sci USA 106:19545–19550

Lowy FD (1998) *Staphylococcus aureus* infections. New Engl J Med 339:520–532

Macintosh RL, Brittan JL, Bhattacharya R, Jenkinson HF, Derrick J, Upton M, Handley PS (2009) The terminal A domain of the fibrillar accumulation-associated protein (Aap) of *Staphylococcus epidermidis* mediates adhesion to human corneocytes. J Bacteriol 191:7007–7016

Mack D, Fischer W, Krokotsch A, Leopold K, Hartmann R, Egge H, Laufs R (1996) The intercellular adhesin involved in biofilm accumulation of *Staphylococcus epidermidis* is a linear beta-1,6-linked glucosaminoglycan: purification and structural analysis. J Bacteriol 178:175–183

Madhusoodanan J, Seo KS, Remortel B, Park JY, Hwang SY, Fox LK, Park YH, Deobald CF, Wang D, Liu S, Daugherty SC, Gill AL, Bohach GA, Gill SR (2011) An enterotoxin-bearing pathogenicity island in *Staphylococcus epidermidis*. J Bacteriol 193:1854–1862

Maiques E, Ubeda C, Tormo MA, Ferrer MD, Lasa I, Novick RP, Penades JR (2007) Role of staphylococcal phage and SaPI integrase in intra- and interspecies SaPI transfer. J Bacteriol 189:5608–5616

Malachowa N, DeLeo FR (2010) Mobile genetic elements of *Staphylococcus aureus*. Cell Mol Life Sci 67:3057–3071

Marchesini B, Bruttin A, Romailler N, Moreton RS, Stucchi C, Sozzi T (1992) Microbiological events during commercial meat fermentations. J Appl Bacteriol 73:203–209
Marraffini LA, Sontheimer EJ (2008) CRISPR interference limits horizontal gene transfer in staphylococci by targeting DNA. Science 322:1843–1845
Marshall JH, Wilmoth GJ (1981) Pigments of *Staphylococcus aureus*, a series of triterpenoid carotenoids. J Bacteriol 147:900–913
Marti M, Trotonda MP, Tormo-Mas MA, Vergara-Irigaray M, Cheung AL, Lasa I, Penades JR (2010) Extracellular proteases inhibit protein-dependent biofilm formation in *Staphylococcus aureus*. Microb Infect 12:55–64
Mazmanian SK, Liu G, Jensen ER, Lenoy E, Schneewind O (2000) *Staphylococcus aureus* sortase mutants defective in the display of surface proteins and in the pathogenesis of animal infections. Proc Natl Acad Sci USA 97:5510–5515
Mazmanian SK, Skaar EP, Gaspar AH, Humayun M, Gornicki P, Jelenska J, Joachmiak A, Missiakas DM, Schneewind O (2003) Passage of heme-iron across the envelope of *Staphylococcus aureus*. Science 299:906–909
McCrea KW, Hartford O, Davis S, Eidhin DN, Lina G, Speziale P, Foster TJ, Hook M (2000) The serine-aspartate repeat (Sdr) protein family in *Staphylococcus epidermidis*. Microbiology 146(Pt 7):1535–1546
McDevitt D, Francois P, Vaudaux P, Foster TJ (1994) Molecular characterization of the clumping factor (fibrinogen receptor) of *Staphylococcus aureus*. Mol Microbiol 11:237–248
Meehl M, Herbert S, Götz F, Cheung A (2007) Interaction of the GraRS two-component system with the VraFG ABC transporter to support vancomycin-intermediate resistance in *Staphylococcus aureus*. Antimicrob Agents Chemother 51:2679–2689
Mehlin C, Headley CM, Klebanoff SJ (1999) An inflammatory polypeptide complex from *Staphylococcus epidermidis*: isolation and characterization. J Exp Med 189:907–918
Merino N, Toledo-Arana A, Vergara-Irigaray M, Valle J, Solano C, Calvo E, Lopez JA, Foster TJ, Penades JR, Lasa I (2009) Protein A-mediated multicellular behavior in *Staphylococcus aureus*. J Bacteriol 191:832–843
Molkanen T, Tyynela J, Helin J, Kalkkinen N, Kuusela P (2002) Enhanced activation of bound plasminogen on *Staphylococcus aureus* by staphylokinase. FEBS Lett 517:72–78
Mwangi MM, Wu SW, Zhou Y, Sieradzki K, de Lencastre H, Richardson P, Bruce D, Rubin E, Myers E, Siggia ED, Tomasz A (2007) Tracking the in vivo evolution of multidrug resistance in *Staphylococcus aureus* by whole-genome sequencing. Proc Natl Acad Sci USA 104:9451–9456
Neoh HM, Cui L, Yuzawa H, Takeuchi F, Matsuo M, Hiramatsu K (2008) Mutated response regulator graR is responsible for phenotypic conversion of *Staphylococcus aureus* from heterogeneous vancomycin-intermediate resistance to vancomycin-intermediate resistance. Antimicrob Agents Chemother 52:45–53
Neubauer H, Götz F (1996) Physiology and interaction of nitrate and nitrite reduction in *Staphylococcus carnosus*. J Bacteriol 178:2005–2009
Neubauer H, Pantel I, Götz F (1999) Molecular characterization of the nitrite-reducing system of *Staphylococcus carnosus*. J Bacteriol 181:1481–1488
Niinivaara FP, Pohja MS (1956) Über die Reifung der Rohwurst. I. Mitt: Die Veränderung der Bakterienflora während der Reifung. Z Lebensm Unters Forsch 104:413–422
Nilsson M, Bjerketorp J, Wiebensjo A, Ljungh A, Frykberg L, Guss B (2004) A von Willebrand factor-binding protein from *Staphylococcus lugdunensis*. FEMS Microbiol Lett 234:155–161
Novick RP (2003) Autoinduction and signal transduction in the regulation of staphylococcal virulence. Mol Microbiol 48:1429–1449
Novick RP, Geisinger E (2008) Quorum sensing in staphylococci. Annu Rev Genet 42:541–564
Novick RP, Subedi A (2007) The SaPIs: mobile pathogenicity islands of *Staphylococcus*. Chem Immunol Allergy 93:42–57
Novick RP, Christie GE, Penades JR (2010) The phage-related chromosomal islands of gram-positive bacteria. Nat Rev Microbiol 8:541–551

Nubel U, Dordel J, Kurt K, Strommenger B, Westh H, Shukla SK, Zemlickova H, Leblois R, Wirth T, Jombart T, Balloux F, Witte W (2010) A timescale for evolution, population expansion, and spatial spread of an emerging clone of methicillin-resistant *Staphylococcus aureus*. PLoS Path 6:e1000855

Otto M (2010a) Basis of virulence in community-associated methicillin-resistant *Staphylococcus aureus*. Annu Rev Microbiol 64:143–162

Otto M (2010b) *Staphylococcus* colonization of the skin and antimicrobial peptides. Expert Rev Dermatol 5:183–195

Otto M (2012) Molecular basis of *Staphylococcus epidermidis* infections. Semin Immunopathol 34:201–214

Otto M, O'Mahoney DS, Guina T, Klebanoff SJ (2004) Activity of *Staphylococcus epidermidis* phenol-soluble modulin peptides expressed in *Staphylococcus carnosus*. J Infect Dis 190:748–755

Palma M, Shannon O, Quezada HC, Berg A, Flock JI (2001) Extracellular fibrinogen-binding protein, Efb, from *Staphylococcus aureus* blocks platelet aggregation due to its binding to the alpha-chain. J Biol Chem 276:31691–31697

Park PW, Rosenbloom J, Abrams WR, Mecham RP (1996) Molecular cloning and expression of the gene for elastin-binding protein (*ebpS*) in *Staphylococcus aureus*. J Biol Chem 271:15803–15809

Pelz A, Wieland KP, Putzbach K, Hentschel P, Albert K, Götz F (2005) Structure and biosynthesis of staphyloxanthin from *Staphylococcus aureus*. J Biol Chem 280:32493–32498

Place RB, Hiestand D, Gallmann HR, Teuber M (2003) *Staphylococcus equorum* subsp. *linens*, subsp. nov., a starter culture component for surface ripened semi-hard cheeses. Syst Appl Microbiol 26:30–37

Postma B, Poppelier MJ, van Galen JC, Prossnitz ER, van Strijp JA, de Haas CJ, van Kessel KP (2004) Chemotaxis inhibitory protein of *Staphylococcus aureus* binds specifically to the C5a and formylated peptide receptor. J Immunol 172:6994–7001

Proctor RA, von Eiff C, Kahl BC, Becker K, McNamara P, Herrmann M, Peters G (2006) Small colony variants: a pathogenic form of bacteria that facilitates persistent and recurrent infections. Nat Rev Microbiol 4:295–305

Qin Z, Ou Y, Yang L, Zhu Y, Tolker-Nielsen T, Molin S, Qu D (2007) Role of autolysin-mediated DNA release in biofilm formation of *Staphylococcus epidermidis*. Microbiology 153:2083–2092

Rahimpour R, Mitchell G, Khandaker MH, Kong C, Singh B, Xu L, Ochi A, Feldman RD, Pickering JG, Gill BM, Kelvin DJ (1999) Bacterial superantigens induce down-modulation of CC chemokine responsiveness in human monocytes via an alternative chemokine ligand-independent mechanism. J Immunol 162:2299–2307

Rainard P (2007) *Staphylococcus aureus* leucotoxin LukM/F' is secreted and stimulates neutralising antibody response in the course of intramammary infection. Vet Res 38:685–696

Rautenberg M, Joo HS, Otto M, Peschel A (2011) Neutrophil responses to staphylococcal pathogens and commensals via the formyl peptide receptor 2 relates to phenol-soluble modulin release and virulence. FASEB J 25:1254–1263

Rehm SJ, Tice A (2010) *Staphylococcus aureus*: methicillin-susceptible S. *aureus* to methicillin-resistant *S. aureus* and vancomycin-resistant S. *aureus*. Clin Infect Dis 51 Suppl 2:S176–182

Resch A, Rosenstein R, Nerz C, Götz F (2005) Differential gene expression profiling of *Staphylococcus aureus* cultivated under biofilm and planktonic conditions. Appl Environ Microbiol 71:2663–2676

Resch A, Leicht S, Saric M, Pasztor L, Jakob A, Götz F, Nordheim A (2006) Comparative proteome analysis of *Staphylococcus aureus* biofilm and planktonic cells and correlation with transcriptome profiling. Proteomics 6:1867–1877

Roche FM, Massey R, Peacock SJ, Day NP, Visai L, Speziale P, Lam A, Pallen M, Foster TJ (2003a) Characterization of novel LPXTG-containing proteins of *Staphylococcus aureus* identified from genome sequences. Microbiology 149:643–654

Roche FM, Meehan M, Foster TJ (2003b) The *Staphylococcus aureus* surface protein SasG and its homologues promote bacterial adherence to human desquamated nasal epithelial cells. Microbiology 149:2759–2767

Rohde H, Burdelski C, Bartscht K, Hussain M, Buck F, Horstkotte MA, Knobloch JK, Heilmann C, Herrmann M, Mack D (2005) Induction of *Staphylococcus epidermidis* biofilm formation via proteolytic processing of the accumulation-associated protein by staphylococcal and host proteases. Mol Microbiol 55:1883–1895

Rohde H, Burandt EC, Siemssen N, Frommelt L, Burdelski C, Wurster S, Scherpe S, Davies AP, Harris LG, Horstkotte MA, Knobloch JK, Ragunath C, Kaplan JB, Mack D (2007) Polysaccharide intercellular adhesin or protein factors in biofilm accumulation of *Staphylococcus epidermidis* and *Staphylococcus aureus* isolated from prosthetic hip and knee joint infections. Biomaterials 28:1711–1720

Rooijakkers SH, Ruyken M, Roos A, Daha MR, Presanis JS, Sim RB, van Wamel WJ, van Kessel KP, van Strijp JA (2005a) Immune evasion by a staphylococcal complement inhibitor that acts on C3 convertases. Nat Immunol 6:920–927

Rooijakkers SH, van Wamel WJ, Ruyken M, van Kessel KP, van Strijp JA (2005b) Anti-opsonic properties of staphylokinase. Microb Infect 7:476–484

Rosenstein R, Götz F (2000) Staphylococcal lipases: biochemical and molecular characterization. Biochimie 82:1005–1014

Rosenstein R, Götz F (2010) Genomic differences between the food-grade *Staphylococcus carnosus* and pathogenic staphylococcal species. Int J Med Microbiol 300:104–108

Rosenstein R, Nerz C, Biswas L, Resch A, Raddatz G, Schuster SC, Götz F (2009) Genome analysis of the meat starter culture bacterium *Staphylococcus carnosus* TM300. Appl Environ Microbiol 75:811–822

Sakinc T, Kleine B, Gatermann SG (2006) SdrI, a serine-aspartate repeat protein identified in *Staphylococcus saprophyticus* strain 7108, is a collagen-binding protein. Infect Immun 74:4615–4623

Sakinc T, Kleine B, Gatermann SG (2007) Biochemical characterization of the surface-associated lipase of *Staphylococcus saprophyticus*. FEMS Microbiol Lett 274:335–341

Sakinc T, Kleine B, Michalski N, Kaase M, Gatermann SG (2009a) SdrI of *Staphylococcus saprophyticus* is a multifunctional protein: localization of the fibronectin-binding site. FEMS Microbiol Lett 301:28–34

Sakinc T, Michalski N, Kleine B, Gatermann SG (2009b) The uropathogenic species *Staphylococcus saprophyticus* tolerates a high concentration of D-serine. FEMS Microbiol Lett 299:60–64

Sass P, Berscheid A, Jansen A, Oedenkoven M, Szekat C, Strittmatter A, Gottschalk G, Bierbaum G (2012) Genome sequence of *Staphylococcus aureus* VC40, a vancomycin- and daptomycin-resistant strain, to study the genetics of development of resistance to currently applied last-resort antibiotics. J Bacteriol 194:2107–2108

Savolainen K, Paulin L, Westerlund-Wikstrom B, Foster TJ, Korhonen TK, Kuusela P (2001) Expression of *pls*, a gene closely associated with the *mecA* gene of methicillin-resistant *Staphylococcus aureus*, prevents bacterial adhesion in vitro. Infect Immun 69:3013–3020

Schijffelen MJ, Boel CH, van Strijp JA, Fluit AC (2010) Whole genome analysis of a livestock-associated methicillin-resistant *Staphylococcus aureus* ST398 isolate from a case of human endocarditis. BMC Genomics 11:376

Schlag M, Biswas R, Krismer B, Köhler T, Zoll S, Yu W, Schwarz H, Peschel A, Götz F (2010) Role of staphylococcal wall teichoic acid in targeting the major autolysin Atl. Mol Microbiol 75:864–873

Schleifer KH, Fischer U (1982) Description of a new species of the genus *Staphylococcus*: *Staphylococcus carnosus*. Int J Syst Bacteriol 32:153–156

Schlievert PM, McCormick JK, Bohach GA, Ohlendorf DH (2009) Exotoxins. In: Crossley KB et al (eds) Staphylococci in human disease, 2nd edn. Blackwell-Wiley, Chicester

Schneewind O, Model P, Fischetti VA (1992) Sorting of protein A to the staphylococcal cell wall. Cell 70:267–281

Schroeder K, Jularic M, Horsburgh SM, Hirschhausen N, Neumann C, Bertling A, Schulte A, Foster S, Kehrel BE, Peters G, Heilmann C (2009) Molecular characterization of a novel *Staphylococcus aureus* surface protein (SasC) involved in cell aggregation and biofilm accumulation. PloS One 4:e7567

Schumacher-Perdreau F, Heilmann C, Peters G, Götz F, Pulverer G (1994) Comparative analysis of a biofilm-forming *Staphylococcus epidermidis* strain and its adhesion-positive, accumulation-negative mutant M7. FEMS Microbiol Lett 117:71–78

Seitter M, Nerz C, Rosenstein R, Götz F, Hertel C (2011) DNA microarray based detection of genes involved in safety and technologically relevant properties of food associated coagulase-negative staphylococci. Int J Food Microbiol 145:449–458

Sellman BR, Howell AP, Kelly-Boyd C, Baker SM (2005) Identification of immunogenic and serum binding proteins of *Staphylococcus epidermidis*. Infect Immun 73:6591–6600

Sharp JA, Echague CG, Hair PS, Ward MD, Nyalwidhe JO, Geoghegan JA, Foster TJ, Cunnion KM (2012) *Staphylococcus aureus* surface protein SdrE binds complement regulator factor H as an immune evasion tactic. PloS One 7:e38407

Shore AC, Rossney AS, Brennan OM, Kinnevey PM, Humphreys H, Sullivan DJ, Goering RV, Ehricht R, Monecke S, Coleman DC (2011) Characterization of a novel arginine catabolic mobile element (ACME) and staphylococcal chromosomal cassette mec composite island with significant homology to Staphylococcus epidermidis ACME type II in methicillin-resistant Staphylococcus aureus genotype ST22-MRSA-IV. Antimicrob Agents Chemother 55:1896–1905

Sinha B, Francois PP, Nusse O, Foti M, Hartford OM, Vaudaux P, Foster TJ, Lew DP, Herrmann M, Krause KH (1999) Fibronectin-binding protein acts as *Staphylococcus aureus* invasin via fibronectin bridging to integrin alpha5beta1. Cell Microbiol 1:101–117

Skaar EP, Humayun M, Bae T, DeBord KL, Schneewind O (2004) Iron-source preference of *Staphylococcus aureus* infections. Science 305:1626–1628

Sloot N, Thomas M, Marre R, Gatermann S (1992) Purification and characterisation of elastase from *Staphylococcus epidermidis*. J Med Microbiol 37:201–205

Smeltzer MS, Lee CY, Harik N, Hart ME (2009) Molecular basis of pathogenicity. In: Crossley KB et al (eds) Staphylococci in human disease, 2nd edn. Blackwell-Wiley, Chichester

Stegger M, Price LB, Larsen AR, Gillece JD, Waters AE, Skov R, Andersen PS (2012) Genome sequence of Staphylococcus aureus strain 11819–97, an ST80-IV European community-acquired methicillin-resistant isolate. J Bacteriol 194:1625–1626

Szabados F, Kleine B, Anders A, Kaase M, Sakinc T, Schmitz I, Gatermann S (2008) *Staphylococcus saprophyticus* ATCC 15305 is internalized into human urinary bladder carcinoma cell line 5637. FEMS Microbiol Lett 285:163–169

Szmigielski S, Prevost G, Monteil H, Colin DA, Jeljaszewicz J (1999) Leukocidal toxins of staphylococci. Zentralbl Bakteriol Int J Med Mikrobiol 289:185–201

Takeuchi F, Watanabe S, Baba T, Yuzawa H, Ito T, Morimoto Y, Kuroda M, Cui L, Takahashi M, Ankai A, Baba S, Fukui S, Lee JC, Hiramatsu K (2005) Whole-genome sequencing of *Staphylococcus haemolyticus* uncovers the extreme plasticity of its genome and the evolution of human-colonizing staphylococcal species. J Bacteriol 187:7292–7308

Teufel P, Götz F (1993) Characterization of an extracellular metalloprotease with elastase activity from *Staphylococcus epidermidis*. J Bacteriol 175:4218–4224

Thomas D, Chou S, Dauwalder O, Lina G (2007) Diversity in *Staphylococcus aureus* enterotoxins. Chem Immunol Allergy 93:24–41

Tormo MA, Knecht E, Götz F, Lasa I, Penades JR (2005) Bap-dependent biofilm formation by pathogenic species of *Staphylococcus*: evidence of horizontal gene transfer? Microbiology 151:2465–2475

Tormo MA, Ubeda C, Marti M, Maiques E, Cucarella C, Valle J, Foster TJ, Lasa I, Penades JR (2007) Phase-variable expression of the biofilm-associated protein (Bap) in *Staphylococcus aureus*. Microbiology 153:1702–1710

Tse H, Tsoi HW, Leung SP, Lau SK, Woo PC, Yuen KY (2010) Complete genome sequence of *Staphylococcus lugdunensis* strain HKU09-01. J Bacteriol 192:1471–1472

Tse H, Tsoi HW, Leung SP, Urquhart IJ, Lau SK, Woo PC, Yuen KY (2011) Complete genome sequence of the veterinary pathogen *Staphylococcus pseudintermedius* strain HKU10-03, isolated in a case of canine pyoderma. J Bacteriol 193:1783–1784

Tung H, Guss B, Hellman U, Persson L, Rubin K, Ryden C (2000) A bone sialoprotein-binding protein from *Staphylococcus aureus*: a member of the staphylococcal Sdr family. Biochem J 345(Pt 3):611–619

Turlej A, Hryniewicz W, Empel J (2011) Staphylococcal cassette chromosome *mec* (Sccmec) classification and typing methods: an overview. Pol J Microbiol 60:95–103

Ubeda C, Tormo MA, Cucarella C, Trotonda P, Foster TJ, Lasa I, Penades JR (2003) Sip, an integrase protein with excision, circularization and integration activities, defines a new family of mobile *Staphylococcus aureus* pathogenicity islands. Mol Microbiol 49:193–210

Uhlemann AC, Porcella SF, Trivedi S, Sullivan SB, Hafer C, Kennedy AD, Barbian KD, McCarthy AJ, Street C, Hirschberg DL, Lipkin WI, Lindsay JA, DeLeo FR, Lowy FD (2012) Identification of a highly transmissible animal-independent *Staphylococcus aureus* ST398 clone with distinct genomic and cell adhesion properties. mBio 3:e00027–12

Valentin-Weigand P, Timmis KN, Chhatwal GS (1993) Role of fibronectin in staphylococcal colonisation of fibrin thrombi and plastic surfaces. J Med Microbiol 38:90–95

van de Guchte M, Penaud S, Grimaldi C, Barbe V, Bryson K, Nicolas P, Robert C, Oztas S, Mangenot S, Couloux A, Loux V, Dervyn R, Bossy R, Bolotin A, Batto JM, Walunas T, Gibrat JF, Bessieres P, Weissenbach J, Ehrlich SD, Maguin E (2006) The complete genome sequence of *Lactobacillus bulgaricus* reveals extensive and ongoing reductive evolution. Proc Natl Acad Sci USA 103:9274–9279

van Wamel WJ, Rooijakkers SH, Ruyken M, van Kessel KP, van Strijp JA (2006) The innate immune modulators staphylococcal complement inhibitor and chemotaxis inhibitory protein of *Staphylococcus aureus* are located on beta-hemolysin-converting bacteriophages. J Bacteriol 188:1310–1315

Vandenesch F, Etienne J, Reverdy ME, Eykyn SJ (1993) Endocarditis due to *Staphylococcus lugdunensis*: report of 11 cases and review. Clin Infect Dis 17:871–876

Vazquez V, Liang X, Horndahl JK, Ganesh VK, Smeds E, Foster TJ, Hook M (2011) Fibrinogen is a ligand for the *Staphylococcus aureus* microbial surface components recognizing adhesive matrix molecules (MSCRAMM) bone sialoprotein-binding protein (Bbp). J Biol Chem 286:29797–29805

Voggu L, Schlag S, Biswas R, Rosenstein R, Rausch C, Götz F (2006) Microevolution of cytochrome *bd* oxidase in staphylococci and its implication in resistance to respiratory toxins released by *Pseudomonas*. J Bacteriol 188:8079–8086

von Eiff C, Heilmann C, Proctor RA, Woltz C, Peters G, Götz F (1997) A site-directed *Staphylococcus aureus hemB* mutant is a small-colony variant which persists intracellularly. J Bacteriol 179:4706–4712

von Eiff C, Peters G, Heilmann C (2002) Pathogenesis of infections due to coagulase-negative staphylococci. Lancet Infect Dis 2:677–685

Wagner E, Götz F, Brückner R (1993) Cloning and characterization of the *scrA* gene encoding the sucrose-specific enzyme II of the phosphotransferase system from *Staphylococcus xylosus*. Mol Gen Genet 241:33–41

Wagner E, Doskar J, Götz F (1998) Physical and genetic map of the genome of *Staphylococcus carnosus* TM300. Microbiology 144(Pt 2):509–517

Wallmark G, Arremark I, Telander B (1978) *Staphylococcus saprophyticus*: a frequent cause of acute urinary tract infection among female outpatients. J Infect Dis 138:791–797

Walsh EJ, O'Brien LM, Liang X, Hook M, Foster TJ (2004) Clumping factor B, a fibrinogen-binding MSCRAMM (microbial surface components recognizing adhesive matrix molecules) adhesin of *Staphylococcus aureus*, also binds to the tail region of type I cytokeratin 10. J Biol Chem 279:50691–50699

Wang R, Braughton KR, Kretschmer D, Bach TH, Queck SY, Li M, Kennedy AD, Dorward DW, Klebanoff SJ, Peschel A, DeLeo FR, Otto M (2007) Identification of novel cytolytic peptides as key virulence determinants for community-associated MRSA. Nat Med 13:1510–1514

Wann ER, Gurusiddappa S, Hook M (2000) The fibronectin-binding MSCRAMM FnbpA of *Staphylococcus aureus* is a bifunctional protein that also binds to fibrinogen. J Biol Chem 275:13863–13871

Weinstein MP, Mirrett S, Van Pelt L, McKinnon M, Zimmer BL, Kloos W, Reller LB (1998) Clinical importance of identifying coagulase-negative staphylococci isolated from blood cultures: evaluation of MicroScan Rapid and Dried Overnight Gram-Positive panels versus a conventional reference method. J Clin Microbiol 36:2089–2092

Wertheim HF, Vos MC, Ott A, van Belkum A, Voss A, Kluytmans JA, van Keulen PH, Vandenbroucke-Grauls CM, Meester MH, Verbrugh HA (2004) Risk and outcome of nosocomial *Staphylococcus aureus* bacteraemia in nasal carriers versus non-carriers. Lancet 364:703–705

Wertheim HF, Melles DC, Vos MC, van Leeuwen W, van Belkum A, Verbrugh HA, Nouwen JL (2005) The role of nasal carriage in *Staphylococcus aureus* infections. Lancet Infect Dis 5:751–762

Wiedenheft B, Sternberg SH, Doudna JA (2012) RNA-guided genetic silencing systems in bacteria and archaea. Nature 482:331–338

Wieland B, Feil C, Gloria-Maercker E, Thumm G, Lechner M, Bravo JM, Poralla K, Götz F (1994) Genetic and biochemical analyses of the biosynthesis of the yellow carotenoid 4,4'-diaponeurosporene of *Staphylococcus aureus*. J Bacteriol 176:7719–7726

Williams RJ, Henderson B, Sharp LJ, Nair SP (2002) Identification of a fibronectin-binding protein from *Staphylococcus epidermidis*. Infect Immun 70:6805–6810

Wright JA, Nair SP (2012) The lipoprotein components of the Isd and Hts transport systems are dispensable for acquisition of heme by *Staphylococcus aureus*. FEMS Microbiol Lett 329:177–185

Zell C, Resch M, Rosenstein R, Albrecht T, Hertel C, Götz F (2008) Characterization of toxin production of coagulase-negative staphylococci isolated from food and starter cultures. Int J Food Microbiol 127:246–251

Zhang YQ, Ren SX, Li HL, Wang YX, Fu G, Yang J, Qin ZQ, Miao YG, Wang WY, Chen RS, Shen Y, Chen Z, Yuan ZH, Zhao GP, Qu D, Danchin A, Wen YM (2003) Genome-based analysis of virulence genes in a non-biofilm-forming *Staphylococcus epidermidis* strain (ATCC 12228). Mol Microbiol 49:1577–1593

Ziebuhr W, Krimmer V, Rachid S, Lossner I, Götz F, Hacker J (1999) A novel mechanism of phase variation of virulence in *Staphylococcus epidermidis*: evidence for control of the polysaccharide intercellular adhesin synthesis by alternating insertion and excision of the insertion sequence element IS256. Mol Microbiol 32:345–356

Zoll S, Patzold B, Schlag M, Götz F, Kalbacher H, Stehle T (2010) Structural basis of cell wall cleavage by a staphylococcal autolysin. PLoS Pathog 6:e1000807

Zoll S, Schlag M, Shkumatov AV, Rautenberg M, Svergun DI, Götz F, Stehle T (2012) Ligand-binding properties and conformational dynamics of autolysin repeat domains in staphylococcal cell wall recognition. J Bacteriol 194:3789–3802

Microevolution of *Pseudomonas aeruginosa* to a Chronic Pathogen of the Cystic Fibrosis Lung

Michael Hogardt and Jürgen Heesemann

Abstract *Pseudomonas aeruginosa* is the leading pathogen of chronic cystic fibrosis (CF) lung infection. Life-long persistance of *P. aeruginosa* in the CF lung requires a sophisticated habitat-specific adaptation of this pathogen to the heterogeneous and fluctuating lung environment. Due to the high selective pressure of inflamed CF lungs, *P. aeruginosa* increasingly experiences complex physiological and morphological changes. Pulmonary adaptation of *P. aeruginosa* is mediated by genetic variations that are fixed by the repeating interplay of mutation and selection. In this context, the emergence of hypermutable phenotypes (mutator strains) obviously improves the microevolution of *P. aeruginosa* to the diverse microenvironments of the CF lung. Mutator phenotypes are amplified during CF lung disease and accelerate the intraclonal diversification of *P. aeruginosa*. The resulting generation of numerous subclonal variants is advantegous to prepare *P. aeruginosa* population for unpredictable stresses (insurance hypothesis) and thus supports long-term survival of this pathogen. Oxygen restriction within CF lung environment further promotes persistence of *P. aeruginosa* due to increased antibiotic tolerance, alginate production and biofilm formation. Finally, *P. aeruginosa* shifts from an acute virulent pathogen of early infection to a host-adapted chronic virulent pathogen of end-stage infection of the CF lung. Common changes that are observed among chronic *P. aeruginosa* CF isolates include alterations in surface antigens, loss

M. Hogardt (✉)
Department of Infectiology, Bavarian Health and Food Safety Authority,
Veterinärstraße 2, 85764 Oberschleissheim, Germany
e-mail: Michael.Hogardt@lgl.bayern.de

J. Heesemann
Max von Pettenkofer-Institut für Hygiene and Medical Microbiology,
Ludwig-Maximilians-Universitity Munich, Pettenkoferstraße 9a,
80336 Munich, Germany
e-mail: heesemann@mvp.uni-muenchen.de

Current Topics in Microbiology and Immunology (2013) 358: 91–118 91
DOI: 10.1007/82_2011_199

Published Online: 8 February 2011

of virulence-associated traits, increasing antibiotic resistances, the overproduction of the exopolysaccharide alginate and the modulation of intermediary and micro-aerobic metabolic pathways (Hogardt and Heesemann, Int J Med Microbiol 300(8):557–562, 2010). Loss-of-function mutations in *mucA* and *lasR* genes determine the transition to mucoidity and loss of quorum sensing, which are hallmarks of the chronic virulence potential of *P. aeruginosa*. Metabolic factors that are positively selected in response to the specific environment of CF lung include the outer membrane protein OprF, the microaerophilic oxidase Cbb3-2, the blue copper protein azurin, the cytochrome c peroxidase c551 and the enzymes of the arginine deiminase pathway ArcA-ArcD. These metabolic adaptations probably support the growth of *P. aeruginosa* within oxygen-depleted CF mucus. The deeper understanding of the physiological mechanisms of niche specialization of *P. aeruginosa* during CF lung infection will help to identify new targets for future anti-pseudomonal treatment strategies to prevent the selection of mutator isolates and the establishment of chronic CF lung infection.

Contents

1 Introduction

Cystic fibrosis (CF) is an autosomal-recessive genetic disorder that affects about 70,000 people worldwide. The disease is based on mutations in the cystic fibrosis transmembrane conductance regulator (CFTR) gene, encoding a chloride channel that controls transepithelial electrolyte transport. Impaired CFTR function results in the accumulation of thick dehydrated mucus in all exocrine (secretory) glands. This abnormality causes a plethora of clinical symptoms especially in the respiratory, gastrointestinal and reproductive systems. In the last decades, a continuous amendment in the treatment options for CF patients has greatly improved their quality of life and life expectancies. Nonetheless, the clinical picture of CF is dictated by chronic lung disease and recurrent pulmonary infections that lead to inexorable decline of lung function, respiratory failure and still premature death of patients (Dodge et al. 2007). Although there are several lines of evidence that CF lung infection is of polymicrobial nature, *Pseudomonas aeruginosa* is considered to be the major determinant of bacterial lung infection, morbidity and mortality of diseased individuals (Sibley and Surette 2011). *P. aeruginosa* is a Gram-negative microorganism commonly found in environmental reservoirs, such as soil, plants and water. Occasionally, *P. aeruginosa* is transmitted from its natural habitat to humans resulting in opportunistic infections such as pneumonia, wound infections and urinary tract infections, particularly in the immunocompromised host. Albeit the immune system of CF patients is generally intact, this patient group is especially vulnerable to infections with *P. aeruginosa* likely as a consequence of defective pulmonary defenses (Burns et al. 2001). Despite intensive research, the exact molecular mechanisms causing the high susceptibility of CF patients to *P. aeruginosa* infections following CFTR dysfunction are still not fully understood (Grassmé et al. 2010). However, once entering the CF lung, *P. aeruginosa* has to adapt its pathogenic strategy to this special environment, initially by sensing the new environment and by responding with an appropriate regulation of gene expression. This early phase of colonization by *P. aeruginosa* as well as the subsequent acute infection correlates with the expression of an armament of cell-associated as well as secreted virulence factors. At this time, colonization of CF lung is often temporary and eradication of *P. aeruginosa* still possible. In case of ineffective eradication efforts, nearly all CF patients are chronically colonized and *P. aeruginosa* becomes a persisting pathogen up to the end of the patient's life. According to the current concept, chronic lung infection is latest established when *P. aeruginosa* forms mucoid phenotypes and/or anaerobic biofilms. However, during the survival in the hostile and fluctuating microenvironments of CF airways, *P. aeruginosa* undergoes dramatic genetic, physiological and morphological changes. A striking result regarding the genomic structure of *P. aeruginosa* CF isolates is that they are in general genotypically indistinguishable from their environmental counterparts, indicating that no habitat-specific pathotypes exist (Wiehlmann et al. 2007). Lastly, lung adaptation of *P. aeruginosa* is characterized by the transition from an acute environmental pathogen to a chronic habitat-adapted pathogen with reduced cell-cytotoxicity and the emergence of a population of heterogenous phenotypes and pathotypes typically for the chronic infection (Hogardt and Heesemann 2010). This adaptation process can be accelerated by the occurrence of mutator phenotypes of *P. aeruginosa* with high mutation rates that effectively hitchhike beneficial mutations (Oliver et al. 2000). Interestingly, similar evolutionary patterns are observed among independent clonal lineages of *P. aeruginosa* CF isolates. A common finding in CF is the selection of *P. aeruginosa* variants that are characterized by mucoidity, multi-resistance to antibiotics and the loss of various classical virulence-associated traits. In conclusion, during chronic lung disease *P. aeruginosa* obviously no longer benefits from acute virulence factors that directly damage host tissue. Obviously, the offensive strategy of *P. aeruginosa* switches to a defensive strategy effective to withstand antibacterial host response and to thrive within the suppurative and hypoxic (micro-aerobic to anaerobic) milieu of CF airway secretions. This dramatic change in the persistence strategy of *P. aeruginosa* emerges in response to prevailing environmental signals and progressive lung destruction. Obviously, the characterization of *P. aeruginosa* CF isolates of chronically infected patients expands our understanding of chronic bacterial virulence and the mechanisms of persistance in a given host. Interestingly, a significant percentage of genes differentially expressed in end-stage CF isolates comprise metabolic functions underscoring that metabolic adaptability of *P. aeruginosa* is a leading factor of chronic virulence. In order to improve the prognosis of CF patients, future emphasis should be layed on the identification and the therapeutic targeting of specific persistence factors that are active during late stages of CF lung disease.

2 Challenges Within the Stressful Environment of CF Lungs

CF Lung disease is characterized by alterations in the ionic composition and hydration state of the airway surface fluid (ASF), accumulation of viscous mucus, endobronchial mucus plugging, recurrent bacterial infections, recurrent exacerbations, excessive pulmonary inflammation and finally irreversible decline in lung function due to bronchiolar obliteration, bronchiectasis, atelectasis, lung bleeding, cell damage and fibrotic remodeling of lung tissue. Accordingly, management of chronic CF lung disease requires periodic hospitalization and a variety of therapeutic interventions, such as agents for sputum liquefaction, antibiotics, anti-inflammatory agents and bronchodilatators. Thus, the dynamics of progressive lung disease promote the formation of multiform microenvironments within the respiratory tract; *P. aeruginosa* has to cope with (Fig. 1a).

CF airways chronically infected by *P. aeruginosa* exhibit an abundant inflammatory response supported by host and bacterial factors. Activation of epithelial cells together with *P. aeruginosa* antigens such as surface components and secreted exoproduct initiate a sequence of immune cell recruitment (neutrophils, macrophages) and release of immune mediators (cytokines, chemoattractants) amplifying the inflammatory reaction (Elizur et al. 2008). Toll-like receptors (TLRs) recognize mainly lipopolysaccharide and/or flagella from *P. aeruginosa* and activate the epithelial cells to produce inflammatory mediators. Neutrophil levels recovered from CF lungs are reported to be about 1,500 times higher than those from uninfected individuals (Konstan et al. 1994). Neutrophil activation and degranulation leads to the release of neutrophil elastase, myeloperoxidase, reactive oxygen species (ROS) and reactive nitrogen species (RNS) which are considered to significantly contribute to lung tissue damage. The free proteolytic activity in airway secretions can degrade protease inhibitors, immunoglobulins, surfactant proteins, antibacterial peptides, complement components and structural proteins such as elastin and fibronectin (Doring and Gulbins 2009; Gaggar et al. 2011). These observations lead to the concept that irreversible damage of *P. aeruginosa*-infected CF lungs results mainly from an overwhelming inflammation preserved by repeated cycles of pro- and anti-inflammatory responses as well as from

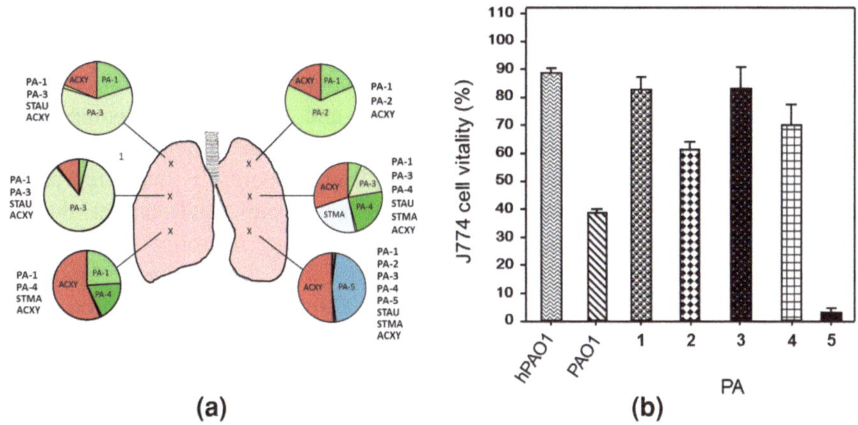

Fig. 1 **a** Diverse microbial communities recovered from six different areas of explanted CF-lungs of an individual CF patient by standard quantitative microbiological culture of lung biopsies (unpublished data). Different *P. aeruginosa* morphotypes (PA-1, mucoid; PA-2, non-mucoid; PA-3, mucoid, multi-drug-resistant; PA-4, non-mucoid, small-colony-variant; and PA-5, non-mucoid), *Staphylococcus aureus* (STAU), *Stenotrophomonas maltophilia* (STMA) and *Achromobacter xylosoxydans* (ACXY) were simultaneously recovered in variable compositions. The species composition and/or recovered *P. aeruginosa* morphotypes are given aside each area. The relative abundance (CFU) of each species and/or *P. aeruginosa* morphotype is indicated by the color code of each circle. Cultures for anaerobes, fungi and non-tuberculous mycobacteria were negative. **b** Cell cytotoxicity of different *P. aeruginosa* (PA) morphotypes, namely PA-1, PA-2, PA-3, PA-4 and PA-5), toward J774 murine macrophage cell line ATCC TIB-67. J774 cell cytotoxicity assay demonstrated that coexisting *P. aeruginosa* variants represent different cytotoxic pathotypes, while *P. aeruginosa* PA-5 exhibited striking cytotoxic effect. *P. aeruginosa* PAO1 and heat-inactivated *P. aeruginosa* PAO1 cells are shown as controls. Methyltiazole tetrazolium (MTT) cytotoxicity assay was performed as described (Hogardt et al. 2007)

imbalances between protease/antiprotease and oxidants/antioxidants activities (Doring 1999; Ratjen et al. 2002). Thus, in the CF lung *P. aeruginosa* has to protect itself against oxidative stresses that especially derive from the liberation of ROS by neutrophils or that are endogenously produced during incomplete oxidation reactions or the redox-active molecule pyocyanin. Pyocyanin, a cyclic phenazine of *P. aeruginosa*, is detected in high concentrations in CF sputum where it may deplete cellular antioxidants (Lau et al. 2004).

Besides the challenges imposed by host defence such as phagocytosis and oxidative stress, *P. aeruginosa* is regularly faced with high doses of alternate antibiotics (aminoglycosides, ß-lactams, fluorochinolones and polymyxins) and has to compete with other diverse bacterial species present in CF secretions (Gilligan et al. 1985; Tunney et al. 2008) (Fig. 1a). Nevertheless, *P. aeruginosa* is capable to resist these environmental stresses and may even thrive lifelong within CF lungs. Strikingly, purulent CF sputum represents an excellent growth substrate for *P. aeruginosa* (Harrison 2007; Barth and Pitt 1996; Palmer et al. 2005). Basically, airway secretions are an aqueous mixture of electrolytes, glycoproteins (e.g. mucins), proteins and lipids (Galabert et al. 1987; Rose and Voynow 2006).

Mucins are highly glycosylated macromolecules with ≥50% carbohydrate content. CF sputum contains airway mucus but also loads cellular debris from decaying neutrophils and microbes. This results in an endobronchial deposition of large amounts of DNA, F-actin, membrane lipids and other cellular components (Whitchurch et al. 2002). Thus, CF airway secretions are enriched in degraded proteins, amino acids, DNA and phospholipids. Furthermore, CF mucus represents an oxygen-depleted (micro-aerobic or even anaerobic) environment due to impaired ventilation and rapid oxygen consumption by microbes and invading neutrophils, above all the high burden of up to 1 x 10^{10} *P. aeruginosa* cells per g sputum (Worlitzsch et al. 2002; Hassett et al. 2009). The oxygen depletion in CF sputum and the steep oxygen gradients in CF mucus have been directly determined in situ. Moreover, the recovery of high amounts of obligate anaerobic bacteria from CF lung secretions is a further strong indication for an anaerobic environment (Rogers et al. 2004; Harris et al. 2007; Tunney et al. 2008; Worlitzsch et al. 2009).

In conclusion, the CF lung and its secretions are predestinated to present various microenvironments due to variations regarding inflammation, hypoxia, pH, oxidative stress and selected nutrient availability. Chronic exposure to the special and diverse niches of the CF airways probably serves as the biological matrix that triggers microevolution of *P. aeruginosa* in CF lung disease. For instance, the hypoxic microenvironments of inflammed areas of CF lung provide a strong selection pressure on *P. aeruginosa* to acquire anaerobic adaptations. Moreover, there are several lines of evidence that in CF lungs *P. aeruginosa* forms anaerobic biofilm-like microcolonies within pockets of thick airway mucus. Chronic exposure to the diverse niches of the CF airways probably serves as the biological matrix that fuels microevolution of *P. aeruginosa* in CF lung disease.

3 Hypermutability of *P. aeruginosa*: The Driving Force of Niche Specialization

Hypermutability has been predicted to be beneficial in the selection of adapted niche specialist and survival of bacterial pathogens in hostile environments such as chronic CF lung infection. Hypermutable bacterial isolates (also termed hypermutators or mutators) have increased spontaneous mutation rates as a result of defects in DNA repair and/or error avoidance systems (Jolivet-Gougeon et al. 2011). *H. influenzae*, *S. aureus* and *P. aeruginosa* isolates with increased mutation rates were frequently isolated from chronically infected CF lungs but not from infections of non-CF patients. Excessive inflammation within CF airways promotes oxidative stress e.g. by reactive oxygen species (ROS) and is considered to be associated with the emergence of hypermutable phenotypes (Ciofu et al. 2005). ROS includes superoxide (O_2^-), hydrogen peroxide (H_2O_2) and hydroxyl radical (HO•) that excerts mutagenic stresses on bacterial DNA. Among hypermutable CF isolates *P. aeruginosa* mutators are most frequently recovered, ranging from 15 to 65% among chronic-infecting isolates. *P. aeruginosa* possesses two major systems involved in the repair of mutagenic DNA lesions: the methyl-directed DNA mismatch repair (MMR) system and the guanine 8-oxo-2-deoxyguanosine (GO) system. Inactivation of both the MMR and the GO system can lead to the generation of hypermutable phenotypes due to the inability to repair genetic mismatches efficiently (Oliver and Mena 2010). The MMR system is specified by a set of proteins encoded by the genes *mutS*, *mutL* and *uvrD*. The GO system in turn depends on the action of proteins encoded by the genes *mutT*, *mutY* and *mutM*. Isolates with defects in MMR genes exhibit a strong mutator phenotype (100 to 1,000-fold increased mutation frequency), while defects in the GO system genes result in only weak mutator phenotypes (Ciofu et al. 2010). In CF, *P. aeruginosa* mutators emerge mainly as a result of defects in the MMR system, while *mutS*-mutators are most often recovered followed by *mutL*-mutators. Mutator genes of *E. coli* have been at best functionally characterized (Kunkel and Erie 2005). In *E. coli* MutS recognizes basepair mismatches as well as short insertions or deletions, interacts with MutL in order to bind to DNA and finally removes mismatches. DNA helicase II encoded by *uvrD* is required for separation of DNA strands. Thus, MMR deficiency increases not only the mutation frequency of affected isolates but also the rate of recombination events between homologous DNA sequences.

Interestingly, among chronic *P. aeruginosa* CF isolates also mutants defective in the GO system have been reported (Mandsberg et al. 2009). The GO system is especially involved in the repair of oxidatively damaged DNA. One common oxygen-related DNA lesion is the oxidation of the guanine 8-oxo-2-deoxyguanosine (8-oxodG or GO) by ROS. GO promotes the incorrect insertion of adenine during DNA replication, leading to GC → TA transversions (Maki and Sekiguchi 1992). The main pathway for the removal of GO from DNA involves MutM a formamidopyrimidine DNA glycosylase. Alternatively, mispaired adenine may be removed by the action of the glycosylase MutY. Oxidized GTP (8-oxodGTP) a potential source for GO may be cleared from the nucleotide pool by MutT. Thus, mutations in genes of the GO system increase the occurrence of GO and thus mutation frequencies in *P. aeruginosa*.

Strikingly, several in vitro experiments have underscored the evolutionary advantage of mutator strains, whereas in particular the investigation of *P. aeruginosa* mutators from CF lung established the importance of hypermutability during bacterial persistence in vivo (Mena et al. 2008). The co-existence of *P. aeruginosa* mutator/non-mutator subpopulations and the amplification of mutator variants in CF lungs over time indicated a selective advantage of mutators during chronic infections. Further, it has been shown that during the course of chronic CF lung infection *P. aeruginosa* mutators increasingly accumulated adaptive gene mutations. The most striking characteristic of *P. aeruginosa* mutators in comparison to nonmutators is their high antibiotic resistance rate that per se should result in negative clinical impact on the outcome of CF patients (Oliver et al. 2000; Hogardt et al. 2007). In conclusion, under selective conditions hypermutable isolates more effectively acquire mutations in antibiotic resistance determinants. Under the selection pressure of the CF-lung, *P. aeruginosa* mutators more rapidly acquire adaptations in virulence genes and metabolic functions as compared to isogenic nonmutators (Oliver and Mena 2010). Finally, patient-to-patient transmission of mutators has not yet been reported. This may be explained by the reduced survival of mutator strains in environmental habitats, indicating that mutators reach niche specialization on the expense of overall fitness. Nevertheless, further studies are needed for an in-depth evaluation of the clinical impact of *P. aeruginosa* mutators for the progression of CF lung disease.

4 Multicellular Growth of *P. aeruginosa* Promotes Chronic CF Lung Infection

The growth of *P. aeruginosa* in complex microbial communities also known as biofilms is considered to mediate several advantages as compared to free-floating planktonic cells, both in the environment or during biofilm-driven human infections, in particular chronic CF lung disease. First evidence that the environment of CF airways promote biofilm formation of *P. aeruginosa* has been derived from microscopic examinations of CF sputum that displayed a high burden of multicellular bacterial aggregates. Later, it has been shown that the release of *P. aeruginosa* quorum sensing (QS) molecules in CF sputum indicates biofilm growth and the induction of a QS-dependent virulence program. Finally, CF-lung infection has become the major setting to investigate multicellular growth and signaling of *P. aeruginosa*.

The *P. aeruginosa* QS circuit is a complex network of intercellular communication that coordinates various cellular activities in a cell-density-dependent manner. QS targets seem to be more effective when conducted by a larger population. The core of the *P. aeruginosa* QS network consists of the LasIR and RhlIR systems that in addition interacts with the Pseudomonas quinolone signal system, several global regulators, two-component systems and sigma factors. The action of the LasIR and RhlIR systems depends on the production, accumulation and sensing of small freely diffusible hormone-like molecules, the acyl-homoserine lactones (HSL). *P. aeruginosa* primarily release the 3-oxo-C12-HSL (produced by LasI synthase) and C4-HSL (produced by RhlI synthase) that bind to both their cognate regulators LasR and RhlR. While the synthesis of the HSL is constitutive, the transcriptional regulators are activated when reaching a critical AHL concentration or respective cell densities (quorum sensing) (Venturi 2006).

By contrast, the Pseudomonas quinolone signal system is functioning by a chemically distinct class of signal molecules, the 2-alkyl-4-quinolones. The *pqsABCDE* locus encodes the enzymes needed to generate quinolones including 2-heptyl-3-hydroxy-4-quinolone (PQS) and its precursor 2-heptyl-4-quinolone (HHQ). Released HHQ is taken up by adjacent bacterial cells and converted into PQS by PqsH, a putative mono-oxygenase. Further, there is growing evidence that another signal molecule, the second messenger cyclic di-guanylate (c-di-GMP), influences several cellular processes such as swarming, twitching, surface attachment and thus biofilm formation. The current concept is that accumulation of c-di-GMP promotes biofilm behavior of *P. aeruginosa*, while its breakdown favors motility, dispersal from biofilms and planctonic lifestyle. C-di-GMP is produced from GTP by di-guanylate cyclases (DGC), and degraded to pGpG by phosphodiesterases (PDE). These enzymes contain conserved GGDEF and EAL domains, while it is thought that the GGDEF domain excerts DGC activity and the EAL domain accounts for PDE activity (Lory et al. 2009).

The action of the QS cascade is highly regulated and is suspected to start with the 3-oxo-C12-HSL/LasR complex that drives the expression of LasI (autoinduction loop) as well as that of RhlR and RhlI. After this signal amplification step QS may alter gene expression of about 10% of the genes of the *P. aeruginosa* genome including genes encoding virulence, motility, chemotaxis and metabolism-associated factors (QS regulon). Briefly, the LasRI system hierarchically controls not only the RhlRI system but also the gene expression of virulence factors such as elastase (*lasB*), staphylolytic protease (*lasA*), Exotoxin A (*exoA*) and components of xcp secretion machinery (*xcpP* and *xcpR*). The RhlRI system in turn regulates via C4-HSL/RhlR the expression of RhlI (autoinduction loop), sigma factor RpoS, rhamnolipid (*rhlAB*), LasB, LasA, hydrogen cyanide (*hcnABC*), pyocyanin (*phzABCDEFG*), lipase and alkaline protease (*aprA*). PQS regulates its own production by driving the expression of *pqsABCDE* operon through the interaction with PqsR (autoinduction loop) and the expression of various target genes encoding, for example, elastase, rhamnolipid, PA-IL lectin (*lecA*) and pyocyanin. Not surprisingly, mutation of either system results in the attenuation of *P. aeruginosa* virulence. However, RhlR is able to overcome the absence the Las system in a lasR mutant by activating specific LasR-controlled functions, including the production of 3-oxo-C12-HSL and PQS (Dekimpe and Deziel 2009). Finally, QS molecules may interact with host cells and modulate host responses. 3-oxo-C12-HSL has been reported to induce apoptosis in neutrophils, affect the production of cytokines by immune cells and stimulate IL-8 production in bronchial epithelial cells (Smith et al. 2001). Finally, QS by HSL and quinolones is not restricted to *P. aeruginosa* enabling a cross-talk with other bacterial species recovered from CF secretions.

Although the precise role of QS for biofilm formation is still a matter of debate, it is established that QS signaling modifies the architecture and density of biofilms as well as the dynamics of biofilm maturation. In agreement, inhibitory drugs of QS signaling, for example, furanones significantly reduce the ability of *P. aeruginosa* to form biofilms and advise these substances as new anti-pseudomonal treatment options (Christensen et al. 2007). So far, several data support the hypothesis that biofilms are a major cause of ineffective host immune response and *P. aeruginosa*'s persistence in CF airways. In general, it is not clear how long after the onset of colonization it takes *P. aeruginosa* to switch to a biofilm mode of growth. Obviously, CF sputum provides an excellent milieu for microcolony formation, while especially the presence of extracellular DNA (eDNA) probably derived mainly from decayed cells induce the set-up of a bacterial aggregates (Barken et al. 2008; Whitchurch et al. 2002). Additional eDNA may be released from PQS-dependent cell lysis of a subpopulation of biofilm-grown bacteria during biofilm development (attachment, division, expression of matrix polymers, maturation and dispersal) (Allesen-Holm et al. 2006). Within mature biofilms *P. aeruginosa* cells are embedded in a surrounding polymeric and partly self-produced matrix of polysaccharides, alginate, DNA and protein. Obviously, the complexity of the biofilm matrix enables *P. aeruginosa* to withstand the continuous challenges imposed by recurrent antibiotic treatments, inflammatory molecules and phagocytes (Spasenovski et al. 2010). Biofilms are characterized by a striking tenacity against various environmental stresses. For instance, they are up to 1,000 times more resistant to antimicrobial agents than planktonic cells likely due to various mechanisms, such as restricted diffusion of the antibiotic agent, activation of efflux pumps, selection of persister phenotypes and a reduced growth rate (Drenkard 2003). Further, the deeper layers of *P. aeruginosa* biofilms demonstrate complete oxygen depletion due to oxygen consumption of bacteria in upper layers of the biofilm (Werner et al. 2004). Thus, *P. aeruginosa* biofilms are organized in different subpopulations while a subset of bacteria exist under anaerobic conditions and in a reduced metabolic state. Both factors contribute to the antibiotic tolerance of biofilms e.g. against aminoglycosides and ß-lactam antibiotics (Borriello et al. 2004a, b; Fux et al. 2005; Martínez and Rojo 2011).

In summary, the bacteria in *P. aeruginosa* biofilms are in close contact to each other rather than to epithelial cells of CF lungs and exist in various physiological states depending on their localization within a biofilm (Whitchurch et al. 2002; Worlitzsch et al. 2002; Borlee et al. 2010). For example, quorum-sensing genes were reported to be expressed specifically in the stalk portion of the mushroom-shaped *P. aeruginosa* biofilms (De Kievit et al. 2001; Lequette and Greenberg 2005; Yang et al. 2009). In conclusion, the growth of *P. aeruginosa* in biofilms may directly contribute to genetic diversity of *P. aeruginosa* recovered from CF lung, by imposing selective physiological conditions, by driving the exchange of eDNA and thus the emergence of specifically adapted subpopulations (Molin and Tolker-Nielsen 2003; Friedman and Kolter 2004; Jackson et al. 2004; Ma et al. 2009).

5 Anaerobic Survival Strategies of *P. aeruginosa*

The recognition that CF airway secretions are more or less micro-aerobic or even anaerobic represents one of the most striking results of clinical CF microbiology in the last decade (Worlitzsch et al. 2002; Yoon et al. 2002). Not surprisingly, within metabolically active multicellular bacterial aggregates that are embedded in an extracellular matrix and located inside abscess-like lesions and/or thick mucus plugs oxygen consumption exceeds its supply.

At first glance, *P. aeruginosa* prefers aerobic respiration using oxygen as terminal electron acceptor for its growth because of the highest energy yield.

To transfer electrons to molecular oxygen, the aerobic respiratory chain of *P. aeruginosa* posesses five terminal oxidases: the bo(3) oxidase (Cyo), the cyanide-insensitive oxidase (CIO), the cbb(3)-1 oxidase (Cbb3-1), the cbb(3)-2 oxidase (Cbb3-2) and the aa(3) oxidase (CoxBA), all with different biochemical characteristics and/or oxygen affinities. Moreover, *P. aeruginosa* is also able to perfectly survive, grow and thrive under oxygen limitation up to anoxic conditions. *P. aeruginosa* may in fact release compounds that reduce the oxygen transfer into growth medium and thus lead to restricted oxygen availability (Kim et al. 2003). *P. aeruginosa* may further grow better under reduced oxygen concentration in vitro (Sabra et al. 2002). This leads to the suggestion that *P. aeruginosa* indeed prefers micro-aerobic to anaerobic conditions. Anaerobic growth of *P. aeruginosa* is possible using the sequential eight-electron reduction of the alternative electron acceptor nitrate (NO_3) via nitrite (NO_2), nitric oxide (NO) and nitrous oxide (N_2O) to N_2, a process called anaerobic respiration or denitrification. Denitrification is catalyzed by different enzymes termed Nar (nitrate reductase), Nir (nitrite reductase), Nor (nitric oxide reductase) and Nos (nitrous oxide reductase) that are located within the inner membrane or periplasm (Zumft 1997). Like aerobic respiration of oxygen, nitrate reduction generates protons that are translocated into the periplasm to build up a proton motive force that drives ATP production. Electrons enter denitrification by different dehydrogenases before the transfer to respective reductases via ubiquinone, cytochrome bc1 complex, azurin or c-type cytochromes NirM and NirC (for details see Schobert and Tielen 2010). Interestingly, nitrate has been detected in CF sputum in concentrations up to 700 μM, which is sufficient to support anaerobic growth of *P. aeruginosa* (Grasemann et al. 1998). Moreover, three out of five *P. aeruginosa* oxidases, namely Cbb3-1, Cbb3-2 and CIO seem to be important for growth under micro-aerobic conditions (Alvarez-Ortega and Harwood 2007). Interestingly, only a triple mutant but not single mutants of CIO, Cbb3-1 and Cbb3-2 failed to grow at 2% oxygen and forms atypical flat biofilms with decreased stress resistance (Alvarez-Ortega and Harwood 2007). Cbb3-1 is expressed at high oxygen tension, while Cbb3-2 is expressed under oxygen-limiting conditions and CIO is active at high cyanide levels. Cyanide has been detected in CF sputum at concentrations up to 72 μM, why CIO might be relevant under the increasing cyanide concentrations found under anaerobic conditions.

Under the absence of oxygen and N-oxides *P. aeruginosa* may use arginine fermentation as alternative, namely the conversion of arginine to ornithine with concomitant ATP formation, for survival or slow growth (Schobert and Jahn, 2010). Since CF mucus is generally rich in amino acids this pathway may be also relevant in vivo. Arginine fermentation requires the arginine-ornithine antiporter (ArcD), the arginine deiminase (ArcA), the ornithine carbamoyltransferase (ArcB) and the carbamate kinase (ArcC), each encoded by the *arcDABC* operon. If arginine is unavailable *P. aeruginosa* may survive by generating ATP from the conversion of pyruvate to acetate (Eschbach et al. 2004). It has been suggested that pyruvate fermentation support survival in the deep anaerobic regions of *P. aeruginosa* biofilms (Price-Whelan et al. 2007). Expression of genes involved in anaerobic respiration and arginine fermentation are regulated by ANR (anaerobic nitrate regulator), an oxygen sensitive FNR-type (fumarate/nitrate regulator) transcriptional activator. ANR in turn is required for transcription of the downstream regulator DNR (dissimilatory nitrate respiration regulator), which in the presence of N-oxides promotes the expression of *nir*, *nor* and *nos* genes. In addition, anaerobic energy metabolism of *P. aeruginosa* is coordinated by the two-component regulatory system NarX-NarL. When nitrate is present, NarL represses the arginine-dependent activation of the *arcDABC* operon, explaining the preferred use of denitrification over arginine fermentation (Benkert et al. 2008).

In conclusion, oxygen limited conditions of the CF lung environment result in profound changes of *P. aeruginosa* metabolism and physiology, which should have important implications for antibiotic treatment (Trunk et al. 2010). Anaerobiosis significantly increases antibiotic tolerance of *P. aeruginosa* and robustness of biofilms by mechanisms that are still only partly understood (Yoon et al. 2002; Borriello et al. 2004a, b; Field et al. 2005). Denitrification plays an important role in *P. aeruginosa* biofilm development and prevents accumulation of toxic NO (Worlitzsch et al. 2002; Yoon et al. 2002). In agreement, the *nirCMSQ* and *napEF* (periplasmic nitrate reductase NAP) genes that are required for denitrification are upregulated in *P. aeruginosa* during biofilm formation. Oxygen depletion and NO, a product of anaerobic metabolism, in turn trigger dispersal of cells during late stages of biofilm development. Reduced oxygen tension also result in increased alginate production by *P. aeruginosa* (Bragonzi et al. 2005). Alginate as the main component of biofilm matrix provides a physical barrier to host defense and antibiotics (Govan and Deretic 1996; Gordon et al. 1988; Nichols et al. 1988; Hentzer et al. 2001). The anaerobic increase in alginate production in non-mucoid strains ahead of the selection of stable mucoid variants could explain the survival of *P. aeruginosa* during early infection. Anaerobic growth conditions also promote the production of cyanide that is expressed in dependence of ANR and of exotoxin A (Zimmermann et al. 1991; Gaines et al. 2005). Recently, it has been shown that components of the denitrification machinery are important for the expression of type III secretion system (Van Alst et al. 2009). These data indicate that anaerobe metabolism also interacts with the production of virulence factors by *P. aeruginosa*.

Finally, it is still under debate whether in the CF lung *P. aeruginosa* grow rather by aerobic respiration, anaerobic respiration or both pathways (Alvarez-Ortega and Harwood 2007). Interestingly, it has been demonstrated that *P. aeruginosa* exhibit denitrification already in the presence of low oxygen concentrations (Chen et al. 2006). The CF lung environment is heterogeneous and may present both, aerobic regions because of normal air and blood supply, microaerobic and anaerobic regions due to mucus plugging and formation of abscess-like lesions. Finally, the low oxygen tension found in the CF lung is also demanding for immune cells such as recruited neutrophils (Sica et al. 2011). Neutrophils sense and respond to low oxygen by the major regulators hypoxia-inducible factors (HIFs) and AMP-activated protein kinase (AMPK). Low oxygen conditions promote the metabolic switch of neutrophils toward anaerobic pathways including glycolysis, increase neutrophil inflammatory functions such as tissue infiltration, protease degranulation, ROS generation and survival. However, anoxic conditions may cause also a major reduction in the respiratory burst activity of neutrophils which distinctly impair killing of *S. aureus* but not of *P. aeruginosa* (Mizgerd and Brain 1995; McGovern 2011). Although neutrophils are extremely well adapted to operate successfully under low oxygen conditions under lasting anaerobiosis and carbon starvation neither macrophages nor neutrophils may survive, suggesting that oxygen limitation may account for the death and decay of immune cells resulting in the release of DNA, elastase and myeloperoxidase (Döring et al. 2011).

6 Transition of *P. aeruginosa* from Acute to Chronic Infecting Pathogen

Driven by the selective pressure in CF airways, clonal expansion of *P. aeruginosa* leads to a variety of phenotypic and genotypic variants that are considered to be specifically adapted to the requirements of the CF lung environment (Nguyen and Singh 2006). Common characteristics of chronic *P. aeruginosa* phenotypes are multiple morphological changes (e.g. mucoidity, small colony variants, non-pigmentation), antibiotic resistances, attenuation of acute virulence and adaptations in metabolic pathways. This evolutionary change of *P. aeruginosa*, also termed adaptive radiation, has been interpreted as a continuous selection process toward a population of niche specialists. Various stress conditions that occur in the *P. aeruginosa*-infected CF lung such as high osmolarity, low pH, dehydration, inflammation, oxidative stress, oxygen limitation and antibiotics may select for *P. aeruginosa* adaptations (Mathee et al. 1999; Bragonzi et al. 2005). The diversification of *P. aeruginosa* e.g. accelerated by hypermutability may ensure bacterial survival under extreme environmental fluctuations (insurance hypothesis), while physiological characteristics that are maintained or increased during survival in the CF lung are putative chronic virulence determinants of *P. aeruginosa*.

6.1 Morphological Characteristics of Chronic P. aeruginosa Phenotypes

A consistent finding in CF airways is the presence of mucoid *P. aeruginosa* that mark the transition from acute to chronic lung infection. Mucoid *P. aeruginosa* overproduce the exopolysaccharide alginate, a polymer of D-mannuronic and L-guluronic acid (Deretic et al. 1995; Boucher et al. 1997). Alginate promotes the encapsulation of *P. aeruginosa* and biofilm formation (Worlitzsch et al. 2002). Further, alginate protects *P. aeruginosa* from antibiotics, ROS, opsonizing antibodies and keeps phagocytes from entering the biofilms. (Hatch and Schiller 1998;

Oliver and Weir 1985; Pedersen et al. 1990; Simpson et al. 1989), altogether favoring lung persistence. Consequently, the occurrence of mucoid phenotypes is associated with a remarkably diminished likelihood of eradication of *P. aeruginosa*, deterioration of lung function and increased tissue damage (Hoffmann et al. 2005). Mucoidity is thus a major determinant of chronic virulence and poor prognosis of CF lung disease.

The production of alginate by *P. aeruginosa* is highly regulated. The alginate biosynthesis operon (*algD–algA*) of *P. aeruginosa* is controlled by the *algD* promoter. The *algD* gene encodes GDP-mannose dehydrogenase, the first enzyme involved in alginate biosynthesis. The alternative sigma factor AlgT (also termed AlgU), induces the expression of *algD* and that of regulatory proteins that enhance *algD* transcription (Ramsey and Wozniak 2005). The *algT* gene belongs to an operon with four other genes, *mucA-mucB-mucC-mucD*. MucA a cytoplasmatic membrane bound protein acts as an anti-sigma factor that negatively controls *algD* expression by binding to AlgT. MucB and MucD also negatively control transcription of the *algD* gene. The anti-sigma factor MucA inhibits AlgT; thus, inactivation of *mucA* leads to upregulation of AlgT and conversion to mucoidity. Mutations in *mucB* or *mucD* can also lead to mucoidity (Govan and Deretic 1996), while in clinical CF isolates loss-of-function mutations are predominantly located in *mucA* and rarely in *mucB* or *mucD* (Boucher et al. 1997). *MucA* mutants are highly mucoid, *mucB* and *mucD* mutants have only a weak mucoid phenotype.

Moreover, AlgT negatively regulates flagella-mediated motility explaining the sessile phenotype of mucoid isolates. In turn, non-mucoid *P. aeruginosa* cells are able to regain motility e.g. to move to aerobic regions that confer faster growth rates. Alginate expression is up-regulated under micro-aerobic conditions of viscous CF mucus that are already found during early infection and possibly promotes lung persistence (Bragonzi et al. 2005). Mucoidity is initially unstable but may be fixed by above mentioned mutations during chronic infection. In agreement, mutator strains show more frequently a mucoid phenotype than non-mutators. However, during chronic infection mucoid and non-mucoid *P. aeruginosa* isolates may occur simultaneously in CF sputum, whilst during end-stage disease again non-mucoid isolates dominated (Bragonzi et al. 2009). Non-mucoid isolates may be sustained wild-type *P. aeruginosa* or revertants from mucoid phenotypes that only occur intermittently in the CF lungs. It is known that mucoid phenotypes may revert to non-mucoid forms in vitro in the absence of a selective pressure. Reversion to non-mucoidity may result from repair of the defect *mucA* alleles or by secondary-site mutations. Several non-mucoid CF isolates carry *mucA* mutation (Bragonzi et al. 2006) suggesting the latter and indicating that the selective pressure of the CF lung milieu changed. If *P. aeruginosa* cells spend a lot of energy to produce alginate, it is not surprising that during ongoing CF lung disease mucoidity is unstable. Inactivation of *algT* is the most frequent mechanism that has been demonstrated for revertants selected in vitro. In contrast, in the CF lung in the majority of non-mucoid revertants no secondary *algT* mutations are found. Thus, *algT* mutants are probably less proficient to survive in the CF lung environment indicating that secondary-site mutations must be located in genes other than *algT*

(Ciofu et al. 2008). Moreover, it seems that an anaerobic environment conserves this mucoid phenotype by lowering the rate of revertants (Hassett 1996). Interestingly, mucoid *P. aeruginosa* isolates typically have no flagellum, probably since motility is not needed during static non-gradient growth conditions within anaerobic biofilms. However, AlgT represses the biosynthesis of the flagellum in *P. aeruginosa* and thus non-mucoid revertants with mutations in *algT* become motile. The fact that motility is not favored in the CF lung, may account for the low frequency of *algT*-dependent revertants.

Another intriguing variant of chronic *P. aeruginosa* CF isolates are small colony variants (SCV). SCVs are selected during prolonged antibiotic treatment and become apparent due to their small colony size of 1–3 mm (Haussler et al. 1999). SCVs may exhibit a hyperpiliated and hyperadherent phenotype with autoaggregative behavior as well as an elevated capacity to form biofilm (Haussler 2004; Deziel et al. 2001). Further, it has been demonstrated that the increased expression of the *pel* and *psl* polysaccharide gene loci and elevated intracellular c-di-GMP levels contribute to SCV morphotype (Starkey et al. 2009). The *psl* (polysaccharide synthesis locus) genes are responsible for the production of mannose-rich exopolysaccharide and the *pel* locus in pellicle formation that both favor biofilm formation. Thus, SCV show increased PEL and PSL exopolysaccharide production, enhanced capacity to form biofilms and may partially contribute to increased persistence of *P. aeruginosa* in the airways of CF lungs (Lory et al. 2009).

6.2 Loss of Acute Virulence Factors Indicates Activation of Chronic Virulence

P. aeruginosa possesses many virulence factors, which can be divided into cell-associated and extracellular factors. Almost all virulence factors are regulated by the quorum-sensing circuit of *P. aeruginosa* and/or by other regulatory proteins. Cell appendages such as flagellum, type IV pili, Cup fimbria and Psl polysaccharide are known to mediate adhesion of *P. aeruginosa* to host epithelial cell surfaces (Vallet et al. 2001; Ma et al. 2009). The pilin- and flagellum-dependent adherence is particularly needed for the initial colonization step and may further promote cell injury due to translocation of type III exotoxins during tight host cell contact of *P. aeruginosa*. Generally, the type III secretion system (T3SS) injects the exoenzymes S, T, Y and U into host cells to impair cell integrity and actin cytoskeleton.

However, chronic CF isolates are characterized by the disability to move by twitching and swimming due to nonpiliation and loss of the flagellum, respectively. The binding and recognition of these appendages to Toll-like receptor 5 (TLR5) can induce a strong pro-inflammatory response. Thus, the down-regulation of the flagellum and pili probably counteract this response. Similarly, during ongoing lung infection T3SS of *P. aeruginosa* is turned off leading to distinct decrease in cell cytotoxicity (Hogardt et al. 2007). Strikingly, beside motility and T3SS many other classical virulence traits of *P. aeruginosa* are targets of adaptations resulting in reduced production of homoserine lactones, proteases, phospholipase C, loss of pyoverdine, pyocyanin, pyocins and elastase. Therefore, also type-I-dependent (alkaline protease) secretion and type-II-dependent (LasB and LasA elastases, protease IV, phospholipase C) secretion of anti-host factors are affected during lung adaptation (D'Argenio et al. 2002, 2007; Ramsey and Wozniak 2005; Lee et al. 2011; Williams and Camara 2009; Winstanley and Fothergill 2009). A further characteristic of chronic *P. aeruginosa* CF isolates is that they typically show a rough colony phenotype due to modification of the lipid A moiety of lipopolysaccharide (LPS). LPS-rough strains contain few, short, or no O side chains and are therefore non-typeable (Hancock et al. 1983). Moreover, isolates that lack O-specific antigens are sensitive to in vitro killing by serum complement and are readily permeabilized by surfactant-protein A and are more tolerant to the gentamicin (Kadurugamuwa et al. 1993). In contrast to environmental *P. aeruginosa* with penta-acylated LPS, CF-isolates synthesize hexa-acylated LPS due to addition of palmitate and aminoarabinose residues (Ernst et al. 1999). These CF specific lipid A structures show a higher proinflammatory activity and an increase in IL-8 expression due to a differential recognition of the altered LPS structure by human TLR4 receptor (Hajjar et al. 2002). CF-LPS is often associated with increased resistance to antimicrobial peptides. However, the definite selective advantage of an amplified immune response to highly-acylated CF-LPS is unclear.

In conclusion, during the chronic course of lung infections in patients with CF several virulence factors of *P. aeruginosa* are part of its chronic survival strategy and either switched off or produced in reduced quantities or modified appearance. The emergence of pyocyanin-negative isolates may reduce auto-oxidative stress of *P. aeruginosa*. First, these adaptive changes may reduce immune recognition by the host possibly due to the reduction or modification of immunostimulatory components. For instance, *P. aeruginosa* isolates lacking the flagellum are less effectively phagocytosed by alveolar macrophages and neutrophils. Thus, the loss of flagellar expression may enable *P. aeruginosa* to better evade immune recognition and airway clearance by phagocytosis by avoiding activation of TLR5 expressing epithelial cells, neutrophils and macrophages. During the evolution of *P. aeruginosa* from acute to chronic infection, the lipid A moiety changes its acylation pattern to one of less immunogenicity and weaker stimulation through TLR4. Further, it has been shown that CF sera may contain high amounts of antibodies against elastase, exotoxin A, exotoxin S, alkaline protease, LPS and flagella structural proteins. Although obviously not effective in preventing colonization or infection by *P. aeruginosa,* these immunoglobulins may disturb the correct function of bacterial targets and trigger inflammation by the formation of immune complexes. Further, during this matrix-enclosed sessile growth *P. aeruginosa* cells downregulate the flagellum, Typ IV pili and the type III secretory pathway. Due to the inability of the type III secretion system to translocate its effector proteins (ExoS or ExoU, ExoT and

ExoY) into target cells, the flagellum and pili probably are no longer needed to move along epithelial cell surfaces (Hogardt et al. 2004). Finally, downregulation is needed to bypass the high energy costs to keep these systems in stand-by position (retention of the flagellum, chemotactic components, cytoplasmic exotoxins and secretion machineries).

To fully understand the physiological adaptation process of *P. aeruginosa*, the knowledge of their genetic background and interactions is a fundamental prerequisite. One common genetic change in chronic CF isolates that account for several phenotypes is the mutational inactivating of the *lasR* gene encoding the QS transcriptional regulator LasR. Several studies have shown that LasR mutants emerged within individual patients from existing chronically-infecting lineages (D'Argenio et al. 2007; Smith et al. 2006), but also occur frequently in the natural environment (Cabrol et al. 2003), in intubated patients (Denervaud et al. 2004) and patients suffering from wound infection (Hamood et al. 1996). Loss-of-function mutations in *lasR* may explain the loss or reduced production in AHL and several virlence factors such as LasB in chronic *P. aeruginosa* CF isolates, but also confer autolysis and iridescent shine of *P. aeruginosa* colonies as well as growth advantage on amino acids. The occurrence of LasR mutants during CF lung infection indicate a strong selective pressure against its function and that the role of the total QS circuit for the pathogenesis of CF lung infection is limited or even restricted to the early stages of the disease (Winstanley et al. 2009). Mutations in both *mucA* and *lasR* gene may occur ahead of the selection of mutator phenotypes, suggesting that rather the selective environment than hypermutability is necessary for effective lung adaptation (Ciofu et al. 2010). A further link of genotype and phenotype was found for loss of twitching and swimming motility. In general, nonpiliation may derive from disruption of *pilB*, encoding an ATPase that is needed for the extention and retraction of pili, or from a defect in the gene of PilQ that is required to extrude the pilus fibre through the bacterial outer membrane (Chang et al. 2007). Mutations of the *rpoN* gene (required for both pilus and flagellum assembly) account for the majority of pili and flagellum losses in CF isolates (Mahenthiralingam et al. 1994; Kresse et al. 2003). Whole genome comparison between an isogenic early and late *P. aeruginosa* pair recovered from one patient 90 months apart showed that the late isolate accumulated a total 68 mutations, while intermediate isolates of the same clone carried a subset of these mutations (Smith et al. 2006). Interestingly, examinations of explanted lungs suggested that bacterial aggregates from anaerobic purulent sputum plugs prevail in the conductive lung zone of the lung and may serve as a reservoir for *P. aeruginosa,* whereas the respiratory zone may be effectively protected over long periods from massive biofilm infection (Bjarnholt et al. 2009). We found that after successful lung-transplantation cytotoxic variants may reemerge likely from reservoirs of the upper respiratory tract (Hogardt et al. 2007) and that in explanted lungs non-cytotoxic and cytotoxic *P. aeruginosa* variants coexist (unpublished), indicating that does *P. aeruginosa* not completely loose its acute virulence potential during chronic CF disease (Fig. 1b).

6.3 Metabolic Adaptations Toward Chronic CF Lung Persistence

P. aeruginosa is known for its environmental versatility and metabolic flexibility that likely determines its outstanding capacity to thrive in diverse habitats such as CF lungs. Beside host defenses, *P. aeruginosa* is subjected to various nutritional stresses during colonization, early infection and long-term persistence in CF airways. First of all, nutrient acquisition in the lungs takes place in competition with commensal microorganisms and host cells. For the production of virulence factors (e.g. alginate), growth in high cell densities and survival in biofilm-like aggregates *P. aeruginosa* has to acquire sufficient carbon and energy sources and several cofactors including iron. Not surprisingly, several studies provide evidence that also metabolic pathways of *P. aeruginosa* are major targets of lung adaptation and thus potential markers for the activation of chronic virulence strategies.

P. aeruginosa is known to preferentially metabolize organic acids and TCA cycle intermediates before metabolizing nonorganic acids, such as glycerol, or glucose. Expression of genes involved in catabolism of hexoses is repressed by the catabolite repression control protein (Crc) in the presence of preferred carbon sources. Generally, CF sputum is nutritional rich in mucin, DNA, lipids, amino acids and proteins that likely promote efficient *P. aeruginosa* growth and energy metabolism. Considerable data showed that peptides, amino acids and fatty acids fuelled the growth of *P. aeruginosa* within CF lungs (Barth and Pitt 1996; Palmer et al. 2005; Son et al. 2007). Further, in vivo and in vitro transcriptome/proteome analysis of chronic *P. aeruginosa* CF isolates showed that various metabolic genes responsible for the uptake and the utilization of different nutrients are positively selected (Son et al. 2007; Hoboth et al. 2009; Oberhardt et al. 2010). Several metabolic enzymes (e.g. malat synthase GlvB/glcB etc.), transporters (e.g. ArcD, AotJQMOP, BraC, PA0119 a homolog of DctA etc.) and regulators (e.g. ANR) were found to be up-regulated (Hoboth et al. 2009). The frequent recovery of mutants that are auxotroph for different amino acids among chronic *P. aeruginosa* CF isolates additionally underlined their availability in sufficient amounts within CF airway secretions (Thomas et al. 2000; Barth et al. 1996; Taylor et al. 1993). Interestingly, methionine auxotroph isolates have been most frequently recovered and in particular during acute exacerbations (Taylor et al. 1993). However, so far no clear survival benefit can be attributed to amino acid auxotrophy. From CF sputum *P. aeruginosa* preferentially catabolize L-alanine, L-arginine, L-glutamate over other carbon sources like lactic acid and glucose (Palmer et al. 2007).

In vivo metabolism of amino acids and fatty acids by *P. aeruginosa* is further indicated by high in vivo expression of genes encoding uptake and/or catabolism of these compounds (Son et al. 2007; Hoboth et al. 2009). Likewise, phenylalanine and tyrosine are important carbon sources for *P. aeruginosa* in CF sputum, but also enhance PQS production. Consistently, when grown in aromatic amino acids containing synthetic CF sputum *P. aeruginosa* produces increased amounts of the PQS signal molecule 2-heptyl-3-hydroxy-4-quinolone. Since PQS increases the formation of protective biofilms it may represent an important adaptive behavior of *P. aeruginosa* in CF lung (Palmer et al. 2005; Palmer et al. 2007; Guina et al. 2003). Loss-of-function mutations of *lasR* confer reduced expression of virulence genes but also a growth advantage with particular amino acids such as phenylalanine (D'Argenio et al. 2007). Moreover, mutations in *rhlR* are frequently found among CF isolates, while *P. aeruginosa rhlR* mutants die under denitrifying growth conditions (Heurlier et al. 2006). Since a distinct increase in the nitrate and nitrite reductase activities but only a moderate upregulation of the NO reductase was found, intoxication by NO has been suggested to account for this phenotype (Yoon et al. 2002). NO accumulation due to imbalance of nitrite reductase and NO reductase activity result from PQS production that requires molecular oxygen and is downregulated under anaerobiosis. This mechanism could explain NO accumulation in upper biofilm layers. Further, it has been demonstrated that a *rpoN* mutation is able to rescue the *P. aeruginosa rhlR* mutant, which accumulates NO under anaerobic conditions, by downregulation of the nitrite reductase activity (Yoon et al. 2002).

Another striking phenotype of *P. aeruginosa* CF isolates is mucoidity due to derepression of alginate biosynthesis. Mucoid *P. aeruginosa* exhibit increased levels of G6PDH activity and expression of the corresponding *zwf* gene that does not result from mutations of its repressor Crc. This metabolic variation enables *P. aeruginosa* to supply sufficient carbohydrates for alginate biosynthesis. Moreover, expression of high-level G6PDH activity is also important for the optimal production of cyanide, probably by the conversion of glyoxylate to glycine, a preferred substrate for cyanide synthase (Silo-Suh et al. 2005). Besides G6PDH also activity of isocitrate lyase (ICL) is needed for optimal alginate production by *P. aeruginosa* (Hagins et al. 2009) and increased in chronic CF isolates (Son et al. 2007; Hoboth et al. 2009). *P. aeruginosa* ICL mutant was found to be less virulent in a rat lung infection model (Lindsey et al. 2008). ICL (PA2634) is encoded by *aceA* gene and specific to the glyoxylate shunt pathway that allows growth on fatty acids or acetate as a sole carbon source (Díaz-Pérez et al. 2007; Honer zu Bentrup et al. 1999) again indicate catabolism of fatty acids from CF sputum. Moreover, ICL that is absent in human cells may be an interesting target for the development of anti-pseudomonal agents.

Regarding metabolism of *P. aeruginosa* in the CF lung, the most striking result is that the oxygen tension within CF mucus is significantly reduced or even absent. Under anaerobic conditions *P. aeruginosa* may utilize either denitrification or alternatively arginine fermentation for energy generation. In line with this, microarray analysis of CF sputum samples showed anaerobic gene expression, including denitrifying *nar*, *nir*, *nor* genes and *arc* genes needed for arginine fermentation. Further, several additional data point to the fact that *P. aeruginosa* adapt to the relative oxygen-depletion found within CF mucus. With respect to anaerobic metabolism the outer membrane protein, OprF of *P. aeruginosa* seems to represent an important cellular factor, if OprF may function as a permease for NO_3^-/NO_2^- uptake in order to sustain denitrification (Yoon et al. 2002). It has been shown that OprF is indispensible for the growth in anaerobic biofilms, is increasingly

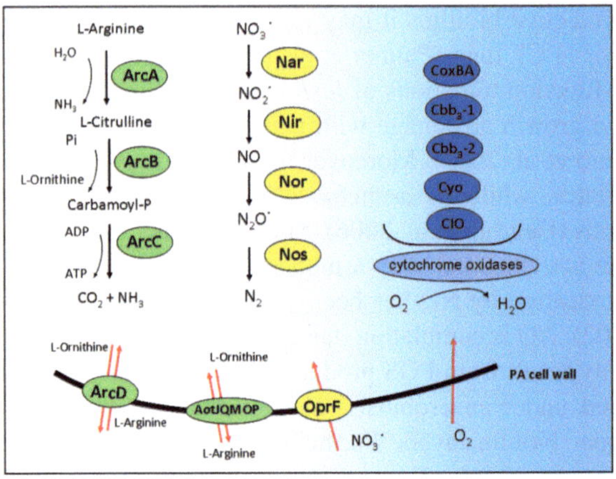

Fig. 2 Arginine fermentation, aerobic and anaerobic respiration (denitrification) of *P. aeruginosa* may ensure survival of this pathogen in the oxygen restricted CF lung milieu. Under aerobic conditions electrons may be transferred to O_2 by five terminal oxidases of *P. aeruginosa* (bo(3) oxidase, Cyo; cyanide-insensitive oxidase, CIO; cbb(3)-1 oxidase, Cbb3-1; cbb(3)-2 oxidase (Cbb3-2) and aa(3) oxidase (CoxBA)). The anaerobic reduction of NO_3 to N_2 is catalyzed by the sequential activity of nitrate reductase (Nar), nitrite reductase (Nir), nitric oxide reductase (Nor) and nitrous oxide reductase. Electrons are transferred to the respective oxidases/reductases via ubiquinone, the cytochrome bc_1 complex, azurin, NirM and NirC. Various dehydrogenases feed electrons into the electron transport chain while the protons are pumped into the periplasm to generate a proton-motive force, used for ATP generation by ATPase (not shown). Whether in the microaerobic CF lung environment *P. aeruginosa* grow rather by aerobic respiration, anaerobic respiration or both pathways is still unknown. Arginine fermentation depends on the activity of the arginine deiminase (ArcA), the ornithine carbamoyltransferase (ArcB) and the carbamate kinase (ArcC). For arginine uptake *P. aeruginosa* possesses the arginine-ornithine antiporters ArcD and AotJQMOP, while OprF may function as a permease for NO_3^-/NO_2^- uptake

expressed in lung-selected *P. aeruginosa* mutator strains and is required for full virulence factor expression (Yoon et al. 2002; Hoboth et al. 2009; Fito-Boncompte et al. 2011). Finally, OprF raises antibody production in CF patients with chronic *P. aeruginosa* infection altogether confirming its in vivo expression (Hassett et al. 2009). Likewise, sera of CF patients contained antibodies specific to two *P. aeruginosa* nitrate reductases, NapA and NarG. While OprF contributes to anaerobic growth of *P. aeruginosa,* NapA and NarG are important for anaerobic respiration using the alternative electron acceptor nitrate (Yoon et al. 2002; Zumft 1997). *NarG* was shown to be essential for growth in synthetic CF sputum. The anaerobic metabolism of *P. aeruginosa* is regulated by ANR and DNR. ANR-regulated genes such as (i) the blue copper protein azurin and the cytochrome c peroxidase c551 (NirS) both relevant to transfer electrons to denitrifying N-oxide reductases, (ii) the anaerobic oxidase Cbb3-2 typically expressed under oxygen-limiting conditions and (iii) enzymes of the arginine-deiminase pathway are found to be increased in CF isolates from end-stage disease and thus seem to be additional biomarkers for the anaerobic CF lung adaptation (Hoboth et al. 2009) (Fig. 2).

In summary, nutrients such as amino acids, fatty acids and carbohydrates are important carbon sources for *P. aeruginosa* but their metabolism also may influence its virulence and persistence in CF lung. Although the impact of specific carbon sources on host colonization, virulence and progression of lung disease are widely unknown, the understanding the nutrient requirements of *P. aeruginosa* for high-density growth may identify new potential treatment options against this pathogen.

7 Concluding Remarks

Genotypes and phenotypes of *P. aeruginosa* isolates from chronic CF lung disease differ substantially from those that initially have colonized the CF lung. The transition from environmental phenotypes to specialized lung-adapted phenotypes of *P. aeruginosa* results from an ongoing adaptation process based on the lasting selection of mutations that are beneficial for survival and/or growth within CF airways. This diversification of *P. aeruginosa* may be accelerated by the emergence of hypermutable isolates with up to 1,000 -fold increased mutation rates. Similar adaptation characteristics among independent clonal lineages of lung-adapted *P. aeruginosa* from different CF patient's point to comparable selective requirements of chronically-infected CF lungs. Many factors such as progressive airway destruction, host immune response, antibiotic therapy, interspecies competition, hypoxia, biofilm growth, starvation and the supply of nutrients may trigger the acquisition of adaptations by *P. aeruginosa*, while the exact mechanism of this selection process and its benefits still have to be defined in more detail. The formation of biofilms protects bacteria from environmental stresses such as antibiotics. Loss or reduction of traditional virulence determinants important for acute infection and/or reduction of immunostimulatory factors are common features of chronic *P. aeruginosa* isolates. Thus, these traits should not play any important role in maintaining lung persistence or are required solely in decreased quantities. By contrast, the presence of characteristics positively selected in chronic *P. aeruginosa* CF isolates should confer lung survival. *P. aeruginosa lasR* and *mucA* mutants are typical indicators of chronic CF lung infection. Hypermutability is not strictly necessary for effective lung adaptation, but that mutators are more frequently multidrug resistant than non-mutators is of great clinical relevance. Chronically persisting *P. aeruginosa* obviously adapt metabolically to microaerobiosis and nutrient environments of mucous CF secretions. Under microaerobiosis/anaerobiosis of CF mucus *P. aeruginosa* is able to grow and survive via denitrification and arginine fermentation. As a consequence, several enzymes, transporters and regulators for different metabolic purposes are up-regulated. Denitrification may be further important for detoxification of NO generated during the infection process. The consumption of fatty acids that belong to host defense such as prostaglandins and phosphatidylcholine may promote lung colonization and/or persistence. Likewise exploiting oxygen from CF mucus and generation of anaerobic conditions may impair ROS production by neutrophils and the antibacterial activity of antibiotics. Chronic virulence determinants of *P. aeruginosa* that are active within CF lungs should be relevant for chronic pathogenesis and lung survival. The identification of the persistence strategies will provide important clues to develop new approaches for effectively combat chronic lung disease. In this context, improved simulation of CF lung conditions and meta-transcriptomic or -proteomic analysis of in vivo infection requires further attention. Finally, metabolic adaptations of *P. aeruginosa* seem to play a key role during lung survival and may provide promising targets for new treatment options against CF lung infection and possibly other chronic infections that follow similar mechanisms. A good example for such efforts may be the inhibition of ICL, anaerobic energy metabolism and the increased killing of *P. aeruginosa* by nitrite.

References

Allesen-Holm M et al (2006) A characterization of DNA release in *Pseudomonas aeruginosa* cultures and biofilms. Mol Microbiol 59(4):1114–1128

Alvarez-Ortega C, Harwood CS (2007) Responses of *Pseudomonas aeruginosa* to low oxygen indicate that growth in the cystic fibrosis lung is by aerobic respiration. Mol Microbiol 65(1):153–165

Barken KB et al (2008) Roles of type IV pili, flagellum-mediated motility and extracellular DNA in the formation of mature multicellular structures in *Pseudomonas aeruginosa* biofilms. Environ Microbiol 10(9):2331–2343

Barth AL, Pitt TL (1996) The high amino-acid content of sputum from cystic fibrosis patients promotes growth of auxotrophic *Pseudomonas aeruginosa*. J Med Microbiol 45(2):110–119

Benkert B et al (2008) Nitrate-responsive NarX-NarL represses arginine-mediated induction of the *Pseudomonas aeruginosa* arginine fermentation arcDABC operon. Microbiology 154(Pt 10): 3053–3060

Bjarnsholt T et al (2009) *Pseudomonas aeruginosa* biofilms in the respiratory tract of cystic fibrosis patients. Pediatr Pulmonol 44(6):547–558

Borlee BR et al (2010) *Pseudomonas aeruginosa* uses a cyclic-di-GMP-regulated adhesin to reinforce the biofilm extracellular matrix. Mol Microbiol 75(4):827–842

Borriello G et al (2004a) Oxygen limitation contributes to antibiotic tolerance of *Pseudomonas aeruginosa* in biofilms. Antimicrob Agents Chemother 48(7):2659–2664

Borriello G et al (2004b) Oxygen limitation contributes to antibiotic tolerance of *Pseudomonas aeruginosa* in biofilms. Antimicrob Agents Chemother 48(7):2659–2664

Boucher C et al (1997) Mucoid *Pseudomonas aeruginosa* in cystic fibrosis: characterization of muc mutations in clinical isolates and analysis of clearance in a mouse model of respiratory infection. Infect Immun 65(9):3838–3846

Bragonzi A et al (2005) Nonmucoid *Pseudomonas aeruginosa* expresses alginate in the lungs of patients with cystic fibrosis and in a mouse model. J Infect Dis 192(3):410–419

Bragonzi A, Wiehlmann L, Klockgether J, Cramer N, Worlitzsch D, Döring G, Tümmler B (2006) Sequence diversity of the mucABD locus in *Pseudomonas aeruginosa* isolates from patients with cystic fibrosis. Microbiology 152(Pt 11):3261–3269

Bragonzi A et al (2009) *Pseudomonas aeruginosa* microevolution during cystic fibrosis lung infection establishes clones with adapted virulence. Am J Respir Crit Care Med 180(2):138–145

Burns JL et al (2001) Longitudinal assessment of *Pseudomonas aeruginosa* in young children with cystic fibrosis. J Infect Dis 183(3):444–452

Cabrol S et al (2003) Transcription of quorum-sensing system genes in clinical and environmental isolates of *Pseudomonas aeruginosa*. J Bacteriol 185(24):7222–7230
Chang YS et al (2007) An intragenic deletion in pilQ leads to nonpiliation of a *Pseudomonas aeruginosa* strain isolated from cystic fibrosis lung. FEMS Microbiol Lett 270(2):201–206
Chen F, Xia Q, Ju LK (2006) Competition between oxygen and nitrate respirations in continuous culture of *Pseudomonas aeruginosa* performing aerobic denitrification. Biotechnol. Bioeng 93(6):1069–1078
Christensen LD et al (2007) Impact of *Pseudomonas aeruginosa* quorum sensing on biofilm persistence in an in vivo intraperitoneal foreign-body infection model. Microbiology 153(Pt 7): 2312–2320
Ciofu O, Riis B, Pressler T, Poulsen HE, Høiby N (2005) Occurrence of hypermutable *Pseudomonas aeruginosa* in cystic fibrosis patients is associated with the oxidative stress caused by chronic lung inflammation. Antimicrob Agents Chemother 49(6):2276–2282
Ciofu O et al (2008) Investigation of the algT operon sequence in mucoid and non-mucoid *Pseudomonas aeruginosa* isolates from 115 Scandinavian patients with cystic fibrosis and in 88 in vitro non-mucoid revertants. Microbiology 154(Pt 1):103–113
Ciofu O et al (2010) Genetic adaptation of *Pseudomonas aeruginosa* during chronic lung infection of patients with cystic fibrosis: strong and weak mutators with heterogeneous genetic backgrounds emerge in mucA and/or lasR mutants. Microbiology 156(Pt 4):1108–1119
D'Argenio DA et al (2002) Autolysis and autoaggregation in *Pseudomonas aeruginosa* colony morphology mutants. J Bacteriol 184(23):6481–6489
D'Argenio DA et al (2007) Growth phenotypes of *Pseudomonas aeruginosa* lasR mutants adapted to the airways of cystic fibrosis patients. Mol Microbiol 64(2):512–533
Dekimpe V, Déziel E (2009) Revisiting the quorum-sensing hierarchy in *Pseudomonas aeruginosa*: the transcriptional regulator RhlR regulates LasR-specific factors. Microbiology 155(Pt 3):712–723
De Kievit TR, Gillis R, Marx S, Brown C, Iglewski BH (2001) Quorum-sensing genes in *Pseudomonas aeruginosa* biofilms: their role and expression patterns. Appl Environ Microbiol 67(4):1865–1873
Denervaud V et al (2004) Characterization of cell-to-cell signaling-deficient *Pseudomonas aeruginosa* strains colonizing intubated patients. J Clin Microbiol 42(2):554–562
Deretic V, Schurr MJ, Yu H (1995) *Pseudomonas aeruginosa*, mucoidy and the chronic infection phenotype in cystic fibrosis. Trends Microbiol 3(9):351–356
Deziel E et al (2001) Initiation of biofilm formation by *Pseudomonas aeruginosa* 57RP correlates with emergence of hyperpiliated and highly adherent phenotypic variants deficient in swimming, swarming and twitching motilities. J Bacteriol 183(4):1195–1204
Díaz-Pérez AL, Román-Doval C, Díaz-Pérez C, Cervantes C, Sosa-Aguirre CR, López-Meza JE, Campos-García J (2007) Identification of the aceA gene encoding isocitrate lyase required for the growth of *Pseudomonas aeruginosa* on acetate, acyclic terpenes and leucine. FEMS Microbiol Lett 269(2):309–316
Dodge JA et al (2007) Cystic fibrosis mortality and survival in the UK: 1947–2003. Eur Respir J 29(3):522–526
Doring G (1999) Serine proteinase inhibitor therapy in alpha(1)-antitrypsin inhibitor deficiency and cystic fibrosis. Pediatr Pulmonol 28(5):363–375
Doring G, Gulbins E (2009) Cystic fibrosis and innate immunity: how chloride channel mutations provoke lung disease. Cell Microbiol 11(2):208–216 [Epub: 2 Dec, 2008]
Döring G et al (2011) Differential adaptation of microbial pathogens to airways of patients with cystic fibrosis and chronic obstructive pulmonary disease. FEMS Microbiol Rev 35(1): 124–146
Drenkard E (2003) Antimicrobial resistance of *Pseudomonas aeruginosa* biofilms. Microbes Infect 5(13):1213–1219
Elizur A et al (2008) Airway inflammation in cystic fibrosis. Chest 133(2):489–495
Ernst RK et al (1999) Specific lipopolysaccharide found in cystic fibrosis airway *Pseudomonas aeruginosa*. Science 286(5444):1561–1565

Eschbach M et al (2004) Long-term anaerobic survival of the opportunistic pathogen *Pseudomonas aeruginosa* via pyruvate fermentation. J Bacteriol 186(14):4596–4604
Field TR et al (2005) Effect of oxygen limitation on the in vitro antimicrobial susceptibility of clinical isolates of *Pseudomonas aeruginosa* grown planktonically and as biofilms. Eur J Clin Microbiol Infect Dis 24(10):677–687
Fito-Boncompte L et al (2011) Full virulence of *Pseudomonas aeruginosa* requires OprF. Infect Immun 79(3):1176–1186
Friedman L, Kolter R (2004) Genes involved in matrix formation in *Pseudomonas aeruginosa* PA14 biofilms. Mol Microbiol 51(3):675–690
Fux CA, Costerton JW, Stewart PS, Stoodley P (2005) Survival strategies of infectious biofilms. Trends Microbiol 13(1):34–40
Gaggar A et al (2011) The role of matrix metalloproteases in cystic fibrosis lung disease. Eur Respir J 38(3):721–727
Gaines JM et al (2005) Effect of static growth and different levels of environmental oxygen on *toxA* and *ptxR* expression in the *Pseudomonas aeruginosa* strain PAO1. Microbiology 151(Pt 7): 2263–2275
Galabert C et al (1987) Relationships between the lipid content and the rheological properties of airway secretions in cystic fibrosis. Clin Chim Acta 164(2):139–149
Gilligan PH et al (1985) Isolation medium for the recovery of *Pseudomonas cepacia* from respiratory secretions of patients with cystic fibrosis. J Clin Microbiol 22(1):5–8
Gordon CA, Hodges NA, Marriott C (1988) Antibiotic interaction and diffusion through alginate and exopolysaccharide of cystic fibrosis-derived *Pseudomonas aeruginosa*. J Antimicrob Chemother 22(5):667–674
Govan JR, Deretic V (1996) Microbial pathogenesis in cystic fibrosis: mucoid *Pseudomonas aeruginosa* and *Burkholderia cepacia*. Microbiol Rev 60(3):539–574
Grasemann H et al (1998) Nitric oxide metabolites in cystic fibrosis lung disease. Arch Dis Child 78(1):49–53
Grassmé H et al (2010) CFTR-dependent susceptibility of the cystic fibrosis-host to *Pseudomonas aeruginosa*. Int J Med Microbiol 300(8):578–583
Guina T et al (2003) Quantitative proteomic analysis indicates increased synthesis of a quinolone by *Pseudomonas aeruginosa* isolates from cystic fibrosis airways. Proc Natl Acad Sci USA 100(5):2771–2776
Hagins JM et al (2009) Isocitrate lyase supplies precursors for hydrogen cyanide production in a cystic fibrosis isolate of *Pseudomonas aeruginosa*. J Bacteriol 191(20):6335–6339
Hajjar AM et al (2002) Human Toll-like receptor 4 recognizes host-specific LPS modifications. Nat Immunol 3(4):354–359
Hamood AN et al (1996) Characterization of elastase-deficient clinical isolates of *Pseudomonas aeruginosa*. Infect Immun 64(8):3154–3160
Hancock RE et al (1983) *Pseudomonas aeruginosa* isolates from patients with cystic fibrosis: a class of serum-sensitive, nontypable strains deficient in lipopolysaccharide O side chains. Infect Immun 42(1):170–177
Harris JK, De Groote MA, Sagel SD, Zemanick ET, Kapsner R, Penvari C, Kaess H, Deterding RR, Accurso FJ, Pace NR (2007) Molecular identification of bacteria in bronchoalveolar lavage fluid rom children with cystic fibrosis. Proc Natl Acad Sci U S A 104(51):20529–20533 [Epub 2007 Dec 11]
Harrison F (2007) Microbial ecology of the cystic fibrosis lung. Microbiology 153(Pt 4):917–923
Hassett DJ (1996) Anaerobic production of alginate by *Pseudomonas aeruginosa*: alginate restricts diffusion of oxygen. J Bacteriol 178(24):7322–7325
Hassett DJ et al (2009) *Pseudomonas aeruginosa* hypoxic or anaerobic biofilm infections within cystic fibrosis airways. Trends Microbiol 17(3):130–138
Hatch RA, Schiller NL (1998) Alginate lyase promotes diffusion of aminoglycosides through the extracellular polysaccharide of mucoid *Pseudomonas aeruginosa*. Antimicrob Agents Chemother 42(4):974–977

Haussler S (2004) Biofilm formation by the small colony variant phenotype of *Pseudomonas aeruginosa*. Environ Microbiol 6(6):546–551

Haussler S et al (1999) Small-colony variants of *Pseudomonas aeruginosa* in cystic fibrosis. Clin Infect Dis 29(3):621–625

Hentzer M, Teitzel GM, Balzer GJ, Heydorn A, Molin S, Givskov M, Parsek MR (2001) Alginate overproduction affects *Pseudomonas aeruginosa* biofilm structure and function. J Bacteriol 183(18):5395–5401

Heurlier K et al (2006) Impact of quorum sensing on fitness of *Pseudomonas aeruginosa*. Int J Med Microbiol 296(2–3):93–102

Hoboth C et al (2009) Dynamics of adaptive microevolution of hypermutable *Pseudomonas aeruginosa* during chronic pulmonary infection in patients with cystic fibrosis. J Infect Dis 200(1):118–130

Hoffmann NR et al (2005) Novel mouse model of chronic *Pseudomonas aeruginosa* lung infection mimicking cystic fibrosis. Infect Immun 73(4):2504–2514

Hogardt M, Heesemann J (2010) Adaptation of *Pseudomonas aeruginosa* during per-sistence in the cystic fibrosis lung. Int J Med Microbiol 300(8):557–562

Hogardt M et al (2004) Expression of *Pseudomonas aeruginosa* exoS is controlled by quorum sensing and RpoS. Microbiology 150(Pt 4):843–851

Hogardt M et al (2007) Stage-specific adaptation of hypermutable *Pseudomonas aeruginosa* isolates during chronic pulmonary infection in patients with cystic fibrosis. J Infect Dis 195(1):70–80

Honer Zu, Bentrup K, Miczak A, Swenson DL, Russell DG (1999) Characterization of activity and expression of isocitrate lyase in Mycobacterium avium and Mycobacterium tuberculosis. J Bacteriol 181(23):7161–7167

Jackson KD, Starkey M, Kremer S, Parsek MR, Wozniak DJ (2004) Identification of psl, a locus encoding a potential exopolysaccharide that is essential for *Pseudomonas aeruginosa* PAO1 biofilm formation. J Bacteriol 186(14):4466–4475

Jolivet-Gougeon A et al (2011) Bacterial hypermutation: clinical implications. J Med Microbiol 60(Pt 5):563–573

Kadurugamuwa JL et al (1993) Interaction of gentamicin with the A band and B band lipopolysaccharides of *Pseudomonas aeruginosa* and its possible lethal effect. Antimicrob Agents Chemother 37(4):715–721

Kim EJ et al (2003) Iron deficiency leads to inhibition of oxygen transfer and enhanced formation of virulence factors in cultures of *Pseudomonas aeruginosa* PAO1. Microbiology 149(Pt 9): 2627–2634

Konstan MW, Hilliard KA, Norvell TM, Berger M (1994) Bronchoalveolar lavage findings in cystic fibrosis patients with stable, clinically mild lung disease suggest ongoing infection and inflammation. Am J Respir Crit Care Med 150(2):448–454

Kresse AU et al (2003) Impact of large chromosomal inversions on the adaptation and evolution of *Pseudomonas aeruginosa* chronically colonizing cystic fibrosis lungs. Mol Microbiol 47(1):145–158

Kunkel TA, Erie DA (2005) DNA mistmatch repair. Annu Rev Biochem 74:681–710

Lau GW et al (2004) The role of pyocyanin in *Pseudomonas aeruginosa* infection. Trends Mol Med 10(12):599–606

Lee KM, Yoon MY, Park Y, Lee JH, Yoon SS (2011) Anaerobiosis-induced loss of cytotoxicity is due to inactivation of quorum sensing in *Pseudomonas aeruginosa*. Infect Immun 79(7): 2792–2800

Lequette Y, Greenberg EP (2005) Timing and localization of rhamnolipid synthesis gene expression in *Pseudomonas aeruginosa* biofilms. J Bacteriol 187(1):37–44

Lindsey TL et al (2008) Virulence determinants from a cystic fibrosis isolate of *Pseudomonas aeruginosa* include isocitrate lyase. Microbiology 154(Pt 6):1616–1627

Lory S et al (2009) Multiple activities of c-di-GMP in *Pseudomonas aeruginosa*. Nucleic Acids Symp Ser 53:51–52

Ma L, Conover M, Lu H, Parsek MR, Bayles K, Wozniak DJ (2009) Assembly and development of the *Pseudomonas aeruginosa* biofilm matrix. PLoS Pathog 5(3):e1000354 [Epub 2009 Mar 27]
Mahenthiralingam E et al (1994) Nonmotility and phagocytic resistance of *Pseudomonas aeruginosa* isolates from chronically colonized patients with cystic fibrosis. Infect Immun 62(2):596–605
Maki H, Sekiguchi M (1992) MutT protein specifically hydrolyses a potent mutagenic substrate for DNA synthesis. Nature 355(6357):273–275
Mandsberg LF et al (2009) Antibiotic resistance in *Pseudomonas aeruginosa* strains with increased mutation frequency due to inactivation of the DNA oxidative repair system. Antimicrob Agents Chemother 53(6):2483–2491
Martínez JL, Rojo F (2011) Metabolic regulation of antibiotic resistance. FEMS Microbiol Rev 35(5):768–789
Mathee K et al (1999) Mucoid conversion of *Pseudomonas aeruginosa* by hydrogen peroxide: a mechanism for virulence activation in the cystic fibrosis lung. Microbiology 145(6):1349–1357
McGovern NN (2011) Hypoxia selectively inhibits respiratory burst activity and killing of *Staphylococcus aureus* in human neutrophils. J Immunol 186(1):453–463
Mena A et al (2008) Genetic adaptation of *Pseudomonas aeruginosa* to the airways of cystic fibrosis patients is catalyzed by hypermutation. J Bacteriol 190(24):7910–7917
Mizgerd JP, Brain JD (1995) Reactive oxygen species in the killing of *Pseudomonas aeruginosa* by human leukocytes. Curr Microbiol 31(2):124–128
Molin S, Tolker-Nielsen T (2003) Gene transfer occurs with enhanced efficiency in biofilms and induces enhanced stabilisation of the biofilm structure. Curr Opin Biotechnol 14(3):255–261
Nguyen D, Singh PK (2006) Evolving stealth: genetic adaptation of *Pseudomonas aeruginosa* during cystic fibrosis infections. Proc Natl Acad Sci U S A 103(22):8305–8306
Nichols WW, Dorrington SM, Slack MP, Walmsley HL (1988) Inhibition of tobramycin diffusion by binding to alginate. Antimicrob Agents Chemother 32(4):518–523
Oberhardt MA et al (2010) Metabolic network anal-ysis of *Pseudomonas aeruginosa* during chronic cystic fibrosis lung infection. J Bacteriol 192(20):5534–5548
Oliver and Mena (2010) Bacterial hypermutation in cystic fibrosis, not only for antibiotic resistance. Clin Microbiol Infect 16(7):798–808
Oliver AM, Weir DM (1985) The effect of Pseudomonas alginate on rat alveolar macrophage phagocytosis and bacterial opsonization. Clin Exp Immunol 59(1):190–196
Oliver A et al (2000) High frequency of hypermutable *Pseudomonas aeruginosa* in cystic fibrosis lung infection. Science 288(5469):1251–1254
Palmer KL et al (2005) Cystic fibrosis sputum supports growth and cues key aspects of *Pseudomonas aeruginosa* physiology. J Bacteriol 187(15):5267–5277
Pedersen SS, Kharazmi A, Espersen F, Høiby N (1990) *Pseudomonas aeruginosa* alginate in cystic fibrosis sputum and the inflammatory response. Infect Immun 58(10):3363–3368
Palmer KL et al (2007) Nutritional cues control *Pseudomonas aeruginosa* multicellular behavior in cystic fibrosis sputum. J Bacteriol 189(22):8079–8087
Price-Whelan A et al (2007) Pyocyanin alters redox homeostasis and carbon flux through central metabolic pathways in *Pseudomonas aeruginosa* PA14. J Bacteriol 189(17):6372–6381
Ramsey DM, Wozniak DJ (2005) Understanding the control of *Pseudomonas aeruginosa* alginate synthesis and the prospects for management of chronic infections in cystic fibrosis. Mol Microbiol 56(2):309–322
Ratjen F, Hartog CM, Paul K, Wermelt J, Braun J (2002) Matrix metalloproteases in BAL fluid of patients with cystic fibrosis and their modulation by treatment with dornase alpha. Thorax 57(11):930–934
Rose MC, Voynow JA (2006) Respiratory tract mucin genes and mucin glycoproteins in health and disease. Physiol Rev 86(1):245–278
Rogers GB, Carroll MP, Serisier DJ, Hockey PM, Jones G, Bruce KD (2004) Characterization of bacterial community diversity in cystic fibrosis lung infections by use of 16s ribosomal DNA terminal restriction fragment length polymorphism profiling. J Clin Microbiol 42(11): 5176–5183

Sabra W et al (2002) Physiological responses of *Pseudomonas aeruginosa* PAO1 to oxidative stress in controlled micro-aerobic and aerobic cultures. Microbiology 148(Pt 10):3195–3202
Schobert M, Jahn D (2010) Anaerobic physiology of *Pseudomonas aeruginosa* in the cystic fibrosis lung. Int J Med Microbiol 300(8):549–556
Schobert M, Tielen P (2010) Contribution of oxygen-limiting conditions to persistent infection of *Pseudomonas aeruginosa*. Future Microbiol 5(4):603–621
Sibley CD, Surette MG (2011) The polymicrobial nature of airway infections in cystic fibrosis: Cangene Gold Medal Lecture. Can J Microbiol 57(2):69–77
Sica A et al (2011) Hypoxia: a double-edged sword of immunity. J Mol Med 89(7):657–665
Silo-Suh L et al (2005) Adaptations of *Pseudomonas aeruginosa* to the cystic fibrosis lung environment can include deregulation of zwf, encoding glucose-6-phosphate dehydrogenase. J Bacteriol 187(22):7561–7568
Simpson JA, Smith SE, Dean RT (1989) Scavenging by alginate of free radicals released by macrophages. Free Radic Biol Med 6(4):347–353
Smith RS et al (2001) IL-8 production in human lung fibroblasts and epithelial cells activated by the *Pseudomonas* autoinducer N-3-oxododecanoyl homoserine lactone is transcriptionally regulated by NF-kB and activator protein-2. J Immunol 167(1):366–374
Smith EE et al (2006) Genetic adaptation by *Pseudomonas aeruginosa* to the airways of cystic fibrosis patients. Proc Natl Acad Sci U S A 103(22):8487–8492
Son MS et al (2007) In vivo evidence of *Pseudomonas aeruginosa* nutrient acquisition and pathogenesis in the lungs of cystic fibrosis patients. Infect Immun 75(11):5313–5324
Spasenovski T et al (2010) Modelling the bacterial communities associated with cystic fibrosis lung infections. Eur J Clin Microbiol Infect Dis 29(3):319–328
Starkey M et al (2009) *Pseudomonas aeruginosa* rugose small-colony variants have adaptations that likely promote persistence in the cystic fibrosis lung. J Bacteriol 191(11):3492–3503
Taylor RF, Hodson ME, Pitt TL (1993) Adult cystic fibrosis: association of acute pulmonary exacerbations and increasing severity of lung disease with auxotrophic mutants of *Pseudomonas aeruginosa*. Thorax 48(10):1002–1005
Thomas SR et al (2000) Increased sputum amino acid concentrations and auxotrophy of *Pseudomonas aeruginosa* in severe cystic fibrosis lung disease. Thorax 55(9):795–797
Trunk K et al (2010) Anaerobic adaptation in *Pseudomonas aeruginosa*: definition of the Anr and Dnr regulons. Environ Microbiol 12(6):1719–1733
Tunney MM, Field TR, Moriarty TF, Patrick S, Doering G, Muhlebach MS, Wolfgang MC, Boucher R, Gilpin DF, McDowell A, Elborn JS (2008) Detection of anaerobic bacteria in high numbers in sputum from patients with cystic fibrosis. Am J Respir Crit Care Med 177(9):995–1001
Vallet I, Olson JW, Lory S, Lazdunski A, Filloux A (2001) The chaperone/usher pathways of *Pseudomonas aeruginosa*: identification of fimbrial gene clusters (cup) and their involvement in biofilm formation. Proc Natl Acad Sci U S A 98(12):6911–6916
Van Alst NE et al (2009) Nitrite reductase NirS is required for Type III secretion system expression and virulence in the human monocyte cell line THP-1 by *Pseudomonas aeruginosa*. Infect Immun 77(10):4446–4454
Venturi V (2006) Regulation of quorum sensing in Pseudomonas. FEMS Microbiol Rev 30(2):274–291
Werner E et al (2004) Stratified growth in *Pseudomonas aeruginosa* biofilms. Appl Environ Microbiol 70(10):6188–6196
Whitchurch CB, Tolker-Nielsen T, Ragas PC, Mattick JS (2002) Extracellular DNA required for bacterial biofilm formation. Science 295(5559):1487
Wiehlmann L et al (2007) Population structure of *Pseudomonas aeruginosa*. Proc Natl Acad Sci U S A 104(19):8101–8106
Williams P, Cámara M (2009) Quorum sensing and environmental adaptation in *Pseudomonas aeruginosa*: a tale of regulatory networks and multifunctional signal molecules. Curr Opin Microbiol 12(2):182–191
Winstanley C, Fothergill JL (2009) The role of quorum sensing in chronic cystic fibrosis *Pseudomonas aeruginosa* infections. FEMS Microbiol Lett 290(1):1–9

Worlitzsch D et al (2002) Effects of reduced mucus oxygen concentration in airway Pseudomonas infections of cystic fibrosis patients. J Clin Invest 109(3):317–325

Worlitzsch D, Rintelen C, Böhm K, Wollschläger B, Merkel N, Borneff-Lipp M, Döring G (2009) Antibiotic-resistant obligate anaerobes during exacerbations of cystic fibrosis patients. Clin Microbiol Infect 15(5):454–460

Yang L, Rybtke MT, Jakobsen TH, Hentzer M, Bjarnsholt T, Givskov M, Tolker-Nielsen T (2009) Computer-aided identification of recognized drugs as *Pseudomonas aeruginosa* quorum-sensing inhibitors. Antimicrob Agents Chemother 53(6):2432–2443

Yoon SS et al (2002) *Pseudomonas aeruginosa* anaerobic respiration in biofilms: relationships to cystic fibrosis pathogenesis. Dev Cell 3(4):593–603

Zimmermann A et al (1991) Anaerobic growth and cyanide synthesis of *Pseudomonas aeruginosa* depend on *anr*, a regulatory gene homologous with *fnr* of *Escherichia coli*. Mol Microbiol 5(6):1483–1490

Zumft WG (1997) Cell biology and molecular basis of denitrification. Microbiol Mol Biol Rev 61(4):533–616

Lactobacillus: Host–Microbe Relationships

John O'Callaghan and Paul W. O'Toole

Abstract Lactobacilli are a subdominant component of the human intestinal microbiota that are also found in other body sites, certain foods, and nutrient-rich niches in the free environment. They represent the types of microorganisms that mammalian immune systems have learned not to react to, which is recognized as a potential driving force in the evolution of the human immune system. Co-evolution of lactobacilli and animals provides a rational basis to postulate an association with health benefits. To further complicate a description of their host interactions, lactobacilli may rarely cause opportunistic infections in compromised subjects. In this review, we focus primarily on human–*Lactobacillus* interactions. We overview the microbiological complexity of this extraordinarily diverse genus, we describe where lactobacilli are found in or on humans, what responses their presence elicits, and what microbial interaction and effector molecules have been identified. The rare cases of *Lactobacillus* septicaemia are explained in terms of the host impairment required for such an outcome. We discuss possibilities for exploitation of lactobacilli for therapeutic delivery and mucosal vaccination.

Contents

J. O'Callaghan · P. W. O'Toole (✉)
Department of Microbiology and Alimentary Pharmabiotic Centre,
University College Cork, Cork, Ireland
e-mail: pwotoole@ucc.ie

Current Topics in Microbiology and Immunology (2013) 358: 119–154 119
DOI: 10.1007/82_2011_187

Published Online: 20 November 2011

List Of Abbreviations

aCGH	Array-based comparative genome hybridization
GIT	Gastrointestinal tract
GRAS	Generally Regarded As Safe
IL10	Interleukin 10
LAB	Lactic acid bacteria
MAMP	Microbe associated molecular pattern
PRR	Pattern recognition receptor
TLR	Toll-like receptor

1 *Lactobacillus* Microbiology

1.1 Introduction to Lactobacillus Microbiology and Phylogeny

The phylogenetically diverse genus *Lactobacillus* currently comprises of approximately 125 species (Dellaglio and Felis 2005). In the context of host–microbe interactions, many of the species are significant; however, the ubiquitous nature of the lactobacilli means that there are many possible points of interaction between *Lactobacillus* species and human and other mammalian hosts.

Lactobacilli are used commercially in the production of fermented foods, including dairy products, fermented vegetables and meat products and as such are consumed as part of the normal diet. In addition, the lactobacilli are an important component of the microbiota of humans and other mammals where they are present in the respiratory tract, gastro-intestinal tract (GIT) and genital tracts. Finally, the lactobacilli are present in sewage and decaying plant material.

The presence of members of the genus in such a diverse range of niches has led to the evolution of a number of different groupings of lactobacilli, each adapted to the environmental niche from which it originated. This review will focus primarily on the niches that involve host–microbe interaction, and begins with a consideration of the microbiology, phylogeny and metabolism of the lactobacilli to provide context and rationale for how host interactions have developed.

The genus *Lactobacillus* belongs to the phylum *Firmicutes*, class *Bacilli*, order *Lactobacillales* and family *Lactobacillus*. The *Lactobacillus* family also contains the genera *Paralactobacillus* and *Pediococcus* (Kandler and Weiss 1986). The family most closely related to the lactobacilli is the *Leuconostocaceae* family (Dellaglio and Felis 2005). The lactobacilli are quite diverse at the genetic level with an extremely wide variation in the G + C content of the genome (Schleifer and Ludwig 1995) ranging from 32 to 54%. It is also notable that even within closely related groups of *lactobacillus* species there may be considerable variation in G + C content (Felis and Dellaglio 2007). The large number of species in the genus *Lactobacillus* form a number of different phylogenetic clades. The traditional classification of the lactobacilli was derived from the original classification of the lactic acid bacteria (LAB) by Orla-Jensen (1919) based mainly on fermentation type and morphology, which divided the lactobacilli into three groups based on the end products they produced from the fermentation of carbohydrates: (a) obligately homofermentative, (b) facultatively heterofermentative and (c) obligately heterofermentative. The use of fermentation profiles to classify lactobacilli remained common until relatively recently (Vandamme et al. 1996), and complicated the unequivocal identification of strains associated with humans.

The development of 16s ribosomal RNA sequencing (Woese 1987) was a significant milestone in the area of bacterial phylogeny, and three distinct phylogenetic clusters were identified in lactobacilli (Collins et al. 1991). These groupings did not correspond to the classical phenotypic ones. In recent years a large number of *Lactobacillus* genomes have been sequenced and annotated. This has allowed the relationships of the lactobacilli to be further dissected by phylogenomic analysis. In 2006, we examined the phylogeny of the lactobacilli based on the 5 *Lactobacillus* genome sequences then available (Canchaya et al. 2006). We detected no long-range synteny between the genome sequences, but we identified 583 orthologs common to all 5 species, with a 50–60% identity range between the conserved orthologues. A supertree was constructed from 383 conserved orthologs that divided the genus into four main divisions, *E. faecalis*, *L. sakei*, *L. plantarum*/*L. salivarius* and *L. johnsonii*/*L. acidophilus*. In addition a 16s rRNA tree was constructed for 111 *Lactobacillus* species which identified 5 divisions that were not dissimilar to the groupings from the supertree in that *L. acidophilus* and *L. johnsonii* formed a single group (A) while *L. plantarum* and *L. salivarius* were in two closely related groups (D and E). The *L. sakei* species was one of a number of strains that could not be assigned to a group because potential groupings had either low bootstrap values or fewer than 10 members. A more recent review of *Lactobacillus* phylogenomics from our group (Claesson et al. 2008) concluded that the use of GroEL (a well conserved stress protein) appeared to offer a more robust phylogenetic marker for the classification of the lactobacilli than did the traditional 16s ribosomal RNA markers. These studies established that there were no strong correlations between phylogenetic groupings and the environmental niches occupied by individual species.

It is clear that niche specificity of the lactobacilli may not be predicted with confidence from taxonomic groupings based on the use of traditional taxonomic markers. In contrast if the gene complement of the various *Lactobacillus* sp. genome sequences is analyzed, it becomes obvious that a considerable amount of niche-driven changes in gene complement have occurred. It may be argued that adaptation to a particular environmental niche by specific *Lactobacillus* species has driven a reductive evolution process in which genes that are unnecessary are progressively eliminated (Bolotin et al. 2004; Makarova et al. 2006). The publication of the genome sequences of *L. plantarum* (Kleerebezem et al. 2003), *L. delbrueckii* subsp. *bulgaricus* (van de Guchte et al. 2006), *L. acidophilus* (Altermann et al. 2005), *L. helveticus* (Callanan et al. 2008) and *L. iners* (Macklaim et al. 2011) have confirmed that a substantial degree of reductive evolution has occurred in the lactobacilli and that this appears to be niche driven. Attempts have been made to identify genes that are diagnostic for the occupation by lactobacilli of environmental niches or body sites, and from 12 genomes, 4 genes related to lactobacilli associated with meat were identified (Claesson et al. 2008). Recently when the gene complements of 11 fully sequenced lactobacilli were analyzed, 9 niche-specific genes were identified, six related to dairy origin and three related to intestinal origin (O'Sullivan et al. 2009).

In the most recent analysis of *Lactobacillus* phylogenomics (Zhang et al. 2011), a division of the LAB into 3 groupings emerged; group 2 included the lactobacilli, pediococci, leuconostocs and oenococci and comprised 2 distinct clades; one containing the *L. acidophilus* complex together with *L. sakei* and *L. casei*, and the other the *L. salivarius* subgroup. As mentioned already, the GIT-associated LAB are not confined to a single clade, as *L. casei, L. gasseri*, *L. johnsonii* and *L. acidophilus* are in the *L. acidophilus* complex and *L. reuterii*, *L. plantarum* and *L. salivarius* belong to the *L. salivarius* group.

Of considerable significance for studying *Lactobacillus*–host interaction is the recent realization that there a significant level of genetic heterogeneity within individual *Lactobacillus* species. A group of 20 strains of *L. plantarum* from different sources were analyzed by array-based comparative genome hybridization (aCGH) and two distinct groups of strains were identified (Molenaar et al. 2005). *L. reuteri* strains isolated from six host species (human, mouse, rat, pig, chicken and turkey) when analyzed by AFLP and MLST, separated into clades that reflected the origin of the isolates (Oh et al. 2010). Our recent analysis of the species *L. salivarius* by aCGH revealed the highest level of genomic diversity seen to date in lactobacilli, within only recent human isolates forming a discrete grouping (Raftis et al. 2011). Thus it is generally not possible to extrapolate phenotypic properties from one strain to another within *L. salivarius*, and presumably other species as well.

1.2 The *Lactobacillus acidophilus* Complex

One of the largest discrete divisions of the lactobacilli that also includes many strains with proven host interaction properties is the *L. acidophilus* complex that includes *L. acidophilus*, *L. helveticus*, *L. gasseri*, *L. johnsonii* and the type strain of the genus *L. delbrueckii*. This clade also contains many species used as probiotics (see below). The closest relatives outside of the complex are *L. sakei* and *L. casei* (Zhang et al. 2011). The initial isolation of *L. acidophilus* in 1900 was from infant feces and the species was included in the genus *Lactobacillus* from 1929. The use of DNA–DNA hybridization led to the identification of six DNA-homology groups in the species that led to its division into *L. acidophilus*, *L. amylovorus*, *L. crispatus*, *L. gallinarum*, *L. gasseri and L. johnsonii* (Fujisawa et al. 1992). The 16s rRNA sequence studies by Collins et al. (1991) identified three groups, one of which was the *L. delbrueckii* group that included all of the 6 species delineated from the original *L. acidophilus* group, including *L. helveticus*, *L. delbreukii* and some other species. This group was subsequently renamed the *L. acidophilus* group as *L. delbrueckii* was not considered representative of the entire group (Felis and Dellaglio 2007). It is noteworthy that the extreme range of G + C values mentioned earlier also applies to the *L. acidophilus* group ranging from 33% in *L. gasseri* to 51% in *L. delbrueckii* (Berger et al. 2007).

The overall relationships within the *L. acidophilus* group established by Berger et al. (2007) are similar to those we established using rRNA gene trees (Canchaya et al. 2006), and by others using concatenated ribosomal protein trees (Makarova et al. 2006). Thus *L. delbrueckii* diverged earliest while the other strains diverged later to form the *L. acidophilus*/*L. helveticus* group and the *L. gasseri*/*L. johnsonii* groups. In addition to these species, the vaginal isolate *Lactobacillus iners* also belongs to the *L. acidophilus* complex (Macklaim et al. 2011) where its closest relatives are *L. gasseri* and *L. johnsonii*. In terms of host interactions therefore
the *L. acidophilus* complex contains three known vaginal commensals (*L. gasseri*, *L. crispatus* and *L. iners*), and three gut commensals (*L. gasseri*, *L. acidophilus* and *L. johnsonii*).

1.3 The *L. salivarius* Clade

The *L. salivarius* clade occupies a peripheral position in the *Lactobacillus* phylogenetic tree, and contains a number of well-characterized species that interact with mammalian hosts (Neville and O'Toole 2010). The closest relatives outside the clade are *L. plantarum* and *L. brevis* (Zhang et al. 2011). The clade currently comprises 25 species of which *L. salivarius* is the best characterized and hence gives its name to the clade. The species in the clade have been isolated from a wide range of environmental niches including vertebrates, the environment, alcohol fermentations and food fermentations (Neville and O'Toole 2010). There seems to be little overlap between strains from different sources, with only *L. agilis* being common to both the environmental and intestinal groups of isolates. However, as the environmental source of *L. agilis* was sewage, the classification of this species as an environmental isolate may be considered tentative. *L. salivarius* was reported in 1953 (Rogosa et al. 1953) and was initially considered to contain two subspecies *salicinus* and *salivarius*, distinguished by the ability to ferment rhamnose and salicin, respectively. However, a more recent polyphasic analysis has led to the discontinuation of the subspecies classification as there was no concordance between the phylogeny and sugar fermentation status (Li et al. 2006). The majority of the species in this clade were reported within the last 10 years, with only *L. salivarius* (Rogosa et al. 1953), *L. mali* (Carr and Davies 1970), *L. ruminis* (Sharpe et al. 1973), *L. agilis* (Weiss et al. 1981), *L. animalis* (Dent and Williams 1982) and *L. aviarus* (Fujisawa et al. 1984) being reported prior to the year 2000. The phylogenetic robustness of the group together with the presence of two groups of strains that represent the vertebrate and environmental niches suggests that study of the *L. salivarius* clade will provide important insights into the differences between environmental and commensal species.

There is a strong correlation between phylogenetic groupings in the *L. salivarius* clade and the source from which the species was isolated. In the most comprehensive recent analysis (Chen et al. 2010), in which 23 members of the *L. salivarius* clade were included, *Lactobacillus ruminus*, *L. agilis* and *L. equi* diverged early to form a distinct group. All of these strains are vertebrate isolates and *L. ruminis* and *L. agilis* are motile. The remaining 20 strains were divided between 2 groups, the first of which contained *L. salivarius*, *L. acidipiscis*, *L. pobuzihi*, *L. aviarius*, *L. hayakitensis*, *L. saerimneri* and *L. ceti*. All of these except *L. pobuzhi* (fermented grain) and *L. acidipiscis* (fermented fish) are vertebrate/mammalian isolates, and none are motile. Both environmental species appeared to diverge early from the mammalian isolates. The second grouping contained 2 distinct subgroups (a) *L. vini*, *L. nagelii*, *L. ghanensis*, *L. satsumensis*, *L. oeni*, *L. uvarum*, *L. capillatus*, *L. cacaonum*, *L. mali* and *L. hordei*, all of which were isolated from environmental locations or from food fermentations and of which all but *L. hordei* and *L. cacaonum* are motile and (b) *L. apodemi*, *L. murinis* and *L. animalis* all of which are mammalian isolates and none of which are motile. The *L. salivarius* clade thus contains phylogenetic groups based on 16s rRNA sequences that correspond very well with the isolation sources of the species. Three of these groupings are associated with vertebrate/mammalian niches and it is particularly noteworthy that motility is far more common in those species isolated from environmental/fermentation niches. As the flagellin component of bacterial flagella is a potent inducer of an inflammatory immune response in eukaryotes (Tallant et al. 2004), it is tempting to speculate that the preponderance of non-motile strains among the vertebrate isolates is because they are relatively nonstimulatory to the host immune system. It is important to bear in mind, however, that while flagellin is strongly immunogenic it is not in itself inimical, indeed the induction of host defensins by the probiotic Nissle 1917 strain of *Escherichia coli* is induced by flagellin

(Schlee et al. 2007). It is quite possible that motile members of the *L. salivarius* clade may be particularly effective probiotic strains through flagellin-mediated interactions with the host immune system, in particular as the isolation of such strains from food fermentations would support the view that they would be considered to be GRAS bacteria. We have recently completed a comparative genomic analysis of *L. ruminis* (Forde et al., submitted), which has facilitated analysis of the genetic basis for carbohydrate utilization in this species (O'Donnell et al., submitted) and for motility in the *L. salivarius* clade (Neville et al., submitted).

1.4 *Lactobacillus* Metabolism in the Context of Host Adaptation

Lactobacilli ferment carbohydrates to produce lactic acid. In the case of the homofermentative and obligately heterofermentative species, hexose sugars are fermented by the Embden–Mayerhof–Parnas pathway (EMP) to lactic acid. Homofermentative species lack the phosphoketolase enzyme necessary for the fermentation of pentoses, while the facultatively heterfermentative strains possess this enzyme and may also produce an aldolase that allows them to utilize gluconate. The heterofermentative species do not use the EMP pathway (Hammes and Vogel 1995).

The main evolutionary changes that have occurred during adapation to the intestine (and other mammalian niches) is that strains and species that originate in the intestine have the ability to utilize a wider range of carbohydrates (Ventura et al. 2009). These include oligosaccharides that are released during digestion of starch and cellulose in the mammalian intestine. In contrast, strains that belong to the dairy groupings where lactose is the main sugar tend to have a reduced carbohydrate metabolism capability. For example, *L. delbreuckii* subsp. *bulgaricus*, long used as a dairy starter, possesses only about 50% of the carbohydrate metabolism and transport genes as closely related species (Bolotin et al. 2004). Strains that are derived from fermented meat tend to have a more specialized carbohydrate metabolism capability that allows them to metabolize nucleosides (Chaillou et al. 2005), while the 'environmental' isolates also have a wide-ranging carbohydrate utilization repertoire (Ventura et al. 2009). A more recent report of niche-driven adaptation of carbohydrate metabolism concerns the *L. iners* species (Macklaim et al. 2011) which is the dominant *Lactobacillus* species in the vaginal microbiota (see next section). The major source of carbohydrate in the vagina is mucus which *L. iners* can utilize as a carbohydrate source (Macklaim et al. 2011). Since mucin levels are constantly replenished in the vaginal niche, there is likely to be an adequate carbohydrate supply for the maintenance of *L. iners* (Macklaim et al. 2011).

Relating to the point noted above about intraspecies genomic diversity, there may also be considerable intraspecies variation in the ability to utilize carbohydrates. For example, when 39 *L. casei* strains originating from cheese, plant, GI tract and blood were examined for the ability to ferment a range of carbohydrates, some niche-specific patterns were observed (Cai et al. 2007). Few of the plant-derived strains were capable of lactose metabolism (22%), while the majority of strains from cheese (83%) and the GIT (71%) could ferment lactose. The GIT (83%) and plant (67%) strains were more likely to ferment inulin than were the cheese strains (27%); fermentation of starch was confined to strains of plant origin only.

The second evolutionary effect on metabolism is that the ability to synthesize amino acids de novo tends to be lost in the dairy and multi-niche strains while the environmental isolates have a wider range of biosynthetic capabilities (Cai et al. 2009). In the dairy environment, the milk protein casein is the main amino acid source, and following hydrolysis, the resulting peptides are transported into the cells where further hydrolysis to free amino acids occurs (Kunji et al. 1996). In this environment, the ability to synthesize amino acids de novo is not advantageous to the cell. However, there is an obvious requirement for proteolytic enzymes, peptide transport systems and peptidases. In the GIT-associated probiotic strain *L. johnsonii* NC533, a similar level of amino acid auxotrophy was observed and the number of peptidases (>25) was similar to what would be expected in a dairy strain (Pridmore et al. 2004).

To conclude this section, we suggest that the metabolic capabilities and the complement of biosynthetic genes are often a better indication of the environmental niche of a particular *Lactobacillus* than purely phylogenetic data because changes to the gene complement of a strain or species occur much more rapidly than do changes to 16s rRNA sequences and other phylogenetic markers in an evolutionary timescale. This is exemplified by the study of *L. casei* cheese subgroups (Cai et al. 2009) which showed that the group A organisms had undergone substantial gene decay to the extent that they could be considered dairy specialists while the group B strains were less decayed and could be considered generalists. It is considered that the main *L. casei* clusters diverged approximately 1.5 million years ago, while the divergence of the dairy strains occurred approximately 10,000 years ago, a timescale that coincides with the commencement of dairying and thus exposure to the dairy niche. High levels of genome decay are not confined to the dairy niche as is demonstrated by the level of gene loss in the vaginal species *L. iners* compared to other species (Macklaim et al. 2011). The methodology of phylogenetic analysis will itself have to 'evolve' if it is to address the challenges posed by the ever-increasing numbers of new species and to utilize the availability of genome sequence data.

2 General Biology of *Lactobacillus*–Host Interaction

2.1 Species Range and Niches of Human-Associated Lactobacilli

Since one of the defining microbiological properties of the lactobacilli is their nutritional fastidiousness (Hammes and Vogel 1995), it follows that they are naturally found in human body sites where nutritional substrates are readily

Table 1 *Lactobacillus* species found in various human body sites

Species	Oral cavity	Stomach	Feces	Vagina
L. acidophilus	+		+	
L. antri		+		
L. brevis			+	
L. casei			+	+
L. buchneri	+			
L. crispatus	+		+	+
L. curvatus			+	
L. delbrueckii		+	+	
L. fermentum		+	+	
L. gallinarium				+
L. gasseri		+	+	
L. gastricus		+		+
L. insers				+
L. jensenii				+
L. johnsonii	+		+	
L. kalixensis		+		
L. paracasei			+	+
L. plantarum				
L. rhamnosus			+	
L. reuteri		+	+	+
L. ruminis		+	+	
L. salivarius	+	+	+	
L. sakei			+	
L. ultunensis		+		
L. vaginalis		+		+

Data compiled from references cited in main text

available. These include the vagina, oral cavity and gastrointestinal tract (Walter 2005). The range of species of lactobacilli reportedly found in or on various human body sites are listed in Table 1.

Lactobacilli represent the dominant bacterial genus that colonizes the healthy vagina of women of reproductive age [reviewed in ref. (Witkin et al. 2007)], where they are thought to promote vaginal health by modulating the environment and other bacteria through production of lactic acid, bacteriocins and by competitive exclusion. The major species present are variably reported as *L. crispatus* and *L. iners* (Zhou et al. 2004; Fredricks et al. 2005) or *L. crispatus* and *L. gasseri* (Verhelst et al. 2004). The additional species *L. jensenii*, *L. gallinarum* and *L. vaginalis* may also be present in many subjects (Pavlova et al. 2002). Although vaginosis may be accompanied by an altered microbiota including absence of lactobacilli, presence of lactobacilli *per se* is not sufficient or strictly required for prevention of vaginosis (Witkin et al. 2007). A pyrosequencing-based microbiota profiling approach recently showed that lactobacilli were approximately ten times less abundant in subjects with bacterial vaginosis than in healthy controls (Ling et al. 2010a, b). Ravel et al. (2011) numerically substantiated the interesting observation that the vagina is the only anatomical site where lactobacilli predominate, constituting >50% of sequence reads in 73% of subjects, and emphasizing the unique aspects of *Lactobacillus* ecology that may be studied in this niche.

Lactobacilli may be isolated from almost all sites in the gastrointestinal tract, starting from the oral cavity through to the rectum (Walter 2005). The numbers in the oral cavity (saliva and mucosae) and small intestine are low, typically 10^3 per ml (Walter 2005). Recent culture-independent metagenomic surveys of the oral cavity revealed around 700 species to be present (Parahitiyawa et al. 2010), of which lactobacilli represent a small proportion. Oral lactobacilli have been linked to caries causation especially in children (see Sect. 5). Lactobacilli are also present in the stomach (Bik et al. 2006), representing approximately 0.2% of the culture-independent gastric microbiota in those subjects. We and others have successfully cultured lactobacilli from the stomach (Roos et al. 2005; Ryan et al. 2008a), in the search for probiotic strains with the ability to suppress *Helicobacter pylori* infection (Ryan et al. 2008b). Although it is relatively easily to successfully culture lactobacilli from the feces of the majority of human subjects [estimated at over 70% in analysis of 62 human subjects; (Finegold et al. 1983)], the proportion of the culturable microbiota that lactobacilli comprise is typically low, from 10^6 to 10^8 organisms per gram in feces, or from 0.01 to 1% (Mitsuoka 1978); also reviewed in Walter 2005). This runs counter to a common misconception that lactobacilli are dominant or abundant members of the intestinal microbiota, possibly because they are relatively easy to culture, and many species are aerotolerant. Molecular methods including in situ hybridization confirmed the relatively low proportions of lactobacilli in human feces (reviewed in Walter 2008). In several pioneering 16S gene clone library studies generating relatively low numbers of sequence reads, lactobacilli were not recorded at all, or only in low numbers (Hayashi et al. 2002; Eckburg et al. 2005). In later studies with more subjects or deeper sequencing, reads assigned to the genus *Lactobacillus* were recorded at 5% (Frank et al. 2007) and 12.9% (Hayashi et al. 2005). However, in the most comprehensive metagenomics survey performed to date, of 124 European subjects, but where genus proportions were not provided, lactobacilli did not feature among the most abundant 155 species that were present in at least one individual at greater than 1% genome sequence coverage (Qin et al. 2010). It is thus clear that lactobacilli are not dominant members of the human distal intestinal microbiota, accepting the caveat that feces is an acceptable surrogate for luminal content.

The *Lactobacillales* constitutes a subdominant order on the human skin. Data from the NIH Intramural Sequencing Center showed that they range from 0.69% to 1.37% of overall reads, aggregated across 20 skin sites sampled in each of 10 individuals (Grice et al. 2009; our analysis of data in Supplementary Online Material). The data of Grice et al. do not give taxonomic assignment below Order level, and obviously these data will include many genera other than *Lactobacillus*. Blaser and colleagues (Gao et al. 2008) assigned 16 out of 3,963 Operational Taxonomic Units (OTUs) from skin to the *Lactobacilli* (0.4% of assigned reads). The species identified were *L. crispatus*, *L. delbrueckii* subsp. *bulgaricus*, *L. jensenii* and an unidentified *Lactobacillus* sp. (Gao et al. 2008). Lactobacilli would be expected to be transiently present on human skin, especially in the perineum area and hands, and such incidental carriage may be important for their dissemination (see also below).

The *Lactobacillus* species and associated body sites listed in Table 1 have been identified primarily by culture-based studies, and one would have expected major revision or amplification of such datasets by culture-independent methods. However, although many high-throughput phylogenetic profiling analyses of the intestinal microbiota have recently been performed, a major caveat is that only partial 16S rRNA gene sequences are usually generated from next-generation sequencing technologies, typically of a single- or two-tandem variable regions, which do not allow assignment below genus level, and not even to genus level in all cases (Claesson et al. 2009). Earlier clone-library sequencing of full-length 16S genes usually could not achieve deep enough sequence coverage to sample the lactobacilli. Reductions in next-generation sequencing run costs, longer read lengths and new technologies, will inevitably make it economically feasible to apply deep shot-gun sequencing, as performed recently on faecal samples (Qin et al. 2010), to samples from other human and animal body sites.

2.2 Transients Versus Residents

The *Lactobacillus* species that colonize the vagina have been identified with reproducibility and confidence, but there is a degree of ambiguity about which lactobacilli are long-term residents of the human intestine, or are simply passing through when detected e.g. in feces. Since some *Lactobacillus* species have historically been used for food preservation or flavor enhancement, and more recently added as functional ingredients e.g. to promote health, these species have probably become more common, and may be incorrectly considered as gut commensals as a consequence. Since many probiotic cultures have been selected based upon robustness in products, survival of gastric transit and human epithelial cell adherence, it is not surprising that consumers of fermented or functional food products may become colonized or at least transient shedders of such cultures (Berg 1996), for which there is ample evidence from surveys and trials (Tannock et al. 2000; Saxelin et al. 2010). Application of highly sensitive molecular methods may actually serve only to further confuse the issue, since low numbers of food-borne lactobacilli or incidental environmental strains will be recorded. Further complications arise from consideration of studies such as ones which showed that some adults who had consumed *L. rhamnosus* GG were apparently colonized in the colon but not at sufficient levels to detect cells in feces (Alander et al. 1999).

A helpful term in this discussion is an "autochthonous" microbe—formed where found, as distinct from an "allochthonous" bacterium that is passing through and does not sustain a replicating population in the sampled locale (Savage 1977). It is widely agreed that many of the lactobacilli cultured from or detected in the gastrointestinal tract (Table 1) are allochthonous, including

L. acidophilus, *L. plantarum*, *L. rhamnosus*, *L. casei* and *L. paracasei*, and do not form stable long-term communities in the human gut. Studies of *Lactobacillus* communities in animals are also helpfully illustrative of the ecological complexities that may pertain (Tannock 2004), such as the colonization of older broiler chicken crops by *L. salivarius* which only occurs as a succession event following other lactobacilli (Guan et al. 2003). Species such as *L. salivarius* (Knarreborg et al. 2002), and *L. reuteri* (Tannock et al. 2005) form biofilms in the chicken crop and murine forestomach, respectively. Shedding of cells from such biofilms into the distal intestine is likely to have ecological consequences not only in term of competition (Tannock 2004), but also complicates determining what lactobacilli are autochthonous in the distal bowel.

One reason why distinguishing between autochthonous and allochthonous lactobacilli in the human intestine is important is because strains for use as probiotic cultures in humans should ideally be of human origin (Dunne et al. 2001). Our group in University College Cork has devoted much research effort to *L. salivarius* (summarized in Neville and O'Toole 2010), but we are also exploiting metagenomic and phylogenetic profiling of the intestinal microbiota to identify other lactobacilli with autochthonous host associations. For example, Gill and colleagues deposited 2,062 near full-length 16S rRNA gene sequences in GenBank, derived from the faecal metagenome of two individuals (Gill et al. 2006). Our unpublished analysis of these sequences shows that, although 99 sequences had identity values greater than 90% to *Lactobacillus* sp. 16S gene sequences, only 21 of these were greater than 99% identity (which indicates a high-confidence species identification). Among these 21, three were *L. casei/paracasei*; two were *L. plantarum* and the remaining sixteen were 99–100% identical to the *L. ruminis* 16S rRNA gene sequence. Coupled with the meta-analyses of Reuter which consistently identified *L. ruminis* in human faecal samples (Reuter 2001), this finding underlines the likelihood that *L. ruminis* is a member of the autochthonous *Firmicutes* microbiota in human subjects, though its location in the human GI tract is not currently known.

2.3 *Lactobacilli* as Probiotics

Probiotics are defined as "live microorganisms that, when administered in adequate amounts, confer a health benefit on the host" (FAO/WHO 2001). Some of the probiotic applications of lactobacilli are listed in Table 2. A comprehensive description of probiotic microbes including lactobacilli, and their application, is outside the scope of this review, but excellent recent reviews are available (Rastall et al. 2005; O'Toole and Cooney 2008; Kleerebezem and Vaughan 2009; Rijkers et al. 2010). Many different microorganisms have been used as probiotics, with *Lactobacillus* sp., *Bifidobacterium* sp., *Enterococcus* sp., *E. coli*, *Propionibacterium* sp., *Leuconostoc*, *Lactococcus*, *Pediococcus*, *Streptococcus*, *Saccharomyces cerevisiae* and *S. boulardii* being among the organisms most frequently used (Dunne et al. 1999). A diverse range of benefits have been attributed to probiotic

Table 2 Examples of probiotic properties attributed to *Lactobacillus* sp

Organism	Benefit attributed	Comments	References*
L. acidophilus NCFM	Relief of lactose intolerance	Linked to betagalactosidase production	(Sanders and Klaenhammer 2001)
L. casei Shirota	Modulation of innate immune response	Effector unknown	(Spanhaak et al. 1998)
L. casei Immunitass	Reduction in antibiotic-associated diarrhea	Linked to suppression of *Clostridium difficile*	(Hickson et al. 2007)
L. fermentum RC-14	Reduction in vaginal yeast and coliforms	In combination with *L. rhamnosus* GR-1	(Reid et al. 2003)
L. gasseri SBT2055	Reduction in obesity	Reduced adiposity in DBRPCT[a]	(Kadooka et al. 2010)
L. johnsonii *La1*	Reduced gastritis in 50 subjects with *H. pylori*	Mechanism unknown	(Felley et al. 2001)
L. plantarum	Alleviation of birch pollen allergy	Strain specific effect; pre-screen for IL10 induction	(Snel et al. 2010)
L. reuteri	Alleviation of colic in infants	Mechanism unknown	(Savino et al. 2010)
L. rhamnosus GG	Reduced severity of pediatric diarrhea	Competitive exclusion?	(Isolauri et al. 1991)
L. salivarius UCC118	Attenuation of mild-moderate colitis	Indirect report of EU research project outcomes	(Mattila-Sandholm et al. 1999)

* Representative property/reference only; many others typically exist for these and other strains
[a] Double-blind, randomized, placebo-controlled trial

consumption, many of which are not confined to the gut, and which enjoy greater or lesser degrees of rigorous scientific proof, from strong to nonexistent (Rijkers et al. 2010). The emphasis of Sect. 3 will be to describe *Lactobacillus*–host interactions where the interaction and outcome have been characterized at a molecular level. In this section, we will briefly describe why lactobacilli have attracted so much interest as probiotics, and we will advance some integrative hypotheses for why such interactions might be beneficial.

As explained above, lactobacilli are not dominant organisms in the intestine, so the fact that they are so widely used as probiotics may seem counter-intuitive. The Nobel laureate Eli Metchnikoff had concluded that aging was accelerated by toxic products of microbial origin in the large bowel (Metchnikoff 1907), and he had linked the increased longevity of some Balkan residents to consumption of fermented milk (Metchnikoff 1908). The "Bulgarian bacillus" was proposed to elaborate products that prevented the proliferation of the (postulated) putrefactive bacteria that accelerated aging, and consumption of fermented foodstuffs was more attractive than Metchnikoff's other idea of surgical removal of the large bowel. Lactobacilli are used as starters or secondary fermenters in a wide range of food products, most of which have traditional roots in food preservation, and some of which are retained as household or artisanal food fermentation agents throughout the world (Stiles 1996). The fact that lactobacilli had been consumed safely for centuries made their "generally regarded as safe" status (GRAS) easy to recognize (Bernardeau et al. 2006). Coupled with their historical record for health promotion, lactobacilli thus became natural targets for commercial exploitation as probiotic ingredients in functional foods, particularly in recent decades (Vaughan et al. 2002).

The investigation of the mechanisms of action of probiotics in general, and lactobacilli in particular, has entered a new phase with the recent realization of the complexity and biological significance of the intestinal microbiota (reviewed in Bik 2009; Sekirov et al. 2010). This was aided to a large degree by development of methods for profiling the composition and function of complex bacterial communities in the environment that were the adapted for intestinal samples (Zoetendal et al. 2008; O'Toole and Claesson 2010). The intestinal microbiota plays essential roles in health, aiding metabolism of dietary polysaccharides (Xu et al. 2003), metabolising nutrients [particularly lipids and fibre (Fava et al. 2006)], regulating energy balance (Ley et al. 2006; Duncan et al. 2008), producing vitamins (Hill 1997) and both priming and regulating the immune reaction to self- and non-self antigens (Round and Mazmanian 2009; Atarashi et al. 2011). The beneficial effects of probiotics (and the difficulty in elucidating their mechanisms, especially in reductionist models) could be due to indirect action, mediated by effects upon the microbiota (reviewed in O'Toole and Cooney 2008). Thus, administration of probiotic lactobacilli, or natural variations of lactobacilli in the intestine, could affect the composition and function of the microbiota because of competition for nutritional substrates in the diet, conversion of substrates into compounds that alter the physical environment such as pH or by direct antagonism through production of antimicrobial compounds such as bacteriocins (O'Toole and Cooney 2008). In the case of lactobacilli, effects upon immune function could be due to direct interaction with immune cells, such as dendritic cells, as illustrated for *L. plantarum* (Meijerink et al. 2010) and several other *Lactobacillus* species (Mohamadzadeh et al. 2005).

Loss from the microbiota of commensal bacteria involved in tolerance and regulatory T cell function has been advanced as an explanation for the recent surge in the incidence of diseases related to immune dysregulation in developed countries (Guarner et al. 2006). A key driver in this development would be increased sanitation—the so-called "hygiene hypothesis revisited" (Guarner et al. 2006). Alterations in the diet of citizens of Western industrialized countries in particular, with higher glycaemic index and lower fibre, are also plausibly responsible for alterations in the composition and function of the microbiota (Maslowski and Mackay 2011), so that a combination of sanitation and diet changes might explain the increase in diseases related to immune dysfunction. Clearly other organisms such as *Bacteroidetes* sp. (Mazmanian et al. 2008), *Faecalibacterium prausnitzii* (Sokol et al. 2008) and the *Clostridia* (Atarashi et al. 2011) are also likely to be major agents, but one can rationalize the probiotic properties of lactobacilli as having developed as part of a broader spectrum of mutualistic associations between host and intestinal bacteria.

Consistent with their hypothetical role as components of the intestinal microbiota that have co-evolved with the mammalian intestine and immune system, lactobacilli have a range of known or proposed interactions with the innate and adaptive immune response (reviewed in Lebeer et al. 2008). In addition to anti-inflammatory effects which improve barrier function by remodelling tight junctions (Miyauchi et al. 2009), lactobacilli have been reported to interact with or to regulate the activity of epithelial cells (O'Hara et al. 2006), macrophages (Miettinen et al. 1998), dendritic cells (Christensen et al. 2002); see also above), and regulatory T cells (von der Weid et al. 2001). The broad consequences of these interactions with gut-associated lymphoid tissue amount to potentially profound modulating influences on immune response, for which there are varying levels of proof in humans (Lebeer et al. 2008). Indeed well-designed human ingestion trials (see Sect. 4) are critical for extrapolating from cell culture systems and animal models.

In concluding this somewhat speculative perspective on the evolution of beneficial interactions between lactobacilli and host, the *L. reuteri* paradigm uncovered by Walter and colleagues may be illustrative. It was demonstrated that monophyletic clades of *L. reuteri* are associated with distinct mammalian host lineages, and competition experiments suggest that this was an adaptive process that enhanced competition and survival in that host (Oh et al. 2010). The authors decry the fact that most probiotic strains are not selected in the light of evolutionary criteria, and they argue cogently that the *L. reuteri*–mammal association can be used as a model to identify essential host–microbe interaction mechanisms (Walter et al. 2011). This is undoubtedly true, but the lack of a diverse range of verifiable host-species tropisms for most other *Lactobacillus* species makes it hard to extend this approach to the many other interesting probiotic mechanisms in those species.

3 Molecular Characterization of Host Interaction Factors

3.1 Genetic and Genomic Analysis of Lactobacillus–Host Interaction Macromolecules

As noted in Sect. 1, the genome sequence has been determined for representatives of many *Lactobacillus* species, and there are multiple genome sequences available for some species (see the GOLD database at www.genomesonline.org for current information). At the time of writing, there were 27 completed *Lactobacillus* genomes from 16 species on the GOLD database, with a further 55 draft *Lactobacillus* genomes from 23 species emanating from the Human Microbiome Project (www.hmpdacc.org). Comparative genomics has provided powerful insights into the biology of commensal lactobacilli (Klaenhammer et al. 2005; Makarova and Koonin 2007; Kleerebezem and Vaughan 2009; Ventura et al. 2009; Kleerebezem et al. 2010). As well as providing superior datasets for defining phylogenetic divisions in this very diverse genus (Canchaya et al. 2006; Claesson et al. 2007; Claesson et al. 2008), comparative genomics has also defined core genes for particular species and niches (O'Sullivan et al. 2009). The current scale of *Lactobacillus* genome sequence availability presents temporary practical challenges such as automated genome annotation quality and computational resources for bioinformatics, but will ultimately facilitate definitive genome-wide analysis of the genus *Lactobacillus*.

It is generally agreed that surface structures of bacteria are the primary players in initial host interaction, and the *Lactobacillus* secretome has been the study of several trans-species studies. Siezen and colleagues established a carefully curated database of predicted extracellular proteins in 26 lactic acid bacteria that included 15 *Lactobacillus* genomes from 12 different species (Zhou et al. 2010). The *Lactobacillales*-specific clusters of orthologous groups of proteins (LaCOGs) that had been previously established (Makarova et al. 2006) were extended, and pseudogenes and false annotations were identified. A striking finding was that the function was not known for some 31% of the LAB secretome analyzed in this study (Zhou et al. 2010), while only 4% of the secretome was devoted to binding protein function. Enzyme function, transport and cell wall turn-over were the largest LaCOG functional classes, representing about 15% each. Drawing upon previous approaches (Boekhorst et al. 2006), Zhou and colleagues expanded their definition of mucin-binding proteins into seven families, and also categorized the predicted mannose-specific adhesins, collagen-binding proteins, chitin-binding proteins and unclassified adhesins. The partly overlapping study by Kleerebezem et al. (2010) extended the comparative analysis to the genes for the important surface components peptidoglycan, teichoic acid and exopolysaccharide.

Teichoic acid in an anionic polymer that is not produced by all lactobacilli (Kandler and Weiss 1986), which may now be rationalized upon the basis of the variable presence of the respective biosynthesis genes in the genome (e.g. lacking in *L. casei*, *L. reuteri* and *L. fermentum*; see (Kleerebezem et al. 2010) for a genomic overview of 12 lactobacilli). Similarly to peptidoglycan-derived peptides (Asong et al. 2009), teichoic acid from lactobacilli (and other gram-positive bacteria) is recognized by host toll-like receptor 2 (TLR2) (Matsuguchi et al. 2003), which may contribute to a broad range of immunomodulatory phenotypes, exemplified by enhancing antiviral activity of dendritic cells (Weiss et al. 2010). Teichoic acid, at least in some lactobacilli, also contributes synergistically to induction of interleukin 10 (IL10) (Kaji et al. 2010). The lipid-anchored form of teichoic acid, lipoteichoic acid or LTA, is D-alaninylated by the product of the *dltA* gene, and ablation of this gene rendered *L. rhamnosus* LGG more sensitive to human gastric juice and beta-defensin (Perea Velez et al. 2007). However, in addition to this structural/defensive role for LTA, D-alanine substitution is also important for interaction with immune cells, since a *dltA* mutant of *L. plantarum* was less proinflammatory than the wild type, and more efficacious in ameliorating murine colitis (Grangette et al. 2005). D-alanine substitution of LTA is also required for efficient biofilm formation and fore-stomach colonization by *L. reuteri* (Walter et al. 2007), which may be a stress-resistance-related phenotype judging from the in vivo morphology of the mutant cells. Notwithstanding these fragmented insights, the overall contribution of teichoic acids to *Lactobacillus* biology and host interaction is poorly understood, and warrants much greater research attention.

The ability to produce exopolysaccharide (EPS) is widespread in lactobacilli, and EPS production by lactobacilli is a desirable thickening ingredient in some foods (van Kranenburg et al. 1999). EPS may be bound by the cell or released into the medium, in proportions that vary by strain and carbon source, and in some studies, the cell bound form is referred to as capsular polysaccharide. EPS of commensal bacteria attracted much greater interest when it was shown that a zwitterionic polysaccharide of *Bacteroides fragilis* could down-regulate inflammatory responses in the gut (Mazmanian et al. 2005, 2008). The mechanism for this was recently shown to involve increased development of Foxp3+ T-cells, such that *B. fragilis* effectively co-opts the Treg lineage differentiation pathway in the gut toward tolerance (Round and Mazmanian 2010). EPS produced by lactobacilli appears primarily to mask potential host interaction mechanisms. A mutant of *L. johnsonii* deficient in EPS production had a slightly increased persistence time in orally fed mice compared to the wildtype (Denou et al. 2008). Attachment of some strains of *L. acidophilus* to immobilized type I collagen and fibronectin, abundant proteins in intestinal basement membrane, was also reduced when capsule was produced (Lorca et al. 2002). Exposure of the adhesive pili of *L. rhamnosus* GG (see below) was enhanced upon removal of EPS (Kankainen et al. 2009). Biofilm formation is a putative colonization mechanism, and we have recently shown that the ability of *L. salivarius* strains to form biofilms is also inversely related to the levels of EPS produced (Raftis et al. 2011). There are some reports of reduced biofilm production in EPS-deficient strains, or adhesion activity associated with EPS (reviewed in Lebeer et al. 2008).

EPS produced by lactobacilli is increasingly being recognized as being capable of interacting with innate immune cells. Undefined high molecular weight polysaccharide from *L. casei* Shirota suppressed the inflammatory response (to LPS) in murine macrophage cell lines and spleen cells (Yasuda et al. 2008). A positive role for EPS production in immunomodulation has also recently been described in *L. reuteri*. Strain 100-23 produces a levan-containing EPS, lack of which reduces the proportion of Foxp3+ T-cells in mouse spleen compared to the wildtype (Sims et al. 2011). EPS produced by lactobacilli is generally under-characterized, however, and the role of this macromolecule in the dialog between lactobacilli and the host also deserves greater scrutiny.

3.2 Surface Proteins

Lactobacillus rhamnosus strain GG (often called LGG) is employed as a probiotic ingredient in many functional foods, and despite having been the subject of more than 500 published studies, it was only recently subjected to genome sequencing (Kankainen et al. 2009). Comparison of the LGG genome to that of strain LC705, a starter adjunct strain with persistence ability and mucin-adhesion activity that are both inferior to those of LGG, identified a locus encoding sortase-dependent pili. Experiments with a gene knock-out and anti-pilin antibody confirmed that the pilin subunit bound to immobilized human mucins (Kankainen et al. 2009). A second pilin locus in the LGG genome encodes another pilin whose recombinant form also bound mucins, and which may contribute to the adhesive properties of LGG, or to competitive exclusion of related adhesins on pathogens (von Ossowski et al. 2010). Sortase-dependent pilin genes were also annotated in the *L. johnsonii* NCC533 genome (Pridmore et al. 2004), and this kind of adhesin could prove to be a more general host-interaction mechanism in this genus.

The *L. plantarum* genome was the first among the lactobacilli to be sequenced (Kleerebezem et al. 2003), and strain WCFS1 remains among the best characterized. Biodiversity-based screening of multiple *L. plantarum* strains, comparing array-comparative genomic hybridization (aCGH) data to mannose-sensitive hemagglutination phenotypes (Pretzer et al. 2005), resulted in the identification of a gene encoding the mannose-sensitive haemagglutinin Msa. This protein contains both Mub mucin-binding repeats and a ConA lectin-like motif, and might contribute to the ability of *L. plantarum* to compete with pathogens such as enterotoxigenic *E. coli* that target mannose, but has not yet been demonstrated experimentally (Kleerebezem and Vaughan 2009).

Polk and colleagues identified two proteins encoded by LGG, designated p75 and p40, that had the potential to modulate epithelium proliferation because they inhibited cytokine-induced apoptosis (Yan et al. 2007). Recent characterization of the homologous proteins in *L. casei* showed them to be two predicted cell wall hydrolases with CHAPS and NLPC domains, respectively, which bound to mucin, collagen, and intestinal epithelial cells (Bauerl et al. 2010). Both proteins were secreted from *L. casei*, but partly associated with the cell envelope, and the broader biological significance for host interaction of these proteins is still open to question pending animal or human trials.

Targetted analysis of proteins that are predicted from annotated genomes to be surface located has been a productive way of identifying host interaction partners in lactobacilli. The majority of these studies have used in vitro/cell culture models. Surface proteins in gram-positive bacteria have a limited number of secretion and anchoring strategies, which for lactobacilli have been reviewed in detail elsewhere (Boekhorst et al. 2005; Kleerebezem et al. 2010; Zhou et al. 2010). Sortase-anchored proteins, covalently attached by a transpeptidation linkage to peptidoglycan, have been cataloged in many lactobacilli e.g. *L. reuteri* (Bath et al. 2005), and have been the target of functional genomic studies including in *L. acidophilus* (Buck et al. 2005) and *L. salivarius* (van Pijkeren et al. 2006). Insertional inactivation of each of five putative surface protein genes in *L. acidophilus* resulted in decreased adhesion to Caco-2 cells in tissue culture (Buck et al. 2005). A smaller reduction of adhesion was observed in our laboratory when either of three genes for predicted sortase-dependent surface proteins were knocked out in *L. salivarius* strain UCC118 (van Pijkeren et al. 2006). The greatest reduction in adhesion to

HT29 cells (ca. 40% compared to wild-type) was achieved when the single sortase gene was deleted, confirming the role of adhesins anchored by this enzyme in interacting with cultured human cells. A 2,419 residue protein, anchored by sortase, was recently identified in LGG, that contributes to biofilm formation, epithelial cell adhesion, and binding to murine intestinal tissue (Velez et al. 2010). It is therefore highly likely that functional analysis of more sortase-anchored proteins, in more species, will identify a broad range of functions and ligands. The real challenge will be to relate in vitro or ex vivo assays to in vivo functions.

Inactivation of genes encoding proteins with predicted mucin binding (Mub) repeats reduced adhesion by *L. acidophilus* and *L. salivarius* in the studies described above. The first *Lactobacillus* Mub-repeat protein to be functionally characterized was that of *L. reuteri* strain 1063 (Roos and Jonsson 2002), and the Msa adhesin of *L. plantarum*, as noted above, also contains Mub repeats. The original *L. reuteri* Mub protein was elegantly identified by screening a lambda gene library with antiserum raised against an extracellular protein extract from the *L. reuteri* strain (Roos and Jonsson 2002). The 358 kDa protein contained 14 copies of a ca. 200 amino acid repeat, and was found to bind to mucin. Many proteins with homologous Mub repeats have since been identified in the genomes of lactic acid bacteria. Zhou and colleagues identified 7 discernible sub-families of Mub-containing proteins in the genomes of 26 lactic acid bacteria (Zhou et al. 2010), but for the majority of these, there is no functional investigation. That such analysis is warranted was dramatically illustrated when the crystal structure of the original *L. reuteri* type 2 Mub was determined (MacKenzie et al. 2009), revealing structural homology to the Ig-binding Protein L of *Peptostreptoccus magnus* (now *Finegoldia magna*). This binding of immunoglobulin could have important implications for how commensal lactobacilli expressing Mub-containing proteins interact with the immune system.

Surface layer (S-layers) are paracrystalline arrays formed from a single protein (or glycoprotein) that cover the surface of certain bacteria (Sleytr and Beveridge 1999). In the same study of sortase-anchored adhesins of *L. acidophilus* described above, it was shown that a mutant lacking the S-layer protein A (SlpA) had significantly reduced binding to Caco-2 cells (Buck et al. 2005). However, the *L. acidophilus* S-layer has a greater and more complex role in host interaction, as it governs the nature of interaction with dendritic cells (DCs). When the SlpA protein is produced, the corresponding S-layer functions to promote specific binding to DC-specific ICAM-3-grabbing nonintegrin (DC-SIGN), a DC- specific receptor (Konstantinov et al. 2008). As a result, the DCs produce IL-10. If an S-layer containing the SlpB protein was produced, however, there was a shift to higher DC production of proinflammatory cytokines including IL-12p70, TNF, and IL-1β (Konstantinov et al. 2008). This represents one of the clearest examples whereby a defined component of a probiotic bacterium can modulate immune cell functions.

4 Molecular Dialog Between Lactobacilli and Host

Although lactobacilli may interact with an animal host at many sites, the substantial level of interest in probiotic bacteria has meant that the vast majority of work on immune system interactions concerns the GIT. As discussed elsewhere in this review, a distinction must be made between commensal lactobacilli that are resident in the GIT and transient lactobacilli that enter the GIT fortuitously or following consumption of specific probiotic products. The intestinal mucosa represents the primary point at which commensal or transient lactobacilli interact with the host immune system. The beneficial effects of the lactobacilli may result from modulation of the host immune system or by prevention of pathogen infection.

4.1 Interaction of Commensal Bacteria with the Intestinal Epithelia

One of the most intriguing aspects of mucosal immunology is the tolerance of the system to commensal bacteria. It is generally recognized that systemic tolerance to commensal bacteria does not occur but rather the immune response is confined to the mesenteric lymph nodes, thus avoiding a systemic immune response. Bacteria administered in the presence of mesenteric lymph nodes are detectable both in dendritic cells from Peyer's patches, and in the mesenteric lymph nodes. However in the absence of intact lymph nodes, administered commensals will enter the systemic immune system and lead to their reaching the spleen (Macpherson et al. 2005). It is thought unlikely that tolerance to commensal bacteria is due to evasion or suppression of the mucosal immune response, as this suppression would impair the ability of the host to mount an immune response to any pathogen and the genetic determinants necessary for immune evasion could be easily transferred to pathogenic bacteria in the intestine (Duerkop et al. 2009). Recent experiments with *E. coli* strains engineered to lack colonization ability have shown that there is attrition of memory with regard to IgA production against commensals (Hapfelmeier et al. 2010), which is relevant for understanding the duration and specificity of recognition of lactobacilli by the adaptive immune response. This is similar to the inflammatory response and cytokine production that occurred six days after *L. reuterii* was administered to mice that had not previously encountered this species; the inflammation subsided after 21 days even though the lactobacilli were still present (Livingston et al. 2010).

A further reason for the failure of commensal intestinal bacteria to elicit chronic intestinal inflammation is because their interaction with epithelial cells is restricted by the mucus layer that coats the intestinal epithelium (Johansson et al. 2008). The mucus layer is a relatively complex structure composed of two distinct sub-layers, an outer layer, relatively permeable to and colonized by bacteria, and an inner layer that contains far lower numbers of bacteria (Johansson et al. 2008; Hansson and Johansson 2010). The major mucin component of both layers is the mucin MUC2; the differences in density and permeability to bacteria is due to proteolytic action that reduces the number of cross-links, thus allowing a volume expansion of the outer layer (Johansson et al. 2011). It has been reported that the inner mucus layer contains defensins and other antimicrobial compounds secreted by epithelial cells, and that these molecules contribute to the absence of bacteria from the inner mucus layer (Meyer-Hoffert et al. 2008). Two mouse strains with separate spontaneous mutations in the MUC2 gene develop diarrhea and colitis, while MUC2-/- mice are prone to developing adenocarcinoma of the small intestine, colon and rectum (Velcich et al. 2002). Persistence of *L. acidophilus* and *L. rhamnosus* in the colonic mucus was observed in rats following administration of a multi-bacterial probiotic mixture even after bacteria were no longer being shed in the feces (Lee et al. 2009). Indeed an earlier study reported a similar observation for *L. rhamnosus* following administration to humans (Alander et al. 1999). The presence of probiotic lactobacilli in the intestinal mucus layer may be a significant factor in the inhibition of colonization by pathogens; for example the adherence of *Enterobacter sakazaki* to human colonic mucus was inhibited by *L. rhamnosus*, *L. paracasei* and other probiotic species (Collado et al. 2008).

The sampling of bacteria in the mucus layer by the immune system effectively allows the immune system to monitor bacterial numbers in the mucus and therefore to mount a proportionate response to the microbial challenge that is present. It has been suggested that sampling of luminal bacteria by dendritic cells at Peyer's patches in the epithelium has a significant role in the exclusion of commensal bacteria from the intestinal surface. The sampled bacteria trigger an IgA response that coats the bacteria that are outside the inner mucus layer and hinders their penetration of the mucus layer (Macpherson et al. 2005). It is estimated that only 0.0001% of a bacterial challenge dose will reach the Peyer's patches and elicit IgA production, and that the response is modulated by bacterial cell numbers, i.e. the more bacteria present, the more IgA is produced (Macpherson and Uhr 2004). In this way, the permeability of the mucosa to bacteria is regulated in response to the numbers of bacterial that encounter the mucosal immune system.

In addition to providing a binding site, the mucus layer also provides a source of nutrients as commensal bacteria may remove monosaccharides one by one from the mucus glycans, and in doing this may mediate turnover of the mucus layer, since it is known that germ-free mice have much thicker, looser mucus layers compared to conventional mice (Johansson et al. 2008). *L. iners* uses mucin as the preferred carbon source (Macklaim et al. 2011).

4.2 Transcriptional Studies of In Vivo Responses to Probiotic Bacteria

The interaction between the immune system and *Lactobacillus plantarum* has been investigated using human intervention studies in healthy volunteers whereby *L. plantarum* was administered and the transcriptional changes in the duodenal mucosa were studied. Troost et al. (2008) administered the bacteria to a group of healthy volunteers by continuous intraduodenal infusion in two separate studies, in which infusion times of 1 and 6 h were used. Doudenal mucosal tissue was sampled by gastroduodenoscopy before and after infusion of the bacteria. Gene transcription in the tissue samples was measured by microarray analysis and proteomic analysis by MALDI-TOF was also performed. Following the one-hour infusion, 669 genes were differentially expressed while 424 genes were differentially expressed after the 6 h infusion. Only 54 genes had altered expression levels in both the 1 and 6 h infusions subjects, and of these, 19 were immune system related (Troost et al. 2008). The effects of the probiotic administration evolved over time in that the host response after 1 h exposure was to down-regulate genes for fatty acid metabolism and cell cycle progression, whereas after 6 h exposure, the response had changed to the up-regulation of genes for lipid metabolism and cellular growth and development (Troost et al. 2008)

Two major recent studies have examined the transcriptional changes in the intestinal epithelium of healthy human subjects in response to the administration of probiotic cultures. In the first of these studies, suspensions of either dead cells, live stationary phase cells and exponentially growing cells were administered orally at intervals for a total time of 6 h, and biopsy samples of the proximal duodenum were taken before and after exposure. A number of different cellular pathways were identified as being affected, and these clustered around the NF-kappaB complex which is an important modulator of the inflammatory responses (van Baarlen et al. 2009). Stationary-phase cells and dead cells showed little difference in their ability to elicit NF-kappaB-related responses, but exponential phase cells were more potent inducers of anti-inflammatory activities (e.g. BCL3, IkB and SADM). In the second study, *L. plantarum* was compared to two other probiotic bacteria, *L. casei* and *L. rhamnosus*, and to placebo, for their effect on gene transcription in duodenal biopsies (van Baarlen et al. 2011). Two interesting outcomes to this study were that there were differences in the transcriptional effects elicited by each species, but also that there were very significant differences in response between each individual subject in the study (van Baarlen et al. 2011). The inter subject differences were greater than differences between the bacterial species. It was notable however that the transcription of genes with a more central role in regulatory networks differed less between individuals than did genes with more peripheral roles in the processes. However the differences in responses between each species were distinct, e.g. *L. acidophilus* had a distinct effect on regulation of the immune response, tissue growth and development hormones and ion homeostasis, whereas *L. casei* affected genes involved in cell proliferation, Th1–Th2 balance and blood pressure regulating hormones. The differences in the effect of the strains on the expression of genes that regulate the immune system were quite marked, with far more genes being differentially transcribed in response to *L. acidophilus*. It was reassuring that there was a good correlation between previous reports of probiotic effects for the three species and the observed patters of gene expression. The differences in transcriptional responses between the individual subjects underline the importance of other factors such as diet, lifestyle and resident microbiota in gauging the effects of probiotic bacteria on the host.

5 Lactobacilli as Incidental or Opportunistic Pathogens?

5.1 Lactobacilli as Oral Pathogens

Lactobacilli have been associated with dental caries for nearly a century (Kligler 1915) and prior to the 1950s they were the leading cariogenic candidate (Beighton 2005). Two classic studies of Leverett and coworkers (Leverett et al. 1993) clearly show lactobacilli as a co-factor and strong predictor of caries in children, even more so than *Streptococcus mutans*. Interestingly, they found lactobacilli to be a good predictor of caries before cavitation occurred. Similarly, longitudinal studies by Loesche et al. (1984) and Burt et al. (1985) support an association of lactobacilli with the caries process, but not to the same extent as for the mutans streptococci. The lactobacilli became epidemiologically linked to dental caries when salivary counts of lactobacilli were shown to be highly correlated to caries levels, at least at the population level (Marsh and Martin 1999). One of the earliest 16S rRNA gene sequencing studies identified *L. fermentum* in samples from caries but not healthy controls (Becker et al. 2002). A particularly aggressive form of dental caries is termed rampant or severe early childhood caries (S-ECC) (Drury et al. 1999). S-ECC is associated with prolonged exposure to carbohydrates, a history of maternal malnutrition, and early infection by, and dominance of mutans streptococci (Li et al. 1994). Page Caufield and colleagues at NYU have shown that the microbiota of children with S-ECC differ significantly from those of their caries-free counterparts (Li et al. 1994; Yang et al. 2010). While the mutans streptococci are considered the main initiators of caries, lactobacilli are likely the major determinants of the severity and rate of caries progression. Table 3 summarizes studies linking oral lactobacilli to caries, and lists the main *Lactobacillus* species implicated. Four to six species, including *L. fermentum*, *L. rhamnosus*, *L. gasseri* and *L. salivarius*, seem most consistently detected in caries subjects. A recent study detected *L. gasseri*, *L. paracasei*, *L. vaginalis* and *L. fermentum* in the oral microbiota of 41 caries-free children, but only one of which (*L. gasseri*) was detected in 39 SECC subjects (Kanasi et al. 2010). Another study applying greater sequencing depth failed to detect lactobacilli in association with childhood caries in 60 subjects in China (Ling et al. 2010a, b). As noted by the authors, their sampling of saliva and supragingival plaques may explain the near absence of lactobacilli (and mutans streptococci), as these organisms might be preferentially localized to deep-dentin lesions with lower redox potential. Other oral microbiota profiling and microbiome sequencing projects currently underway in a number of centers will shed further light on lactobacilli as causative agents or accomplices in dental caries.

Table 3 Selected studies linking oral lactobacilli to caries formation, and species implicated

14 S-ECC extracted teeth (Marchant et al. 2001)	65 adult extracted teeth (Byun et al. 2004)	5 adults caries lesions (Munson et al. 2004)	3 adults saliva and feces (Dal Bello and Hertel 2006)	10 caries-active mothers saliva (Caufield et al. 2007)	7 S-ECC saliva and caries lesions (Yang et al. 2010)
L. casei	*L. gasseri*	*L. gasseri*	*L. gasseri*	*L. fermentum*	*L. rhamnosus*
L. fermentum	*L. rhamnosus*	*L. rhamnosus*	*L. paracasei*	*L. gasseri*	*L. gasseri*
L. rhamnosus	*L. crispatus*	*L. casei*	*L. rhamnosus*	*L. vaginalis*	*L. fermentum*
	Unknown *Lactobacillus*	*L. plantarnum*	*L. vaginalis*	*L. salivarius*	*L. casei*
	L. casei	*L. reuteri*		*L. oris*	*L. oris*
	L. ultunensis	*L. vaginalis*			*L. salivarius*
	L. salivarius	*L. fermentum*			*L. vaginalis*

5.2 Lactobacilli as Opportunistic Pathogens Emanating from the Gut

Lactobacilli have been clearly demonstrated to cause septicemia, rheumatic heart disease and infective endocarditis (Harty et al. 1994), but this typically occurs only in patients that are compromised in some way (reviewed in Sanders et al. 2010). A diagnosis of *Lactobacillus* bacteraemia usually requires two positive blood cultures, and the duration for such infections ranged from 1 to 14 days in a clinical survey of 45 cases (Husni et al. 1997). In a 2005 review of 200 cases of *Lactobacillus*-associated infections, a diverse spectrum of species was recorded, but with a clear preponderance of species associated with the intestine (Cannon et al. 2005). Endocarditis and bacteremia were the most common *Lactobacillus*-associated infections. The overall mortality rate was very high, at 30%, but there was a significant association between mortality and polymicrobial infections (Cannon et al. 2005). Focussing specifically on *Lactobacillus*-associated endocarditis, a meta-analysis reported that mortality rates had improved between 1992 and 2006, possibly due to more aggressive antibiotic regimens (Salvana and Frank 2006). Treatment durations of four to six weeks, and including various combinations of vancomycin, gentamicin, clindamycin, ciprofloxacin, chloramphenicol and amoxicillin *inter alia* were required to produce clean blood cultures.

Lactobacillus sepsis associated with probiotic therapy has received special attention, because there is the potential that subjects with the highest risk factors for opportunistic infection might be selected because they stand to gain most from probiotic intervention. For example, a 6-week old infant undergoing right ventricle surgery developed diarrhea, and was administered *L. rhamnosus* that was linked to bacteremia 20 days later (Land et al. 2005). The rep-PCR profiles of the blood isolate and the administered probiotic culture were indistinguishable, which is strong (but not inconclusive) evidence for strain identity. In a high-profile study, probiotic administration was associated with a higher mortality rate than the control group (16% compared to 6%), in a cohort of severe acute pancreatitis patients (Besselink et al. 2008). The probiotic mixture comprised four *Lactobacillus* strains and two bifidobacteria at a combined total of 10^{10} colony forming units per day. Decreased barrier function appears not to have been the reason for the increased mortality, but rather an increased oxidative stress burden in an already critically affected redox system (Sanders et al. 2010). The design and implementation of the study was called into question by Dutch regulatory authorities, and a number of ways in which best practice was not followed were highlighted (Lancet 2010).

It has been suggested that bacterial factors considered desirable for probiotic strains, such as adhesion, might contribute to potential to cause bacteremia (Apostolou et al. 2001). However, only three of eight clinical isolates tested in one study aggregated platelets (which has been associated with endocarditis in other gram-positive bacteria) and this trait did not appear to be an indicative risk factor for association with *Lactobacillus*-associated endocarditis (Kirjavainen et al. 1999). In a study of 55 strains comprising isolates from bacteremia cases, from feces of healthy subjects, and from dairy foods, adhesion levels to intestinal mucus were significantly higher in clinical isolates compared to non-clinical isolates (Apostolou et al. 2001). Higher adhesion levels were not confined to strains considered intestinal (see Sect. 1), and so-called "dairy" strains were also capable of high-level adhesion. Further investigation of a broader range of virulence properties failed to detect significantly different phenotypes in clinical isolates, that were tested for adhesion to extracellular matrix proteins, hemolysis, resistance to monocyte respiratory burst killing and resistance to killing by whole serum (Vesterlund et al. 2007). The significance of the linkage to increased mucus binding was also weaker than in a previous study ($P = 0.07$).

The current consensus is that major risk factors should be used as exclusion criteria for usage of probiotics, including lactobacilli, in clinical practice. Such risk factors (Boyle et al. 2006) include immune compromise, and premature birth infants, as major factors, and catheterization, impaired barrier function, broad spectrum antibiotics, administration by jejunostomy or presence of cardiac valvular disease (for *Lactobacillus* probiotics only). Interestingly, usage of probiotics with "high mucosal adhesion or known pathogenicity" was also listed as a risk factor for probiotic sepsis (Boyle et al. 2006), although the literature is vague as to what constitutes pathogenicity in lactobacilli. More recent commentary (Sanders et al. 2010) has pointed out that for the vast majority of healthy subjects, the historical rates of lactobacillemia are "unequivocally negligible" (Borriello et al. 2003), and that even in many immunocompromised patients (including patients with HIV, premature infants, elderly and patients with Crohn's Disease, evidence for opportunistic pathogenicity is weak (reviewed in Sanders et al. 2010). Pending more convincing evidence to the contrary, the balance of evidence suggests that *Lactobacillus*-associated infections are due to opportunistic pathogenesis by normally innocuous strains.

5.3 *Lactobacilli* as Delivery Agents for Vaccines or Therapeutics

Because of the fact that they are Generally Regarded As Safe, by virtue of being consumed in large quantities in fermented foods, and because they appear not to trigger inflammation, lactic acid bacteria in general have attracted attention as delivery vehicles for vaccines and therapeutics (Wells et al. 1996; Steidler and Neirynck 2005). It is technically facile to express heterologous antigens in many lactobacilli, and it was shown that mice administered *L. casei* cells expressing the flagellin protein FliC from *Salmonella enterica* serovar Enteriditis were afforded protective immunity against subsequent *Salmonella* challenge (Kajikawa et al. 2007). This system was adapted for additional antigens, but because the *Lactobacillus* cells are poorly immunogenic, the desired antigen was fused to the FliC protein, which acted as an adjuvant (Kajikawa and Igimi 2010). A protective antigen from *Bacillus anthracis* was also successfully expressed in *L. gasseri* and the recombinant afforded protection against anthrax in mice (Mohamadzadeh et al. 2010), underlining the diversity of mucosal vaccination possibilities using *Lactobacillus* vehicles.

In another exciting application, two-domain CD4 domain proteins were expressed and secreted in *L. jensenii*, and shown to inhibit HIV entry into cultured HeLa cells (Chang et al. 2003). Subsequent genome sequencing and refinement of expression and anchoring technology has been carried out to optimize this system, with a view to developing topical microbicides for the mitigation of heterosexual transmission of HIV and other mucosally transmitted viral pathogens (Liu et al. 2008).

6 Concluding Perspectives

It is open to question if the US Surgeon General really ever did declare in the late 1960s that the war on infectious disease had been won (Spellberg 2008), but thankfully we still witness the application of powerful analytical approaches to microbes in the following four decades, that enabled us to understand how many pathogens cause disease, circumvent the immune response, and express resistance to antibiotics. In the last twenty years, similar approaches have been applied to the commercially useful traits of lactic acid bacteria, in many cases drawing closely upon experience and methodologies developed in microbial pathogenesis laboratories. The fact that many lactic bacteria are representatives of, and in many cases isolates from, the intestinal microbiota, has been one factor in drawing increasing research attention to their host interaction properties. However, despite recent advances, significant questions remain unanswered. There is a surfeit of descriptive analyses of in vitro tests and even trials in animals and humans, but a dearth of knowledge of molecular effectors and biologically significant host responses. Are the responses so far described (even the better characterized examples) enduring, or will a subject respond less, or at all, upon repeated exposure to a given

Lactobacillus? Do lactobacilli present as sub-dominant commensals in the GIT of healthy subjects elicit the same responses in the host as probiotic cultures do when ingested, perhaps at a lower level? And does the change in the intestinal microbiota of people in developed industrialized countries mean we may have already sanitized or starved to extinction the commensal bacteria, including lactobacilli, that potentially co-evolved with us as key factors in immune regulation and bystander function? There are many chapters in the book on *Lactobacillus*-host interaction still to be written.

Acknowledgments Work in PWOT's laboratory is supported in part by awards from Science Foundation Ireland (Principal Investigator program, and a CSET grant to the Alimentary Pharmabiotic Centre), and from the Dept. Agriculture Fisheries and Food/Health Research Board FHRI program. We thank Page Caufield for helpful discussions.

References

Alander M, Satokari R et al (1999) Persistence of colonization of human colonic mucosa by a probiotic strain, *Lactobacillus rhamnosus* GG, after oral consumption. Appl Environ Microbiol 65(1):351–354

Altermann E, Russell WM et al (2005) Complete genome sequence of the probiotic lactic acid bacterium *Lactobacillus acidophilus* NCFM. Proc Natl Acad Sci U S A 102(11):3906–3912

Apostolou E, Kirjavainen PV et al (2001) Good adhesion properties of probiotics: a potential risk for bacteremia? FEMS Immunol Med Microbiol 31(1):35–39

Asong J, Wolfert MA et al (2009) Binding and cellular activation studies reveal that toll-like receptor 2 can differentially recognize peptidoglycan from gram-positive and gram-negative bacteria. J Biol Chem 284(13):8643–8653

Atarashi K, Tanoue T et al (2011) Induction of colonic regulatory T cells by indigenous clostridium species. Science 331:337–341

Bath K, Roos S et al (2005) The cell surface of *Lactobacillus reuteri* ATCC 55730 highlighted by identification of 126 extracellular proteins from the genome sequence. FEMS Microbiol Lett 253(1):75–82

Bauerl C, Perez-Martinez G et al (2010) Functional analysis of the p40 and p75 proteins from *Lactobacillus casei* BL23. J Mol Microbiol Biotechnol 19(4):231–241

Becker MR, Paster BJ et al (2002) Molecular analysis of bacterial species associated with childhood caries. J Clin Microbiol 40(3):1001–1009

Beighton D (2005) The complex oral microflora of high-risk individuals and groups and its role in the caries process. Comm Dent Oral Epidemiol 33(4):248–255

Berg RD (1996) The indigenous gastrointestinal microflora. Trends Microbiol 4(11):430–435

Berger B, Pridmore RD et al (2007) Similarity and differences in the *Lactobacillus acidophilus* group identified by polyphasic analysis and comparative genomics. J Bacteriol 189(4):1311–1321

Bernardeau M, Guguen M et al (2006) Beneficial lactobacilli in food and feed: long-term use, biodiversity and proposals for specific and realistic safety assessments. FEMS Microbiol Rev 30(4):487–513

Besselink MG, van Santvoort HC et al (2008) Probiotic prophylaxis in predicted severe acute pancreatitis: a randomised, double-blind, placebo-controlled trial. Lancet 371(9613):651–659

Bik EM (2009) Composition and function of the human-associated microbiota. Nutr Rev 67(2):S164–S171

Bik EM, Eckburg PB et al (2006) Molecular analysis of the bacterial microbiota in the human stomach. Proc Natl Acad Sci U S A 103(3):732–737

Boekhorst J, de Been MW et al (2005) Genome-wide detection and analysis of cell wall-bound proteins with LPxTG-like sorting motifs. J Bacteriol 187(14):4928–4934
Boekhorst J, Helmer Q et al (2006) Comparative analysis of proteins with a mucus-binding domain found exclusively in lactic acid bacteria. Microbiology 152(Pt 1):273–280
Bolotin A, Quinquis B et al (2004) Complete sequence and comparative genome analysis of the dairy bacterium *Streptococcus thermophilus*. Nat Biotechnol 22(12):1554–1558
Borriello SP, Hammes WP et al (2003) Safety of probiotics that contain lactobacilli or bifidobacteria. Clin Infect Dis 36(6):775–780
Boyle RJ, Robins-Browne RM et al (2006) Probiotic use in clinical practice: what are the risks? Am J Clin Nutr 83(6):1256–1264; quiz 1446–1257
Buck BL, Altermann E et al (2005) Functional analysis of putative adhesion factors in *Lactobacillus acidophilus* NCFM. Appl Environ Microbiol 71(12):8344–8351
Burt BA, Loesche WJ et al (1985) Stability of selected plaque species and their relationship to caries in a child population over 2 years. Caries Res 19(3):193–200
Byun R, Nadkarni MA et al (2004) Quantitative analysis of diverse *Lactobacillus* species present in advanced dental caries. J Clin Microbiol 42(7):3128–3136
Cai H, Rodriguez BT et al (2007) Genotypic and phenotypic characterization of *Lactobacillus casei* strains isolated from different ecological niches suggests frequent recombination and niche specificity. Microbiology 153(Pt 8):2655–2665
Cai H, Thompson R et al (2009) Genome sequence and comparative genome analysis of *Lactobacillus casei*: insights into their niche-associated evolution. Genome Biol Evol 1:239–257
Callanan M, Kaleta P et al (2008) Genome sequence of *Lactobacillus helveticus*, an organism distinguished by selective gene loss and insertion sequence element expansion. J Bacteriol 190(2):727–735
Canchaya C, Claesson MJ et al (2006) Diversity of the genus *Lactobacillus* revealed by comparative genomics of five species. Microbiology 152:3185–3196
Cannon JP, Lee TA et al (2005) Pathogenic relevance of *Lactobacillus*: a retrospective review of over 200 cases. Eur J Clin Microbiol Infect Dis 24(1):31–40
Carr JG, Davies PA (1970) Homofermentative lactobacilli of ciders including *Lactobacillus mali* nov. spec. J Appl Bacteriol 33(4):768–774
Caufield PW, Li Y et al (2007) Diversity of lactobacilli in the oral cavities of young women with dental caries. Caries Res 41(1):2–8
Chaillou S, Champomier-Verges MC et al (2005) The complete genome sequence of the meat-borne lactic acid bacterium *Lactobacillus sakei* 23K. Nat Biotechnol 23(12):1527–1533
Chang TL, Chang CH et al (2003) Inhibition of HIV infectivity by a natural human isolate of *Lactobacillus jensenii* engineered to express functional two-domain CD4. Proc Natl Acad Sci U S A 100(20):11672–11677
Chen YS, Miyashita M et al (2010) *Lactobacillus pobuzihii* sp. nov., isolated from *pobuzihi* (fermented cummingcordia). Int J Syst Evol Microbiol 60(Pt 8):1914–1917
Christensen HR, Frokiaer H et al (2002) Lactobacilli differentially modulate expression of cytokines and maturation surface markers in murine dendritic cells. J Immunol 168(1):171–178
Claesson MJ, Van Sinderen D et al (2008) *Lactobacillus* phylogenomics—towards a reclassification of the genus. Int J Syst Evol Microbiol 58(12):2945–2954
Claesson MJ, O'Sullivan O et al (2009) Comparative analysis of pyrosequencing and a phylogenetic microarray for exploring microbial community structures in the human distal intestine. PLoS One 4(8):e6669
Claesson MJ, van Sinderen D et al (2007) The genus *Lactobacillus*—a genomic basis for understanding its diversity. FEMS Microbiol Lett 269(1):22–28
Collado MC, Isolauri E et al (2008) Specific probiotic strains and their combinations counteract adhesion of *Enterobacter sakazakii* to intestinal mucus. FEMS Microbiol Lett 285(1):58–64
Collins MD, Rodrigues U et al (1991) Phylogenetic analysis of the genus *Lactobacillus* and related lactic-acid bacteria as determined by reverse-transcriptase sequencing of 16s ribosomal-RNA. FEMS Microbiol Lett 77(1):5–12

Dal Bello F, Hertel C (2006) Oral cavity as natural reservoir for intestinal lactobacilli. Syst Appl Microbiol 29(1):69–76

Dellaglio F, Felis GE (2005) Taxonomy of lactobacilli and bifidobacteria. In: Tannock GW (ed) Probiotics and prebiotics: scientific aspects. Caister Academic Press, Norfolk, pp 25–49

Denou E, Pridmore RD et al (2008) Identification of genes associated with the long-gut-persistence phenotype of the probiotic *Lactobacillus johnsonii* strain NCC533 using a combination of genomics and transcriptome analysis. J Bacteriol 190(9):3161–3168

Dent VE, Williams RAD (1982) *Lactobacillus animalis* sp-nov, a new species of *Lactobacillus* from the alimentary canal of animals. Zentralblatt Bakteriol. Mikrobiol. Hyg 3(3):377–386

Drury TF, Horowitz AM et al (1999) Diagnosing and reporting early childhood caries for research purposes. A report of a workshop sponsored by the National Institute of Dental and Craniofacial Research, the Health Resources and Services Administration, and the Health Care Financing Administration. J Public Health Dent 59(3):192–197

Duerkop BA, Vaishnava S et al (2009) Immune responses to the microbiota at the intestinal mucosal surface. Immunity 31(3):368–376

Duncan SH, Lobley GE et al (2008) Human colonic microbiota associated with diet, obesity and weight loss. Int. J. Obes. 32(11):1720–1724

Dunne C, Murphy L et al (1999) Probiotics: from myth to reality. Demonstration of functionality in animal models of disease and in human clinical trials. Ant Van Leeuwenh 76(1–4):279–292

Dunne C, O'Mahony L et al (2001) In vitro selection criteria for probiotic bacteria of human origin: correlation with in vivo findings. Am J Clin Nutr 73(2):386S–392S

Eckburg PB, Bik EM et al (2005) Diversity of the human intestinal microbial flora. Science 308(5728):1635–1638

FAO/WHO (2001). Health and nutritional properties of probiotics in food including powder milk with live lactic acid bacteria. Cordoba, Argentina

Fava F, Lovegrove JA et al (2006) The gut microbiota and lipid metabolism: implications for human health and coronary heart disease. Curr Med Chem 13(25):3005–3021

Felis GE, Dellaglio F (2007) "Taxonomy of *Lactobacilli* and *Bifidobacteria*. Curr Issues Intest Microbiol 8(2):44–61

Felley CP, Corthesy-Theulaz I et al (2001) Favourable effect of an acidified milk (LC-1) on *Helicobacter pylori* gastritis in man. Eur J Gastroenterol Hepatol 13(1):25–29

Finegold SM, Sutter VL et al (1983) Normal indigenous intestinal microbiota. In: Heneges DJ (ed) Human intestinal microbiota in health and disease. Academic Press, New York, pp 3–31

Frank DN, St Amand AL et al (2007) Molecular-phylogenetic characterization of microbial community imbalances in human inflammatory bowel diseases. Proc Natl Acad Sci U S A 104(34):13780–13785

Fredricks DN, Fiedler TL et al (2005) Molecular identification of bacteria associated with bacterial vaginosis. N Engl J Med 353(18):1899–1911

Fujisawa T, Shirasaka S et al (1984) *Lactobacillus aviarius* sp. nov.—a new species Isolated from the intestine of chickens. Syst Appl Microbiol 5(3):414–420

Fujisawa T, Benno Y et al (1992) Taxonomic study of the *Lactobacillus acidophilus* group, with recognition of *Lactobacillus gallinarum* sp. nov. and *Lactobacillus johnsonii* sp. nov. and synonymy of *Lactobacillus acidophilus* group A3 (Johnson et al. 1980) with the type strain of *Lactobacillus* amylovorus (Nakamura 1981). Int J Syst Bacteriol 42(3):487–491

Gao Z, Tseng CH et al (2008) Substantial alterations of the cutaneous bacterial biota in psoriatic lesions. PLoS One 3(7):e2719

Gill SR, Pop M et al (2006) Metagenomic analysis of the human distal gut microbiome. Science 312(5778):1355–1359

Grangette C, Nutten S et al (2005) Enhanced antiinflammatory capacity of a *Lactobacillus plantarum* mutant synthesizing modified teichoic acids. Proc Natl Acad Sci U S A 102(29): 10321–10326

Grice EA, Kong HH et al (2009) Topographical and temporal diversity of the human skin microbiome. Science 324(5931):1190–1192

Guan LL, Hagen KE et al (2003) Detection and identification of *Lactobacillus* species in crops of broilers of different ages by using PCR-denaturing gradient gel electrophoresis and amplified ribosomal DNA restriction analysis. Appl Environ Microbiol 69(11):6750–6757

Guarner F, Bourdet-Sicard R et al (2006) Mechanisms of disease: the hygiene hypothesis revisited. Nat Clin Pract Gastroenterol Hepatol 3(5):275–284

Hammes WP, Vogel RF (1995) The genus *Lactobacillus*. In: Wood BJB, Holzapfel WH (eds) The genera of lactic acid bacteria, vol 2. Blackie Academic and Professional, Glasgow, pp 19–54

Hansson GC, Johansson ME (2010) The inner of the two Muc2 mucin-dependent mucus layers in colon is devoid of bacteria. Gut Microbes 1(1):51–54

Hapfelmeier S, Lawson MA et al (2010) Reversible microbial colonization of germ-free mice reveals the dynamics of IgA immune responses. Science 328(5986):1705–1709

Harty DW, Oakey HJ et al (1994) Pathogenic potential of lactobacilli. Int J Food Microbiol 24(1–2):179–189

Hayashi H, Sakamoto M et al (2002) Phylogenetic analysis of the human gut microbiota using 16S rDNA clone libraries and strictly anaerobic culture-based methods. Microbiol Immunol 46(8):535–548

Hayashi H, Takahashi R et al (2005) Molecular analysis of jejunal, ileal, caecal and recto-sigmoidal human colonic microbiota using 16S rRNA gene libraries and terminal restriction fragment length polymorphism. J Med Microbiol 54(Pt 11):1093–1101

Hickson M, D'Souza AL et al (2007) Use of probiotic *Lactobacillus* preparation to prevent diarrhoea associated with antibiotics: randomised double blind placebo controlled trial. Brit Med J 335(7610):80

Hill MJ (1997) Intestinal flora and endogenous vitamin synthesis. Eur J Cancer Prev 6(1): S43–S45

Husni RN, Gordon SM et al (1997) *Lactobacillus* bacteremia and endocarditis: review of 45 cases. Clin Infect Dis 25(5):1048–1055

Isolauri E, Juntunen M et al (1991) A human *Lactobacillus* strain (*Lactobacillus casei* sp strain GG) promotes recovery from acute diarrhea in children. Pediatrics 88(1):90–97

Johansson ME, Holmen Larsson JM et al (2011) Microbes and health sackler colloquium: The two mucus layers of colon are organized by the MUC2 mucin, whereas the outer layer is a legislator of host-microbial interactions. Proc Natl Acad Sci U S A 108(Suppl 1): 4659–4665

Johansson ME, Phillipson M et al (2008) The inner of the two Muc2 mucin-dependent mucus layers in colon is devoid of bacteria. Proc Natl Acad Sci U S A 105(39):15064–15069

Kadooka Y, Sato M et al (2010) Regulation of abdominal adiposity by probiotics (*Lactobacillus gasseri* SBT2055) in adults with obese tendencies in a randomized controlled trial. Eur J Clin Nutr 64(6):636–643

Kaji R, Kiyoshima-Shibata J et al (2010) Bacterial teichoic acids reverse predominant IL-12 production induced by certain *Lactobacillus* strains into predominant IL-10 production via TLR2-dependent ERK activation in macrophages. J Immunol 184(7):3505–3513

Kajikawa A, Igimi S (2010) Innate and acquired immune responses induced by recombinant *Lactobacillus casei* displaying flagellin-fusion antigen on the cell-surface. Vaccine 28(19): 3409–3415

Kajikawa A, Satoh E et al (2007) Intragastric immunization with recombinant *Lactobacillus casei* expressing flagellar antigen confers antibody-independent protective immunity against *Salmonella enterica* serovar enteritidis. Vaccine 25(18):3599–3605

Kanasi E, Dewhirst FE et al (2010) Clonal analysis of the microbiota of severe early childhood caries. Caries Res 44(5):485–497

Kandler O, Weiss N (1986) Regular, nonsporing Gram-positive rods. In: Sneath PHA, Mair NS, Sharpe ME, Holt JG (eds) Bergey's manual of systematic bacteriology, vol 2. Williams and Wilkins, Baltimore, pp 1208–1234

Kankainen M, Paulin L et al (2009) Comparative genomic analysis of *Lactobacillus rhamnosus* GG reveals pili containing a human-mucus binding protein. Proc Natl Acad Sci U S A 106(40):17193–17198

Kirjavainen PV, Tuomola EM et al (1999) In vitro adhesion and platelet aggregation properties of bacteremia-associated lactobacilli. Infect Immun 67(5):2653–2655
Klaenhammer TR, Barrangou R et al (2005) Genomic features of lactic acid bacteria effecting bioprocessing and health. EMS Microbiol Rev 29(3):393–409
Kleerebezem M, Vaughan EE (2009) Probiotic and gut lactobacilli and bifidobacteria: molecular approaches to study diversity and activity. Annu Rev Microbiol 63:269–290
Kleerebezem M, Boekhorst J et al (2003) Complete genome sequence of *Lactobacillus plantarum* WCFS1. Proc Natl Acad Sci U S A 100(4):1990–1995
Kleerebezem M, Hols P et al (2010) The extracellular biology of the lactobacilli. FEMS Microbiol Rev 34(2):199–230
Kligler IJ (1915) Chemical studies of the relations of oral microorganisms to dental caries. J Allied Dental Soc 10:141–166
Knarreborg A, Simon MA et al (2002) Effects of dietary fat source and subtherapeutic levels of antibiotic on the bacterial community in the ileum of broiler chickens at various ages. Appl Environ Microbiol 68(12):5918–5924
Konstantinov SR, Smidt H et al (2008) S layer protein A of *Lactobacillus acidophilus* NCFM regulates immature dendritic cell and T cell functions. Proc Natl Acad Sci U S A 105(49): 19474–19479
Kunji ER, Mierau I et al (1996) The proteolytic systems of lactic acid bacteria. Ant Van Leeuwenh 70(2–4):187–221
Lancet EoT (2010) Expression of concern—probiotic prophylaxis in predicted severe acute pancreatitis: a randomised, double-blind, placebo-controlled trial. Lancet 375(9718):875–876
Land MH, Rouster-Stevens K et al (2005) *Lactobacillus* sepsis associated with probiotic therapy. Pediatrics 115(1):178–181
Lebeer S, Vanderleyden J et al (2008) Genes and molecules of lactobacilli supporting probiotic action. Microbiol Mol Biol Rev 72(4):728–764
Lee HH, Orlovich DA et al (2009) Detection and specific enumeration of multi-strain probiotics in the lumen contents and mucus layers of the rat intestine after oral administration. Probiotics Antimicrob Prot 1:113–120
Leverett DH, Featherstone JD et al (1993) Caries risk assessment by a cross-sectional discrimination model. J Dent Res 72(2):529–537
Ley RE, Turnbaugh PJ et al (2006) Microbial ecology: human gut microbes associated with obesity. Nature 444(7122):1022–1023
Li Y, Navia JM et al (1994) Colonization by mutans streptococci in the mouths of 3- and 4-year-old Chinese children with or without enamel hypoplasia. Arch Oral Biol 39(12):1057–1062
Li Y, Raftis E et al (2006) Polyphasic analysis indicates that *Lactobacillus salivarius* subsp. *salivarius* and *Lactobacillus salivarius* subsp. *salicinius* do not merit separate subspecies status. Int J Sys Evol Microbiol 56:2397–2403
Ling Z, Kong J et al (2010a) Analysis of oral microbiota in children with dental caries by PCR-DGGE and barcoded pyrosequencing. Microb Ecol 60(3):677–690
Ling Z, Kong J et al (2010b) Molecular analysis of the diversity of vaginal microbiota associated with bacterial vaginosis. BMC Genom 11:488
Liu X, Lagenaur LA et al (2008) Engineering of a human vaginal *Lactobacillus* strain for surface expression of two-domain CD4 molecules. Appl Environ Microbiol 74(15):4626–4635
Livingston M, Loach D et al (2010) Gut commensal *Lactobacillus reuteri* 100-23 stimulates an immunoregulatory response. Immunol Cell Biol 88(1):99–102
Loesche WJ, Eklund S et al (1984) Longitudinal investigation of bacteriology of human fissure decay: epidemiological studies in molars shortly after eruption. Infect Immun 46(3):765–772
Lorca G, Torino MI et al (2002) Lactobacilli express cell surface proteins which mediate binding of immobilized collagen and fibronectin. FEMS Microbiol Lett 206(1):31–37
MacKenzie DA, Tailford LE et al (2009) Crystal structure of a mucus-binding protein repeat reveals an unexpected functional immunoglobulin binding activity. J Biol Chem 284(47): 32444–32453

Macklaim JM, Gloor GB et al (2011) Microbes and health sackler colloquium: at the crossroads of vaginal health and disease, the genome sequence of *Lactobacillus iners* AB-1. Proc Natl Acad Sci U S A 108(Suppl 1):4688–4695
Macpherson AJ, Uhr T (2004) Induction of protective IgA by intestinal dendritic cells carrying commensal bacteria. Science 303(5664):1662–1665
Macpherson AJ, Geuking MB et al (2005) Immune responses that adapt the intestinal mucosa to commensal intestinal bacteria. Immunology 115(2):153–162
Makarova KS, Koonin EV (2007) Evolutionary genomics of lactic acid bacteria. J Bacteriol 189(4):1199–1208
Makarova K, Slesarev A et al (2006) Comparative genomics of the lactic acid bacteria. Proc Natl Acad Sci U S A 103(42):15611–15616
Marchant S, Brailsford SR et al (2001) The predominant microflora of nursing caries lesions. Caries Res 35(6):397–406
Marsh P, Martin MV (1999) The resident oral microflora: oral microbiology. Reed Educational and Professional, Woburn, pp 17–33
Maslowski KM, Mackay CR (2011) Diet, gut microbiota and immune responses. Nat Immunol 12(1):5–9
Matsuguchi T, Takagi A et al (2003) Lipoteichoic acids from *Lactobacillus* strains elicit strong tumor necrosis factor alpha-inducing activities in macrophages through toll-like receptor 2. Clin Diagn Lab Immunol 10(2):259–266
Mattila-Sandholm T, Blum S et al (1999) Probiotics: towards demonstrating efficiency. Trends Food Sci Technol 10:393–399
Mazmanian SK, Liu CH et al (2005) An immunomodulatory molecule of symbiotic bacteria directs maturation of the host immune system. Cell 122(1):107–118
Mazmanian SK, Round JL et al (2008) A microbial symbiosis factor prevents intestinal inflammatory disease. Nature 453(7195):620–625
Meijerink M, van Hemert S et al (2010) Identification of genetic loci in *Lactobacillus plantarum* that modulate the immune response of dendritic cells using comparative genome hybridization. PLoS One 5(5):e10632
Metchnikoff E (1907) The prolongation of life: optimistic studies. William Heinemann, London
Metchnikoff E (1908) The nature of man: studies in optimistic philosophy. William Heinemann, London
Meyer-Hoffert U, Hornef MW et al (2008) Secreted enteric antimicrobial activity localises to the mucus surface layer. Gut 57(6):764–771
Miettinen M, Matikainen S et al (1998) Lactobacilli and streptococci induce interleukin-12 (IL-12), IL-18, and gamma interferon production in human peripheral blood mononuclear cells. Infect Immun 66(12):6058–6062
Mitsuoka T (1978) Intestinal bacteria and health. Harcourt Brace Jovanovich, Tokyo
Miyauchi E, Morita H et al (2009) *Lactobacillus rhamnosus* alleviates intestinal barrier dysfunction in part by increasing expression of zonula occludens-1 and myosin light-chain kinase in vivo. J Dairy Sci 92(6):2400–2408
Mohamadzadeh M, Olson S et al (2005) Lactobacilli activate human dendritic cells that skew T cells toward T helper 1 polarization. Proc Natl Acad Sci U S A 102(8):2880–2885
Mohamadzadeh M, Durmaz E et al (2010) Targeted expression of anthrax protective antigen by *Lactobacillus gasseri* as an anthrax vaccine. Future Microbiol 5(8):1289–1296
Molenaar D, Bringel F et al (2005) Exploring *Lactobacillus plantarum* genome diversity by using microarrays. J Bacteriol 187(17):6119–6127
Munson MA, Banerjee A et al (2004) Molecular analysis of the microflora associated with dental caries. J Clin Microbiol 42(7):3023–3029
Neville BA, O'Toole PW (2010) Probiotic properties of *Lactobacillus salivarius* and closely related Lactobacillus species. Future Microbiol 5(5):759–774
O'Hara AM, O'Regan P et al (2006) Functional modulation of human intestinal epithelial cell responses by *Bifidobacterium infantis* and *Lactobacillus salivarius*. Immunology 118(2): 202–215

O'Sullivan O, O'Callaghan J et al (2009) Comparative genomics of lactic acid bacteria reveals a niche-specific gene set. BMC Microbiol 9:50

O'Toole PW, Claesson MJ (2010) Gut microbiota: changes throughout the lifespan from infancy to elderly. Internat Dairy J 20:281–291

Oh PL, Benson AK et al (2010) Diversification of the gut symbiont *Lactobacillus reuteri* as a result of host-driven evolution. ISME J 4(3):377–387

Orla-Jensen S (1919) The lactic acid bacteria. Mem Acad Roy Sci Denmark Sect Sci 8 Ser 5:81–197

O'Toole PW, Cooney JC (2008) Probiotic bacteria influence the composition and function of the intestinal microbiota. Interdiscip Perspect Infect Dis 2008:175285

Parahitiyawa NB, Scully C et al (2010) Exploring the oral bacterial flora: current status and future directions. Oral Dis 16(2):136–145

Pavlova SI, Kilic AO et al (2002) Genetic diversity of vaginal lactobacilli from women in different countries based on 16S rRNA gene sequences. J Appl Microbiol 92(3):451–459

Perea Velez M, Verhoeven TL et al (2007) Functional analysis of D-alanylation of lipoteichoic acid in the probiotic strain *Lactobacillus rhamnosus* GG. Appl Environ Microbiol 73(11): 3595–3604

Pretzer G, Snel J et al (2005) Biodiversity-based identification and functional characterization of the mannose-specific adhesin of *Lactobacillus plantarum*. J Bacteriol 187(17):6128–6136

Pridmore RD, Berger B et al (2004) The genome sequence of the probiotic intestinal bacterium *Lactobacillus johnsonii* NCC 533. Proc Natl Acad Sci U S A 101(8):2512–2517

Qin J, Li R et al (2010) A human gut microbial gene catalogue established by metagenomic sequencing. Nature 464(7285):59–65

Raftis EJ, Salvetti E et al (2011) Genomic diversity of *Lactobacillus salivarius*. Appl Environ Microbiol 77(3):954–965

Rastall RA, Gibson GR et al (2005) Modulation of the microbial ecology of the human colon by probiotics, prebiotics and synbiotics to enhance human health: an overview of enabling science and potential applications. FEMS Microbiol Ecol 52(2):145–152

Ravel J, Gajer P et al (2011) Microbes and health sackler colloquium: vaginal microbiome of reproductive-age women. Proc Natl Acad Sci U S A 108(Suppl 1):4523–4530

Reid G, Charbonneau D et al (2003) Oral use of *Lactobacillus rhamnosus* GR-1 and *L. fermentum* RC-14 significantly alters vaginal flora: randomized, placebo-controlled trial in 64 healthy women. FEMS Immunol Med Microbiol 35(2):131–134

Reuter G (2001) The *Lactobacillus* and *Bifidobacterium* microflora of the human intestine: composition and succession. Curr Issues Intest Microbiol 2(2):43–53

Rijkers GT, Bengmark S et al (2010) Guidance for substantiating the evidence for beneficial effects of probiotics: current status and recommendations for future research. J Nutr 140(3): 671S–676S

Rogosa M, Wiseman RF et al (1953) Species differentiation of oral lactobacilli from man including description of *Lactobacillus salivarius* nov spec and *Lactobacillus cellobiosus* nov spec. J Bacteriol 65(6):681–699

Roos S, Jonsson H (2002) A high-molecular-mass cell-surface protein from *Lactobacillus reuteri* 1063 adheres to mucus components. Microbiology 148(Pt 2):433–442

Roos S, Engstrand L et al (2005) *Lactobacillus gastricus* sp. nov. *Lactobacillus antri* sp. nov. *Lactobacillus kalixensis* sp. nov. and *Lactobacillus ultunensis* sp. nov. isolated from human stomach mucosa. Int J Syst Evol Microbiol 55(Pt 1):77–82

Round JL, Mazmanian SK (2009) The gut microbiota shapes intestinal immune responses during health and disease. Nat Rev Immunol 9(5):313–323

Round JL, Mazmanian SK (2010) Inducible Foxp3+ regulatory T-cell development by a commensal bacterium of the intestinal microbiota. Proc Natl Acad Sci U S A 107(27): 12204–12209

Ryan KA, Jayaraman T et al (2008a) Isolation of lactobacilli with probiotic properties from the human stomach. Lett Appl Microbiol 47: 269–274

Ryan KA, Daly P et al (2008b) Strain-specific inhibition of *Helicobacter pylori* by *Lactobacillus salivarius* and other lactobacilli. J Antimicrob Chemother 61(4):831–834
Salvana EM, Frank M (2006) *Lactobacillus* endocarditis: case report and review of cases reported since 1992. J Infect 53(1):e5–e10
Sanders ME, Klaenhammer TR (2001) Invited review: the scientific basis of *Lactobacillus acidophilus* NCFM functionality as a probiotic. J Dairy Sci 84(2):319–331
Sanders ME, Akkermans LMA et al (2010) Safety assessment of probiotics for human use. Gut Microbes 1:164–185
Savage DC (1977) Microbial ecology of the gastrointestinal tract. Annu Rev Microbiol 31: 107–133
Savino F, Cordisco L et al (2010) *Lactobacillus reuteri* DSM 17938 in infantile colic: a randomized, double-blind, placebo-controlled trial. Pediatrics 126(3):e526–e533
Saxelin M, Lassig A et al (2010) Persistence of probiotic strains in the gastrointestinal tract when administered as capsules, yoghurt, or cheese. Int J Food Microbiol 144(2):293–300
Schlee M, Wehkamp J et al (2007) Induction of human beta-defensin 2 by the probiotic *Escherichia coli* Nissle 1917 is mediated through flagellin. Infect Immun 75(5):2399–2407
Schleifer KH, Ludwig W (1995) Phylogenetic relationships of lactic acid bacteria. In: Wood BJB, Holzapfel WH (eds) The genera of Lactic Acid Bacteria, Vol 2. Blackie Academic and Professional, pp 7–18
Sekirov I, Russell SL et al (2010) Gut microbiota in health and disease. Physiol Rev 90(3): 859–904
Sharpe ME, Latham MJ et al (1973) Two new species of *Lactobacillus* isolated from the bovine rumen, *Lactobacillus ruminis* sp.nov. and *Lactobacillus vitulinus* sp.nov. J Gen Microbiol 77(1):37–49
Sims IM, Frese SA et al (2011) Structure and functions of exopolysaccharide produced by gut commensal *Lactobacillus reuteri* 100-23 ISME J 5(7):1115–1124
Sleytr UB, Beveridge TJ (1999) Bacterial S-layers. Trends Microbiol 7(6):253–260
Snel J, Vissers YM et al (2010) Strain-specific immunomodulatory effects of *Lactobacillus plantarum* strains on birch-pollen-allergic subjects out of season. Clin Exp Allergy 41(2):232–242
Sokol H, Pigneur B et al (2008) *Faecalibacterium prausnitzii* is an anti-inflammatory commensal bacterium identified by gut microbiota analysis of Crohn disease patients. Proc Natl Acad Sci U S A 105(43):16731–16736
Spanhaak S, Havenaar R et al (1998) The effect of consumption of milk fermented by *Lactobacillus casei* strain Shirota on the intestinal microflora and immune parameters in humans. Eur J Clin Nutr 52(12):899–907
Spellberg B (2008) Dr. William H. Stewart: mistaken or maligned? Clin Infect Dis 47(2):294
Steidler L, Neirynck S (2005) Genetically modified probiotics. In: Tannock GW (ed) Probiotics and prebiotics: scientific aspects. Caister Academic Press, Norfolk, pp 155–170
Stiles ME (1996) Biopreservation by lactic acid bacteria. Ant Van Leeuwenh 70(2–4):331–345
Tallant T, Deb A et al (2004) Flagellin acting via TLR5 is the major activator of key signaling pathways leading to NF-kappa B and proinflammatory gene program activation in intestinal epithelial cells. BMC Microbiol 4:33
Tannock GW (2004) A special fondness for lactobacilli. Appl Environ Microbiol 70(6): 3189–3194
Tannock GW, Munro K et al (2000) Analysis of the fecal microflora of human subjects consuming a probiotic product containing *Lactobacillus rhamnosus* DR20. Appl Environ Microbiol 66(6):2578–2588
Tannock GW, Ghazally S et al (2005) Ecological behavior of *Lactobacillus reuteri* 100-23 is affected by mutation of the luxS gene. Appl Environ Microbiol 71(12):8419–8425
Troost FJ, van Baarlen P et al (2008) Identification of the transcriptional response of human intestinal mucosa to *Lactobacillus plantarum* WCFS1 in vivo. BMC Genom 9:374
van Baarlen P, Troost F et al (2011) Microbes and health sackler colloquium: human mucosal in vivo transcriptome responses to three lactobacilli indicate how probiotics may modulate human cellular pathways. Proc Natl Acad Sci U S A 108(Suppl 1):4562–4569

van Baarlen P, Troost FJ et al (2009) Differential NF-kappaB pathways induction by *Lactobacillus plantarum* in the duodenum of healthy humans correlating with immune tolerance. Proc Natl Acad Sci U S A 106(7):2371–2376

van de Guchte M, Penaud S et al (2006) The complete genome sequence of *Lactobacillus bulgaricus* reveals extensive and ongoing reductive evolution. Proc Natl Acad Sci U S A 103(24):9274–9279

van Kranenburg R, Boels IC et al (1999) Genetics and engineering of microbial exopolysaccharides for food: approaches for the production of existing and novel polysaccharides. Curr Opin Biotechnol 10(5):498–504

van Pijkeren J-P, Canchaya C et al (2006) Comparative and functional analysis of sortase-dependent proteins in the predicted secretome of *Lactobacillus salivarius* UCC118. Appl Env Microbiol 72:4143–4153

Vandamme P, Pot B et al (1996) Polyphasic taxonomy, a consensus approach to bacterial systematics. Microbiol Rev 60(2):407–438

Vaughan EE, de Vries MC et al (2002) The intestinal LABs. Ant Van Leeuwenh 82(1–4): 341–352

Velcich A, Yang W et al (2002) Colorectal cancer in mice genetically deficient in the mucin Muc2. Science 295(5560):1726–1729

Velez MP, Petrova MI et al (2010) Characterization of MabA, a modulator of *Lactobacillus rhamnosus* GG adhesion and biofilm formation. FEMS Immunol Med Microbiol 59(3): 386–398

Ventura M, O'Flaherty S et al (2009) Genome-scale analyses of health-promoting bacteria: probiogenomics. Nat Rev Microbiol 7(1):61–71

Verhelst R, Verstraelen H et al (2004) Cloning of 16S rRNA genes amplified from normal and disturbed vaginal microflora suggests a strong association between *Atopobium vaginae*, *Gardnerella vaginalis* and bacterial vaginosis. BMC Microbiol 4:16

Vesterlund S, Vankerckhoven V et al (2007) Safety assessment of *Lactobacillus* strains: presence of putative risk factors in faecal, blood and probiotic isolates. Int J Food Microbiol 116(3): 325–331

von der Weid T, Bulliard C et al (2001) Induction by a lactic acid bacterium of a population of CD4(+) T cells with low proliferative capacity that produce transforming growth factor beta and interleukin-10. Clin Diagn Lab Immunol 8(4):695–701

von Ossowski I, Reunanen J et al (2010) Mucosal adhesion properties of the probiotic *Lactobacillus rhamnosus* GG SpaCBA and SpaFED pilin subunits. Appl Environ Microbiol 76(7):2049–2057

Walter J (2005) The microecology of *Lactobacilli* in the gastrointestinal tract. In: Tannock GW (ed) Probiotics and prebiotics: scientific aspects. Caister Academic Press, Norfolk, pp 51–82

Walter J (2008) Ecological role of lactobacilli in the gastrointestinal tract: implications for fundamental and biomedical research. Appl Environ Microbiol 74(16):4985–4996

Walter J, Britton RA et al (2011) Microbes and health sackler colloquium: host-microbial symbiosis in the vertebrate gastrointestinal tract and the *Lactobacillus reuteri* paradigm. Proc Natl Acad Sci U S A 108(Suppl 1):4645–4652

Walter J, Loach DM et al (2007) D-alanyl ester depletion of teichoic acids in *Lactobacillus reuteri* 100-23 results in impaired colonization of the mouse gastrointestinal tract. Environ Microbiol 9(7):1750–1760

Weiss N, Schillinger U et al (1981) *Lactobacillus-sharpeae* sp-nov and *Lactobacillus-agilis* sp. nov. 2 new species of homofermentative, meso-diaminopimelic acid-containing lactobacilli isolated from sewage. Zentralblatt Fur Bakteriologie Mikrobiologie Und Hygiene I Abteilung Originale C-Allgemeine Angewandte Und Okologische Mikrobiologie 2(3):242–253

Weiss G, Rasmussen S et al (2010) *Lactobacillus acidophilus* induces virus immune defence genes in murine dendritic cells by a toll-like receptor-2-dependent mechanism. Immunology 131(2):268–281

Wells JM, Robinson K et al (1996) Lactic acid bacteria as vaccine delivery vehicles. Ant Van Leeuwenh 70(2–4):317–330

Witkin SS, Linhares IM et al (2007) Bacterial flora of the female genital tract: function and immune regulation. Best Pract Res Clin Obstet Gynaecol 21(3):347–354
Woese CR (1987) Bacterial evolution. Microbiol Rev 51(2):221–271
Xu J, Bjursell MK et al (2003) A genomic view of the human–bacteroides thetaiotaomicron symbiosis. Science 299(5615):2074–2076
Yan F, Cao H et al (2007) Soluble proteins produced by probiotic bacteria regulate intestinal epithelial cell survival and growth. Gastroenterology 132(2):562–575
Yang R, Argimon S et al (2010) Determining the genetic diversity of lactobacilli from the oral cavity. J Microbiol Methods 82(2):163–169
Yasuda E, Serata M et al (2008) Suppressive effect on activation of macrophages by *Lactobacillus casei* strain *Shirota* genes determining the synthesis of cell wall-associated polysaccharides. Appl Environ Microbiol 74(15):4746–4755
Zhang ZG, Ye ZQ et al (2011) Phylogenomic reconstruction of lactic acid bacteria: an update. BMC Evol Biol 11:1
Zhou X, Bent SJ et al (2004) Characterization of vaginal microbial communities in adult healthy women using cultivation-independent methods. Microbiology 150(Pt 8):2565–2573
Zhou M, Theunissen D et al (2010) LAB-secretome: a genome-scale comparative analysis of the predicted extracellular and surface-associated proteins of lactic acid bacteria. BMC Genom 11:651
Zoetendal EG, Rajilic-Stojanovic M et al (2008) High-throughput diversity and functionality analysis of the gastrointestinal tract microbiota. Gut 57(11):1605–1615

Bacterial Moonlighting Proteins and Bacterial Virulence

Brian Henderson and Andrew Martin

Abstract Implicit in the central dogma is the hypothesis that each protein gene product has but one function. However, over the past decade, it has become clear that many proteins have one or more unique functions, over-and-above the principal biological action of the specific protein. This phenomenon is now known as protein moonlighting and many well-known proteins such as metabolic enzymes and molecular chaperones are now recognised as moonlighting proteins. A growing number of bacterial species are being found to have moonlighting proteins and the moonlighting activities of such proteins can contribute to bacterial virulence behaviour. The glycolytic enzymes, glyceraldehyde-3-phosphate dehydrogenase (GAPD) and enolase, and the cell stress proteins: chaperonin 60, Hsp70 and peptidyl prolyl isomerase, are among the most common of the bacterial moonlighting proteins which play a role in bacterial virulence. Moonlighting activities include adhesion and modulation of cell signalling processes. It is likely that only the tip of the bacterial moonlighting iceberg has been sighted and the next decade will bring with it many new discoveries of bacterial moonlighting proteins with a role in bacterial virulence.

B. Henderson (✉)
Department of Microbial Diseases, UCL-Eastman Dental Institute,
University College London, 256 Gray's Inn Road,
London, WC1X 8LD, UK
e-mail: Brian.Henderson@ucl.ac.uk

A. Martin
Division of Biosciences, Institute of Structural and Molecular Biology,
University College London, London, UK

Current Topics in Microbiology and Immunology (2013) 358: 155–213
DOI: 10.1007/82_2011_188

Published Online: 6 December 2011

Abbreviations

AFB	American foulbrood
AMF	Autocrine motility factor
ATP	Adenosine triphosphate
CD	Cluster of differentiation
Cpn	Chaperonin
DC-SIGN	Dendritic cell-specific intercellular adhesion molecule
ECM	Extracellular matrix
EMT	Epithelial-to-mesenchymal transition
ER	Endoplasmic reticulum
FBA	Fructose bisphosphate aldolase
FBP	Fibronectin-binding protein
FHL	Factor H-like (protein)
GAPD	Glyceraldehyde-3-phosphate dehydrogenase
GH-RH	Growth hormone releasing hormone
HDL	High-density lipoprotein
HGFR	Hepatocyte growth factor receptor
HSP	Heat shock protein
HtrA	High-temperature requirement
IFN-γ	Interferon-gamma
IRE	Iron-responsive elements
LAP	*Listeria* adhesion protein.
LCV	*Legionella*-containing vacuole
LDL	Low density lipoprotein
LPS	Lipopolysaccharide
MAP	Mitogen-activated protein (kinase)
MHC	Major histocompatibility complex
MIP	Macrophage infectivity potentiator
MyD	Myeloid differentiation factor
NFAT	Nuclear factor of activated T cells
ORF	Open reading frame
PFC	Protein-folding catalyst
6PGD	Phosphogluconate dehydrogenase
PGI	Phosphoglucoisomerase
PPD	Purified protein derivative
PPI	Peptidyl prolyl isomerase
SDH	Streptococcal surface dehydrogenase
STAT	Signal transducer and activator of transcription
TCA	Tricarboxylic acid cycle
TLR	Toll-like receptor
TPI	Triose phosphate isomerase
uPAR	Urokinase plasminogen activator receptor
VDAC	Voltage dependent anion channel

Contents

1 Introduction

Francis Crick's Central Dogma which states

$$\text{Gene(DNA)} \rightarrow \text{RNA} \rightarrow \text{Protein}$$

and defines (incorrectly) the direction of information flow in gene-to-protein synthesis, also carries with it the unspoken assumption that each protein has a single function. The human genome is now believed to encode 23–25,000 proteins and up until the 1980s there was no evidence for the alternative hypothesis—that individual proteins can have more than one biological activity. However, in 1988, Joram Piatigorsky, working at the National Eye Institute in Bethesda, USA, reported that the lens crystallin protein in the duck was the metabolic enzyme, argininosuccinate lyase (Piatigorsky et al. 1988). He proposed that this phenomenon, in which one gene product has more than one function, be called *gene sharing* (Piatigorsky et al. 1988). Subsequent work showed that a whole range of metabolic enzymes, and other proteins, can generate transparent lens structures in a variety of animal species. Piatigorsky also realised that this gene sharing had consequences for the transcriptional control of the shared genes (Piatigorsky 1998). Although the pioneer of this field of protein molecular biology (see Piatigorsky 2007), Piatigorsky's gene sharing hypothesis failed to attract the attention it deserved and the term 'gene sharing' is rarely seen in the current literature.

The term 'protein' moonlighting can be traced back to the British biochemist, Robert Freedman, who used the term 'moonlighting' in a report on a Biochemical Society meeting on protein function in 1978 (Freedman 1978). However, it was two scientists working for Hoffman-La-Roche who reported that the well-established neuropeptides, somatostatin and growth hormone releasing hormone

(GH-RH), also exhibited immunological activity and termed this activity 'moonlighting' (Campbell and Scanes 1995). The next use of the term 'moonlighting' was in an editorial describing a paper that reported that yeast enzymes involved in repairing double strand breaks in DNA are also involved in the maintenance of telomeres (Weaver 1998). It has been Constance Jeffery who has publicised the concept of protein moonlighting and attempted to bring some definition to this novel area of protein biology (Jeffery 1999). While this is a simplistic version of the history of protein moonlighting, it should be noted that there are earlier descriptions of protein moonlighting, but with the phenomenon not being explicitly recognised. Glyceraldehyde-3-phosphate dehydrogenase (GAPD) is one of the most accomplished moonlighting proteins, now well recognised for having functions in the nucleus (Sirover 2005). That GAPD could bind single, but not double, stranded DNA was reported as early as 1980 (e.g. Perucho et al. 1980).

The term moonlighting means to have a second job, at night, in addition to the daytime occupation. In Jeffery's first review (1999) she attempted to categorise the various functional facets of moonlighting proteins and to limit their definition. Thus proteins generated by gene fusions, homologous but non-identical proteins, splice variants, protein decoration variants, protein fragments and proteins operating in different locations or utilising different substrates are not considered to be moonlighting proteins (Jeffery 1999). Enzymes which have two metabolic functions or utilise two different substrates are categorised as bifunctional enzymes (Moore 2004). The term 'catalytic promiscuity' has also been applied to the situation of an enzyme which has an active site able to catalyse two different reactions (Copley 2003). This form of 'moonlighting' will not be discussed in this article. The key attributes of moonlighting proteins are the expression of clearly distinct biological activities. This may be an attribute of the original active site of the protein or may be due to the evolution of other sites on the protein with biological function (Fig. 1). Often the exhibition of moonlighting activity is dependent on where the protein is found. Many proteins have one originally defined activity within a cell or a cell compartment and another biological action when the protein is present in another cell compartment or, indeed, is secreted from the cell. Thus moonlighting activity is partially dependent on the environment the protein finds itself in and this may be thought of as 'geographical' moonlighting. As will be seen, as this review develops, a growing number of proteins have not just one moonlighting activity but multiple such activities and there is evidence that moonlighting can contribute to pathology in eukaryotes. In prokaryotes it is emerging that an increasing number of moonlighting proteins are involved in bacterial virulence. One of the most fascinating aspects of protein moonlighting is its apparent spanning of the kingdoms of life. It is now accepted that mitochondria originated from a eubacterium (Gray et al. 1999). Two examples of moonlighting proteins involving mitochondria/eukaryotic cell interactions are: (1) the role of the cytokine-induced transcription factor Stat (signal transducer and activator of transcription) 3, which can enter into mitochondria and control mitochondrial respiration (Wegrzyn et al. 2009) and; (2) the role of the mitochondrial F_1F_o ATP synthase as a cell surface high affinity receptor for apoA-I,

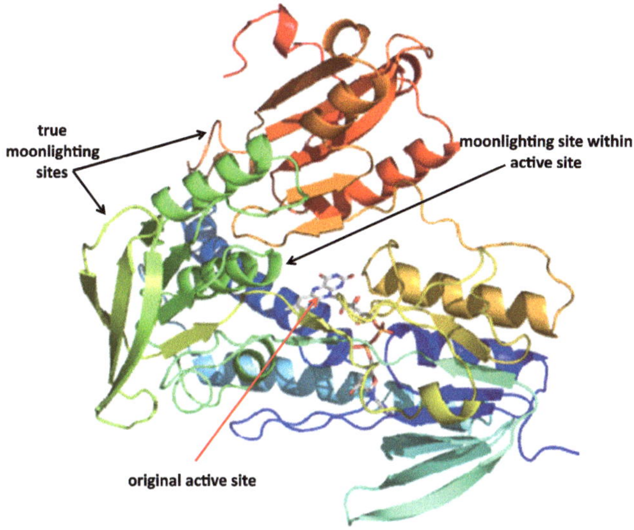

Fig. 1 A cartoon illustration of the potential positions of the moonlighting sites in proteins. This cartoon is the crystal structure of human thioredoxin deposited by Debreczeni et al. (2006) and obtained from (http://www.thesgc.org/structures/structure_description/2CFY/). The diagram suggests that moonlighting sites in moonlighting proteins may be associated with the true active site of the protein and/or may be completely different sites which would likely be found on the protein surface. The definition of active and moonlighting sites on this cartoon bears no relationship to the thioredoxin active site and putative thioredoxin moonlighting sites of this protein

which is the main protein in high-density lipoproteins (HDL) (Vantourout et al. 2010). It is even speculated that this cell surface ATP synthase could be a therapeutic target in heart disease. The reader should be aware that if protein moonlighting proves to be a property of the majority of proteins it has major ramifications for evolutionary theory, cellular network complexity, genetics and medicine. This will be briefly discussed at the conclusion of this review.

2 Eukaryotic Moonlighting Proteins and Human Disease

This article will focus on prokaryotic moonlighting proteins but, at the moment, more is known about moonlighting in eukaryotes, and many of the eukaryotic moonlighting proteins have homologues in bacteria. Thus it is possible that these bacterial proteins might be able to mimic the moonlighting actions of eukaryotic proteins. This could be important, as a number of human moonlighting proteins have pathological activity.

There are probably well over a hundred eukaryotic moonlighting proteins described in the literature. One of the surprising findings is that many of these proteins are well-known metabolic proteins such as the glycolytic and the tricarboxylic acid (TCA) cycle enzymes. Another growing family of moonlighting proteins are the molecular chaperones and protein-folding catalysts involved in protein folding and the cell stress response. These are all ancient gene products, and this leads to the question—are all moonlighting proteins early-evolved gene products? Unfortunately, it is too early to answer this question definitively. Now one of the major surprises in protein moonlighting is the number of moonlighting actions that individual proteins have acquired. For example, the glycolytic enzyme, phosphoglucoisomerase (PGI) has four distinct extracellular functions which have been identified and individually named: (1) a neurotrophic activity termed neuroleukin (Chaput et al. 1988); (2) a factor which promotes cell motility, and is involved in tumour malignancy, called autocrine motility factor (AMF—Watanabe et al. 1996); (3) differentiation and maturation mediator, which promotes myeloid cell differentiation and may play some role in leukaemia (Xu et al. 1996) and (4) an implantation factor activity in the ferret (Schulz and Bahr 2003). More recently, another potential moonlighting function of AMF has been described—regulation of endoplasmic reticulum (ER) stress by a novel mechanism involving controlling ER calcium levels (Fu et al. 2011). Most studies of PGI's moonlighting activity have concentrated on its AMF activity. It is now established that AMF is significantly correlated with breast cancer progression and with a poor prognosis of this condition. This appears to be due to the ability of secreted PGI/AMF to promote epithelial-to-mesenchymal transition (EMT), which is a precursor to tumour metastasis (Funasaka et al. 2009). Now one of the fascinating aspects about the biology of PGI, when it functions as AMF, is the nature of the receptor. It turns out to be another moonlighting protein called gp78, an endoplasmic reticulum (ER) membrane-anchored ubiquitin ligase involved in endoplasmic reticulum-associated protein degradation (Fairbank et al. 2009). Secreted PGI clearly has important pathophysiological moonlighting actions which could be mimicked by secreted bacterial homologues of this enzyme and contribute to bacterial pathogenicity. The available evidence is contradictory with one group, who cloned, expressed and crystallised the PGI from *Bacillus stearothermophilus* (a Gram-positive environmental thermophile) finding the bacterial protein to have both AMF and neuroleukin activity (Sun et al. 1999). In contrast, another group, using a commercially available source of this same enzyme, found the protein enzymically active but lacking in moonlighting activity (Amraei and Nabi 2002). The most sensible explanation for these opposite findings is that the commercially available enzyme has some alteration in the PGI moonlighting site. A recent study has proposed that the *Mycobacterium tuberculosis* PGI structure is similar to that of the human enzyme (Anand et al. 2010) suggesting this bacterial protein could mimic human PGI in patients with tuberculosis. Other well-known eukaryotic proteins with moonlighting functions are listed in Table 1. In case the reader is sceptical about the possibility that glycolytic enzymes could have biological actions other than driving glycolysis the recent report that cancer cells ulitilise an alternative glycolytic pathway (Vander-Heiden et al. 2010) reveals that we are still far away from completely understanding how eukaryotic cells metabolise glucose. To strengthen this message, it was a well-known finding that

Table 1 Some eukaryotic proteins with protein moonlighting actions

Protein	Source	Original function	Moonlighting functions
Aldehyde dehydrogenase	Cow	Alcohol metabolism	Lens protein
Thymidine phosphorylase	Human	DNA metabolism	Platelet-derived endothelial cell growth factor
Fumarate hydratase	Human	TCA cycle	Tumour suppressor
Gelsolin	Human	Structural protein	Controlling apoptosis
Glycogen synthase kinase 3β	Rat	Sugar metabolism	Role in embryonic development
Lactate dehydrogenase	Human	Glycolysis	Protein translation factor
Lactate dehydrogenase	Rat	Glycolysis	DNA maintenance
Citrate synthase	Tetrahymena	TCA cycle	Structural filament-forming protein
Hexokinase	Human	Glycolysis	Controlling apoptosis
Thioredoxin	Multiple	Redox enzyme	Multiple moonlighting functions
Xanthine oxidoreductase	Mouse	Oxidase	Structural role in milk secretion
Cytochrome C	Many	Electron transport chain	Controlling apoptosis
Phosphoglycerate kinase	Human	Glycolysis	Controlling angiogenesis
Quinone oxidoreductase	Guinea pig	Electron transport chain	Lens protein
Succinate dehydrogenase	Human	TCA cycle	Tumour suppressor gene
Aconitase	Yeast	TCA cycle	DNA maintenance
Enolase	Yeast	Glycolysis	Molecular chaperone
Isocitrate dehydrogenase	Yeast	TCA cycle	RNA metabolism
STAT3	Rat	Signalling protein	Electron transport chain
Chaperonin 10	Human	Molecular chaperone	Immunosuppressant
Chaperonin 60	Human	Molecular chaperone	Immunomodulator

References can be found in Piatigorsky (2007) or in this text

tumour cells secreted a reductase that reduced and activated plasmin allowing the production of the blood vessel inhibitor, angiostatin. It is now known that this secreted reductase is the glycolytic enzyme, phosphoglycerate kinase, which is secreted in large amounts by tumour cells (Lay et al. 2000).

Now it is easy to make unwarranted assumptions in biology. This applies to moonlighting proteins where it would be sensible to suppose that if one protein exhibits particular moonlighting actions then all homologues of this protein should also have these additional activities. This turns out not to be the case. Only one example of this will be provided at this stage and others will be briefly mentioned later in the text. Sperm has to undergo physiological and biochemical changes after ejaculation to be able to fertilise the oocyte. Collectively, these changes are termed 'capacitation' and result in functionally competent sperm. It would be natural to assume that the crucial importance of the capacitation phenomenon would limit evolutionary change in this process. It was surprising to find that capacitation of mouse sperm required the active participation of two molecular chaperones—Hsp60 and Hsp90 (Asquith et al. 2004) with a potential additional involvement of Hsp10 (Walsh et al. 2008). The Hsp60 and Hsp90 proteins are located on the sperm surface and require to be tyrosine phosphorylated to achieve capacitation. An even bigger surprise is the report that human sperm have none of these molecular chaperones on their surface and there is no evidence of cell surface tyrosine phosphorylation (Mitchell et al. 2007). Thus the evolutionary development of these moonlighting chaperones in the mouse has clearly occurred within the past 75 million years since the mouse diverged from the precursors of *Homo sapiens* (Stillman and Stewart 2004). This reveals a fairly rapid evolutionary dynamic in the evolution of these molecular chaperone moonlighting sites. Studies of bacterial moonlighting proteins should be able to probe, in more detail, the evolutionary dynamics of moonlighting sites.

3 An Introduction to Protein Moonlighting in Bacteria

The first moonlighting bacterial protein described was the glycolytic enzyme, glyceraldehyde-3-phosphate dehydrogenase (GAPD), which was found on the cell surface of group A streptococci (*Streptococcus pyogenes*) in studies that were conducted before the moonlighting concept was introduced (Pancholi and Fischetti 1992). This protein was shown to be tightly attached to the cell surface by a mechanism which is still not defined. The isolated purified GAPD from the surface of this organism was found to bind to a variety of ligands including the major host adhesive glycoprotein, fibronectin and to lysozyme, as well as the cytoskeletal proteins, myosin and actin. Thus GAPD is truly the prototypic bacterial moonlighting protein, as many bacterial moonlighting proteins are cytoplasmic proteins found on the cell surface and endowed with adhesive functionality. Indeed, Jeffery (2005) has proposed that proteomic analysis of proteins in 'unusual' sites is one way of identifying moonlighting proteins. Clearly, the mere presence of a protein on, say, the cell surface, does not mean that it is necessarily a moonlighting protein. However, it is a start for such analyses. Over the past decade a number of bacteria have been subject to proteomic analysis and a surprising number of cytosolic proteins have been found to exist on the outer bacterial surface. Various different streptococci have been studied in this regard and there is interesting variation and similarity in the populations of secreted cell surface proteins. In *Streptococcus agalactiae* the following cytosolic proteins were identified on the outer surface of the cell: PGI, non-phosphorylating GAPD, enolase, purine nucleotide phosphorylase, ornithine carbamoyltransferase, cysteine synthase, superoxide dismutase, chaperonin 60 and DnaK (Hughes et al. 2002). In *Streptococcus oralis*, 27 cell surface proteins were identified including the glycolytic enzymes: fructose bisphosphate aldolase, triosephosphate isomerase (TPI), GAPD, enolase and phosphoglycerate kinase. However chaperonin 60 and DnaK were absent. In addition, the population of proteins on the cell surface changed when bacteria were grown under acidic conditions (Wilkins et al. 2003). Another streptococcal species, the swine pathogen, *Streptococcus suis,* has another set of cell surface proteins including: phosphoglycerate mutase, 6-phosphofructokinase, phosphopentomutase, amylase-binding protein, acetylserine lyase, chaperonin 10, chaperonin 60 and GrpE (Hsp90) (Wu et al. 2008). *Streptococcus pneumoniae* also has a number of glycolytic enzymes on its cell surface including lactate dehydrogenase. Interestingly, antibodies to many of these pneumococcal proteins are found in healthy individuals (Ling et al. 2004).

Surprisingly, the papers on the cell surface location of enzymes in the streptococci suggest that the whole of the glycolytic pathway enzyme cohort could be present on the bacterial cell surface. This raises the speculation that some bacteria may be able to drive glycolysis at the cell surface. To do this would require a source of glucose and ADP/ATP. It is possible that levels of these essential glycolytic substrates, present within specific environments of the bacterial host, may be high enough to drive this pathway. Indeed, a recent study has concluded that tethering glycolytic enzymes closely together on two-dimensional matrices dramatically increases the efficiency of glycolysis (Mukai et al. 2009). What would be the advantage to cell surface glycolysis? The products of glycolysis include a range of sugar metabolites and ATP. It is now appreciated that ATP is a potent signal, acting through ionotropic P2X receptors and metabotropic P2Y receptors (Corriden and Insel 2010). In addition, it is possible that extracellular glycolytic intermediates may have intercellular signalling actions. If extracellular glycolysis does exist, and has cell signalling activity, this would be the first example of a moonlighting pathway rather than a moonlighting protein.

It is expected that the glycolytic pathway and its individual enzymes will present us with many surprises in the years to come and some of these surprises will be due to moonlighting actions. As an example, the glycolytic pathway in the model organism, *Bacillus subtilis*, is revealing an interesting pattern of protein–protein interactions. Two hybrid analyses have revealed interactions between phosphofructokinase, phosphoglyceromutase and enolase, suggesting these enzymes may form a complex within the cytoplasm. In addition, such interatomic analysis has established interactions between glycolytic enzymes and proteins involved in RNA degradation, suggesting the presence of additional moonlighting functions for these proteins (Commichau et al. 2009).

It is still early days in the study of bacterial protein moonlighting. However, it is already clear that a theme is emerging in the role of such proteins. Most of them seem to play some role in bacterial pathogenesis, either by acting as secreted cell signalling agents and/or cell surface bacterial adhesins. A confusing factor in the literature is the separation between the adhesive function and the signalling function of individual moonlighting proteins and it appears that many bacterial moonlighting proteins can both act as ligands for host receptors and as adhesins for either host matrix components and/or receptor proteins on the host cell.

The remainder of this article will deal separately with the signalling and the adhesive actions of bacterial moonlighting proteins and the contribution such moonlighting makes to bacterial virulence behaviour. Bacteria are not the only microbes to utilise moonlighting proteins and examples will also be given of moonlighting in eukaryotic microbes where appropriate.

4 Bacterial Moonlighting Proteins that Signal to the Host

It is now well-recognised that signalling between commensal or pathogenic bacteria and their hosts is essential for mutual bacterial-host survival and for bacterial pathogenesis. Understanding such signalling has given rise to the discipline of cellular microbiology (Henderson et al. 1999). One of the most prevalent classes of bacterial moonlighting proteins, which function as extracellular signals for the host, are molecular chaperones and protein-folding catalysts. As will be seen, this type of protein is also implicated in moonlighting bacterial adhesion and cell invasion. Some of the earlier literature has been reviewed in Henderson et al. (2006).

4.1 Molecular Chaperones and Protein-Folding Catalysts

Each cell contains a huge amount of protein [between 80 and 400 g/L (Ellis and Minton 2006)] concentrated into a small volume. Such large protein concentrations favour the misfolding of nascent proteins and the aggregation of proteins generally. Such protein misfolding and protein aggregation is enhanced when cells are subject to stress (now termed proteotoxic stress). The evolutionary solution to protein misfolding and protein aggregation is the molecular chaperone (Gregersen and Bross 2010) and protein-folding catalyst (PFC) (Gething 1997). These proteins aid in: (1) the correct folding of proteins, (2) the inhibition of protein aggregation and (3) the solubilisation of existing protein aggregates. Moreover, many of these proteins (collectively termed cell stress proteins) dramatically increase in concentration in stressed cells (Richter et al. 2010). There are now around 25 families of molecular chaperones and PFCs (Gething 1997; Makarow and Braakman 2010) with probably a few hundred proteins (many being homologues) now identified (Table 2). New molecular chaperone types are still being found. For example, analysis of protein folding in *Escherichia coli* has discovered a novel molecular chaperone termed Spy (Quan et al. 2011). Many of the molecular chaperones and PFCs are essential for cell viability and individual cell stress protein gene inactivation leads to non-viable cells.

The signalling activity of molecular chaperones and PFCs, which will, collectively, be termed cell stress proteins in this review, was first discovered with eukaryotic proteins, and a growing number of these proteins have been reported to

Table 2 Prokaryotic Molecular Chaperones[1]

Chaperonin (Hsp)10
Thioredoxin family
Glutaredoxin
Trigger factor
Peroxiredoxins
Small heat shock proteins
Peptidylprolyl isomerases
DnaJ/Hsp40
GrpE
Protein disulphide isomerases
Chaperonin (Hsp)60
DnaK/Hsp70
HtpG/Hsp90
ClpA/Hsp100 family
Spy

[1] Names highlighted are the molecular chaperones and protein-folding catalysts from bacteria currently known to moonlight. In eukaryotes most of the homologues of these prokaryotic proteins have moonlighting actions

be secreted by cells and to function as intercellular signalling proteins or cell surface receptors (Henderson and Pockley 2010). While there are many individual prokaryotic molecular chaperones, only four of these proteins, in order of increasing molecular mass: chaperonin (Hsp)10, peptidylprolyl isomerase (PPI), chaperonin (Cpn) (sometime heat shock protein (Hsp 60) and DnaK (Hsp70) have been reported to have the ability to signal to human cells. This does not mean that other bacterial cell stress proteins do not act as cell signals, only that they have not yet been tested for this ability. This compares with around 16 eukaryotic cell stress proteins with cell signalling activity (Henderson and Pockley 2011). Most studies of moonlighting cell stress proteins in bacteria have focused on three pathogens: *M. tuberculosis*, *Chlamydia pneumoniae* and *Helicobacter pylori* with a few other organisms studied in far less detail. The signalling actions of the cell stress proteins of these individual organisms will be dealt with in turn. The different bacteria that utilise moonlighting proteins in bacteria-host interactions are delineated in Table 3. Note that there is some confusion with the nomenclature of the 60 kDa protein which in bacteria is generally known as Cpn60 (sometimes GroEL) while in humans the more common term is Hsp60.

4.1.1 Mycobacterium tuberculosis

Chaperonin (Cpn) also known as heat shock protein (Hsp)10 is a heptamer composed of 10 kDa subunits and forms a cap for the chaperonin (Cpn)60 protein to aid in protein folding (Horwich et al. 2009). The human Cpn10 protein was originally identified as an immunosuppressant protein termed early pregnancy

Table 3 The bacteria currently known to employ moonlighting proteins

Bacterium	Moonlighting proteins
Aeromonas hydrophila	Enolase
Aeromonas salmonicida	Phosphoglycerate kinase
Aggregatibacter actinomycetemcomitans	Cpn60
Bacillus anthracis	GAPD
Bacillus stearothermophilus	PGI
Bacillus subtilis	Glutamate dehydrogenase
Bartonella bacilliformis	Cpn60
Bifidobacterium animalis	Hsp70
Bifidobacteria spp	Enolase
Subsp. lactis	
Borrelia burgdorferi	Cpn60
Brucella abortis	Cpn60
Candida albicans[a]	Phosphoglycerate kinase, phosphoglycerate mutase
Chlamydia pneumoniae	Cpn60.1
Chlamydia trachomatis	Cpn60.2, PPI
Clostridium difficile	Cpn60
Coxiella burnetti	Hsp70
Entamoeba invadens[b]	Enolase
Enterobacter aerogenes	Cpn60
Enteropathogenic *E. coli*	Hsp70, GAPD
Enterohaemorrhagic *E. coli*	GAPD
Francisella tularensis	Cpn60
Haemophilus ducreyi	Cpn60
Haemophilus influenza	Hsp70
Helicobacter pylori	Hsp20, Hsp60, Hsp70, PPI, HtrA
Histoplasma capsulatum[a]	Cpn60, PPI
Lactobacillus crispatus	GAPD, enolase
Lactobacillus jensenii	Enolase
Lactobacillus johnsonii	Cpn60
Lactococcus lactis	Cpn60, Hsp70, pyruvate kinase
Lactobacillus plantarum	GAPD, enolase
Legionella dumoffi	DnaK (DjlA)
Legionella pneumophila	Cpn60, Hsp70, PPI
Leishmania donovani[b]	Hexokinase
Listeria monocytogenes	GAPD, InlB, alcohol acetaldehyde dehydrogenase, IspC
Mycobacterium avium	Cpn60, Hsp70
Mycobacterium bovis BCG	Cpn60.1
Mycobacterium leprae	Cpn60.2
Mycobacterium smegmatis	Cpn60.1
Mycobacterium tuberculosis	Cpn10, Cpn60.1, Cpn60.2, Hsp70, GAPD, glutamate racemase, mycosyltransferases, malate synthase, phosphodiesterase, phosphoglycerate mutase, superoxide dismutase, aconitase

(continued)

Table 3 (continued)

Bacterium	Moonlighting proteins
Mycobacterium ulcerans	Small heat shock protein (18 kDa)
Mycoplasma pneumoniae	Ef-TU, β-subunit of pyruvate dehydrogenase
Mycoplasma suis	GAPD
Neisseria gonorrhoeae	PPI
Neisseria meningitiditis	Peroxiredoxin,Hsp70, GAPD, fructose-bisphosphate aldolase
Paenibacillus larvae	Enolase
Phycomyces blakesleeanus[a]	GAPD
Plesiomonas shigelloides	Cpn60
Rhizobium leguminosarum	Cpn60.1
Rickettsia prowazekii	PPI
Salmonella enterica	Cpn60
Serovar Typhimurium	
Schistosoma bovis[b]	GAPD
Staphylococcus aureus	GAPD, triose phosphate isomerase, Aaa
Staphylococcus epidermidis	GAPD, Aae,
Streptococcus agalactiae	Cpn60, Hsp70, GAPD, phosphoglycerate kinase, enolase, C5a peptidase
Streptococcus caprae	Atlc
Streptococcus equisimilis	GAPD, enolase
Streptococcus gordoni	Enolase
Streptococci oral	Phosphoglycerate mutase
Streptococcus mutans	AtlA
Streptococcus pneumoniae	PPI, GAPD, fructose-bisphosphate aldolase, enolase, 6-phosphogluconate dehydrogenase
Streptococcus pyogenes	GAPD, enolase, serum opacity factor, SpnA
Streptococcus sobrinus	Enolase
Streptococcus saprophyticus	Aas
Streptococcus suis	Enolase, 6-phosphogluconate dehydrogenase
Toxoplasma gondii[b]	Fructose-bisphosphate aldolase
Trichomoas vaginalis[b]	GAPD, pyruvate:ferredoxin oxidoreductase
Trypanosoma brucei[b]	GAPD
Xanthomonas oryzae pv. oryzae	Chorismate mutase
Xenorhabdus nematophila	Cpn60
In red-fungus	
In blue-protozoan	

Fungus and protozoan are designated with superscript letters 'a' and 'b' respectively.

factor (Noonan et al. 1979) and has been shown to inhibit both macrophage (Johnson et al. 2005) and T cell function (Zhang et al. 2003) and is being used in clinical trials for a variety of diseases including rheumatoid arthritis (Vanags et al. 2006) and psoriasis (Williams et al. 2008). It has also recently been reported to be an endothelial-derived differentiation factor (Dobocan et al. 2009). Thus the human homologue has some interesting moonlighting actions.

The mycobacterial Cpn10 protein is a powerful inducer of Th1 responses in patients with leprosy (Launois et al. 1995). Administration of *M. tuberculosis* Cpn10 to rats with adjuvant arthritis (Ragno et al. 1996) or mice with experimental allergic asthma (Riffo-Vasquez et al. 2004) inhibited these distinct experimental lesions. Curiously, in spite of these various findings, it has also been reported that *M. tuberculosis* Cpn10 is a potent inducer of osteoclast formation and bone resorption in in vitro bone cell and organ cultures, with activity being associated with the flexible loop and residues 65–70, suggesting there is a single moonlighting conformational unit which encompasses the bone-resorbing activity (Meghji et al. 1997).

Most information on moonlighting cell stress proteins in *M. tuberculosis* comes from the study of the chaperonin 60 proteins of this bacterium. Most mycobacteria encode two Cpn60 proteins termed Cpn60.1 and Cpn60.2 (Kong et al. 1993). The Cpn60.2 protein was the first such chaperonin discovered in *M. tuberculosis* and is also known as Hsp65. This has been long recognised as a potent immunomodulatory protein influencing T cell function (van Eden et al. 2005). One of the first reports that molecular chaperones have extracellular signalling activity with human leukocytes was of the stimulation of human monocyte cytokine synthesis by *M. tuberculosis* Cpn60.2 (Friedland et al. 1993). This has led on to the study of a range of Cpn60 proteins from other bacteria and from rodents and humans (Henderson and Pockley 2010). The assumption from Friedland's study was that *M. tuberculosis* Cpn60.2 stimulated so-called 'classic activation' of monocytes. This is the type of activation induced by bacterial lipopolysaccharide (LPS) or the cytokine, gamma-interferon (γ-IFN), and primes macrophages to ingest bacteria and present bacterial antigens to T lymphocytes. Markers of this form of activation include increases in cell surface MHC class II and Fc receptor expression and upregulation of macrophage free radical production. The possibility that macrophages could undergo other patterns of activation, now termed alternative macrophage activation, was only introduced by Siamon Gordon in 1992 [(Stein et al. 1992) see review by Martinez et al. (2009)]. When macrophages exposed to *M. tuberculosis* Cpn60.2 were examined for markers of classic macrophage activation, such markers were not present. Thus the influence of this molecular chaperone on human macrophages is to induce proinflammatory cytokine synthesis without, at the same time, inducing other cellular changes found in classically-activated macrophages (Peetermans et al. 1994). The exact alternative activation state induced by *M. tuberculosis* Cpn60.2 has not been identified. Thus this particular Cpn60.2 protein appears to be inducing an alternative macrophage activation state. Indeed, it has been proposed that many molecular chaperones are able to induce such alternative macrophage activation states (Henderson and Henderson 2009). As will be discussed in the next section, *M. tuberculosis* Cpn60.2 also acts as an adhesin, binding to CD43 on the macrophage surface (Hickey et al. 2009, 2010). Such receptor binding may be responsible for this particular form of macrophage activation induced by Cpn60.2.

There is a growing realisation that the normal and pathological turnover of the largest organ system of the body, the skeleton, is controlled by cells of the immune system. This has given rise to a new discipline termed osteoimmunology (Takayanagi 2009). Work from the principal author's lab revealed that the Cpn60 protein from the oral pathogen, *Aggregatibacter actinomycetemcomitans*, was a potent inducer of bone destruction in vitro (Kirby et al. 1995). The homologous *E. coli* protein, GroEL, was also a potent stimulator of bone resorption (Kirby et al. 1995) and turned out to be a very active promoter of the formation of the multi-nucleated osteoclast population of bone—the cells responsible for bone breakdown (Reddi et al. 1998). Curiously, the *M. tuberculosis* and *Mycobacterium leprae* Cpn60.2 proteins failed to stimulate bone breakdown (Kirby et al. 1995) as did the recombinant *M. tuberculosis* Cpn60.1 protein (Meghji et al. 1997). The two *M. tuberculosis* Cpn60 proteins have >60% sequence identity and it was assumed they would have identical biological activity. Direct comparison of their effects on human monocytes revealed significant differences in their potency as cytokine inducers (Lewthwaite et al. 2001) and direct competition studies have shown that neither protein competes with the other for binding to human monocytes (Cehovin et al. 2010). Thus inspite of their sequence similarities, these proteins act as distinct cellular ligands and signalling proteins (Henderson et al. 2010). This was clearly shown when the effects of the two *M. tuberculosis* chaperonin 60 proteins were re-examined for their influence on bone resorption and on the formation of osteoclasts. This work showed that the Cpn60.2 protein had no positive or negative effects on bone breakdown or osteoclast formation. In contrast, the Cpn60.1 protein proved to be a potent inhibitor of bone breakdown and osteoclast formation in vitro and blocked the massive osteoclast-driven bone destruction found in the joints of rats with adjuvant arthritis, without inhibiting joint inflammation. This inhibition of osteoclast formation was associated with the repression of transcription of the key osteoclast transcription factor NFATc1 (Winrow et al. 2008).

The Cpn60.1 protein from *M. tuberculosis* is a potent inhibitor of the eosinophilia and bronchial hyperreactivity found in mice with experimental allergic asthma. The homologous Cpn60.2 protein was completely inactive in this model (Riffo-Vasquez et al. 2004). This has proved to be a very intriguing finding as an earlier study had shown that the *M. tuberculosis* Cpn60.2 protein was also inactive in this model, yet the *M. leprae* Cpn60.2 protein was a potent inhibitor of this allergic asthma model (Rha et al. 2002). What makes this finding intriguing is that the *M. leprae* and *M. tuberculosis* Cpn60.2 proteins exhibit >95% sequence identity and comparison of the Cpn60.1 and Cpn60.2 sequences from these two bacteria reveals only a handful of residues that could account for this major difference in biological activity of the Cpn60.2 proteins. These few different residues presumably identify the sequence of the moonlighting site that accounts for the anti-asthmatic effects of these proteins. Thus, only minor changes in sequence seem capable of producing distinct biologically active sites in, atleast, this protein. The *M. tuberculosis* Cpn60.1 protein has also been shown to inhibit macrophage activation by the proinflammatory component of this bacterium, PPD (purified protein derivative). Such inhibition involves modulation of cell activity via TLR2 signalling (Khan et al. 2008).

The evidence suggests that the chaperonins of *M. tuberculosis* are important signalling proteins with some role to play in the interaction between the bacterium and the host macrophage. Curiously, the *M. tuberculosis* Cpn60 proteins are structurally and enzymatically distinct from the prototypic Cpn60 protein, the *E. coli* GroEL. GroEL is a tetradecamer comprising two 7-membered rings stacked back to back, giving a molecular mass of around 860 kDa, with significant ATPase activity (Krishna et al. 2007). In contrast, the Cpn60 proteins from *M. tuberculosis* appear not to form tetradecameric structures at the concentrations at which GroEL does and their ATPase activity is very low (Qamra et al. 2004). Further, the crystal structure of the *M. tuberculosis* Cpn60.2 protein is a dimer and not a tetradecamer (Qamra and Mande 2004). It is unclear how these 60 kDa chaperonins will fold proteins if they cannot form at least one ring structure. To address the importance of the chaperonins in the virulence of *M. tuberculosis* efforts were made to inactivate the genes encoding Cpn10, Cpn60.1 and Cpn60.2 in the virulent *M. tuberculosis* strain H37Rv (Hu et al. 2008). As *cpn10* and *cpn60.1* appeared to form an operon, it was assumed that these were the genes coding for the proteins essential for survival and therefore that only the gene encoding Cpn60.2 would be able to be inactivated. In fact, the only one of the three chaperonin genes able to be inactivated was *cpn60.1*. The *Δcpn60.1* isogenic mutant grew normally in culture and within quiescent and activated macrophages and responded to major stresses in an identical manner to the wild type organism. Thus it looked like the Cpn60.1 protein was not acting as a stress protein. This was confirmed in complementation experiments, where it was shown that the *cpn10* and *cpn60.2* genes would complement an *E. coli* mutant with conditional inactivation of the *groES* (*cpn10*) and *groEL* (*cpn60*) genes. However, the *cpn60.1* gene failed to complement this mutant, supporting the hypothesis that this protein has evolved away from protein folding to some other function. This is also the conclusion from study of the *Mycobacterium smegmatis* Cpn60 proteins (Rao and Lund 2010) and may be a more general finding [reviewed by Lund (2009)]. With no in vitro phenotype it was assumed that the *Δcpn60.1* mutant would behave normally when used to infect animals. However, in both infected mice and guinea pigs, although the *Δcpn60.1* mutant grew as well as the wildtype organism, it failed to induce a granulomatous inflammation in the lungs (Hu et al. 2008). This finding was supported by in vitro studies of human granuloma formation from whole blood. Again, the *Δcpn60.1* mutant failed to induce the formation of multinucleate giant cells from monocyte precursors (Cehovin et al. 2010). This reveals some very interesting effects of the *M. tuberculosis* Cpn60.1 protein on myeloid cell differentiation. Thus this protein prevents the formation of osteoclasts, but its absence is associated with the failure to generate multinucleate giant cells, suggesting that this protein is somehow inducing giant cell formation and the formation of granulomas (Fig. 2). As the granuloma is the hallmark of mycobacterial infection, this finding suggests that the Cpn60.1 protein is an important contributor to this form of inflammation and is therefore a key virulence factor of *M. tuberculosis*.

In this context, the differences in the biological actions of the Cpn60 proteins of the plant symbiotic bacterium, *Rhizobium leguminosarum*, are relevant. This

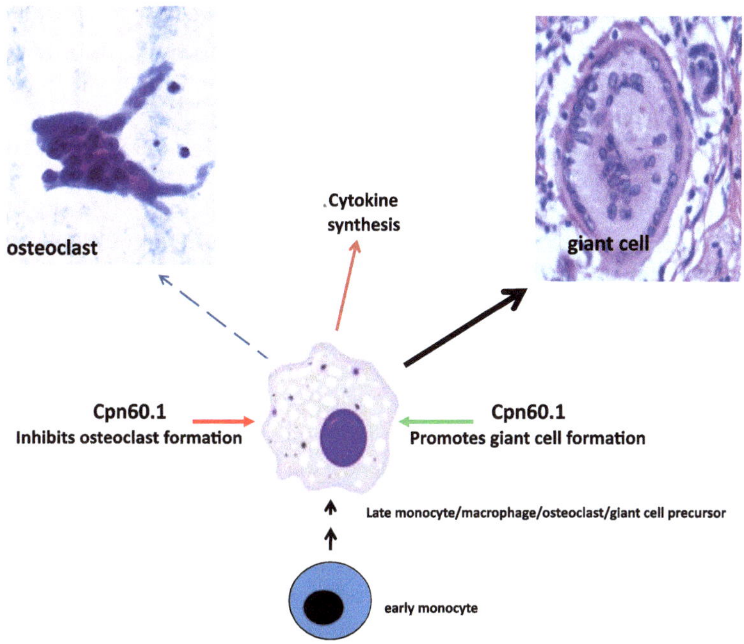

Fig. 2 Interactions of *M. tuberculosis* chaperonin 60.1 with myeloid Cells. The Cpn60.1 protein can stimulate monocyte/macrophage cytokine synthesis but it is not known if this protein can cause classical activation of macrophages. However, the Cpn60.1 protein is able to inhibit the development of osteoclasts from osteoclast precursor cells (both in vitro and in vivo) while also being able to promote the formation of multinucleate giant cells from their precursors. This means that the *M. tuberculosis* Cpn60.1 protein is a molecular probe to dissect the differences between osteoclasts and multinucleate giant cells. The mechanism of inhibition of osteoclastogenesis and giant cell formation are currently unknown

bacterium encodes three Cpn60 proteins of which the Cpn60.1 is the main housekeeping chaperonin and present in the highest amount while the other two are expressed at low levels and are not required for protein folding. Comparison of the human monocyte cytokine-inducing activities of the Cpn60.1 and Cpn60.3 proteins of *R. leguminosarum* revealed that in spite of around 80% sequence identity, only the Cpn60.3 protein was able to induce monocyte cytokine synthesis. The major chaperonin, Cpn60.1, was without any monocyte activating activity. It is not known what sequence or structural differences between these two proteins are responsible for this major difference in biological activity. It also shows that even bacteria, which have no normal interactions with mammals, may have proteins able to influence immune functions (Lewthwaite et al. 2002).

The other major molecular chaperone that functions as a signalling molecule in *M. tuberculosis* is Hsp70 or DnaK. There are at least 13 human Hsp70 genes, of which at least three have been shown to act as intercellular signals, and one of these Hsp70 proteins, BiP (an immunosuppressive protein), is now in clinical trial for the treatment of rheumatoid arthritis (Henderson and Pockley 2010). As has been touched upon, tuberculosis can be thought of as a disease of the macrophage/dendritic cell populations (Mortellaro et al. 2009) in which the production of specific chemokines is important in the immunopathology of the disease (Mendez-Samperio 2008). Lehner's group in London, UK, were the first to show that the *M. tuberculosis* Hsp70 protein signalled to leukocytes, in this case primate CD8 lymphocytes, causing the release of the CC chemokines CCL3-5 (Lehner et al. 2000). Studies of the effects of *M. tuberculosis* Hsp70 on myeloids cells has shown that this protein stimulates monocytes to secrete CCL3-5 through a mechanism involving the transmembrane receptor of the TNFα gene superfamily, CD40 (Wang et al. 2001). This is an interesting finding in light of the recent report that CCL5 participates in protection against *M. tuberculosis* in the early phase of infection (Vesovsky et al. 2010). Most reports of Hsp70 binding to monocytes supports the hypothesis that signalling is through TLR4, although a number of other receptors have been implicated (Henderson and Pockley 2010). There is one study of the human Hsp70 protein that concludes that this protein does bind to CD40. However the receptor binding site in the human Hsp70 protein is within the N-terminal ATP-binding domain (Becker et al. 2002), which is distinct from the identified receptor binding site in the *M. tuberculosis* Hsp70 protein (Wang et al. 2002). This clearly indicates that the moonlighting activity of these two Hsp70 proteins evolved independently. Truncation mutagenesis and peptide mapping identified the receptor binding site of *M. tuberculosis* Hsp70 to a 20-mer peptide sequence within the C-terminus (Wang et al. 2002, 2005). Later studies from the same group also revealed that *M. tuberculosis* Hsp70 also bound the HIV co-receptor CCR5 (Whittall et al. 2006; Floto et al. 2006). Given the known synergy between infection with tuberculosis and HIV this is a very interesting finding and raised the question as to whether *M. tuberculosis* Hsp70 can block HIV binding to target cells? This experiment has been done and the answer is that *M. tuberculosis* Hsp70 can block HIV uptake, suggesting that this protein has some therapeutic potential (Babaahmady et al. 2007).

Since these studies were conducted it has been shown that *M. tuberculosis* releases large amounts of Hsp70 (Hickey et al. 2009). Indeed, an earlier paper had reported that *Mycobacterium bovis* released two ATPases which could inhibit ATP-induced monocyte apoptosis. One of these two enzymes turned out to be the Hsp70 protein of this bacterium (Zaborina et al. 1999).

4.1.2 Helicobacter pylori

This organism colonises the stomach, which must be one of the most stressful environments for any bacterium to occupy. Thus it would be expected to have an active cell stress response. It is therefore not surprising that *H. pylori* Cpn60 is a dominant antigen in patients with gastric disease due to this bacterium and that this is a useful diagnostic marker (Macchia et al. 1993; Yunoki et al. 2000). As with a number of other bacteria (to be discussed) Cpn60 is also found on the cell surface of *H. pylori*, which may account for its immunogenicity (Yamaguchi et al. 1996). Surprisingly, a monoclonal antibody to *H. pylori* Cpn60, when cultured with *H. pylori*, inhibited the growth of the organism, suggesting that, somehow, the surface location of this molecular chaperone controls intracellular mechanisms involved in cell division (Yamaguchi et al. 1997a). It is not known if this is related to the report that the cell surface Cpn60 of *H. pylori* binds to lactoferrin, which might have effects on cell growth (Amini et al. 1996).

Like the *M. tuberculosis* Cpn60 proteins, it has been reported that *H. pylori* Cpn60 can stimulate monocytes (Lin et al. 2005) and epithelial cells (Yamaguchi et al. 1999) to secrete cytokines. There is confusion in the literature in respect of the receptors required for such activation and the nature of the activating ligand. Most studies claim that the recombinant *H. pylori* Cpn60 protein works by binding to TLR2 or TLR4 (e.g. Takenaka et al. 2004; Zhao et al. 2007). However, using a non-recombinant form of the *H. pylori* Cpn60 protein (i.e. purified directly from *H. pylori* cultures) no requirement for TLR2/4 or myeloid differentiation factor (MyD)88 was identified (Gobert et al. 2004). It is unclear what is responsible for the differences in these results. The *H. pylori* Cpn60 protein, like that of the *M. tuberculosis* Cpn60 proteins, in isolation, is a dimer or tetramer and not a tetradecamer. The activity of this protein in stimulating cytokine synthesis is reported to depend on the oligomeric status of the chaperone (Lin et al. 2009) so this may contribute to these differences in receptor requirement. This conflicts with the finding that the isolated equatorial domain of the *M. tuberculosis* Cpn60.1 protein is as active as the native protein (Tormay et al. 2005) and, again, reveals that moonlighting actions of homologous proteins can exhibit promiscuity of mechanism. In addition to acting as a proinflammatory stimulus, the *H. pylori* Cpn60 protein is proposed to be involved in the process of gastric carcinoma formation (e.g. Lin et al. 2010).

The C-terminus of the *M. tuberculosis* Cpn60.1 protein contains 6 histidine residues and this protein can be directly purified on a Ni–NTA affinity column (Kong et al. 1993). It has recently been reported that *H. pylori* contains a protein, HspA, that is a homologue of *E. coli* GroES. It has now been found that HspA contains a unique histidine-rich C-terminal extension that binds nickel. Deletion of the gene encoding HspA caused a decrease in intracellular nickel content and reduced nickel tolerance suggesting this Cpn60 homologue is a moonlighting protein involved in nickel sequestration and detoxification (Schauer et al. 2010).

The other molecular chaperone which acts as a *H. pylori* moonlighting signalling protein is the peptidyl prolyl isomerase (PPI 1075). Like Cpn60, this protein is secreted and is immunogenic in patients with gastric ulceration (Atanassov et al. 2002). There is growing evidence that human PPIs of the cyclophilin class are secreted and function as proinflammatory factors or cell growth factors (Henderson and Pockley 2010). In contrast, the *H. pylori* PPI (1075), when cultured with a gastric epithelial cell line, induced apoptosis of these cells by a mechanism involving TLR4 and apoptosis signal-regulating kinase 1 (Basak et al. 2005). Inactivation of the gene resulted in a mutant with impaired ability to induce epithelial cell apoptosis. In addition to gastric epithelial cell destruction, the gastropathy associated with *H. pylori* infection involves an inflammatory response with overexpression of cytokines, particularly IL-6. Again, the PPI of *H. pylori* is a major inducer of monocyte-induced IL-6 production. Inactivation of the gene encoding this PPI results in an isogenic mutant with attenuated IL-6-inducing activity (Pathak et al. 2006).

The so-called high-temperature requirement A (HtrA) proteins of eukaryotes and prokaryotes are chaperones and serine proteases which monitor intracellular protein quality control. It has been found that *H. pylori* secretes this protein, which can proteolytically cleave E-cadherin allowing it to be shed and causing epithelial cell dissociation and, it is assumed, changes in cell signalling. This presumably can allow the bacterium to access the intercellular space in the gut (Hoy et al. 2010).

4.1.3 Chlamydia spp

The Chlamydia are obligate intracellular bacteria and the two best known in terms of human disease are *C. pneumoniae*, which causes approximately 10% of cases of community-acquired pneumonia and 5% of cases of bronchitis (Burrilo and Bouzo 2010) and *Chlamydia trachomatis* which is best known for its infection of the genital tract (Darville and Hiltke 2010) and for trachoma (Burton and Mabey 2009). The Chlamydia are also implicated in the pathogenesis of atherosclerosis (Watson and Alp 2008). Most attention has focused on the moonlighting actions of the Cpn60 protein of the Chlamydiae due to the perceived role of this protein in the pathogenesis of atheroma. It has already been pointed out that some bacteria have more than one Cpn60 protein. The Chlamydiae have three *cpn60* genes (Karunakaran et al. 2003) with all work on the signalling actions of these proteins being done on the Cpn60.1/GroEL1 protein. Of interest, in this context, is the report that the gene encoding Cpn60.2 in one serovar of *C. trachomatis* is under the response of the extracellular iron content, being greatly increased when the bacterium is maintained in iron-deficient conditions (LeaRue et al. 2007). The first report of the signalling actions of *C. pneumoniae* Cpn60.1 was the ability of this protein to stimulate monocytes to secrete proinflammatory cytokines and metalloproteinases (Kol et al. 1998). It was then demonstrated that recombinant *C. pneumoniae* Cpn60.1 stimulated murine monocytes and human microvascular endothelial cells through, what is perceived to be, a conventional TLR4/MD-2/Myd88-dependent pathway (Bulut et al. 2009). Activity was heat labile and blocked by antibodies to *C. pneumoniae* Cpn60.1, thus controlling for LPS contamination. Recombinant *C. pneumoniae* Cpn60.1 also stimulated maturation of murine bone-marrow-derived dendritic cells in a TLR-2/4-dependent manner (Costa et al. 2002). A similar effect has been reported with *C. pneumoniae* Cpn60.1 as an inducer of human monocyte-derived dendritic cell maturation, which involved induction of expression of IL-12 and IL-23 (Ausiello et al. 2006).

It has already been described that administration of mycobacterial Cpn60.1 proteins can have therapeutic effects. However, these proteins have not been administered to healthy animals. In vivo administration of purified chlamydial Cpn60.1 to the peritoneal cavities of mice resulted in increased serum levels of the CXC chemokines CXCL1 and CXCL2 (both neutrophil chemoattractants) and marked accumulation of neutrophils. Significantly, Cpn60.1 was a more potent neutrophil attractant than was endotoxin or the CpG oligonucleotide 1668 (Da Costa et al. 2004) and required participation of both TLR2 and TLR4. Intratracheal administration of recombinant *C. pneumoniae* Cpn60.1 in wild type mice resulted in local accumulation of inflammatory cells and up-regulation of cytokine levels (Bulut et al. 2009). These findings support the hypothesis that at least this bacterial Cpn60 protein is proinflammatory.

In addition to stimulating cellular cytokine synthesis, it has been reported that *C. pneumoniae* Cpn60, but not Cpn10, is capable of inducing the oxidation of low density lipoprotein (LDL) (Kalayoglu et al. 2000). It also promotes the proliferation of human vascular smooth muscle cells by a mechanism dependent on TLR4 binding and the activation of p44/42 MAP kinase (Sasu et al. 2001).

Finally, a peptidyprolyl isomerase of *C. trachomatis* has been shown to be involved in the invasion of the targets cells by this bacterium (Lundermose et al. 1993).

4.1.4 Other Bacteria

There are a number of sporadic reports of molecular chaperones with various cell–cell signalling actions not covered in the previous sections. Most of these have used bacterial Cpn60 proteins. Some of these have multiple effects such as the Cpn60 protein of the oral bacterium *A. actinomycetemcomitans* which promote bone breakdown (Kirby et al. 1995), has effects on epithelial cell viability and cell turnover which may be relevant in situ (Zhang et al. 2004a) and stimulates epithelial cell migration (Zhang et al. 2004b). The Cpn60 of *Bartonella bacilliformis* is reported to induce apoptosis in cultured vascular endothelial cells (Smitherman and Minnick 2005). The probiotic organism *Lactobacillus johnsonii* La1 (NCC 533) has Cpn60 on its cell surface and like some other bacterial Cpn60 proteins it is able stimulate epithelial cell IL-8 synthesis (Bergonzelli et al. 2006). With most Gram-negative bacteria the LPS is the most proinflammatory component of the organism. However, the causative agent of Tularaemia, *Francisella tularensis*, induces a potentially fatal systemic disease, but has an LPS with virtually no proinflammatory activity. In this organism it is the Cpn60 protein that is the major proinflammatory signal and this chaperone actually synergises with the LPS to activate human macrophages (Noah et al. 2010).

There are a small, but growing, number of examples of bacterial moonlighting proteins being involved in bacterial invasion and post-invasion events. The gene encoding a DnaJ-like protein (DjlA) in *Legionella dumoffi* was found in a transposon mutagenesis analysis to identify genes involved in intracellular growth.

This has identified DjlA as a molecular chaperone involved in inhibiting lysosome-phagolysosome fusion in macrophages invaded by this organism (Ohnishi et al. 2004).

In addition to bacterial molecular chaperones signalling to human or rodent cells, there are a few examples of signalling to lower organisms. One of the most fascinating reports is of the insect known as the antlion or doodlebug (*Myrmeleon bore*) which bites and paralyses its prey through the action of an insect neurotoxin. This neurotoxin was isolated and turned out to be the Cpn60 protein of the bacterium, *Enterobacter aerogenes*, which is a commensal in the saliva of this insect. Curiously, the sequence of this Cpn60 protein was virtually identical to that of *E. coli* GroEL, and single residue mutations were sufficient to turn GroEL, which has no neurotoxic properties, into a potent insect neurotoxin. One such mutation was threonine at position 347, which conferred no neurotoxic activity on GroEL (Yoshida et al. 2001). Another member of the enterobacteriaceae is *Xenorhabdus nematophila*, a virulent insect pathogen and also a symbiotic organism (Herbert and Goodrich-Blair 2007). This bacterium also secretes a Cpn60 protein with insecticidal activity. Structure:function studies suggest that all three domains of the protein are needed for insecticidal activity and that this can be blocked by N-acetylglucosamine and chito-oligosaccharides. Generation of protein mutants identified the surface-exposed residues Thr347 (which was not involved in the neurotoxicity of GroEL) and Ser356 as essential for binding to the target insect gut epithelium and for insecticidal activity (Joshi et al. 2008). Clearly, these two Cpn60 turned insect neurotoxins are dramatically different in their structure:function relationships. However nothing is yet known of the mechanism through which Cpn60 exerts its neurotoxic effects. Yet again, this exemplifies the enormous variation that can occur in moonlighting proteins that have evolved to have the same moonlighting functions.

Another example of the role of the Cpn60 protein in cell–cell communication concerns biofilm formation in *M. smegmatis*. Most bacteria form biofilms, which have to be seen as equivalent to eukaryotic organ systems as they develop particular three-dimensional shapes to optimise oxygen and nutrient uptake and depend upon cell–cell interactions as well as secreted soluble signals (Hall-Stoodley et al. 2004). In *M. smegmatis*, biofilm formation does not occur in an isogenic mutant in which the *cpn60.1* gene has been inactivated (Ohja et al. 2005). In the absence of Cpn60.1, cells show normal planktonic growth but fail to form biofilms due to the role that this chaperonin plays in the formation of cell wall mycolic acids. In contrast, inactivation of the same gene in *M. tuberculosis* has no influence on biofilm formation (Hu et al. 2008). Another mycobacterial molecular chaperone involved in biofilm formation is the small (18 kDa) heat shock protein of *Mycobacterium ulcerans* (Pidot et al. 2010).

To summarise this section, the four bacterial cell stress proteins that have been identified have a staggering range of biological actions. It is expected that the many other cell stress proteins employed by bacteria will also exhibit moonlighting actions whose role can only be imagined.

4.2 Metabolic Enzymes

In eukaryotes, the preponderant proteins showing protein moonlighting activity are enzymes involved in metabolic processes, such as the glycolytic pathway or the TCA cycle. However, there are only a few such moonlighting proteins, identified to date, that act as cell signals. The same is true for prokaryotes, with only a small number of examples of metabolic enzymes functioning as intercellular signals. The potential cell signalling activity of the PGI of *B. stearothermophilus* has already been described (Sun et al. 1999). Another possible signalling glycolytic enzyme is the fructose-bisphosphate aldolase of *S. pneumoniae* which binds to a member of the cadherin superfamily member, flamingo (Blau et al. 2007). The cadherins are a well-established class of cell surface signalling receptor involved in a variety of cellular effects including leukocyte extravasation. Whether binding of this bacterial aldolase induces cell signaling remains to be determined.

The best example of a metabolic enzyme acting as a cellular signal is the glyceraldehyde 3-phosphate dehydrogenase (GAPD) of the group A streptococci. Indeed, the cell signalling properties of this streptococcal surface enzyme were identified before the introduction of the term protein moonlighting. A major surface protein of group A streptococci, termed streptococcal surface dehydrogenase (SDH), was identified as having homology, with high sequence similarity, to GAPD (Pancholi and Fischetti 1992). This protein was tightly bound to the cell surface, being unable to be removed by 2 M NaCl or 2% SDS and, as will be described, bound to a number of host proteins. This cell surface GAPD is found on virtually all streptococcal groups and in all group A streptococcal M strains tested (Pancholi and Fischetti 1992). It was then shown that this streptococcal GAPD could auto-ADP ribosylate itself at a cysteine residue, resulting in inhibition of the GAPD activity (Pancholi and Fiscetti 1993). This ADP ribosylation is also catalysed by eukaryotic GAPD enzymes (e.g. Dimmeler et al. 1992). ADP ribosylation was one of the first enzymic actions identified as being caused by bacterial toxins (Henkel et al. 2010) and it is now recognised that ADP ribosylation, caused either by endogenous or exogenous proteins, is a key mechanism in controlling cell functionality (Hottiger et al. 2010).

Did the cell surface GAPD on group A streptococci function to control the activity of target cells? Incubation of a human pharyngeal cell line with purified GAPD from *S. pyogenes* led to particular patterns of protein tyrosine phosphorylation in what is assumed to be membrane proteins. Such phosphorylation was related to the ability of group A streptococci to invade pharyngeal cells suggesting that the cell surface GAPD plays a major role in the invasiveness of this organism (Pancholi and Fischetti 1997). It turns out that streptococcal GAPD is a rather sticky protein and it took until 2005 for Pancholi to identify that there is a specific cell surface receptor for this protein—urokinase plasminogen activator receptor (uPAR Smith and Marshall, 2010)—on the surface of pharyngeal cells (Jin et al. 2005). Binding to uPAR can promote a wide range of cell signalling events and thus this could account for the significant tyrosine phosphorylation seen in earlier studies.

While these findings are extremely interesting, they really failed to capture the collective bacteriological imagination. This is largely due to the general feeling that GAPD on the cell surface is simply an artefact due to the death and dissolution of bacteria and the sticking of the released GAPD onto the surface of living bacteria. Indeed, this is often the general opinion of the finding of proteins 'where they should not be'—to ignore the finding. One of the great strengths of modern molecular microbiology is the capacity, in most organisms, to inactivate selected genes for the purpose of testing hypotheses about the said gene product. However, in group A streptococci, the gene encoding GAPD is essential and therefore cannot be inactivated, and so the hypothesis that GAPD is an evolved cell surface moonlighting protein could not be tested. In what can only be described as a brilliant innovation, Pancholi devised an alternative 'knockout strategy'. Instead of inactivating the *gapd* gene, his group generated a modified gene encoding a GAPD protein with a hydrophobic C-terminus in the hope that this protein would be unable to be secreted (Boel et al. 2005). This gene was used to replace the wild type gene and the isogenic mutant so generated was shown to grow normally and contain a functionally active GAPD. However, levels of cell surface GAPD were extremely low as assessed by surface enzyme activity or with immunofluorescence. Now this cell surface mutant bound much less to pharyngeal cells, but the level of invasiveness was not reported. What was surprising was that this GAPD cell surface isogenic mutant was extremely sensitive to killing by culturing the bacteria in whole blood. In other words, the isogenic mutant had lost its innate anti-phagocytic capacity. How this anti-phagocytic activity was produced by the cell surface GAPD was not clear. However, in 2006 it was shown that the GAPD of group A streptococci interacts with the key chemotactic complement component and anaphylotoxin, C5a. Such interaction can form a non-functional complex, or the GAPD can enhance the cell surface proteolysis of C5a (Terao et al. 2006). Such inhibition of C5a is an unexpected moonlighting action of this cell surface GAPD and reveals the richness of the moonlighting landscape of this one protein. Another example of the cell signalling actions of GAPD is the immunological activity of the *S. agalactiae* protein (Madureira et al. 2007). The recombinant GAPD from this bacterium stimulated the formation of B lymphocytes in mice in a non-antigenic manner. Moreover, a *S. agalactiae* strain overexpressing GAPD was more virulent than the wild type strain in mice and this virulence was minimised in IL-10-deficient mice and in mice treated with an anti-GAPD antiserum. Thus this particular GAPD is an interesting immunomodulatory factor contributing, as a virulence factor, to the immunopathogenesis of infection (Madureira et al. 2007).

Further support for the evolution of specific binding of GAPD to the outer bacterial cell wall comes from studies of the extracellular GAPD of *Lactobacillus plantarum*. Thus soluble GAPD from this bacterium does not bind to the organism's surface. Moreover, the presence of GAPD on the surface of this bacterium relates to the cell wall permeability of the organism (Saad et al. 2009).

There are only a few other examples of GAPD being involved in controlling microbial cellular function or in microbe–host interactions. Phagocytosis of bacteria which normally survive intracellularly requires control over the fusion of the phagosome and the endosomal/lysosomal compartments. A key element in such control is the small GTPase Rab5, and this can be subverted by the action of the intracellular pathogen, *Listeria monocytogenes* (Alvarez-Dominguez et al. 1996). The *Listeria* protein controlling Rab5 has been found to be GAPD and this protein showed similarity in activity to the ExoS toxin of *Pseudomonas aeruginosa*. This reveals another putative novel virulence activity of the GAPD protein (Alvarez-Dominguez et al. 2008). One of the key virulence protein families of *L. monocytogenes* are the internalins. This organism uses internalin (Inl)A and InlB to bind in a species-specific manner to the adhesion molecule E-cadherin and the hepatocyte growth factor receptor (HGFR) Met, respectively, to aid internalisation (Bonazzi et al. 2009). It is known that InlB mimics the intracellular action of Met in causing triggering of clathrin-dependent endocytosis and lysosomal degradation of Met. In other words InlB can be thought of as a moonlighting form of Met (Li et al. 2005). A completely distinct biological function for the ADP-ribosylating activity of GAPD is its involvement in controlling spore germination in the fungus *Phycomyces blakesleeanus* (Deveze-Alverez et al. 2001).

So at the present time only a few bacterial metabolic proteins are recognised to moonlight as cell signalling agonists. This contrast with a very much larger number of eukaryotic proteins which have such functionality. Does this reveal a major difference between prokaryotes and eukaryotes or is it simply that there are more scientists working on eukaryotic proteins than on prokaryotic proteins. Only time will solve this conundrum.

5 Bacterial Moonlighting Proteins that Function as Adhesins

Colonisation of host organisms with commensals or pathogens requires that the bacteria selectively adhere to some component of the host or to some other bacterium, which is itself attached, selectively, to the host. In the latter case the bacteria are forming a mixed species biofilm. It is now recognised that bacteria have evolved a very large number of molecules which have affinity for binding to one or other components of the host. These are generically termed adhesins and can be: sugars, lipids, glycolipids, peptides, peptidolipids, proteins, protein aggregates and so on (Ofek et al. 2003). However, most of the high affinity adhesins of bacteria are proteins and the story emerging is that many bacterial adhesins are moonlighting proteins. There is now a growing literature on bacterial moonlighting adhesins and it is rather difficult to subdivide the growing number of such proteins into rationale categories. So there will be a somewhat arbitrary division of these proteins into groupings such as molecular chaperones, glycolytic enzymes, other metabolic enzymes, moonlighting fibronectin adhesins and so on.

5.1 Molecular Chaperones Moonlighting as Bacterial Adhesins

At the present time, a large number of bacteria (and other microbes), both Gram-negative and Gram-positive, have been reported to have one or other molecular chaperones, or protein-folding catalysts, on their cell surface (Table 4). For a number of these organisms it has been shown that the cell surface molecular chaperone/protein-folding catalyst functions as an adhesin which is important in the colonisation of the microorganism. For a growing number of such cell surface cell stress proteins the identity of the host receptor is becoming known. The early literature suggested that Cpn60 and Hsp70 recognised a variety of glycolipids (Table 4). Indeed, it has been shown that both prokaryotic and eukaryotic Hsp70 proteins contain a specific sulphogalactolipid binding site (Mamelak et al. 2001). In the more recent literature there is evidence for these surface cell stress proteins acting to bind to known cell surface protein receptors. These receptors include CD43, DC-SIGN, CCR5, CD40 etc.

For some important pathogenic bacteria there is evidence that cell surface molecular chaperones and protein-folding catalysts contribute significantly to colonisation and bacterial invasion of host cells. Thus *Legionella pneumophila*, which causes Legionnaire's disease (Cianciotto 2001) has a key virulence factor termed macrophage infectivity potentiator (MIP) which is a member of the FKBP family of the peptidyl prolyl isomerases. Interestingly, the active site of this PPI is required for the ability of this bacterium to infect the target cells (Helbig et al. 2003). In addition to binding to cells and aiding invasion, the *L. pneumophila* PPI can also bind to a variety of collagens and this binding has been shown to be important in in vivo tissue invasion (Wagner et al. 2007). These findings make MIP an important therapeutic target and small-molecule MIP inhibitors are starting to be produced (Juli et al. 2010).

The chaperonin 60 protein is increasingly being recognised as a bacterial/microbial adhesin as well as being a signalling ligand. Early studies suggested that the causative agent of chancroid, *Haemophilus ducreyi*, used a cell surface-bound Cpn60 as an adhesin for binding target cells (Frisk et al. 1998). More detailed study of this binding phenomenon (Pantzar et al. 2006) revealed that the Cpn60 protein from *H. ducreyi* can bind to a variety of glycosphingolipids including lactosylceramide, gangliotriaosylceramide and GM3 ganglioside. In Chlamydiae species there are three genes encoding Cpn60 proteins (Karunakaran et al. 2003) with the Cpn60.1 protein appearing to be the major cell stress protein. This protein is also found on the cell surface of the organism and mediates the adhesion of this bacterium to host cells. Indeed, coating recombinant latex beads with *C. pneumoniae* Cpn60.1 allows these beads to attach to target cells. Surprisingly, the other two Cpn60 proteins of this bacterium have no cell adherent properties (Wuppermann et al. 2008). This, yet again, reveals the specific nature of moonlighting protein activity.

Returning to *L. pneumophila*, it has been recognised since the late 1990s that this bacterium has a cell surface Cpn60 (Garduno et al. 1998a, b). It is established from a number of studies that this cell surface Cpn60 protein is involved in both

Table 4 Microbial molecular chaperones and protein-folding catalysts present on the cell surface with the identified host receptor proteins where known

Bacterium	Molecular chaperone	Host receptor	Reference
A. actinomycetemcomitans	Cpn60	?	Goulhen et al. 1998
Bifidobacterium animalis Subsp. lactis	Hsp70	Plasminogen	Candela et al. 2010
Borrelia burgdorferi	Cpn60	Glycosphingolipid	Kaneda et al. 1997
Brucella abortis	Cpn60	?	Watarai et al. 2003
Chlamydia pneumoniae	Cpn60.1	?	Wuppermann et al. 2008
Chlamydia trachomatis	PPI	?	Lundermose et al. 1993
Clostridium difficile	Cpn60	?	Hennequin et al. 2001
Coxiella burnetti	Hsp70	?	Macellaro et al. 1998
Enteropathogenic *E. coli*	Hsp70	Sulphogalactosylceramide	Jesus et al. 2005
Haemophilus ducreyi	Cpn60	?	Frisk et al. 1998
Haemophilus ducreyi	Cpn60	Glycosphingolipids	Pantzar et al. 2006
Haemophilus influenza	Hsp70	?	Hartmann and Lingwood 1997; Hartmann et al. 2001
Helicobacter pylori	Hsp20	?	Du and Ho 2003
Helicobacter pylori	Cpn60	?	Yamaguchi et al. 1996, 1997a, b
Helicobacter pylori	Cpn60	Lactoferrin	Amini et al. 1996
Helicobacter pylori	Hsp70	Sulphatides	Huesca et al. 1998
Histoplasma capsulatum	Cpn60	CD11/CD18	Long et al. 2003
Histoplasma capsulatum	PPI	VLA5	Gomez et al. 2008
Lactobacillus johnsonii	Cpn60	Mucin	Bergonzelli et al. 2006
Lactobacillus plantarum	Cpn60	?	Saad et al. 2009
Lactobacillus plantarum	Hsp70	?	Saad et al. 2009
Lactobacillus plantarum	ClpB	?	Saad et al. 2009
Lactococcus lactis	Cpn60	Yeast invertase	Katakura et al. 2010
Lactococcus lactis	Hsp70	Yeast invertase	Katakura et al. 2010
Legionella pneumophila	Cpn60	?	Garduno et al. 1998a, b
Legionella pneumophila	Hsp70	?	Hoffman and Garduno 1999
Legionella pneumophila	PPI	?	Helbig et al. 2001, 2003
Legionella pneumophila	PPI	Various collagens	Wagner et al. 2007
Leptospira interrogans	Cpn60	?	Natarajaseenivasan et al. 2011
Mycobacterium avium	Cpn60	$\alpha_v\beta_3$	Hayashi et al. 1997
Mycobacterium avium	Hsp70	?	Ratnakar et al. 1996
Mycobacterium bovis BCG	Cpn60.1	DC-SIGN	Carroll et al. 2010
Mycobacterium leprae	Cpn60.2	?	Esaguy and Aguas 1997

(continued)

Table 4 (continued)

Bacterium	Molecular chaperone	Host receptor	Reference
Mycobacterium tuberculosis	Cpn60.2	CD43	Hickey et al. 2010
Mycobacterium tuberculosis	Hsp70	CD40	Wang et al. 2001
Mycobacterium tuberculosis	Hsp70	CCR5	Floto et al. 2006
Mycobacterium smegmatis	Cpn60	?	Esaguy and Aguas 1997
Neisseria gonorrhoeae	PPI	?	Leuzzi et al. 2005
Neisseria meningitiditis	Peroxiredoxin	Plasminogen	Knaust et al. 2007
Neisseria meningitidis	Hsp70	Plasminogen	Knaust et al. 2007
Plesiomonas shigelloides	Cpn60	?	Tsugawa et al. 2007
Rickettsia prowazekii	PPI	?	Emelyanov and Loukianov 2004
Salmonella enterica	Cpn60	Mucus	Ensgraber and Loos 1992
Serovar Typhimurium			
Streptococcus agalactiae	Cpn60	?	Hughes et al. 2002
Streptococcus agalactiae	Hsp70	?	Hughes et al. 2002
Streptococcus pneumoniae	PPI	?	Hermans et al. 2006
Streptococcus suis	Cpn10	?	Wu et al. 2008
Streptococcus suis	Cpn60	?	Wu et al. 2008
Streptococcus suis	GrpE (Hsp90)	?	Wu et al. 2008

the adhesion and invasion of this bacterium (Garduno et al. 1998b). *Legionella pneumophila* is one of the cell invading pathogens that can evade intracellular killing. After internalisation, the organism remains confined within a membrane-bound compartment and forms into a specialized vacuole referred to as the *Legionella*-containing vacuole (LCV). Generation of the LCV involves the recruitment to the site of this vacuole of cellular vesicles and also mitochondria. Using latex beads coated with *L. pneumophila* Cpn60 revealed that these beads acted like the intact bacterium in recruiting mitochondria and inducing changes in the actin microfilaments (Chong et al. 2009). When *E. coli* Cpn60 (GroEL) was attached to beads there was no recruitment of mitochondria showing that this biological activity of the Legionella protein is a unique moonlighting function of this protein (Chong et al. 2009) and again revealing the bewildering evolutionary plasticity of the *cpn60* gene.

There is increasing interest in the moonlighting roles of the molecular chaperones of the mycobacteria, and *M. tuberculosis* uses both Cpn60 proteins and Hsp70 as virulence factors. The Cpn60.2 protein of *M. tuberculosis* is well recognised as a powerful immunogen and also a monocyte cytokine-inducing protein (Henderson et al. 2010). In addition to these functions, it is now recognised that the Cpn60.2 protein is also found on the surface of *M. tuberculosis* where it functions as an adhesin for the binding of this bacterium to human monocytes (Hickey et al. 2009). The receptor on the host macrophages recognising the Cpn60.2 protein is CD43 (Hickey et al. 2010), a protein which has been recognised to be involved in the control of the intracellular growth of *M. tuberculosis* (Randhawa et al. 2008). This raises the question of whether binding of *M. tuberculosis* Cpn60.2 to CD43 controls the intracellular growth of this pathogen?

Like the glycolytic enzymes, it is generally difficult to inactivate the genes encoding molecular chaperones and protein-folding catalysts and so direct testing of the hypothesis that cell surface cell stress proteins are involved in adhesion and subsequent virulence is rarely possible. However, with the fungus, *Histoplasma capsulatum*, the causative agent of histoplasmosis, it has been reported that monoclonal antibodies to the Cpn60 protein of this organism significantly prolong the survival of mice infected with this fungus (Guimarães et al. 2009). This fungus has a cell surface Cpn60 protein which binds to CD11/CD18 on target cells (Long et al. 2003). Treated animals revealed reduced intracellular fungal survival, revealing that the Cpn60 protein of this organism is important in the pathogenesis of infection.

It is therefore clear that many bacteria (and other microbes) contain a number of molecular chaperones and protein-folding catalysts on their cell surfaces and that these proteins can play important roles in adhesion of the organism to host matrices and host cells and such binding can promote virulence. It will now be important to identify, structurally, just how these molecular chaperones function as host receptor ligands.

5.2 Glycolytic Enzymes as Bacterial Adhesins

A surprising finding is that most of the proteins of the glycolytic pathway in bacteria have some adhesive actions. Most attention has thus far focused on bacterial glycolytic enzymes, but there are some reports of eukaryotic glycolytic proteins having adhesive properties and these may be relevant to the prokaryotic situation.

Hexokinase is the first enzyme of the glycolytic pathway and so far it has not been shown to have specific adhesive activity in bacteria. However, in the protozoan, *Leishmania donovani*, hexokinase moonlights as a receptor for haemoglobin (Krishnamurthy et al. 2005). In mammals there are four hexokinase isozymes of which types I and II bind to the mitochondria and can enhance mitochondrial oxidation and decrease mitochondrially-driven apoptosis. This is due to the interaction of hexokinase with the Voltage-Dependent Anion Channel 1 (VDAC1) and enhancing such binding may have anti-cancer potential (Rosano et al. 2011). As the mitochondrion was originally a bacterium, such binding to hexokinase may also occur with true bacteria, but may simply not have been looked for, and so, not reported.

The next two enzymes of glycolysis, phosphoglucoisomerase and phosphofructokinase, have not been reported to function as adhesins although, as described earlier, phosphoglucoisomerase as AMF is involved in tumour cell invasion and so may also interfere with cell adhesion. However, the next enzyme in the pathway, fructose-bisphosphate aldolase (FBA), which is a cell surface protein in streptococci (Wu et al. 2008), has been reported (in *S. pneumoniae*) to bind to the human cadherin homologue of the *Drosophila* Flamingo cadherin receptor (Blau et al. 2007). A number of bacteria utilise these cadherins for cell binding. Of interest, a newly identified group of environmental bacteria that metabolise complex polysaccharides contain a member, *Saccharophagus degradans* strain 2-40, which encodes a number of very large proteins which contain multiple cadherin domains and these proteins are thought to contribute to cell–cell interactions within these organisms (Fraiberg et al. 2010). *Neisseria meningitidis* also contains a cell surface FBA and generation of a mutant deficient in this glycolytic enzyme revealed that there was no effect on bacterial growth but the ability of the organism to bind to target cells was significantly affected revealing that this protein is acting as a selective cell adhesin (Tunio et al. 2010a). A cell surface aldolase is also involved in the invasion of the protozoan *Toxoplasma gondii* and mutational analysis has shown that enzymic activity it not required for cell invasion (Starnes et al. 2009).

The next enzyme of glycolysis is triose-phosphate isomerase (TPI). The only reported example of this enzyme being surface located and involved in adhesion is that the TPI of *Staphylococcus aureus*, which is reported to be on the surface and to recognise glycan components of the fungus *Cryptococcus neoformans*. This work started with the discovery that *S. aureus* would kill *C. neoformans* if these organisms were co-cultured and that capsular polysaccharide from the fungus could inhibit such killing (Saito and Ikeda 2005). The hypothesis is that TPI is a cell surface-bound enzyme in *S. aureus* (Yamaguchi et al. 2010) which recognises and binds to glycans in the capsule of the fungal pathogen (Ikeda et al. 2007; Furuya and Ikeda 2009)

5.2.1 Glyceraldehyde-3-Phosphate Dehydrogenase

The discussion now returns to the subject of GAPD. In the previous section on GAPD as a cell signalling ligand, it was obvious that this protein also enabled group A streptococci to bind to target cells (e.g. Boel et al. 2005). There are also reports of other bacteria using cell surface GAPD to bind to host cells. For example, *Mycoplasma suis* binds to and colonises erythrocytes from a variety of vertebrates. The adhesin for such binding has been identified as GAPD (Hoelze et al. 2007). In *N. meningitidis* there are two genes encoding GAPD proteins (Gap-A-1 and -2). Only one of these, GapA-1, is present on the cell surface. Inactivation of the gene encoding this protein resulted in an isogenic mutant with markedly decreased ability to bind to human epithelial or endothelial cells (Tunio et al. 2010b).

In addition to binding to cells, GAPD from a growing number of bacterial species, and from other microorganisms, has been shown to bind to a variety of host ligands including transferrin, plasminogen and fibronectin. One of the first such ligands identified was the iron-binding protein, transferrin and it was reported that a cell surface GAPD on *S. aureus* and *Staphylococcus epidermidis* bound to transferrin, and also to plasmin (Modun and Williams 1999). This initial report of the transferrin-binding actions of GAPD was refuted by another group who claimed that the cell surface transferrin-binding protein of *S. aureus* was a completely different protein (Taylor and Heinrichs 2002). However, in support of the initial claim, other workers have reported that GAPD is a cell surface transferrin receptor in *Trypanosoma brucei* (Tanaka et al. 2004) and human and mouse macrophages (Raje et al. 2007). Clearly, further work is required to determine if GAPD is a general transferrin-binding protein in bacterial species.

The GAPD from group A streptococci was first identified as a plasminogen-binding protein in the early 1990s (Lottenberg et al. 1992). Recruiting plasminogen to a bacterial surface can promote the formation of the active protease, plasmin, allowing the bacterium to degrade the host's extracellular matrix and gain entry into the tissues (Lähteenmäki et al. 2005). This has been shown with the lung pathogen *S. pneumoniae* (Attali et al. 2008) and *Streptococcus agalactiae* (Magalhães et al. 2007) and in both organisms binding to plasminogen enhances virulence. Since this initial discovery a number of bacteria and other microorganisms have been shown to utilise cell surface GAPD to bind plasminogen. Binding of the *Stretococcus equisimilis* GAPD to plasmin(ogen) has been assessed by surface plasmon resonance and has revealed a Kd of 220nM for binding to plasminogen and 25nM for the binding to plasmin (Gase et al. 1996). These data reveal that the binding between GAPD and plasmin(ogen) is of relatively high affinity and is not simply some non-specific interaction. However, the exact role of GAPD binding to plasminogen in bacterial virulence is still unclear. Thus, Winram and Lottenberg (1998) identified a C-terminal lysyl residue as responsible for high affinity plasmin binding by *S. pyogenes* GAPD. The gene encoding this mutated protein was used to replace the wild type *gapd* gene. However, this site-directed isogenic mutant bound as much plasminogen as the wild type organism. This finding does not square with the study in which the *S. pyogenes* GAPD was mutated such that it could not be secreted. In this study (Boel et al. 2005) there was less plasminogen binding by the mutant.

A number of Gram-positive bacteria including: group B streptococci (Seifert et al. 2003), *S. suis* (Jobin et al. 2004), *S. pneumoniae* (Bergmann et al. 2004), *Lactobacillus crispatus* (Hurmalainen et al. 2007; Antikainen et al. 2007) and *Bacillus anthracis* (Matta et al. 2010) have cell surface GAPDs which bind plasminogen. In the latter organism, immunisation of mice with the bacterial GAPD protein conferred significant protection from infection with the bacterium (Matta et al. 2010). To date there is only one report of GAPD as a cell surface plasminogen-binding protein in a Gram-negative bacterium. Enterohaemorrhagic and enteropathogenic strains of *E. coli* secrete GAPD which binds to plasminogen (Egea et al. 2007). Non-pathogenic strains of this bacterium do not secrete GAPD.

Curiously, analysis of the *E. coli* GAPD has revealed two different electrophoretic variants, with only the more basic form being secreted. Indeed, GAPD binding to plasminogen is not simply confined to bacteria as there are reports that the multicellular parasites *Onchocerca volvulus* (Erttmann et al. 2005) and *Schistosoma bovis* (Ramajo-Hernández et al. 2007) also use this enzyme to bind plasminogen. Finally, an early report of GAPD surface binding in *S. pyogenes* found binding to fibronectin (Pancholi and Fischetti 1992). The protist, *Trichomonas vaginalis* has also been reported to have a surface-associated GAPD which binds to fibronectin (Lama et al. 2009).

5.2.2 Enolase

Enolase is the other major cell surface moonlighting glycolytic enzyme of prokaryotes and eukaryotes. However, before turning to this protein, what is known of the moonlighting actions of the remaining two glycolytic enzymes, phosphoglycerate kinase and phosphoglycerate mutase that lie between GAPD and enolase will be briefly described. Phosphoglycerate kinase has been found on the cell wall of *Candida albicans* (Alloush et al. 1997) and the Gram-negative organism *Aeromonas salmonicida* (Ebanks et al. 2005) as well as in Gram-positive organisms. It is involved in the binding of plasminogen by oral streptococci (Kinnby et al. 2008). In group B streptococci, phosphoglycerate kinase is a cell surface enzyme which appears to be involved in inhibiting binding of bacteria to epithelial cells (Burnham et al. 2005). The role of this protein is still inexplicable. Phosphoglycerate mutase is also found on the surface of oral streptococci [(Kinnby et al. 2008) and by Wu et al. (2008)]. In *M. tuberculosis* this enzyme has been reported to be involved in resistance of the bacterium to oxidative stress (Chaturvedi et al. 2010). In the yeast, *C. albicans*, phosphoglycerate mutase is found on the cell surface where it acts as: (1) a plasminogen-binding protein (Crowe et al. 2003) and a complement binding protein, specifically for factor H and factor H-like (FHL-1) protein (Polterman et al. 2007). Binding of factor H and FHL-1 is a key event in the pathogenesis of streptococcal infection (Oliver et al. 2008) and thus phosphoglycerate mutase may contribute to the virulence of these bacteria.

Returning to the subject of this section, enolase, it is established that this protein has multiple moonlighting actions in both eukaryotes and prokaryotes (Pancholi 2001). For example, enolase is involved in tumour biogenesis and is now seen as a therapeutic target (Capello et al. 2011). Encystation is a survival strategy used by many parasites. Enolase has recently been shown to be involved in the encystation mechanism of *Entamoeba invadens* (Segovia-Gamboa et al. 2010.) A growing number of bacteria and other microbes are reported to express moonlighting enolases on their cell surfaces. Currently, most examples of bacteria with cell surface enolases are Gram positive, but there are also Gram negative (e.g. *Borrelia burgdorferi* (Nowalk et al. 2006)) in this group. Attention has largely focused on the cell surface enolases of

Group A streptococci. Pancholi and Fischetti, whose work on *S. pyogenes* cell surface GAPD was discussed, have also been responsible for much of the work on streptococcal cell surface enolase and its role in bacterial virulence. The complex plasminogen activation system of the vertebrate is essential for survival and is also used by tumour cells to invade tissues (Dano et al. 2005). This capacity of the plasminogen activation system to catalyse tissue invasion has meant that it is an evolutionary target for bacterial plasminogen activators and receptors (Lähteenmäki et al. 2001, 2005). Analysis of the plasminogen-binding characteristics of cell wall proteins of *S. pyogenes* identified enolase as the strongest binder, with this cell surface enolase being enzymically active, and antibodies to it induced opsonisation and enhanced phagocytosis (Pancholi and Fischetti 1998). Enolase is now recognised to be present on the surface of most streptococci (Pancholi and Fischetti 1998) including *S. pneumoniae* (Bergmann et al. 2001). Plasminogen binding to the surface of pneumococci enables the bacteria to penetrate a synthetic basement membrane gel (Matrige^{TM}). It is this process that is believed to be important for the invasion of this organism and the consequence of invasion—meningitis (Eberhard et al. 1999). In a separate study it was shown that soluble recombinant *S. pneumoniae* enolase bound to the surface of the pneumococci even when associated with plasminogen. Treatment of the cell surface with proteases inhibited such re-association, suggesting that binding was due to protein–protein/peptide interactions (Bergmann et al. 2001). Inactivation of the enolase gene in *S. pneumoniae* resulted in non-viable cells, showing the essential nature of this protein, presumably for its role in glycolysis (Bergmann et al. 2001). In mammalian enolases binding to plasminogen is dependent on C-terminal lysyl residues in the enolase which binds to lysine binding sites in the plasminogen (Redlitz et al. 1995). To test if pneumococcal enolase also bound plasminogen through these C-terminal lysines, the enolase was treated with carboxypeptidase or was mutated at Lys-433 and Lys-434 resulting in inhibition of enolase binding to plasminogen (Bergmann et al. 2001). There is another, internal, plasminogen-binding site in Gram-positive enolases which seem to have different roles in different streptococci. Thus the enolase of *S. pneumoniae* has an internal binding site consisting of residues 248–256 (FYDKERKVYD). It turns out that this is the major binding site for plasminogen in *S. pneumoniae* (Bergmann et al. 2003). In *S. pyogenes*, binding of native and C-terminal mutated enolase to native plasminogen, termed Glu-plasminogen and plasminogen after cleavage by plasmin, termed Lys-plasminogen was investigated. Deletion or substitution of the lysines in enolase at positions 434 and 435 resulted in significant decreases in the binding of this glycolytic enzyme to both forms of plasminogen. Moreover, the bacteria encoding the mutated enolase demonstrated a significant decrease in the ability to acquire plasminogen from human plasma and penetrate a synthetic extracellular matrix (Derbise et al. 2004). The lysines at position 252 and 255 also contribute to plasminogen binding (Cork et al. 2009) but this internal site does not seem to be as important as it is in the pneumococci. A similar internal binding site has been identified in *Bifidobacteria* spp enolases (Candela et al. 2009).

The available literature supports the hypothesis that cell surface enolase-plasminogen binding contributes to the virulence of a number of bacteria. The clearest evidence comes from the generation of enolase isogenic mutants in which the enolase-plasminogen binding sites have been ablated. This revealed that in the absence of the internal plasminogen-binding motif, bacteria failed to kill mice (Bergmann et al. 2003). The surface enolase of *S. suis* binds to plasminogen with high affinity (Kd = 21 nM) and antibodies to this protein inhibit the adhesion and invasion of the organism into microvascular endothelial cells (Esgleas et al. 2008). The recombinant enolase of the Gram-negative organism, *Aeromonas hydrophila*, (which also binds to plasminogen) has been used to immunise mice and this markedly decreased the pathology consequent upon infection with this bacterium (Sha et al. 2009). These data confirm that enolase binding is a virulence attribute for these bacteria.

It is not only plasminogen that bacterial enolases bind to. The enolase of *Streptococcus gordonii* binds to the salivary mucin, Muc7 (Kesimer et al. 2009). The *L. plantarum* enolase binds to fibronectin (Castaldo et al. 2009). In addition, the cell surface enolase of the vaginal commensal organism, *Lactobacillus jensenii*, is a potent inhibitor of the adherence of *Neisseria gonorrhoeae* to epithelial cells (Spurbeck and Arvidson 2010).

Other functions have been ascribed to the bacterial cell surface enolases. For example, the enolase of *Streptococcus sobrinus* is an immunosuppressive protein (Veiga-Malta et al. 2004) which can be used, if administered orally, to protect against dental caries in the rat (Dinis et al. 2009). In contrast, with *Paenibacillus larvae*, the Gram-positive causative agent of American Foulbrood (AFB), which affects the larvae of the honeybee, *Apis mellifera*, the enolase is a secreted highly immunogenic protein which is thought to play a role in the virulence of this bacterium (Antúnez et al. 2011).

Surprisingly, little is known about the binding of enolase to the bacterial surface. It has been shown that enolase (and GAPD) associates with the surface lipoteichoic acids of *L. crispatus* at pH 5 but dissociates at alkaline pH (Antikainen et al. 2007).

Of the two remaining enzymes of glycolysis, pyruvate kinase is a cell surface enzyme in *Lactococcus lactis* which can bind the hyper-mannosylated yeast invertase (Katakura et al. 2010). Thus far there are no reports of cell surface lactate dehydrogenase on bacteria.

5.3 Other Bacterial Moonlighting Adhesins

Moonlighting adhesins other than molecular chaperones and glycolytic enzymes exist. However, one of the problems in bringing together the literature on these proteins is attempting to find some common thread to the discussion. Fortunately, the most abundant of the moonlighting bacterial adhesins are proteins that bind to the complex host adhesive glycoprotein, fibronectin. This protein is found in high concentration in the fluids of the body and in the extracellular matrix (ECM) and plays a major role in linking cells to the ECM through specific integrins, which can function as transducers of changes in the matrix (Henderson et al. 2011). Fibronectin has a complex domain structure with different parts of the protein binding to different host components including heparin, collagen, gelatin, fibulin, DNA and so on (Henderson et al. 2011). The binding of *S. pyogenes* GAPD (Pancholi and Fischetti 1992) and the *L. plantarum* cell surface enolase (Castaldo et al. 2009) to fibronectin has already been discussed. Another Gram-negative moonlighting fibronectin-binding protein is the C5a peptidase of group B streptococci. This protein was identified as cleaving and inactivating the complement component and anaphylatoxin, C5a. It also binds fibronectin and inactivation of the gene encoding this protein results in bacteria having 50% less fibronectin-binding capacity than the wild type organism (Beckmann et al. 2002).

Mycobacterium tuberculosis secretes three protein homologues, termed the antigen 85 complex, consisting of proteins 85A, 85B and 85C. These are the products of three different genes located at different loci in the genome and showing significant nucleotide and amino acid sequence identity and marked immune cross-reactivity (Wicker and Harboe 1992). Proteins are in the mass range from 30 to 31 kDa and are all able to bind to fibronectin (Abou-Zeid 1988; Wicker and Harboe 1992). In this disguise these proteins are also termed fibronectin-binding proteins (FBPs)1–3. The site of interaction of the antigen85 complex proteins has been reported variously as the gelatin binding domain for the *M. bovis* protein (Peake et al. 1993), and the heparin and cell wall binding regions for the *Mycobacterium kansasii* protein (Naito et al. 2000). It turns out that in addition to fibronectin binding, the antigen 85 complex proteins contain a carboxylesterase domain and act as mycolyltransferases, which are proteins involved in the final stages of the assembly of the complex mycobacterial cell wall (Belisle et al. 1997). All three proteins appear to have the same, or similar, enzymic roles in terms of transferring mycoloyl residues (Puech et al. 2002). This clearly defined these molecules as moonlighting proteins. The fibronectin-binding motif in the antigen 85 complex proteins forms a helix at the surface of the protein and has no homology to other known prokaryotic and eukaryotic fibronectin-binding features and appears to be unique to the mycobacteria (Naito et al. 1998). It is also argued that a large region of conserved surface residues among antigen85 proteins A, B and C is a probable site for the interaction of these proteins with fibronectin (Ronning et al. 2000). Another mycobacterial fibronectin-binding protein brings us back to the role of metabolic enzymes in protein moonlighting. The malate synthase of *M. tuberculosis*, a cytoplasmic protein involved in the glyoxylate pathway, a cytoplasmic metabolic pathway, has also been found to occur on the bacterial surface, associating by an unknown mechanism, where it can bind both fibronectin and laminin. The binding site in the malate synthase for fibronectin lies in a C-terminal region of the protein that is unique to *M. tuberculosis*, but it is not known to which domain in fibronectin it binds (Kinhikar et al. 2006). This is the first glyoxylate cycle enzyme shown to be present on the bacterial cell surface.

The mycoplasmas are cell wall-less organisms that have evolved from a Gram-positive ancestor, and are probably the smallest living form capable of autonomous growth. Using fibronectin affinity chromatography, two fibronectin-binding proteins, of 30 and 45 kDa were identified in *Mycoplasma pneumoniae* and N-terminal sequencing identified these proteins as elongation factor (EF)-Tu and the β-subunit of pyruvate dehydrogenase (Dallo et al. 2002). Elongation factor (EF)-Tu is normally assumed to be a cytoplasmic protein responsible for critical steps in protein synthesis. Pyruvate dehydrogenase is an enzyme complex formed of two α and one β-subunits which transforms pyruvate into acetyl CoA for mitochondrial oxidation (Dallo et al. 2002). Recombinant versions of these proteins were shown to bind fibronectin. Antibody labelling revealed that both of these proteins were present on the surface of *M. pneumoniae* and both antibodies could inhibit the binding of *M. pneumoniae* to fibronectin. Subsequent studies revealed that a 179 residue region in the C-terminus of EF-Tu is responsible for fibronectin binding. Using C-terminal constructs and truncation mutants, two distinct sites with different fibronectin-binding efficiencies were identified. Immunogold electron microscopy, using antibodies raised against recombinant constructs, demonstrated the surface accessibility of the EF-Tu carboxyl region and fractionation of the mycoplasma confirmed the association of EF-Tu with the mycoplasma outer membrane (Balasubramanian et al. 2008). As has been stated, the rules governing protein moonlighting are not understood. This may explain why the EF-Tu protein of *Mycoplasma genitalium* does not bind to fibronectin even though it shares 96% sequence identity with the *M. pneumoniae* protein. This has enabled the fibronectin binding-moonlighting site in *M. pneumoniae* EF-Tu to be identified. Substitutions of amino acids: serine 343, proline 345 and threonine 357 markedly reduced the Fn binding of the *M. pneumoniae* EF-Tu. Moreover, synthetic peptides corresponding to residues 340–358 in this *M. pneumoniae* EF-Tu protein were able to block the binding of recombinant EF-Tu to fibronectin and also the binding of *M. pneumoniae* to this protein (Balasubramanian et al. 2009).

Autolysins are important peptidoglycan-degrading enzymes. A number of the autolysins of the staphylococci have been shown to also function as fibronectin-binding proteins. These include Aaa (autolysin/adhesin of *S. aureus*) which binds fibronectin with high affinity (Kd = 30nM) and which is involved in bacterial adherence to fibronectin (Heilmann et al. 2005). *Staphylococcus epidermidis* Aae (autolysin/adhesin in *S. epidermidis*) is a homologue of *S. aureus* Aaa and binds to the 29 kDa heparin-binding module of fibronectin (Heilmann et al. 2003). Two other staphylococcal autolysins also function as fibronectin-binding proteins. These are large (155 kDa) homologous proteins—*Streptococcus caprae* Atlc (autolysin caprae) (Allignet et al. 2001) and *Streptococcus saprophyticus* Aas (Hell et al. 1998) which, interestingly, have no obvious cell wall anchor motif. AtlC is the only fibronectin-binding protein so far identified in *S. caprae* and it is a bifunctional enzyme that contains a repeat region (R1–R3), with no recognisable similarity to other proteins, sandwiched between two enzymic domains. The repeat region is responsible for binding to fibronectin, but exactly what binds is still unclear. Using far-western blots, only recombinant R1–R3 and R3 alone bind fibronectin. In contrast, using ELISA or surface plasmon resonance methods, all recombinant domain constructs bind fibronectin (Allignet et al. 2001). The binding site for fibronectin in the *S. saprophyticus* autolysin has been localised as lying between the two enzymic domains, within residues 714–1202, and inactivation of the gene was shown to result in loss of fibronectin binding (Hell et al. 1998). The AtlA autolysin of *Streptococcus mutans* is also a fibronectin-binding protein and inactivation of the gene encoding this protein decreases the virulence of this organism in a rat model of infective endocarditis (Jung et al. 2009). Inactivation of the autolysin gene encoding the IspC protein of *L. monocytogenes* (found on the cell surface) had, contrary to expectation, no effect on rates of cell growth or cell separation but had a major influence on in vivo virulence of the bacterium (Wang and Lin 2008). This appeared to be, at least in part, due to a decreased ability to bind to target cells.

While not likely to be a moonlighting protein, the CagL protein of *H. pylori* is a small cell surface protein (of 26 kDa) with an RGD motif, which mimics the activity of host fibronectin and can modulate host cell signalling (Tegtmeyer et al. 2010).

A number of other bacterial proteins moonlight as adhesins. Again, these are examples of cell surface metabolic enzymes functioning as adhesins. The pentose phosphate pathway enzyme, 6-phosphogluconate dehydrogenase (6PGD) is a cell surface located enzyme which also acts as an adhesin in various pneumococcal strains. Binding of *S. pneumoniae* to lung epithelial cells was inhibited by recombinant 6PGD and immunisation of mice with this protein gave significant protection against infection with *S. pneumoniae* (Daniely et al. 2006). A similar story was found with the 6PGD of *S. suis* with the immunisation of mice with this protein able to provide 80% protection against a lethal dose of the bacterium (Tan et al. 2009). These results clearly show the virulence phenotype that can be induced by a moonlighting protein.

The protozoan, *T. vaginalis*, which causes the sexually transmittted disease Trichomoniasis, binds to the vaginal epithelium. The hydrogenosomal enzyme pyruvate:ferredoxin oxidoreductase is an iron-induced cell surface protein which acts as an epithelial cell adhesin in this organism (Moreno-Brito et al. 2005).

Perhaps the most intriguing moonlighting adhesin is the alcohol acetaldehyde dehydrogenase of *L. monocytogenes*. Listeria adhesion protein (LAP) was identified as a key cell-binding adhesin of this organism (Santiago et al. 2006) allowing the bacterium to bind to intestinal epithelial cells. The surprising finding was then made that the host cell receptor for LAP was human Cpn60 (Wampler et al. 2004). Surprise followed surprise when LAP was identified as the alcohol acetaldehyde dehydrogenase of *L. monocytogenes*—yet another metabolic enzyme involved in moonlighting. Measurement of the kinetics of the interaction of LAP with human Cpn60, using surface plasmon resonance, revealed a Kd value in the low nanomolar range, which is a respectable binding affinity (Kim et al. 2006). So, like the example of PGI as AMF binding to an ubiquitin ligase here we have another example of a moonlighting protein (alcohol acetaldehyde dehydrogenase) binding, with high affinity, to another moonlighting protein (human Cpn60) to allow a bacterium to colonise its host and cause disease. How many other such examples will Nature provide us with? Further analysis of LAP/alcohol acetaldehyde dehydrogenase in non-pathogenic strains of *Listeria* have found that while these strains produce this enzyme there is very little of it on the bacterial surface and so only pathogenic strains bind to target cells via LAP/Hsp60 binding (Jagadeesan et al. 2010). As human Cpn60 is a stress protein, the role of cell stress in *Listeria* infection has been examined. Thus exposure of CaCo-2 cells used for infection assays to various stressors increased intracellular Hsp60 levels and enhanced the adhesion, but not invasion of *L. monocytogenes*. Knock-down of Hsp60 with inhibitory RNA reduced the adhesion and translocation of wild type *L. monocytogenes* but a *lap* mutant showed unchanged adhesion. Overexpression of Hsp60 enhanced wild type adhesion and cellular translocation but there was no change in the *lap* mutant. Of importance, infection with *L. monocytogenes* increased plasma membrane expression of Hsp60. Thus there is a dynamic response between these two moonlighting proteins to enhance *L. monocytogenes* infection (Burkholder and Bhunia 2010).

As final examples of 'moonlighting adhesivity' serum opacity factor, a streptococcal virulence factor which binds to high-density lipoproteins and disrupts them forming large lipid vesicles (Courtney and Pownall 2010) also binds the host ECM protein, fibulin (Courtney et al. 2009). The superoxide dismutase of *M. tuberculosis* has also been reported to function as an adhesin, binding to a number of host moonlighting proteins such as GAPD and cyclophilin A (Reddy and Suleman 2004)—again potentially a moonlighting–moonlighting interaction.

5.4 Other Moonlighting Actions of Bacterial Proteins

Cells have a wide range of other metabolic pathways, and such pathways involve an extremely large number of individual enzymes, most of which we know nothing about in relation to their moonlighting functions. The literature only highlights individual enzymes, in individual organisms, and it is difficult to say much about the relevance of such moonlighting activity in a generic sense. For example, *B. subtilis* encodes a glutamate dehydrogenase (RocG) which, in addition to deaminating glutamate to form α-ketoglutarate, also binds to the transcription factor GltC which functions to regulate glutamate production from α-ketoglutarate and so links these two metabolic pathways. Mutants of RocG have been isolated which have lost their dehydrogenase activity and only retain transcription factor-binding activity (Gunka et al. 2010).

Mycobacterium tuberculosis, and other mycobacteria, have evolved moonlighting actions in many of their proteins. Among these is the enzyme glutamate racemase (MurI) which generates d-glutamate, a key component of the peptidoglycan of the bacterial cell wall. In mycobacteria, including *M. tuberculosis*, MurI also functions as a DNA gyrase. This DNA gyrase activity is not related to the racemase function and overexpression of MurI in vivo results in the bacterium being more resistant to ciprofloxacin, an antibiotic targeting DNA gyrases, thus showing that this protein is important in DNA function in the intact organism (Sengupta et al. 2008). *Mycobacterium tuberculosis* has only one cAMP phosphodiesterase which has recently been shown to play an independent role in controlling cell wall permeability to hydrophobic cytotoxic compounds (Podobnik et al. 2009). Such influence on cell wall functioning is likely to contribute to the survival and virulence of this bacterium. The aconitase of *M. tuberculosis*, as well as being a TCA cycle enzyme, also functions as an iron-responsive protein (IRP). Such proteins interact with iron-responsive elements (IREs) present at untranslated regions of mRNAs and such binding controls the post-transcriptional regulation of the expression of proteins involved in iron homeostasis (Banerjee et al. 2007). The rice pathogenic bacterium *Xanthomonas oryzae pv. oryzae* has a moonlighting chorismate mutase, which is an important enzyme in the shikimate pathway responsible for aromatic amino acid synthesis. Bacteria have two forms of chorismate mutases termed AroQ and AroH, and some pathogenic bacteria are reported to possess a subgroup of these enzymes which have been named AroQ(gamma). Now *X. oryzae pv. oryzae* XKK.12 possesses an AroQ(gamma), and inactivation of the gene coding for this enzyme leads to an isogenic mutant which is hypervirulent, implying an important moonlighting role for this protein in bacterium:rice interactions (Degrassi et al. 2010).

6 Identification of the Moonlighting Sites in Bacterial Moonlighting Proteins

An obvious requirement, if we are to understand protein moonlighting, is to identify the moonlighting site, or sites, in moonlighting proteins to determine how they evolved and how they relate to the initially discovered ‘active site’ of the protein in question. Unfortunately, only a few moonlighting sites have been conclusively identified in bacterial proteins. These are shown in Table 5 and mapped onto structures in Fig. 3. As one would expect, all the moonlighting peptides clearly map to the surface of the proteins, with the partial exception of Hsp70. In this case the moonlighting peptide is at the interface of the homodimer raising the possibility that Hsp70 adopts its moonlighting role when disassociated into monomers although the residues identified by Wang et al. (2005) as the most important are exposed in the dimer. It will only be with the accumulation of more and defined moonlighting sites that this information can be used to ascertain the relationship between moonlighting sequence, structure and function and between the moonlighting sites(s) and the active site which provided the said protein with its original (activity) name. Having mapped moonlighting regions to protein structures, it should be possible to compare the shape and chemical nature of the region with the normal protein. For example, as described above, the Ef-Tu from *M. pneumoniae* binds to fibronectin and comparison of two Ef-Tu homologues have provided clues as to the possible

Table 5 Moonlighting sites in bacterial moonlighting proteins

Bacterium	Moonlighting protein	Moonlighting Sequence	PDB	Structure species	ID
M. tuberculosis	Hsp70	**QPSV**QIQVYQGER**E**IAAHNK[1]	3qnj	*E. coli*	12/20
M. tuberculosis	Antigen 85 complex	FEWYYQ[2]	1sfr	*M. tuberculosis*	5/6
M. tuberculosis	Malate synthase	CGAQQPNGYTEPILHRRRREFKARAAEKPAPSDRAGDD[3]	1n8i	*M. tuberculosis*	26/27[6]
M. pneumoniae	Ef-Tu	GSISLPENTEMVLPGDNTS[4]	2y0u	*T. thermophilus*	10/19
Xen. nematophila	Cpn60	QIRQQIEES[5]	3e76	*E. coli*	8/9

PDB indicates PDB codes for a structure of the protein, or an orthologue from a related species as indicated under 'Structure Species'. ID indicates the number of residues in the structure identical to the 'Moonlighting Sequence' and the length of the matched stretch

[1] Wang et al. (2005) J immunol 174: 3306–3316. [residues highlighted in bold are most important for biological activity]

[2] Naito et al. (1998) J Biol Chem 273:2905–2909

[3] Kinhikar et al. (2006) Mol Microbiol 60:999–1013

[4] Balasubramanian et al. (2009) Infect Immun 77:3533–3541

[5] Joshi et al. (2008) J Biol Chem 283:28287–28296

[6] The PDB structure is truncated after the first 27 residues (i.e. KPAPSDRAGDD missing)

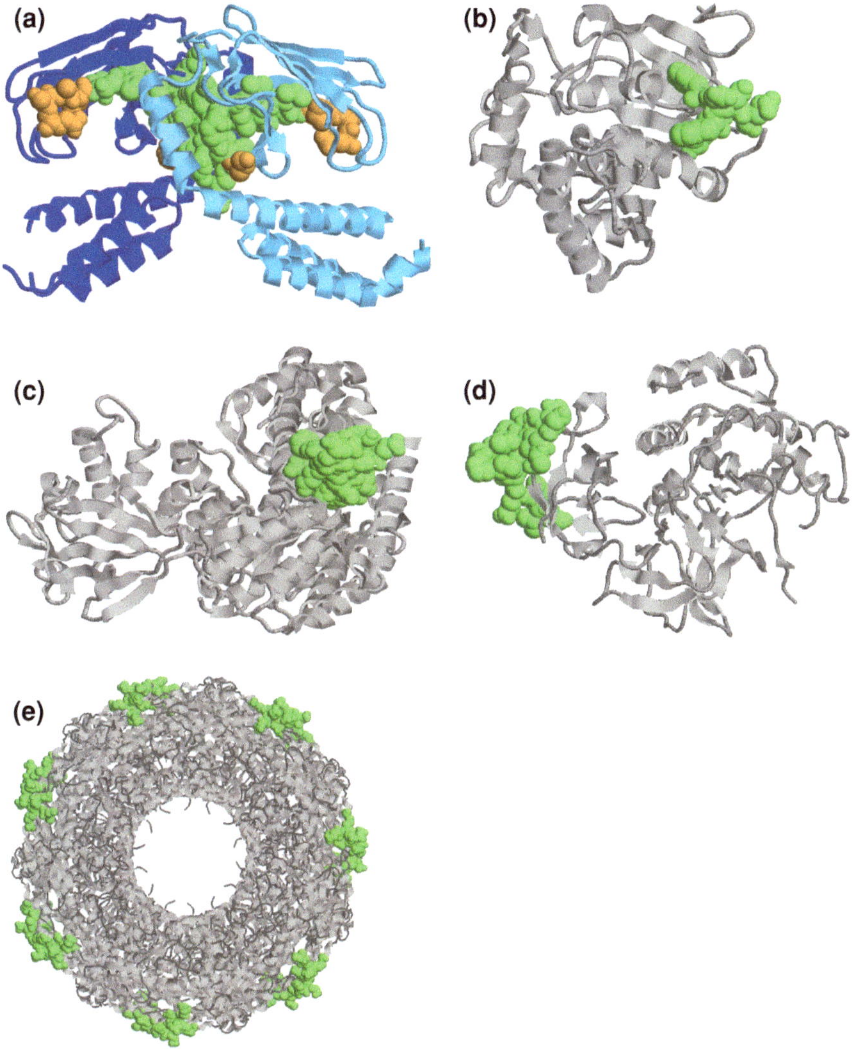

Fig. 3 Known bacterial moonlighting sites mapped to protein structures. Known moonlighting sequences as shown in Table 5 were mapped to protein structures from the same, or related, species to identify the location of the moonlighting site in the structure. Moonlighting sequences are highlighted in green spacefilling with the most important residues in Hsp70 shown in orange. A: Hsp70, B: Antigen 85 Complex, C: Malate Synthase, D: Ef-Tu, E: Cpn60

structural nature of the binding site (Table 5). To determine if this is a known or novel binding site for fibronectin it would be necessary to compare the structural properties of this fibronectin-binding region of Ef-Tu with the surfaces of known fibronectin-binding proteins.

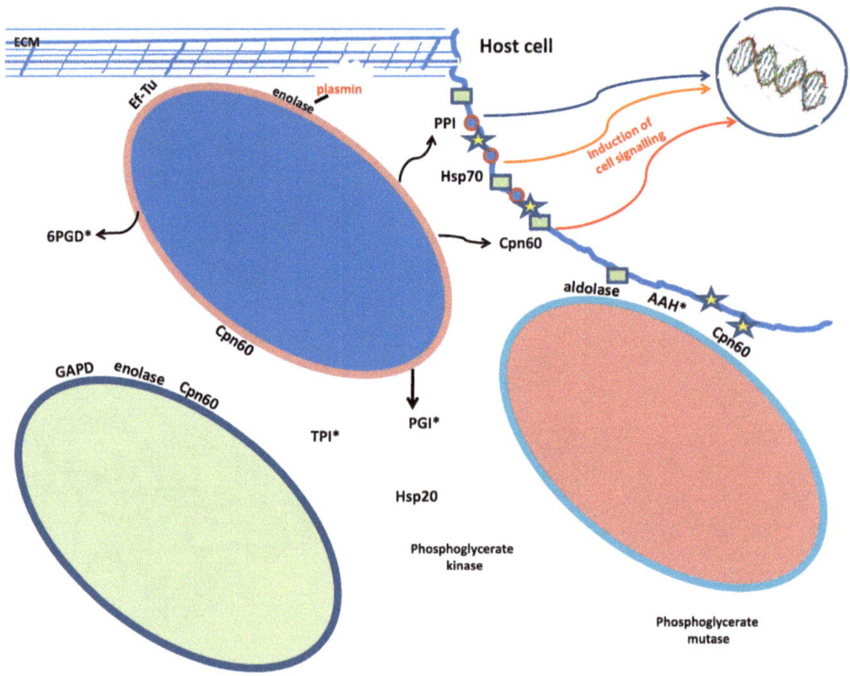

Fig. 4 Schematic diagram of the role of moonlighting proteins in bacterial virulence. On the slide is shown a number of the reported moonlighting proteins in bacteria and the roles that they play, particularly in bacteria–host interactions including induction of cell signalling, adhesion to ECM and cells and degradation of the ECM to allow bacterial invasion. (6PGD—6-phosphogluconate dehydrogenase; AAH—alcohol acetaldehyde dehydrogense; PGI—phosphoglucose isomerase; TPI—triose phosphate isomerase)

7 Conclusions

This review has described a relatively large number of defined moonlighting proteins with actions which are clearly associated with bacterial virulence (Fig. 4). These include proteins such as GAPD, enolase and chaperonin 60 whose moonlighting actions are utilised by a wide range of bacteria and other microorganisms to interact with their hosts. These proteins also have additional functions such as DNA binding or enhancement of biofilm formation. In some cases, although the moonlighting protein seems to be a chaperone or enzyme, it has actually evolved away from this activity and appears to have purely the moonlighting function. A key question is how pervasive is protein moonlighting in bacteria and in bacteria–host interactions. Clearly, other bacterial proteins exist which seem to be moonlighting proteins but, for which, definitive evidence is not yet available. For example, *S. pyogenes* encodes a DNase (SpnA) which is cell surface-associated. Inactivation of the gene encoding SpnA produces a less virulent strain, suggesting that this surface DNase is contributing to virulence (Hasegawa et al. 2010). However, the mechanism by which this protein induces a virulence phenotype is not known. Thus, it is likely that we are just at the beginning of identifying the full panoply of moonlighting bacterial proteins that are contributing to bacterial virulence and definite searches should begin to be made to identify more. The bacterium employing the greatest number of moonlighting proteins is *M. tuberculosis* which, currently, has twelve moonlighting proteins in its armamentarium. It is also interesting as to how many bacteria (and other microbes) have been identified that use either GAPD, enolase or Cpn60 as moonlighting virulence proteins.

Another way in which surface-expressed moonlighting proteins can contribute to pathology is by antibodies to such bacterial proteins cross-reacting with the homologues of the host. Perhaps the most surprising manifestation of this is the proposal that antibody cross-reactivity between glycolytic enymes of group A streptococci and human neuronal surface glycolytic enzymes is responsible for conditions such as Tourette's syndrome and obsessive–compulsive disorder (Dale et al. 2006). Other conditions, such as psoriasis, are also believed to be driven by similar immunological cross-reactivity (McFadden et al. 2009).

Clearly, this review has focused on the role that bacterial moonlighting proteins play in bacterial/host interactions and in the process of bacterial virulence. This is largely because the moonlighting proteins discussed have been discovered by chance and their role in virulence has been the result of additional experimentation. In addition to playing a role in virulence the whole moonlighting phenomenon, which is likely to be much more pervasive than the few examples shown in Table 4, will impinge on our weltanschauung of bacteria and their interaction with their hosts. There are many reasons for this. If every, or a large proportion, protein moonlights then this implies a much more complex cellular system with many more interactions than there would be if each protein had only a single function (see Sriram et al. 2010). One of the problems that has arisen in biology in recent years has resulted from genome sequencing. This has revealed that complex organisms such as *H. sapiens* have unexpectedly small numbers of genes (open reading frames (ORFs)) in the their genomes. Currently, the human genome is estimated to contain only 20–25,000 protein-coding genes. Despite the fact that it is estimated that on average each gene has 3 splice variants, this still seems a very low number of working protein components to generate the complexity of a human being. Initial estimates of the human genome were in the 10^5 range. Moonlighting, which may increase the number of functions each protein has by 2–20 times, would, potentially, provide a system with sufficient number of protein interactors to generate what might be thought of as 'optimal complexity'. This increased complexity of protein–protein interactions would have to be matched by an increased complexity in the control of gene transcription, RNA synthesis/manipulation and protein translation. If the production of each moonlighting protein has to be tailored to match differences in each moonlighting activity then this implies a much more complex cellular network of information control than is currently envisaged.

Moonlighting also has major implications for evolutionary theory which currently envisages only one 'site' which is susceptible to the results of mutation, with most mutations being assumed to be neutral. However, if proteins actually have a much larger proportion of their sequences devoted to biological activity, then a greater proportion of the mutations are likely not to be neutral. This could have major consequences for human genetics, and already there is evidence that this is the case (Sriram et al. 2005). It would be convenient to utilise bacteria to study the evolution and the genetics and biological consequences of protein moonlighting.

References

Abou-Zeid C, Ratliff TL, Wiker HG, Harboe M, Bennedsen J, Rook GA (1988) Characterization of fibronectin-binding antigens released by *Mycobacterium tuberculosis* and *Mycobacterium bovis* BCG. Infect Immun 56:3046–3051

Allignet J, England P, Old I El Solh N (2001) Several regions of the repeat domain of the *Staphylococcus caprae* autolysin, AtlC, are involved in fibronectin binding. FEMS Microbiol Lett 213:193–197

Alloush HM, López-Ribot JL, Masten BJ, Chaffin WL (1997) 3-phosphoglycerate kinase: a glycolytic enzyme protein present in the cell wall of *Candida albicans*. Microbiology 143: 321–330

Alvarez-Dominguez C, Barbieri AM, Berón W, Wandinger-Ness A, Stahl PD (1996) Phagocytosed live *Listeria monocytogenes* influences Rab5-regulated in vitro phagosome-endosome fusion. J Biol Chem 271:13834–13843

Alvarez-Dominguez C, Madrazo-Toca F, Fernandez-Prieto L, Vandekerckhove J, Pareja E, Tobes R, Gomez-Lopez MT, Del Cerro-Vadillo E, Fresno M, Leyva-Cobián F, Carrasco-Marín E (2008) Characterization of a *Listeria monocytogenes* protein interfering with Rab5a. Traffic 9:325–337

Amini HR, Ascencio F, Ruiz-Bustos E, Romero MJ, Wadström T (1996) Cryptic domains of a 60 kDa heat shock protein of *Helicobacter pylori* bound to bovine lactoferrin. FEMS Immunol Med Microbiol 16:247–255

Amraei M, Nabi IR (2002) Species specificity of the cytokine function of phosphoglucose isomerase. FEBS Lett 525:151–155

Anand K, Mathur D, Anant A, Garg LC (2010) Structural studies of phosphoglucose isomerase from *Mycobacterium tuberculosis* H37Rv. Acta Crystallogr Sect F Struct Biol Cryst Commun 66:490–497

Antikainen J, Kuparinen V, Lähteenmäki K, Korhonen TK (2007) pH-dependent association of enolase and glyceraldehyde-3-phosphate dehydrogenase of *Lactobacillus crispatus* with the cell wall and lipoteichoic acids. J Bacteriol 189:4539–4543

Antúnez K, Anido M, Arredondo D, Evans JD, Zunino P (2011) Paenibacillus larvae enolase as a virulence factor in honeybee larvae infection. Vet Microbiol 87:83–89

Asquith KL, Baleato RM, McLaughlin EA, Nixon B, Aitken RJ (2004) Tyrosine phosphorylation activates surface chaperones facilitating sperm-zona recognition. J Cell Sci 117:3645–3657

Atanassov C, Pezennec L, d'Alayer J, Grollier G, Picard B, Fauchère JL (2002) Novel antigens of *Helicobacter pylori* correspond to ulcer-related antibody pattern of sera from infected patients. J Clin Microbiol 40:547–552

Attali C, Durmort C, Vernet T, Di Guilmi AM (2008) The interaction of *Streptococcus pneumoniae* with plasmin mediates transmigration across endothelial and epithelial monolayers by intercellular junction cleavage. Infect Immun 76:5350–5356

Ausiello CM, Fedele G, Palazzo R, Spensieri F, Ciervo A, Cassone A (2006) 60-kDa heat shock protein of *Chlamydia pneumoniae* promotes a T helper type 1 immune response through IL-12/IL-23 production in monocyte-derived dendritic cells. Microbes Infect 8:714–720

Babaahmady K, Oehlmann W, Singh M, Lehner T (2007) Inhibition of human immunodeficiency virus type 1 infection of human CD4+ T cells by microbial HSP70 and the peptide epitope 407–426. J Virol 81:3354–3360

Balasubramanian S, Kannan TR, Baseman JB (2008) The surface-exposed carboxyl region of *Mycoplasma pneumoniae* elongation factor Tu interacts with fibronectin. Infect Immun 76: 3116–3123

Balasubramanian S, Kannan TR, Hart PJ, Baseman JB (2009) Amino acid changes in elongation factor Tu of *Mycoplasma pneumoniae* and *Mycoplasma genitalium* influence fibronectin binding. Infect Immun 77:3533–3541

Banerjee S, Nandyala AK, Raviprasad P, Ahmed N, Hasnain SE (2007) Iron-dependent RNA-binding activity of *Mycobacterium tuberculosis* aconitase. J Bacteriol 189:4046–4052

Basak C, Pathak SK, Bhattacharyya A, Pathak S, Basu J, Kundu M (2005) The secreted peptidyl prolyl cis, trans-isomerase HP0175 of Helicobacter pylori induces apoptosis of gastric epithelial cells in a TLR4- and apoptosis signal-regulating kinase 1-dependent manner. J Immunol 174:5672–5680

Becker T, Hartl FU, Wieland F (2002) CD40, an extracellular receptor for binding and uptake of Hsp70-peptide complexes. J Cell Biol 158:1277–1285

Beckmann C, Waggoner JD, Harris TO, Tamura GS, Rubens CE (2002) Identification of novel adhesins from Group B Streptococci by use of phage display reveals that C5a peptidase mediates fibronectin binding. Infect Immun 70:2869–2876

Belisle JT, Vissart VD, Sievert T, Takayama K, Brennan PJ, Besra GS (1997) Role of the major antigen of *Mycobacterium tuberculosis* in cell wall biogenesis. Science 276:1420–1422

Bergmann S, Rohde M, Chhatwal GS, Hammerschmidt S (2001) A-enolase of *Streptococcus pneumoniae* is a plasmin(ogen)-binding protein displayed on the bacterial cell surface. Mol Microbiol 40:1273–1287

Bergmann S, Wild D, Diekmann O, Frank R, Bracht D, Chhatwal GS, Hammerschmidt S (2003) Identification of a novel plasmin(ogen)-binding motif in surface displayed alpha-enolase of *Streptococcus pneumoniae*. Mol Microbiol 49:411–423

Bergmann S, Rohde M, Hammerschmidt S (2004) Glyceraldehyde-3-phosphate dehydrogenase of *Streptococcus pneumoniae* is a surface-displayed plasminogen-binding protein. Infect Immun 72:2416–2419

Bergonzelli GE, Granato D, Pridmore RD, Marvin-Guy LF, Donnicola D, Corthésy-Theulaz IE (2006) GroEL of *Lactobacillus johnsonii* La1 (NCC 533) is cell surface associated: potential role in interactions with the host and the gastric pathogen *Helicobacter pylori*. Infect Immun 74:425–434

Blau K, Portnoi M, Shagan M, Kaganovich A, Rom S, Kafka D, Chalifa Caspi V, Porgador A, Givon-Lavi N, Gershoni JM, Dagan R, Mizrachi Y, Nebenzahl YM (2007) Flamingo cadherin: a putative host receptor for *Streptococcus pneumoniae*. J Infect Dis 195:1828–1837

Boël G, Jin H, Pancholi V (2005) Inhibition of cell surface export of group A streptococcal anchorless surface dehydrogenase affects bacterial adherence and antiphagocytic properties. Infect Immun 73:6237–6248

Bonazzi M, Lecuit M, Cossart P (2009) *Listeria monocytogenes* internalin and E-cadherin: from bench to bedside. Cold Spring Harb Perspect Biol 1:a003087

Bulut Y, Shimada K, Wong MH, Chen S, Gray P, Alsabeh R, Doherty TM, Crother TR, Arditi M (2009) Chlamydial heat shock protein 60 induces acute pulmonary inflammation in mice via the toll-like receptor 4- and MyD88-dependent pathway. Infect Immun 77:2683–2690

Burillo A, Bouza E (2010) *Chlamydia pneumoniae*. Infect Dis Clin North Am 24:61–71

Burkholder KM, Bhunia AK (2010) *Listeria monocytogenes* uses Listeria adhesion protein (LAP) to promote bacterial transepithelial translocation and induces expression of LAP receptor Hsp60. Infect Immun 78:5062–5073

Burnham C-AD, Shokoples SE, Tyrrell GJ (2005) Phosphoglycerate kinase inhibits epithelial cell invasion by group B streptococci. Microbe Pathog 38:189–200
Burton MJ, Mabey DC (2009) The global burden of trachoma: a review. PLoS Negl Trop Dis 3:e460
Campbell RM, Scanes CG (1995) Endocrine peptides 'moonlighting' as immune modulators: roles for somatostatin and GH-releasing factor. J Endocrinol 147:383–396
Candela M, Biagi E, Centanni M, Turroni S, Vici M, Musiani F, Vitali B, Bergmann S, Hammerschmidt S, Brigidi P (2009) Bifidobacterial enolase, a cell surface receptor for human plasminogen involved in the interaction with the host. Microbiology 155:3294–3303
Candela M, Centanni M, Fiori J, Biagi E, Turroni S, Orrico C, Bergmann S, Hammerschmidt S, Brigidi P (2010) DnaK from *Bifidobacterium animalis* subsp. *lactis* is a surface-exposed human plasminogen receptor upregulated in response to bile salts. Microbiology 156: 1609–1618
Capello M, Ferri-Borgogno S, Cappello P, Novelli F (2011) α-Enolase: a promising therapeutic and diagnostic tumor target. FEBS J [Epub ahead of print]
Carroll MV, Sim RB, Bigi F, Jäkel A, Antrobus R, Mitchell DA (2010) Identification of four novel DC-SIGN ligands on *Mycobacterium bovis* BCG. Protein Cell 1:859–870
Castaldo C, Vastano V, Siciliano RA, Candela M, Vici M, Muscariello L, Marasco R, Sacco M (2009) Surface displaced alfa-enolase of *Lactobacillus plantarum* is a fibronectin-binding protein. Microb Cell Fact 8:14
Cehovin A, Coates AR, Riffo-Vasquez Y, Tormay P, Botanch C, Altare F, Henderson B (2010) Comparison of the moonlighting actions of the two highly homologous chaperonin 60 proteins of *Mycobacterium tuberculosis*. Infect Immun 78:3196–3206
Chaput M, Claes V, Portetelle D, Cludts I, Cravador A, Burny A, Gras H, Tartar A (1988) The neurotrophic factor neuroleukin is 90% homologous with phosphohexose isomerase. Nature 332:454–455
Chaturvedi R, Bansal K, Narayana Y, Kapoor N, Sukumar N, Togarsimalemath SK, Chandra N, Mishra S, Ajitkumar P, Joshi B, Katoch VM, Patil SA, Balaji KN (2010) The multifunctional PE_PGRS11 protein from *Mycobacterium tuberculosis* plays a role in regulating resistance to oxidative stress. J Biol Chem 285:30389–30403
Chong A, Lima CA, Allan DS, Nasrallah GK, Garduño RA (2009) The purified and recombinant *Legionella pneumophila* chaperonin alters mitochondrial trafficking and microfilament organization. Infect Immun 77:4724–4739
Cianciotto NP (2001) Pathogenicity of *Legionella pneumophila*. Int J Med Microbiol 291331–291343
Commichau FM, Rothe FM, Herzberg C, Wagner E, Hellwig D, Lehnik-Habrink M, Hammer E, Völker U, Stülke J (2009) Novel activities of glycolytic enzymes in *Bacillus subtilis*: interactions with essential proteins involved in mRNA processing. Mol Cell Proteomics 8:1350–1360
Copley SD (2003) Enzymes with extra talents: moonlighting functions and catalytic promiscuity. Curr Opin Chem Biol 7:265–272
Cork AJ, Jergic S, Hammerschmidt S, Kobe B, Pancholi V, Benesch JL, Robinson CV, Dixon NE, Aquilina JA, Walker MJ (2009) Defining the structural basis of human plasminogen binding by streptococcal surface enolase. J Biol Chem 284:17129–17137
Corriden R, Insel PA (2010) Basal release of ATP: an autocrine-paracrine mechanism for cell regulation. Sci Signal 3:re1
Costa CP, Kirschning CJ, Busch D, Dürr S, Jennen L, Heinzmann U, Prebeck S, Wagner H, Miethke T (2002) Role of chlamydial heat shock protein 60 in the stimulation of innate immune cells by Chlamydia pneumoniae. Eur J Immunol 32:2460–2470
Courtney HS, Pownall HJ (2010) The structure and function of serum opacity factor: a unique streptococcal virulence determinant that targets high-density lipoproteins. J Biomed Biotechnol 2010:956071
Courtney HS, Li Y, Twal WO, Argraves WS (2009) Serum opacity factor is a streptococcal receptor for the extracellular matrix protein fibulin-1. J Biol Chem 284:12966–12971

Crowe JD, Sievwright IK, Auld GC, Moore NR, Gow NA, Booth NA (2003) Candida albicans binds human plasminogen: identification of eight plasminogen-binding proteins. Mol Microbiol 47:1637–1651

Da Costa CU, Wantia N, Kirschning CJ, Busch DH, Rodriguez N, Wagner H, Miethke T (2004) Heat shock protein 60 from *Chlamydia pneumoniae* elicits an unusual set of inflammatory responses via toll-like receptor 2 and 4 in vivo. Eur J Immunol 34:2874–2884

Dale RC, Candler PM, Church AJ, Wait R, Pocock JM, Giovannoni G (2006) Neuronal surface glycolytic enzymes are autoantigen targets in post-streptococcal autoimmune CNS disease. J Neuroimmunol 172:187–197

Dallo SF, Kannan TR, Blaylock MW, Baseman JB (2002) Elongation factor Tu and E1 β subunit of pyruvate dehydrogenase complex act as fibronectin-binding proteins in *Mycoplasma pneumonia*. Mol Microbiol 46:1041–1051

Daniely D, Portnoi M, Shagan M, Porgador A, Givon-Lavi N, Ling E, Dagan R, Mizrachi Nebenzahl Y (2006) Pneumococcal 6-phosphogluconate-dehydrogenase, a putative adhesin, induces protective immune response in mice. Clin Exp Immunol 144:254–263

Danø K, Behrendt N, Høyer-Hansen G, Johnsen M, Lund LR, Ploug M, Rømer J (2005) Plasminogen activation and cancer. Thromb Haemost 93:676–681

Darville T, Hiltke TJ (2010) Pathogenesis of genital tract disease due to *Chlamydia trachomatis*. J Infect Dis 201(2):S114–S125

de Jesus MC, Urban AA, Marasigan ME, Barnett Foster DE (2005) Acid and bile-salt stress of enteropathogenic *Escherichia coli* enhances adhesion to epithelial cells and alters glycolipid receptor binding specificity. J Infect Dis 192:1430–1440

Degrassi G, Devescovi G, Bigirimana J, Venturi V (2010) Xanthomonas oryzae pv. oryzae XKK.12 contains an AroQgamma chorismate mutase that is involved in rice virulence. Phytopathology 100:262–270

Derbise A, Song YP, Parikh S, Fischetti VA, Pancholi V (2004) Role of the C-terminal lysine residues of streptococcal surface enolase in Glu- and Lys-plasminogen-binding activities of group A streptococci. Infect Immun 72:94–105

Deveze-Alvarez M, García-Soto J, Martínez-Cadena G (2001) Glyceraldehyde-3-phosphate dehydrogenase is negatively regulated by ADP-ribosylation in the fungus Phycomyces blakesleeanus. Microbiology 147:2579–2584

Dimmeler S, Lottspeich F, Brüne B (1992) Nitric oxide causes ADP-ribosylation and inhibition of glyceraldehyde-3-phosphate dehydrogenase. J Biol Chem 267:16771–16774

Dinis M, Tavares D, Veiga-Malta I, Fonseca AJ, Andrade EB, Trigo G, Ribeiro A, Videira A, Cabrita AM, Ferreira P (2009) Oral therapeutic vaccination with *Streptococcus sobrinus* recombinant enolase confers protection against dental caries in rats. J Infect Dis 199:116–123

Dobocan MC, Sadvakassova G, Congote LF (2009) Chaperonin 10 as an endothelial-derived differentiation factor: role of glycogen synthase kinase-3. J Cell Physiol 219:470–476

Du RJ, Ho B (2003) Surface localized heat shock protein 20 (HslV) of *Helicobacter pylori*. Helicobacter 8:257–267

Ebanks RO, Goguen M, McKinnon S, Pinto DM, Ross NW (2005) Identification of the major outer membrane proteins of *Aeromonas salmonicida*. Dis Aquat Organ 68:29–38

Eberhard T, Kronvall G, Ullberg M (1999) Surface bound plasmin promotes migration of *Streptococcus pneumoniae* through reconstituted basement membranes. Microb Pathog 26: 175–181

Egea L, Aguilera L, Giménez R, Sorolla MA, Aguilar J, Badía J, Baldoma L (2007) Role of secreted glyceraldehyde-3-phosphate dehydrogenase in the infection mechanism of entero-hemorrhagic and enteropathogenic *Escherichia coli*: interaction of the extracellular enzyme with human plasminogen and fibrinogen. Int J Biochem Cell Biol 39:1190–1203

Ellis RJ, Minton AP (2006) Protein aggregation in crowded environments. Biol Chem 387:485–497

Emelyanov VV, Loukianov EV (2004) A 29.5 kDa heat-modifiable major outer membrane protein of *Rickettsia prowazekii*, putative virulence factor, is a peptidyl-prolyl cis/trans isomerase. IUBMB Life 56:215–219

Ensgraber M, Loos M (1992) A 66-kilodalton heat shock protein of *Salmonella typhimurium* is responsible for binding of the bacterium to intestinal mucus. Infect Immun 60:3072–3078

Erttmann KD, Kleensang A, Schneider E, Hammerschmidt S, Büttner DW, Gallin M (2005) Cloning, characterization and DNA immunization of an *Onchocerca volvulus* glyceraldehyde-3-phosphate dehydrogenase (Ov-GAPDH). Biochim Biophys Acta 1741:85–94

Esaguy N, Aguas AP (1997) Subcellular localization of the 65-kDa heat shock protein in mycobacteria by immunoblotting and immunogold ultracytochemistry. J Submicrosc Cytol Pathol 29:85–90

Esgleas M, Li Y, Hancock MA, Harel J, Dubreuil JD, Gottschalk M (2008) Isolation and characterization of alpha-enolase, a novel fibronectin-binding protein from *Streptococcus suis*. Microbiology 154:2668–2679

Fairbank M, St-Pierre P, Nabi IR (2009) The complex biology of autocrine motility factor/phosphoglucose isomerase (AMF/PGI) and its receptor, the gp78/AMFR E3 ubiquitin ligase. Mol Biosyst 5:793–801

Floto RA, MacAry PA, Boname JM, Mien TS, Kampmann B, Hair JR, Huey OS, Houben EN, Pieters J, Day C, Oehlmann W, Singh M, Smith KG, Lehner PJ (2006) Dendritic cell stimulation by mycobacterial Hsp70 is mediated through CCR5. Science 314:454–458

Fraiberg M, Borovok I, Weiner RM, Lamed R (2010) Discovery and characterization of cadherin domains in *Saccharophagus degradans* 2–40. J Bacteriol 192:1066–1074

Freedman R (1978) Moonlighting molecules. New Sci 79:560–562

Friedland JS, Shattock R, Remick DG, Griffin GE (1993) Mycobacterial 65-kD heat shock protein induces release of proinflammatory cytokines from human monocytic cells. Clin Exp Immunol 91:58–62

Frisk A, Ison CA, Lagergård T (1998) GroEL heat shock protein of *Haemophilus ducreyi*: association with cell surface and capacity to bind to eukaryotic cells. Infect Immun 66: 1252–1257

Fu M, Li L, Albrecht T, Johnson JD, Kojic LD, Nabi IR (2011) Autocrine motility factor/phosphoglucose isomerase regulates ER stress and cell death through control of ER calcium release. Cell Death Differ [Epub ahead of print]

Funasaka T, Hogan V, Raz A (2009) Phosphoglucose isomerase/autocrine motility factor mediates epithelial and mesenchymal phenotype conversions in breast cancer. Cancer Res 69:5349–5356

Furuya H, Ikeda R (2009) Interaction of triosephosphate isomerase from the cell surface of *Staphylococcus aureus* and alpha-(1-> 3)-mannooligosaccharides derived from glucuronoxylomannan of *Cryptococcus neoformans*. Microbiology 155:2707–2713

Garduño RA, Faulkner G, Trevors MA, Vats N, Hoffman PS (1998a) Immunolocalization of Hsp60 in *Legionella pneumophila*. J Bacteriol 180:505–513

Garduño RA, Garduño E, Hoffman PS (1998b) Surface-associated hsp60 chaperonin of *Legionella pneumophila* mediates invasion in a HeLa cell model. Infect Immun 66:4602–4610

Gase K, Gase A, Schirmer H, Malke H (1996) Cloning, sequencing and functional overexpression of the *Streptococcus equisimilis* H46A gapC gene encoding a glyceraldehyde-3-phosphate dehydrogenase that also functions as a plasmin(ogen)-binding protein. Purification and biochemical characterization of the protein. Eur J Biochem 239:42–51

Gething M-J (1997) Guidebook to Molecular Chaperones and Protein-Folding Catalysts. Oxford University Press, Oxford

Gobert AP, Bambou JC, Werts C, Balloy V, Chignard M, Moran AP, Ferrero RL (2004) *Helicobacter pylori* heat shock protein 60 mediates interleukin-6 production by macrophages via a toll-like receptor (TLR)-2-, TLR-4-, and myeloid differentiation factor 88-independent mechanism. J Biol Chem 279:245–250

Gomez FJ, Pilcher-Roberts R, Alborzi A, Newman SL (2008) Histoplasma capsulatum cyclophilin A mediates attachment to dendritic cell VLA-5. J Immunol 181:7106–7114

Goulhen F, Hafezi A, Uitto VJ, Hinode D, Nakamura R, Grenier D, Mayrand D (1998) Subcellular localization and cytotoxic activity of the GroEL-like protein isolated from *Actinobacillus actinomycetemcomitans*. Infect Immun 66:5307–5313

Gray MW, Burger G, Lang BF (1999) Mitochondrial evolution. Science 283:1476–1481
Gregersen N, Bross P (2010) Protein misfolding and cellular stress: an overview. Methods Mol Biol 648:3–23
Guimarães AJ, Frases S, Gomez FJ, Zancopé-Oliveira RM, Nosanchuk JD (2009) Monoclonal antibodies to heat shock protein 60 alter the pathogenesis of *Histoplasma capsulatum*. Infect Immun 77:1357–1367
Gunka K, Newman JA, Commichau FM, Herzberg C, Rodrigues C, Hewitt L, Lewis RJ, Stülke J (2010) Functional dissection of a trigger enzyme: mutations of the *Bacillus subtilis* glutamate dehydrogenase RocG that affect differentially its catalytic activity and regulatory properties. J Mol Biol 400:815–827
Hall-Stoodley L, Costerton JW, Stoodley P (2004) Bacterial biofilms: from the natural environment to infectious diseases. Nat Rev Microbiol 2:95–108
Hartmann E, Lingwood C (1997) Brief heat shock treatment induces a long-lasting alteration in the glycolipid receptor binding specificity and growth rate of *Haemophilus influenzae*. Infect Immun 65:1729–1733
Hartmann E, Lingwood CA, Reidl J (2001) Heat-inducible surface stress protein (Hsp70) mediates sulfatide recognition of the respiratory pathogen *Haemophilus influenzae*. Infect Immun 69:3438–3441
Hasegawa T, Minami M, Okamoto A, Tatsuno I, Isaka M, Ohta M (2010) Characterization of a virulence-associated and cell-wall-located DNase of *Streptococcus pyogenes*. Microbiology 156:184–190
Hayashi T, Rao SP, Catanzaro A (1997) Binding of the 68-kilodalton protein of *Mycobacterium avium* to alpha(v)beta3 on human monocyte-derived macrophages enhances complement receptor type 3 expression. Infect Immun 65:1211–1216
Heilmann C, Thumm G, Chhatwal GS, Hartleib J, Uekötter A, Peters G (2003) Identification and characterization of a novel autolysin (Aae) with adhesive properties from *Staphylococcus epidermidis*. Microbiology 149:2769–2778
Heilmann C, Hartleib J, Hussain MS, Peters G (2005) The multifunctional *Staphylococcus aureus* autolysin aaa mediates adherence to immobilized fibrinogen and fibronectin. Infect Immun 73:4793–4802
Helbig JH, Lück PC, Steinert M, Jacobs E, Witt M (2001) Immunolocalization of the Mip protein of intracellularly and extracellularly grown *Legionella pneumophila*. Lett Appl Microbiol 32:83–88
Helbig JH, König B, Knospe H, Bubert B, Yu C, Lück CP, Riboldi-Tunnicliffe A, Hilgenfeld R, Jacobs E, Hacker J, Fischer G (2003) The PPIase active site of *Legionella pneumophila* Mip protein is involved in the infection of eukaryotic host cells. Biol Chem 384:125–137
Hell W, Meyer HG, Gatermann SG (1998) Cloning of aas, a gene encoding a *Staphylococcus saprophyticus* surface protein with adhesive and autolytic properties. Mol Microbiol 29: 871–881
Henderson B, Allan E, Coates ARM (2006) Stress wars: the direct role of host and bacterial molecular chaperones in bacterial Infection. Infect Immun 74:3693–3706
Henderson B, Henderson S (2009) Unfolding the relationship between secreted molecular chaperones and macrophage activation states. Cell Stress Chaperones 14:329–341
Henderson B, Pockley AG (2010) Molecular chaperones and protein-folding catalysts as intercellular signaling regulators in immunity and inflammation. J Leukoc Biol 88:445–462
Henderson B, Wilson M, McNab R, Lax AR (1999) Cellular microbiology. Wiley, Chichester
Henderson B, Lund PA, Coates ARM (2010) Multiple moonlighting functions of mycobacterial molecular chaperones. Tuberculosis 90:119–124
Henderson B, Nair S, Pallas J, Williams MA (2011) Fibronectin: a multidomain host adhesin targeted by bacterial fibronectin-binding proteins. FEMS Microbiol Rev 35:147–200
Henkel JS, Baldwin MR, Barbieri JT (2010) Toxins from bacteria. EXS 100:1–29
Hennequin C, Porcheray F, Waligora-Dupriet A, Collignon A, Barc M, Bourlioux P, Karjalainen T (2001) GroEL (Hsp60) of *Clostridium difficile* is involved in cell adherence. Microbiology 147:87–96

Herbert EE, Goodrich-Blair H (2007) Friend and foe: the two faces of *Xenorhabdus nematophila*. Nat Rev Microbiol 5:634–646

Hermans PW, Adrian PV, Albert C, Estevão S, Hoogenboezem T, Luijendijk IH, Kamphausen T, Hammerschmidt S (2006) The streptococcal lipoprotein rotamase A (SlrA) is a functional peptidyl-prolyl isomerase involved in pneumococcal colonization. J Biol Chem 281:968–976

Hickey TB, Thorson LM, Speert DP, Daffé M, Stokes RW (2009) Mycobacterium tuberculosis Cpn60.2 and DnaK are located on the bacterial surface, where Cpn60.2 facilitates efficient bacterial association with macrophages. Infect Immun 77:3389–3401

Hickey TB, Ziltener HJ, Speert DP, Stokes RW (2010) Mycobacterium tuberculosis employs Cpn60.2 as an adhesin that binds CD43 on the macrophage surface. Cell Microbiol 12: 1634–1647

Hoelzle LE, Hoelzle K, Helbling M, Aupperle H, Schoon HA, Ritzmann M, Heinritzi K, Felder KM, Wittenbrink MM (2007) MSG1, a surface-localised protein of *Mycoplasma suis* is involved in the adhesion to erythrocytes. Microbes Infect 9:466–474

Hoffman PS, Garduno RA (1999) Surface-associated heat shock proteins of *Legionella pneumophila* and *Helicobacter pylori*: roles in pathogenesis and immunity. Infect Dis Obstet Gynecol 7:58–63

Horwich AL, Apetri AC, Fenton WA (2009) The GroEL/GroES cis cavity as a passive anti-aggregation device. FEBS Lett 583:2654–2662

Hottiger MO, Hassa PO, Lüscher B, Schüler H, Koch-Nolte F (2010) Toward a unified nomenclature for mammalian ADP-ribosyltransferases. Trends Biochem Sci 35:208–219

Hoy B, Löwer M, Weydig C, Carra G, Tegtmeyer N, Geppert T, Schröder P, Sewald N, Backert S, Schneider G, Wessler S (2010) Helicobacter pylori HtrA is a new secreted virulence factor that cleaves E-cadherin to disrupt intercellular adhesion. EMBO Rep 11: 798–804

Hu Y, Henderson B, Lund PA, Tormay P, Ahmed MT, Gurcha SS, Besra GS, Coates AR (2008) A mycobacterium tuberculosis mutant lacking the groEL homologue cpn60.1 is viable but fails to induce an inflammatory response in animal models of infection. Infect Immun 76:1535–1546

Huesca M, Goodwin A, Bhagwansingh A, Hoffman P, Lingwood CA (1998) Characterization of an acidic-pH-inducible stress protein (hsp70), a putative sulfatide binding adhesin, from *Helicobacter pylori*. Infect Immun 66:4061–4067

Hughes MJ, Moore JC, Lane JD, Wilson R, Pribul PK, Younes ZN, Dobson RJ, Everest P, Reason AJ, Redfern JM, Greer FM, Paxton T, Panico M, Morris HR, Feldman RG, Santangelo JD (2002) Identification of major outer surface proteins of *Streptococcus agalactiae*. Infect Immun 70:1254–1259

Hurmalainen V, Edelman S, Antikainen J, Baumann M, Lähteenmäki K, Korhonen TK (2007) Extracellular proteins of *Lactobacillus crispatus* enhance activation of human plasminogen. Microbiology 153:1112–1122

Ikeda R, Saito F, Matsuo M, Kurokawa K, Sekimizu K, Yamaguchi M, Kawamoto S (2007) Contribution of the mannan backbone of cryptococcal glucuronoxylomannan and a glycolytic enzyme of *Staphylococcus aureus* to contact-mediated killing of *Cryptococcus neoformans*. J Bacteriol 189:4815–4826

Jagadeesan B, Koo OK, Kim KP, Burkholder KM, Mishra KK, Aroonnual A, Bhunia AK (2010) LAP, an alcohol acetaldehyde dehydrogenase enzyme in *Listeria*, promotes bacterial adhesion to enterocyte-like Caco-2 cells only in pathogenic species. Microbiology 156:2782–2795

Jeffery CJ (1999) Moonlighting proteins. Trends Biochem Sci 24:8–11

Jeffery CJ (2005) Mass spectrometry and the search for moonlighting proteins. Mass Spec Revs 24:772–782

Jin H, Youngmia P, Boel G, Kochar J, Pancholi V (2005) Group A streptococcal surface GAPDF, SDH recognises uPAR/CD87 as its receptor on the human pharyngeal cell and mediates bacterial adherence to host cells. J Mol Biol 350:27–41

Jobin MC, Brassard J, Quessy S, Gottschalk M, Grenier D (2004) Acquisition of host plasmin activity by the swine pathogen *Streptococcus suis* serotype 2. Infect Immun 72:606–610

Johnson BJ, Le TT, Dobbin CA, Banovic T, Howard CB, Flores Fde M, Vanags D, Naylor DJ, Hill GR, Suhrbier A (2005) Heat shock protein 10 inhibits lipopolysaccharide-induced inflammatory mediator production. J Biol Chem 280:4037–4047

Joshi MC, Sharma A, Kant S, Birah A, Gupta GP, Khan SR, Bhatnagar R, Banerjee N (2008) An insecticidal GroEL protein with chitin binding activity from *Xenorhabdus nematophila*. J Biol Chem 283:28287–28296

Juli C, Sippel M, Jager J, Thiele A, Weiwad M, Schweimer K, Rosch P, Steinert M, Sotriffer CA, Holzgrabe U (2010) Pipecolic acid derivatives as small-molecule inhibitors of the Legionella MIP protein. J Med Chem [Epub ahead of print]

Jung CJ, Zheng QH, Shieh YH, Lin CS, Chia JS (2009) *Streptococcus mutans* autolysin AtlA is a fibronectin-binding protein and contributes to bacterial survival in the bloodstream and virulence for infective endocarditis. Mol Microbiol 74:888–902

Kalayoglu MV, Indrawati, Morrison RP, Morrison SG, Yuan Y, Byrne GI (2000) Chlamydial virulence determinants in atherogenesis: the role of chlamydial lipopolysaccharide and heat shock protein 60 in macrophage-lipoprotein interactions. J Infect Dis 181(3):S483–489

Kaneda K, Masuzawa T, Yasugami K, Suzuki T, Suzuki Y, Yanagihara Y (1997) Glycosphingolipid-binding protein of *Borrelia burgdorferi sensu lato*. Infect Immun 65:3180–3185

Karunakaran KP, Noguchi Y, Read TD, Cherkasov A, Kwee J, Shen C, Nelson CC, Brunham RC (2003) Molecular analysis of the multiple GroEL proteins of *Chlamydiae*. J Bacteriol 185:1958–1966

Katakura Y, Sano R, Hashimoto T, Ninomiya K, Shioya S (2010) Lactic acid bacteria display on the cell surface cytosolic proteins that recognize yeast mannan. Appl Microbiol Biotechnol 86:319–326

Kesimer M, Kiliç N, Mehrotra R, Thornton DJ, Sheehan JK (2009) Identification of salivary mucin MUC7 binding proteins from *Streptococcus gordonii*. BMC Microbiol 9:163

Khan N, Alam K, Mande SC, Valluri VL, Hasnain SE, Mukhopadhyay S (2008) *Mycobacterium tuberculosis* heat shock protein 60 modulates immune response to PPD by manipulating the surface expression of TLR2 on macrophages. Cell Microbiol 10:1711–1722

Kim KP, Jagadeesan B, Burkholder KM, Jaradat ZW, Wampler JL, Lathrop AA, Morgan MT, Bhunia AK (2006) Adhesion characteristics of Listeria adhesion protein (LAP)-expressing *Escherichia coli* to Caco-2 cells and of recombinant LAP to eukaryotic receptor Hsp60 as examined in a surface plasmon resonance sensor. FEMS Microbiol Lett 256:324–332

Kinhikar AG, Vargas D, Li H, Mahaffey SB, Hinds L, Belisle JT, Laal S (2006) Mycobacterium tuberculosis malate synthase is a laminin-binding adhesin. Mol Microbiol 60:999–1013

Kinnby B, Booth NA, Svensater G (2008) Plasminogen binding by oral streptococci from dental plaque and inflammatory lesions. Microbiology 154:924–931

Kirby AC, Meghji S, Nair SP, White P, Reddi K, Nishihara T, Nakashima K, Willis AC, Sim R, Wilson M, Henderson B (1995) The potent bone resorbing mediator of *Actinobacillus actinomycetemcomitans* is homologous to the molecular chaperone GroEL. J Clin Invest 96:1185–1194

Knaust A, Weber MV, Hammerschmidt S, Bergmann S, Frosch M, Kurzai O (2007) Cytosolic proteins contribute to surface plasminogen recruitment of *Neisseria meningitidis*. J Bacteriol 189:3246–3255

Kol A, Sukhova GK, Lichtman AH, Libby P (1998) Chlamydial heat shock protein 60 localizes in human atheroma and regulates macrophage tumor necrosis factor-alpha and matrix metalloproteinase expression. Circulation 98:300–307

Kong TH, Coates AR, Butcher PD, Hickman CJ, Shinnick TM (1993) *Mycobacterium tuberculosis* expresses two chaperonin-60 homologs. Proc Natl Acad Sci U S A 90:2608–2612

Krishna KA, Rao GV, Rao KR (2007) Chaperonin GroEL: structure and reaction cycle. Curr Protein Pept Sci 8:418–425

Krishnamurthy G, Vikram R, Singh SB, Patel N, Agarwal S, Mukhopadhyay G, Basu SK, Mukhopadhyay A (2005) Hemoglobin receptor in *Leishmania* is a hexokinase located in the flagellar pocket. J Biol Chem 280:5884–5891

Lähteenmäki K, Kuusela P, Korhonen TK (2001) Bacterial plasminogen activators and receptors. FEMS Microbiol Rev 25:531–552

Lähteenmäki K, Edelman S, Korhonen TK (2005) Bacterial metastasis: the host plasminogen system in bacterial invasion. Trends Microbiol 13:79–85

Lama A, Kucknoor A, Mundodi V, Alderete JF (2009) Glyceraldehyde-3-phosphate dehydrogenase is a surface-associated, fibronectin-binding protein of *Trichomonas vaginalis*. Infect Immun 77:2703–2711

LeaRue RW, Dill BD, Giles DK, Whittimore JD, Raulston JE (2007) Chlamydial Hsp60–2 is iron responsive in *Chlamydia trachomatis* serovar E-infected human endometrial epithelial cells in vitro. Infect Immun 75:2374–2380

Launois P, N'Diaye MN, Cartel JL, Mane I, Drowart A, Van Vooren JP, Sarthou JL, Huygen K (1995) Fibronectin-binding antigen 85 and the 10-kilodalton GroES-related heat shock protein are the predominant TH-1 response inducers in leprosy contacts. Infect Immun 63:88–93

Lay AJ, Jiang XM, Kisker O, Flynn E, Underwood A, Condron R, Hogg PJ (2000) Phosphoglycerate kinase acts in tumour angiogenesis as a disulphide reductase. Nature 408:869–873

Lehner T, Bergmeier LA, Wang Y, Tao L, Sing M, Spallek R, van der Zee R (2000) Heat shock proteins generate beta-chemokines which function as innate adjuvants enhancing adaptive immunity. Eur J Immunol 30:594–603

Leuzzi R, Serino L, Scarselli M, Savino S, Fontana MR, Monaci E, Taddei A, Fischer G, Rappuoli R, Pizza M (2005) Ng-MIP, a surface-exposed lipoprotein of *Neisseria gonorrhoeae*, has a peptidyl-prolyl cis/trans isomerase (PPIase) activity and is involved in persistence in macrophages. Mol Microbiol 58:669–681

Lewthwaite JC, Coates AR, Tormay P, Singh M, Mascagni P, Poole S, Roberts M, Sharp L, Henderson B (2001) *Mycobacterium tuberculosis* chaperonin 60.1 is a more potent cytokine stimulator than chaperonin 60.2 (hsp 65) and contains a CD14-binding domain. Infect Immun 69:7349–7355

Lewthwaite J, George R, Lund PA, Poole S, Tormay P, Sharp L, Coates AR, Henderson B (2002) *Rhizobium leguminosarum* chaperonin 60.3, but not chaperonin 60.1, induces cytokine production by human monocytes: activity is dependent on interaction with cell surface CD14. Cell Stress Chaperones 7:130–136

Li N, Xiang GS, Dokainish H, Ireton K, Elferink LA (2005) The *Listeria* protein internalin B mimics hepatocyte growth factor-induced receptor trafficking. Traffic 6:459–473

Lin SN, Ayada K, Zhao Y, Yokota K, Takenaka R, Okada H, Kan R, Hayashi S, Mizuno M, Hirai Y, Fujinami Y, Oguma K (2005) *Helicobacter pylori* heat-shock protein 60 induces production of the proinflammatory cytokine IL8 in monocytic cells. J Med Microbiol 54:225–233

Lin CY, Huang YS, Li CH, Hsieh YT, Tsai NM, He PJ, Hsu WT, Yeh YC, Chiang FH, Wu MS, Chang CC, Liao KW (2009) Characterizing the polymeric status of *Helicobacter pylori* heat shock protein 60. Biochem Biophys Res Commun 388:283–289

Lin CS, He PJ, Tsai NM, Li CH, Yang SC, Hsu WT, Wu MS, Wu CJ, Cheng TL, Liao KW (2010) A potential role for *Helicobacter pylori* heat shock protein 60 in gastric tumorigenesis. Biochem Biophys Res Commun 392:183–189

Ling E, Feldman G, Portnoi M, Dagan R, Overweg K, Mulholland F, Chalifa-Caspi V, Wells J, Mizrachi-Nebenzahl Y (2004) Glycolytic enzymes associated with the cell surface of *Streptococcus pneumoniae* are antigenic in humans and elicit protective immune responses in the mouse. Clin Exp Immunol 138:290–298

Long KH, Gomez FJ, Morris RE, Newman SL (2003) Identification of heat shock protein 60 as the ligand on *Histoplasma capsulatum* that mediates binding to CD18 receptors on human macrophages. J Immunol 170:487–494

Lottenberg R, Broder CC, Boyle MD, Kain SJ, Schroeder BL, Curtiss R (1992) Cloning, sequence analysis, and expression in *Escherichia coli* of a streptococcal plasmin receptor. J Bacteriol 174:5204–5210

Lund PA (2009) Multiple chaperonins in bacteria—why so many? FEMS Microbiol Revs 3:785–800

Lundemose AG, Kay JE, Pearce JH (1993) *Chlamydia trachomatis* Mip-like protein has peptidyl-prolyl cis/trans isomerase activity that is inhibited by FK506 and rapamycin and is implicated in initiation of chlamydial infection. Mol Microbiol 7:777–783

Macchia G, Massone A, Burroni D, Covacci A, Censini S, Rappuoli R (1993) The Hsp60 protein of *Helicobacter pylori*: structure and immune response in patients with gastroduodenal diseases. Mol Microbiol 9:645–652

Macellaro A, Tujulin E, Hjalmarsson K, Norlander L (1998) Identification of a 71-kilodalton surface-associated Hsp70 homologue in *Coxiella burneti*. Infect Immun 66:5882–5888

Madureira P, Baptista M, Vieira M, Magalhães V, Camelo A, Oliveira L, Ribeiro A, Tavares D, Trieu-Cuot P, Vilanova M, Ferreira P (2007) *Streptococcus agalactiae* GAPDH is a virulence-associated immunomodulatory protein. J Immunol 178:1379–1387

Magalhães V, Veiga-Malta I, Almeida MR, Baptista M, Ribeiro A, Trieu-Cuot P, Ferreira P (2007) Interaction with human plasminogen system turns on proteolytic activity in *Streptococcus agalactiae* and enhances its virulence in a mouse model. Microbes Infect 9:1276–1284

Makarow M, Braakman I (2010) Chaperones. Springer, Heidelberg

Mamelak D, Mylvaganam M, Whetstone H, Hartmann E, Lennarz W, Wyrick PB, Raulston J, Han H, Hoffman P, Lingwood CA (2001) Hsp70s contain a specific sulfogalactolipid binding site. Differential aglycone influence on sulfogalactosyl ceramide binding by recombinant prokaryotic and eukaryotic hsp70 family members. Biochemistry 40:3572–3582

Martinez FO, Helming L, Gordon S (2009) Alternative activation of macrophages: an immunologic functional perspective. Annu Rev Immunol 27:451–483

Matta SK, Agarwal S, Bhatnagar R (2010) Surface localized and extracellular Glyceraldehyde-3-phosphate dehydrogenase of *Bacillus anthracis* is a plasminogen binding protein. Biochim Biophys Acta 1804:2111–2120

McFadden JP, Baker BS, Powles AV, Fry L (2009) Psoriasis and streptococci: the natural selection of psoriasis revisited. Br J Dermatol 160:929–937

Meghji S, White PA, Nair SP, Reddi K, Heron K, Henderson B, Zaliani A, Fossati G, Mascagni P, Hunt JF, Roberts MM, Coates AR (1997) *Mycobacterium tuberculosis* chaperonin 10 stimulates bone resorption: a potential contributory factor in Pott's disease. J Exp Med 186:1241–1246

Méndez-Samperio P (2008) Expression and regulation of chemokines in mycobacterial infection. J Infect 57:374–384

Mitchell LA, Nixon B, Aitken RJ (2007) Analysis of chaperone proteins associated with human spermatozoa during capacitation. Mol Hum Reprod 13:605–613

Modun B, Williams P (1999) The staphylococcal transferrin-binding protein is a cell wall glyceraldehyde-3-phosphate dehydrogenase. Infect Immun 67:1086–1092

Moore BD (2004) Bifunctional and moonlighting enzymes: lighting the way to regulatory control. Trends Plant Sci 9:221–228

Moreno-Brito V, Yáñez-Gómez C, Meza-Cervantez P, Avila-González L, Rodríguez MA, Ortega-López J, González-Robles A, Arroyo R (2005) A *Trichomonas vaginalis* 120 kDa protein with identity to hydrogenosome pyruvate:ferredoxin oxidoreductase is a surface adhesin induced by iron. Cell Microbiol 7:245–258

Mortellaro A, Robinson L, Ricciardi-Castagnoli P (2009) Spotlight on mycobacteria and dendritic cells: will novel targets to fight tuberculosis emerge? EMBO Mol Med 1:19–29

Mukai C, Bergkvist M, Nelson JL, Travis AJ (2009) Sequential reactions of surface- tethered glycolytic enzymes. Chem Biol 16:1013–1020

Naito M, Ohara N, Matsumoto S, Yamada T (1998) The novel fibronectin-binding motif and key residues of mycobacteria. J Biol Chem 273:2905–2909

Naito M, Fukuda T, Sekiguchi K, Yamada T, Naito M (2000) The domains of human fibronectin mediating the binding of alpha antigen, the most immunopotent antigen of mycobacteria that induces protective immunity against mycobacterial infection. Biochem J 347:725–731

Natarajaseenivasan K, Artiushin SC, Velineni S, Vedhagiri K, Vijayachari P, Timoney JF (2011) Surface-associated Hsp60 chaperonin of *Leptospira interrogans* serovar *Autumnalis* N2 strain as an immunoreactive protein. Eur J Clin Microbiol Infect Dis [Epub ahead of print]

Noah CE, Malik M, Bublitz DC, Camenares D, Sellati TJ, Benach JL, Furie MB (2010) GroEL and lipopolysaccharide from *Francisella tularensis* live vaccine strain synergistically activate human macrophages. Infect Immun 78:1797–1806

Noonan FP, Halliday WJ, Morton H, Clunie GJ (1979) Early pregnancy factor is immunosuppressive. Nature 278:649–651

Nowalk AJ, Nolder C, Clifton DR, Carroll JA (2006) Comparative proteome analysis of subcellular fractions from *Borrelia burgdorferi* by NEPHGE and IPG. Proteomics 6: 2121–2134

Ofek I, Hasty DL, Doyle RJ (2003) Bacterial adhesion. Washington, ASM Press

Ohnishi H, Mizunoe Y, Takade A, Tanaka Y, Miyamoto H, Harada M, Yoshida S (2004) Legionella dumoffii DjlA, a member of the DnaJ family, is required for intracellular growth. Infect Immun 72:3592–3603

Ojha A, Anand M, Bhatt A, Kremer L, Jacobs WR Jr, Hatfull GF (2005) GroEL1: a dedicated chaperone involved in mycolic acid biosynthesis during biofilm formation in mycobacteria. Cell 123:861–873

Oliver MA, Rojo JM, Rodríguez de Córdoba S, Alberti S (2008) Binding of complement regulatory proteins to group A Streptococcus. Vaccine 26(8):I75–I178

Pancholi V (2001) Multifunctional alpha-enolase: its role in diseases. Cell Mol Life Sci 58:902–920

Pancholi V, Fischetti VA (1992) A major surface protein on group A streptococci is a glyceraldehyde-3-phosphate-dehydrogenase with multiple binding activity. J Exp Med 176: 415–426

Pancholi V, Fischetti VA (1993) Glyceraldehyde-3-phosphate dehydrogenase on the surface of group A streptococci is also an ADP-ribosylating enzyme. Proc Natl Acad Sci U S A 90:8154–8158

Pancholi V, Fischetti VA (1997) Regulation of the phosphorylation of human pharyngeal cell proteins by group A streptoccal surface dehydrogenase: signal transduction between streptococci and pharyngeal cells. J Exp Med 186:1633–1643

Pancholi V, Fischetti VA (1998) Alpha-enolase, a novel strong plasmin(ogen) binding protein on the surface of pathogenic streptococci. J Biol Chem 273:14503–14515

Pantzar M, Teneberg S, Lagergård T (2006) Binding of *Haemophilus ducreyi* to carbohydrate receptors is mediated by the 58.5-kDa GroEL heat shock protein. Microbes Infect 8: 2452–2458

Pathak SK, Basu S, Bhattacharyya A, Pathak S, Banerjee A, Basu J, Kundu M (2006) TLR4-dependent NF-kappaB activation and mitogen- and stress-activated protein kinase 1-triggered phosphorylation events are central to *Helicobacter pylori* peptidyl prolyl cis-, trans-isomerase (HP0175)-mediated induction of IL-6 release from macrophages. J Immunol 177:7950–7958

Peake P, Gooley A, Britton WJ (1993) Mechanism of interaction of the 85B secreted protein of *Mycobacterium bovis* with fibronectin. Infect Immun 61:4828–4832

Peetermans WE, Raats CJ, Langermans JA, van Furth R (1994) Mycobacterial heat-shock protein 65 induces proinflammatory cytokines but does not activate human mononuclear phagocytes. Scand J Immunol 39:613–617

Perucho M, Salas J, Salas ML (1980) Study of the interaction of glyceraldehyde-3-phosphate dehydrogenase with DNA. Biochim Biophys Acta 606:181–195

Piatigorsky J (1998) Multifunctional lens crystallins and corneal enzymes. More than meets the eye. Ann N Y Acad Sci 842:7–15

Piatigorsky J (2007) Gene Sharing and Evolution. Harvard University Press, Cambridge

Piatigorsky J, O'Brien WE, Norman BL, Kalumuck K, Wistow GJ, Borras T, Nickerson JM, Wawrousek EF (1988) Gene sharing by delta-crystallin and argininosuccinate lyase. Proc Natl Acad Sci U S A 85:3479–8343

Pidot SJ, Porter JL, Tobias NJ, Anderson J, Catmull D, Seemann T, Kidd S, Davies JK, Reynolds E, Dashper S, Stinear TP (2010) Regulation of the 18 kDa heat shock protein in *Mycobacterium ulcerans*: an alpha-crystallin orthologue that promotes biofilm formation. Mol Microbiol 78:1216–1231

Podobnik M, Tyagi R, Matange N, Dermol U, Gupta AK, Mattoo R, Seshadri K, Visweswariah SS (2009) A mycobacterial cyclic AMP phosphodiesterase that moonlights as a modifier of cell wall permeability. J Biol Chem 284:32846–32857

Poltermann S, Kunert A, von der Heide M, Eck R, Hartmann A, Zipfel PF (2007) Gpm1p is a factor H-, FHL-1-, and plasminogen-binding surface protein of *Candida albicans*. J Biol Chem 282:37537–37544

Puech V, Guilhot C, Perez E, Tropis M, Armitige LY, Gicquel B, Daffé M (2002) Evidence for a partial redundancy of the fibronectin-binding proteins for the transfer of mycoloyl residues onto the cell wall arabinogalactan termini of *Mycobacterium tuberculosis*. Mol Microbiol 44:1109–1122

Qamra R, Mande SC (2004) Crystal structure of the 65-kilodalton heat shock protein, chaperonin 60.2, of *Mycobacterium tuberculosis*. J Bacteriol 186:8105–8113

Qamra R, Srinivas V, Mande SC (2004) *Mycobacterium tuberculosis* GroEL homologues unusually exist as lower oligomers and retain the ability to suppress aggregation of substrate proteins. J Mol Biol 342:605–617

Quan S, Koldewey P, Tapley T, Kirsch N, Ruane KM, Pfizenmaier J, Shi R, Hofmann S, Foit L, Ren G, Jakob U, Xu Z, Cygler M, Bardwell JC (2011) Genetic selection designed to stabilize proteins uncovers a chaperone called Spy. Nat Struct Mol Biol [Epub ahead of print]

Ragno S, Winrow VR, Mascagni P, Lucietto P, Di Pierro F, Morris CJ, Blake DR (1996) A synthetic 10-kD heat shock protein (hsp10) from *Mycobacterium tuberculosis* modulates adjuvant arthritis. Clin Exp Immunol 103:384–390

Raje CI, Kumar S, Harle A, Nanda JS, Raje M (2007) The macrophage cell surface glyceraldehyde-3-phosphate dehydrogenase is a novel transferrin receptor. J Biol Chem 282: 3252–3261

Ramajo-Hernández A, Pérez-Sánchez R, Ramajo-Martín V, Oleaga A (2007) *Schistosoma bovis*: plasminogen binding in adults and the identification of plasminogen-binding proteins from the worm tegument. Exp Parasitol 115:83–91

Randhawa AK, Ziltener HJ, Stokes RW (2008) CD43 controls the intracellular growth of *Mycobacterium tuberculosis* through the induction of TNF-alpha-mediated apoptosis. Cell Microbiol 10:2105–2117

Rao T, Lund PA (2010) Differential expression of the multiple chaperonins of *Mycobacterium smegmatis*. FEMS Microbiol Lett 310:24–31

Ratnakar P, Rao SP, Catanzaro A (1996) Isolation and characterization of a 70 kDa protein from *Mycobacterium avium*. Microb Pathog 21:471–486

Reddi K, Meghji S, Nair SP, Arnett TR, Miller AD, Preuss M, Wilson M, Henderson B, Hill P (1998) The *Escherichia coli* chaperonin 60 (groEL) is a potent stimulator of osteoclast formation. J Bone Miner Res 13:1260–1266

Reddy VM, Suleman FG (2004) *Mycobacterium avium*-superoxide dismutase binds to epithelial cell aldolase, glyceraldehyde-3-phosphate dehydrogenase and cyclophilin A. Microb Pathog 36:67–74

Redlitz A, Fowler BJ, Plow EF, Miles LA (1995) The role of an enolase-related molecule in plasminogen binding to cells. Eur J Biochem 227:407–415

Rha YH, Taube C, Haczku A, Joetham A, Takeda K, Duez C, Siegel M, Aydintug MK, Born WK, Dakhama A, Gelfand EW (2002) Effect of microbial heat shock proteins on airway inflammation and hyperresponsiveness. J Immunol 169:5300–5307

Richter K, Haslbeck M, Buchner J (2010) The heat shock response: life on the verge of death. Mol Cell 40:253–266

Riffo-Vasquez Y, Spina D, Page C, Desel C, Whelan M, Tormay P, Singh M, Henderson B, Coates ARM (2004) Differential effects of *Mycobacterium tuberculosis* chaperonins on bronchial eosinophilia and hyperresponsiveness in a murine model of allergic inflammation. Clin Exp Allergy 34:712–719

Ronning DR, Klabunde T, Besra GS, Vissa VD, Belisle JT, Sacchettini JC (2000) Crystal structure of the secreted form of antigen 85C reveals potential targets for mycobacterial drugs and vaccines. Nature Struct Biol 7:141–146

Rosano C (2011) Molecular model of hexokinase binding to the outer mitochondrial membrane porin (VDAC1): implication for the design of new cancer therapies. Mitochondrion [Epub ahead of print]

Saad N, Urdaci M, Vignoles C, Chaignepain S, Tallon R, Schmitter JM, Bressollier P (2009) *Lactobacillus plantarum* 299v surface-bound GAPDH: a new insight into enzyme cell walls location. J Microbiol Biotechnol 19:1635–1643

Saito F, Ikeda R (2005) Killing of *cryptococcus neoformans* by *Staphylococcus aureus*: the role of cryptococcal capsular polysaccharide in the fungal–bacteria interaction. Med Mycol 43: 603–612

Santiago NI, Zipf A, Bhunia AK (2006) Influence of temperature and growth phase on expression of a 104-kilodalton Listeria adhesion protein in *Listeria monocytogenes*. Appl Environ Microbiol 65:2765–2769

Sasu S, LaVerda D, Qureshi N, Golenbock DT, Beasley D (2001) *Chlamydia pneumoniae* and chlamydial heat shock protein 60 stimulate proliferation of human vascular smooth muscle cells via toll-like receptor 4 and p44/p42 mitogen-activated protein kinase activation. Circ Res 89:244–250

Schauer K, Muller C, Carrière M, Labigne A, Cavazza C, De Reuse H (2010) The Helicobacter pylori GroES cochaperonin HspA functions as a specialized nickel chaperone and sequestration protein through its unique C-terminal extension. J Bacteriol 192:1231–1237

Schulz LC, Bahr JM (2003) Glucose-6-phosphate isomerase is necessary for embryo implantation in the domestic ferret. Proc Natl Acad Sci U S A 100:8561–8566

Segovia-Gamboa NC, Chávez-Munguía B, Medina-Flores Y, Cázares-Raga FE, Hernández-Ramírez VI, Martínez-Palomo A, Talamás-Rohana P (2010) Entamoeba invadens, encystation process and enolase. Exp Parasitol 125:63–69

Seifert KN, McArthur WP, Bleiweis AS, Brady LJ (2003) Characterization of group B streptococcal glyceraldehyde-3-phosphate dehydrogenase: surface localization, enzymatic activity, and protein–protein interactions. Can J Microbiol 49:350–356

Sengupta S, Ghosh S, Nagaraja V (2008) Moonlighting function of glutamate racemase from *Mycobacterium tuberculosis*: racemization and DNA gyrase inhibition are two independent activities of the enzyme. Microbiology 154:2796–2803

Sha J, Erova TE, Alyea RA, Wang S, Olano JP, Pancholi V, Chopra AK (2009) Surface-expressed enolase contributes to the pathogenesis of clinical isolate SSU of *Aeromonas hydrophila*. J Bacteriol 191:3095–3107

Sirover MA (2005) New nuclear functions of the glycolytic protein, glyceraldehyde-3-phosphate dehydrogenase, in mammalian cells. J Cell Biochem 95:45–52

Smith HW, Marshall CJ (2010) Regulation of cell signalling by uPAR. Nat Rev Mol Cell Biol 11:23–36

Smitherman LS, Minnick MF (2005) *Bartonella bacilliformis* GroEL: effect on growth of human vascular endothelial cells in infected cocultures. Ann N Y Acad Sci 1063:286–298

Spurbeck RR, Arvidson CG (2010) *Lactobacillus jensenii* surface associated proteins inhibit *Neisseria gonorrhoeae* adherence to epithelial cells. Infect Immun 78:3103–3111

Sriram G, Martinez JA, McCabe ER, Liao JC, Dipple KM (2005) Single-gene disorders: what role could moonlighting enzymes play? Am J Hum Genet 76:911–924

Sriram G, Parr LS, Rahib L, Liao JC, Dipple KM (2010) Moonlighting function of glycerol kinase causes systems-level changes in rat hepatoma cells. Metab Eng 12:332–340

Starnes GL, Coincon M, Sygusch J, Sibley LD (2009) Aldolase is essential for energy production and bridging adhesin-actin cytoskeletal interactions during parasite invasion of host cells. Cell Host Microbe 5:353–364

Stein M, Keshav S, Harris N, Gordon S (1992) Interleukin 4 potently enhances murine macrophage mannose receptor activity: a marker of alternative immunologic macrophage activation. J Exp Med 176:287–292

Stillman B, Stewart D (2004) The genome of *Homo Sapiens*. Cold Spring Harbor Laboratory Press, Cold Spring Harbor
Sun YJ, Chou CC, Chen WS, Wu RT, Meng M, Hsiao CD (1999) The crystal structure of a multifunctional protein: phosphoglucose isomerase/autocrine motility factor/neuroleukin. Proc Natl Acad Sci U S A 96:5412–5417
Takayanagi H (2009) Osteoimmunology and the effects of the immune system on bone. Nat Rev Rheumatol 5:667–676
Takenaka R, Yokota K, Ayada K, Mizuno M, Zhao Y, Fujinami Y, Lin SN, Toyokawa T, Okada H, Shiratori Y, Oguma Y (2004) *Helicobacter pylori* heat-shock protein 60 induces inflammatory responses through the toll-like receptor-triggered pathway in cultured human gastric epithelial cells. Microbiology 150:3913–3922
Tan C, Liu M, Liu J, Yuan F, Fu S, Liu Y, Jin M, Bei W, Chen H (2009) Vaccination with *Streptococcus suis* serotype 2 recombinant 6PGD protein provides protection against *S. suis* infection in swine. FEMS Microbiol Lett 296:78–83
Tanaka T, Abe Y, Inoue N, Kim WS, Kumura H, Nagasawa H, Igarashi I, Shimazaki K (2004) The detection of bovine lactoferrin binding protein on *Trypanosoma brucei*. J Vet Med Sci 66:619–625
Taylor JM, Heinrichs DE (2002) Transferrin binding in *Staphylococcus aureus*: involvement of a cell wall-anchored protein. Mol Microbiol 43:1603–1614
Tegtmeyer N, Hartig R, Delahay RM, Rohde M, Brandt S, Conradi J, Takahashi S, Smolka AJ, Sewald N, Backert S (2010) A small fibronectin-mimicking protein from bacteria induces cell spreading and focal adhesion formation. J Biol Chem 285:23515–23526
Terao Y, Yamaguchi M, Hamada S, Kawabata S (2006) Multifunctional glyceraldehyde-3-phosphate dehydrogenase of *Streptococcus pyogenes* is essential for evasion from neutrophils. J Biol Chem 281:14215–14223
Tormay P, Coates AR, Henderson B (2005) The intercellular signaling activity of the *Mycobacterium tuberculosis* chaperonin 60.1 protein resides in the equatorial domain. J Biol Chem 280:14272–14277
Tsugawa H, Ito H, Ohshima M, Okawa Y (2007) Cell adherence-promoted activity of *Plesiomonas shigelloides* groEL. J Med Microbiol 56:23–29
Tunio SA, Oldfield NJ, Berry A, Ala'Aldeen DA, Wooldridge KG, Turner DP (2010a) The moonlighting protein fructose-1, 6-bisphosphate aldolase of *Neisseria meningitidis*: surface localization and role in host cell adhesion. Mol Microbiol 76:605–615
Tunio SA, Oldfield NJ, Ala'Aldeen DA, Wooldridge KG, Turner DP (2010b) The role of glyceraldehyde 3-phosphate dehydrogenase (GapA-1) in *Neisseria meningitidis* adherence to human cells. BMC Microbiol 10:280
Van Eden W, van der Zee R, Prakken B (2005) Heat-shock proteins induce T-cell regulation of chronic inflammation. Nature Revs Immunol 5:318–330
Vanags D, Williams B, Johnson B, Hall S, Nash P, Taylor A, Weiss J, Feeney D (2006) Therapeutic efficacy and safety of chaperonin 10 in patients with rheumatoid arthritis: a double-blind randomised trial. Lancet 368:855–863
Vander Heiden MG, Locasale JW, Swanson KD, Sharfi H, Heffron GJ, Amador-Noguez D, Christofk HR, Wagner G, Rabinowitz JD, Asara JM, Cantley LC (2010) Evidence for an alternative glycolytic pathway in rapidly proliferating cells. Science 329:1492–1499
Vantourout P, Radojkovic C, Lichtenstein L, Pons V, Champagne E, Martinez LO (2010) Ecto-F_1-ATPase: a moonlighting protein complex and an unexpected apoA-I receptor. World J Gastroenterol 16:5925–5935
Veiga-Malta I, Duarte M, Dinis M, Tavares D, Videira A, Ferreira P (2004) Enolase from *Streptococcus sobrinus* is an immunosuppressive protein. Cell Microbiol 6:79–88
Vesosky B, Rottinghaus EK, Stromberg P, Turner J, Beamer G (2010) CCL5 participates in early protection against *Mycobacterium tuberculosis*. J Leukoc Biol 87:1153–1165
Wagner C, Khan AS, Kamphausen T, Schmausser B, Unal C, Lorenz U, Fischer G, Hacker J, Steinert M (2007) Collagen binding protein Mip enables *Legionella pneumophila* to

transmigrate through a barrier of NCI-H292 lung epithelial cells and extracellular matrix. Cell Microbiol 9:450–462

Walsh A, Whelan D, Bielanowicz A, Skinner B, Aitken RJ, O'Bryan MK, Nixon B (2008) Identification of the molecular chaperone, heat shock protein 1 (chaperonin 10), in the reproductive tract and in capacitating spermatozoa in the male mouse. Biol Reprod 78: 983–993

Wampler JL, Kim KP, Jaradat Z, Bhunia AK (2004) Heat shock protein 60 acts as a receptor for the Listeria adhesion protein in Caco-2 cells. Infect Immun 72:931–936

Wang L, Lin M (2008) A novel cell wall-anchored peptidoglycan hydrolase (autolysin), IspC, essential for *Listeria monocytogenes* virulence: genetic and proteomic analysis. Microbiology 154:1900–1913

Wang Y, Kelly CG, Karttunen JT, Whittall T, Lehner PJ, Duncan L, MacAry P, Younson JS, Singh M, Oehlmann W, Cheng G, Bergmeier L, Lehner T (2001) CD40 is a cellular receptor mediating mycobacterial heat shock protein 70 stimulation of CC-chemokines. Immunity 15:971–983

Wang Y, Kelly CG, Singh M, McGowan EG, Carrara AS, Bergmeier LA, Lehner T (2002) Stimulation of Th1-polarizing cytokines, C–C chemokines, maturation of dendritic cells, and adjuvant function by the peptide binding fragment of heat shock protein 70. J Immunol 169:2422–2429

Wang Y, Whittall T, McGowan E, Younson J, Kelly C, Bergmeier LA, Singh M, Lehner T (2005) Identification of stimulating and inhibitory epitopes within the heat shock protein 70 molecule that modulate cytokine production and maturation of dendritic cells. J Immunol 174: 3306–3316

Watanabe H, Takehana K, Date M, Shinozaki T, Raz A (1996) Tumor cell autocrine motility factor is the neuroleukin/phosphohexose isomerase polypeptide. Cancer Res 56:2960–2963

Watarai M, Kim S, Erdenebaatar J, Makino S, Horiuchi M, Shirahata T, Sakaguchi S, Katamine S (2003) Cellular prion protein promotes Brucella infection into macrophages. J Exp Med 198:5–17

Watson C, Alp NJ (2008) Role of *Chlamydia pneumoniae* in atherosclerosis. Clin Sci (Lond) 114:509–531

Weaver DT (1998) Telomeres: moonlighting by DNA repair proteins. Curr Biol 8:R492–R494

Wegrzyn J, Potla R, Chwae YJ, Sepuri NB, Zhang Q, Koeck T, Derecka M, Szczepanek K, Szelag M, Gornicka A, Moh A, Moghaddas S, Chen Q, Bobbili S, Cichy J, Dulak J, Baker DP, Wolfman A, Stuehr D, Hassan MO, Fu XY, Avadhani N, Drake JI, Fawcett P, Lesnefsky EJ, Larner AC (2009) Function of mitochondrial Stat3 in cellular respiration. Science 323:793–797

Whittall T, Wang Y, Younson J, Kelly C, Bergmeier L, Peters B, Singh M, Lehner T (2006) Interaction between the CCR5 chemokine receptors and microbial HSP70. Eur J Immunol 36:2304–2314

Wicker HG, Harboe M (1992) The antigen 85 complex: a major secretion product of *Mycobacterium tuberculosis*. Microbiol Rev 56:648–661

Wilkins JC, Beighton D, Homer KA (2003) Effect of acidic pH on expression of surface-associated proteins of *Streptococcus oralis*. Appl Environ Microbiol 69:5290–5296

Williams B, Vanags D, Hall S, McCormack C, Foley P, Weiss J, Johnson B, Latz E, Feeney D (2008) Efficacy and safety of chaperonin 10 in patients with moderate to severe plaque psoriasis: evidence of utility beyond a single indication. Arch Dermatol 144:683–685

Winram SB, Lottenberg R (1998) Site-directed mutagenesis of streptococcal plasmin receptor protein (Plr) identifies the C-terminal Lys334 as essential for plasmin binding, but mutation of the plr gene does not reduce plasmin binding to group A streptococci. Microbiology 144: 2025–2035

Winrow VR, Mesher J, Meghji S, Morris CJ, Fox S, Coates AR, Tormay P, Blake D, Henderson B (2008) The two homologous chaperonin 60 proteins of *Mycobacterium tuberculosis* have distinct effects on monocyte differentiation into osteoclasts. Cell Microbiol 10:2091–2104

Wu Z, Zhang W, Lu C (2008) Comparative proteome analysis of secreted proteins of *Streptococcus suis* serotype 9 isolates from diseased and healthy pigs. Microb Pathog 45: 159–166

Wuppermann FN, Mölleken K, Julien M, Jantos CA, Hegemann JH (2008) Chlamydia pneumoniae GroEL1 protein is cell surface associated and required for infection of HEp-2 cells. J Bacteriol 190:3757–3767

Xu W, Seiter K, Feldman E, Ahmed T, Chiao JW (1996) The differentiation and maturation mediator for human myeloid leukemia cells shares homology with neuroleukin or phosphoglucose isomerase. Blood 87:4502–4506

Henderson B, Pockley AG. Proteotoxic stress and secreted cell stress proteins in cardiovascular disease (submitted to Circulation)

Yamaguchi H, Osaki T, Taguchi H, Hanawa T, Yamamoto T, Kamiya S (1996) Flow cytometric analysis of the heat shock protein 60 expressed on the cell surface of *Helicobacter pylori*. J Med Microbiol 45:270–277

Yamaguchi H, Osaki T, Taguchi H, Hanawa T, Yamamoto T, Fukuda M, Kawakami H, Hirano H, Kamiya S (1997a) Growth inhibition of *Helicobacter pylori* by monoclonal antibody to heat-shock protein 60. Microbiol Immunol 41:909–916

Yamaguchi H, Osaki T, Kurihara N, Taguchi H, Hanawa T, Yamamoto T, Kamiya S (1997b) Heat-shock protein 60 homologue of *Helicobacter pylori* is associated with adhesion of *H. pylori* to human gastric epithelial cells. J Med Microbiol 46:825–831

Yamaguchi H, Osaki T, Kurihara N, Kitajima M, Kai M, Takahashi M, Taguchi H, Kamiya S (1999) Induction of secretion of interleukin-8 from human gastric epithelial cells by heat-shock protein 60 homologue of *Helicobacter pylori*. J Med Microbiol 48:927–933

Yamaguchi M, Ikeda R, Nishimura M, Kawamoto S (2010) Localization by scanning immunoelectron microscopy of triosephosphate isomerase, the molecules responsible for contact-mediated killing of Cryptococcus, on the surface of Staphylococcus. Microbiol Immunol 54:368–370

Yoshida N, Oeda K, Watanabe E, Mikami T, Fukita Y, Nishimura K, Komai K, Matsuda K (2001) Protein function. Chaperonin turned insect toxin. Nature 411:44

Yunoki N, Yokota K, Mizuno M, Kawahara Y, Adachi M, Okada H, Hayashi S, Hirai Y, Oguma K, Tsuji T (2000) Antibody to heat shock protein can be used for early serological monitoring of *Helicobacter pylori* eradication treatment. Clin Diagn Lab Immunol 7:574–577

Zaborina O, Li X, Cheng G, Kapatral V, Chakrabarty AM (1999) Secretion of ATP-utilizing enzymes, nucleoside diphosphate kinase and ATPase, by *Mycobacterium bovis* BCG: sequestration of ATP from macrophage P2Z receptors? Mol Microbiol 31:1333–1343

Zhang B, Walsh MD, Nguyen KB, Hillyard NC, Cavanagh AC, McCombe PA, Morton H (2003) Early pregnancy factor treatment suppresses the inflammatory response and adhesion molecule expression in the spinal cord of SJL/J mice with experimental autoimmune encephalomyelitis and the delayed-type hypersensitivity reaction to trinitrochlorobenzene in normal BALB/c mice. J Neurol Sci 212:37–46

Zhang L, Pelech S, Uitto VJ (2004a) Long-term effect of heat shock protein 60 from *Actinobacillus actinomycetemcomitans* on epithelial cell viability and mitogen-activated protein kinases. Infect Immun 72:38–45

Zhang L, Koivisto L, Heino J, Uitto VJ (2004b) Bacterial heat shock protein 60 may increase epithelial cell migration through activation of MAP kinases and inhibition of alpha6beta4 integrin expression. Biochem Biophys Res Commun 319:1088–1095

Zhao Y, Yokota K, Ayada K, Yamamoto Y, Okada T, Shen L, Oguma K (2007) *Helicobacter pylori* heat-shock protein 60 induces interleukin-8 via a toll-like receptor (TLR)2 and mitogen-activated protein (MAP) kinase pathway in human monocytes. J Med Microbiol 56: 154–164

Symbionts and Pathogens: What is the Difference?

Vicente Pérez-Brocal, Amparo Latorre and Andrés Moya

Abstract The ecological relationships that organisms establish with others can be considered as broad and diverse as the forms of life that inhabit and interact in our planet. Those interactions can be considered as a continuum spectrum, ranging from beneficial to detrimental outcomes. However, this picture has revealed as more complex and dynamic than previously thought, involving not only factors that affect the two or more members that interact, but also external forces, with chance playing a crucial role in this interplay. Thus, defining a particular symbiont as mutualist or pathogen in an exclusive way, based on simple rules of classification is increasingly challenging if not unfeasible, since new methodologies are providing more evidences that depict exceptions, reversions and transitions within either side of this continuum, especially evident at early stages of symbiotic associations. This imposes a wider and more dynamic view of a complex landscape of interactions.

V. Pérez-Brocal (✉)
Área de Genómica y Salud, Centro Superior de Investigación
en Salud Pública (CSISP), Avenida de Cataluña 21,
46020 Valencia, Spain
e-mail: perez_vicbro@gva.es

A. Latorre
CIBER en Epidemiología y Salud Pública (CIBERESP),
Barcelona, Spain

A. Moya
Institut Cavanilles de Biodiversitat i Biologia Evolutiva,
Universitat de València, Apartado Postal 22085,
46071 Valencia, Spain

Current Topics in Microbiology and Immunology (2013) 358: 215–243
DOI: 10.1007/82_2011_190

Published Online: 11 November 2011

Contents

1 Introduction

All forms of life on Earth interact with their environment that surrounds them. This includes abiotic factors to which organisms respond as well as other living organisms, from the same or different species. Within the enormous diversity of living beings on Earth, one of the groups that undoubtedly have achieved higher success and prosperity is that of the prokaryotes. A feature that characterizes them is their broad spectrum of metabolic capabilities. This has allowed them to colonize virtually the entire planet, from several kilometers inside the lithosphere to the troposphere, and live in all types of terrestrial and aquatic ecosystems, from the most favorable to the most extreme. No wonder, then, that many bacteria have developed physical partnerships with other organisms with more limited metabolic capabilities, such as other bacteria, protozoa, fungi, plants or animals. Their metabolic capabilities allow them to interact with eukaryotes to exploit their resources enabling bacteria to spread at expenses of the eukaryotes, occasionally provoking harm (which, if produces clinical symptoms is called disease) or using them in more than counterbalanced fashion by the contribution of new metabolic capabilities or other benefits for the host. Such beneficial cooperation has enabled eukaryotic organisms and their prokaryotic partners to occupy new ecological niches and diversify in order to be more successful.

The term symbiosis which means 'living together' designates the interdependence of two (or more) organisms of different species that results in a benefit to all of the implied partners (mutualism), or just to some of them. In this case, it can involve harm of one of the species, (parasitism) or not (commensalism). This rather simplistic classification of the symbiosis in discrete categories is being challenged nowadays by a more complex and dynamic view of interaction within a continuum. The spectrum between mutualism and parasitism is continuous and it is often difficult to distinguish one from another since a single association may have positive or negative depending on the environmental circumstances.

The partners that establish a symbiotic association are called host and symbiont. The host organism is defined as the provider of resources or the resource base, while the symbionts are the consumers of such resources, and may or may not provide services in return. This is similar to the definition of Ferrière et al. (2007) who considered the host as the producer of commodities and the symbiont or partner as provider of goods and services. When originally defined by Anton de Bary and Simon Schwendener in 1879, the terms symbiosis and symbiont did not considered whether the effects of the association were beneficial or detrimental for the partners. However, many authors have used these terms in a more restricted way as synonyms of mutualism and mutualist. In order to reduce this ambiguity, throughout this review we use the term symbiosis in the comprehensive meaning to include all kind of interactions (i.e. mutualism, parasitism, commensalism) and only refer to mutualism and mutualists when a benefit to the host is observed.

Symbiosis has acted in the past and still does as a major catalyst of evolution, contributing to the promotion of speciation, diversification and evolutionary novelties such as the development of cell types, tissues and organs to harbor mutualists as well as barriers against pathogens, including the development of the immune system. A classic example of the evolutionary role of symbioses is the endosymbiotic theory (Margulis 1981, 1993) that postulates an initial invasion of the ancestor of the eukaryotic cell by Alphaproteobacteria with a large capacity to consume oxygen, from which the mitochondria would eventually arise, followed by a second colonization by prokaryotes with chlorophyll, believed to be similar to cyanobacteria that gave rise to chloroplasts, resulting in photosynthetic cells such as plants, which have both mitochondria and chloroplasts.

The application of recent technological developments (next-generation sequencing, metagenomics and synthetic biology) has allowed a significant progress in making possible the study of environmental samples and non-cultivable microorganisms, thus offering new opportunities for deciphering the associations between microorganisms and their hosts as well as among different microorganisms within hosts. For example; whole genome amplification methods facilitate genomic studies of host-associated bacteria for which only limited amounts of DNA are available. Metagenomics may reveal how genes interact in bacterial consortia that inhabit the same hosts and tissues and how these interactions affect the outcome of the infection. And synthetic biology holds promise in providing tools to study the function of genes under non-conventional or controlled experimental conditions (Toft and Andersson 2010).

Some examples of symbiotic interactions between bacteria and different eukaryotic hosts are shown in Table 1. A great number of mutualistic interactions have a nutritional character, particularly well studied in arthropods with specialized diets such as sap-sucking, blood-sucking or grain-feeding insects, where bacterial mutualists provide them with essential amino acids and/or vitamins and cofactors. But they can also exert other metabolic roles, in nitrogen recycling, storage and excretion as in omnivorous insects or clams. Finally benefits to the host can be more environmentally related, such as protection against parasitoids, predators and abiotic stress or other functions such as luminescence used by many animals for communication, attraction and other behaviors. On the other hand, the parasitism develops for example between pathogenic bacteria and animals, plants or fungi, causing infectious diseases, reproductive alterations and lesions that can produce the death of the infected host.

Table 1.1 Some examples of bacteria-host symbiotic interactions

Type of interaction	Prokaryotic Partner	Host Partner	Proposed benefit/harm	Type of interaction	Prokaryotic Partner	Host Partner	Proposed benefit/ harm
M	*Symbionidium spp.*	Corals	Nutrient provision	P	*Holospora spp.*	Paramecium	Cell death
M	*Hamiltonella defensa*[a]	Pea aphid	Resistance to parasitoid wasps	P	*Vibrio shiloi*	Coral	Bacterial bleaching
M	*Regiella insecticola*[a]	Pea aphid	Resistance to fungal pathogens	P	*Phormidium corallyticum*	Coral	Black band disease
M	*Serratia symbiotica* [a]	Cedar aphid	Nutrient provision	P	*Pasteuria penetrans*	Root knot nematode	Host sterilization
M	*Wolbachia pipientis*	Nematode	Nutrient provision	P	*Cardinium hertigii a*	Many arthropods	Reproductive parasite
M	*Blochmannia spp.*[a]	Carpenter ant	Nutrient provision-N storage	P	*Wolbachia pipientis*	Many arthropods	Reproductive parasite
M	*Wigglesworthia glossinidia*	Tsetse fly	Nutrient provision	P	*Serratia entomophila*	Grass grub	Septicemia and death
M	*Buchnera aphidicola*	Aphids	Nutrient provision	P	*Rhabdochlamydia crassificans*	Cockroach	Abdominal swelling
M	*Blattabacterium spp*	Cockroach	Nutrient provision-N excretion	P	*Rickettsia spp.*	Fleas, lices, ticks	Insect death
M	*Sulcia muelleri* [a]	Sharpshooter, cicada, spittlebug	Nutrient provision	P	*Photorhabdus spp.*	Many insects	Lethal septicemia
M	*Baumannia cicadellinicola*[a]	Sharpshooter	Nutrient provision	P	*Spiroplasma spp.*	Butterflies	Reproductive parasite
M	*Hodgkinia cicadicola*[a]	Cicada	Nutrient provision	P	*Pasteuria ramosa*	Daphnia	Host sterilization
M	*Zinderia insecticola*[a]	Spittlebug	Nutrient provision	P	*Aeromonas salmonicida*	Salmonids	Furunculosis
M	*"SOPE", "SZPE"*	Weevil	Nutrient provision	P	*Yersinia pestis*	Human	Bubonic plague
M	*Ruthia magnifica*[a]	Deep-sea clam	Nutrient provision-N recycling	P	*Vibrio cholerae*	Human	Cholera
M	*Vibrio fischeri*	Loliginid squids	Chemiluminescence	P	*Xylella fastidiosa*	Grapevine	Pierce's disease
M	*Rhizobium spp.*	Leguminous plants	Nitrogen fixation	P	*Xanthomonas spp.*	Various plants	Necrotic lesions

[a] These bacteria are called Candidatus. M: mutualism, P: parasitism

In this review, we address a number of features that characterize mutualistic symbionts and parasites to try to establish some criteria, if any, that define them, with the purpose of understanding what defines a bacterium as beneficial or harmful. In addition, we address the enormous diversity of interactions that they establish and the complexity that this implies when it comes to classify them.

2 The Plasticity of Symbiotic Associations

The boundaries of parasitism, commensalism and mutualism are vague. By considering the various forms of symbioses as existing throughout a theoretical continuum of fitness outcomes, it is possible to note that mutualism and parasitism may in fact be considered as two sides of the same coin. The symbiotic associations cannot be merely considered as closed-circuits isolated from the external circumstances. In fact, Leung and Poulin (2008) showed how easily symbiotic associations can switch between mutualism and parasitism in response to even the slightest environmental change. Actually, extrinsic factors such as environmental circumstances or time, as well as intrinsic factors such as the transmission mode, can influence the symbiotic relationship, determining that a priory beneficial outcome becomes detrimental if the circumstances change or vice versa.

One aspect to consider when studying symbiotic associations is represented by the fact that the costs and benefits of a symbiosis for a host organism are not always evident or easily measurable because they may exist on many levels (Cushman and Beattie 1991). Parasites may impose with their presence additional costs or mutualist partners may confer multiple benefits to their hosts beyond the immediate physiological effects. For example, bacterial mutualists that confer their hosts novel capabilities and therefore favor the exploitation of novel resources produce the niche expansion for the host, something that has in turn fitness benefits because the hosts find fewer competitors in those unexploited niches (Moran 2007). The pests caused by sap-feeding insects, such as aphids, psyllids, whiteflies, mealybugs, sharpshooters or cicadas illustrate the success of their associations with bacterial obligate endosymbionts. On the contrary, pathogens, in addition to the pathology usually associated with their infection per se, may impose changes in behavior that make them more vulnerable to predators, for example.

Costs of a symbiosis may be hard to detect, possibly even completely hidden due to the host's phenotypic plasticity. If the host eventually evolves to completely tolerate the parasitic infection as a way of mitigating the harm caused by a parasite, the result may superficially appear to be a commensalism when actually it has come about at a significant fitness cost to the host, which has since been masked over evolutionary time (Miller et al. 2006). This happens, for example with the attenuated pathogens, which can be the result of a long history and coevolution with their hosts. At first, bacteria from the environment but sometimes also from the endogenous microbiota can acquire from other strains or species external elements called factors of virulence that render them virulent and able to provoke disease. Susceptible host populations, which are not initially adapted to these newly arisen pathogens, also known as emergent pathogens, may first lead to a dramatic infection and an epidemic. Once a pathogen persists in the new population, the disease tends to become less virulent with time because the hosts are enforced to develop mechanisms of defence able to fight against the interloper as in an "arms race". In this context, if pathogens and hosts coevolve, one possibility is that the pathogen attenuates itself, and eventually a balance is established with the host. This attenuation of the pathogen and increasing dependence on the host may even result in some cases in mutualistic relationships. A possible example of this extreme transition is found in *Wolbachia* spp. which has been described as a reproductive manipulator of many arthropods, but more recently evolved mutualistic relationships with the common bedbug *Cinex lecturalius*, to which it provides with vitamin B, have been discovered (Hosokawa et al. 2010). Even more, *Wolbachia* spp. was transformed in natural populations of *Drosophila simulans* in less than 25 years from a reproductive manipulator to a mutualist that enhanced fecundity (Weeks et al. 2007). Other possibilities are that the pathogen remains virulent or regains virulence due to genetic changes or genetic exchange via horizontal gene transfer leading to new epidemic episodes until the host again responds and another balance is reached. In all these cases, time plays as a key factor, and not only on an evolutionary scale, but even relative short time periods, as demonstrated in *Wolbachia's* example.

Besides the direct costs that pathogens cause to the host fitness in the form of infectious diseases, all parasites take resources from the host, also reducing host fitness in the process as a side effect. But the latter also applies to mutualists which also demand resources from their host. However this cost is usually compensated by the benefit simultaneously conferred on the host by the mutualistic symbiont. This raises the question of whether benefits to host fitness can be surpassed by the costs under certain environments and circumstances or vice versa. The balance between the costs and benefits for the two participants in a symbiosis depends on a range of factors, and often only a small push is needed to shift that balance. An example of the fact that the fitness costs of the association are not always counterbalanced by the benefits is found in the pea aphid *Acyrtosiphon pisum*, which in addition to the obligate endosymbiont *Buchnera aphidicola*, can also harbor a range of facultative secondary symbionts. One of them, "*Ca.* Hamiltonella defensa" was demonstrated to confer to its host resistance against the attack of parasitoid wasps (Oliver et al. 2005). However, aphids with secondary symbionts experience a severe fecundity reduction in comparison with uninfected aphids (Oliver et al. 2006). Symbiont-based fitness differentials are also known for "*Ca.* Serratia symbiotica" and "*Ca.* Regiella insecticola", which affect the ability to withstand heat (Russell and Moran 2006), to use particular host plants (Tsuchida et al. 2004), or to resist pathogenic fungi (Scarborough et al. 2005). Benefits in a natural environment, which can explain the maintenance of certain symbioses, may not be obvious under other conditions, experimental or natural, if the factors that facilitate its persistence are absent. For example, at temperatures above 30°C *B.aphidicola* is eliminated and bacteriocytes are reduced, resulting in the absence of reproduction in aphids, whereas the presence of a secondary symbiont, in conditions of heat stress, resulted in an increase in the reproduction of the host (Montllor et al. 2002), which is also related to the degree of variation in the frequency of secondary symbionts in less stable climates (Haynes et al. 2003).

In addition to the extrinsic factors, an intrinsic key factor in determining the virulence of a symbiont is the mode of transmission (Ewald 1995; Day 2001; Ferdy and Godelle 2005). Thus a vertical transmission implies an alignment of the fitness outcomes of both the symbiont and the host. Cooperation between the two parties or at least lower virulence by the symbiont would be an outcome favored by selection (Ewald 1995). In contrast, if the fitness of the symbiont is not exclusively interweaved with that of its host, then its fitness can be improved by exploiting its host more aggressively while returning fewer benefits, which should push the association on the evolutionary path toward parasitism.

Hosts and bacterial symbionts play an active role in the income-outcome of the symbiotic associations. As for any kind of interaction, in both mutualistic and pathogenic associations there is a conflict of interests between the interacting partners. Thus, they should be viewed as an "arms race" in which changes in one partner must be matched by adaptive changes in the other to maintain equilibrium (Steinert et al. 2000). The need for mechanisms such as host resistance and immunological responses by the hosts has resulted from the conflict of interests that exists within any obligate biological association, and symbiont control and immunological responses perform the role of limiting harm or preventing overexploitation of the host by the symbionts in question. The only difference is that in the case of host–parasite interaction, the relationship is more antagonistic—the host is attempting to completely deny any resources to the symbiont, whereas the latter is attempting to exploit the host while returning nothing of benefit. In this arms raced, which are the weapons that hosts and symbionts brandish? From the host side, defence mechanisms against intruders encompass a broad and varied range of physical barriers (skin, cuticle, mucus, cilia, etc.) and complex immune reactions to cope with the invader. The latter evolved in complexity from simpler responses in the oldest extant metazoan phyla (Müller and Müller 2003) to the most sophisticated ones of vertebrates. The immune system has been particularly well studied in insects (Feldhaar and Gross 2008; Govind 2008; Strand 2008; Gerardo et al. 2010) and vertebrates (Male 2004; Pier et al. 2004; Meyers 2007). Insects rely chiefly on the innate immune system, although new findings challenge this assumption (Kurtz et al. 2006; Schulenburg et al. 2007). This is the oldest one in evolutionary terms, and can be divided into cellular responses, based on haemocytes in insects involving phagocytosis and encapsulation; and the humoral responses resulting in the production of soluble antimicrobial peptides, melanisation and clotting. Responses vary depending on the invader (e.g. antimicrobial peptides against microbes and encapsulation against parasitoids). In insect genomes, there are four pathways that appear well conserved (invasive microbe leads to signal production via four pathways (Toll, immunodeficiency (IMD), c-Jun N-terminal kinase (JNK), and Janus kinase/Signal transducers and activators of transcription (JAK/STAT)). Each pathway is activated in response to particular pathogens (Dionne and Schneider 2008). However, until recently, only holometabolous insects (i.e. with complete metamorphosis) had been analyzed. The study of the immunity and stress in the recently sequenced genome of the hemimetabolous pea aphid *Acyrtosiphon pisum* (Gerardo et al. 2010), which is also the first intimately dependent on obligate and facultative bacterial symbionts for its survival, reveals that many genes central to immune function in other insects are missing (e.g. peptidoglycan receptor proteins, the IMD signaling pathway, defensins, c-type lysozymes) and thus the overall response is more limited. Forces such as reproductive investment to an immune challenge, rather than in a costly immune response, or symbiont-mediated host protection by secondary symbionts, could potentially shape the evolution of aphis stress and immune responses, although more studies characterizing the global aphid response under more conditions are needed. In superior vertebrates, in addition to the innate or nonspecific responses, which include nutritional immunity, the action of the complement and phagocytosis and opsonophagocytosis, an adaptive or specific immunity has developed. This response is based on the recognition of specific antigens by immunoglobulins, T cell receptors and major histocompatibility complex molecules (Davis and Bjorkman 1988; Lebecque and Bearhart 1990; Tonegawa 1983). It comprises humoral responses, mediated by the interaction of B and T helper cells implying the production of antibodies, and cellular responses, triggered by the interaction of macrophages and inflammatory T cells or any infected cell and cytotoxic T cells. Such reactions activate a series of mechanisms to fight the intruders, including the complement activation, toxins neutralization, agglutination reactions, blocking of adhesion of bacteria to epithelia, induction of the free O_2 radicals or liberation of cytotoxins to kill infected cells.

In order to overcome these host mechanisms but also to gain access to the nutrient-rich environment that the host represents the symbionts have developed a range of countermeasures. Since the biological processes needed to successfully infect hosts are largely the same for both types of microorganisms (Gil et al. 2004), some general strategies for symbiont's survival can be applied, no matter if the relationship becomes parasitic or mutualist. In all cases the first stage is the finding of a way of entering the host. Next, the symbionts must find unique niches within hosts. Third, it will be necessary to avoid, circumvent or subvert normal host cell defenses. The goal for the bacteria is to multiply sufficiently either to establish their progeny within the host and/or to move into a new susceptible host. A low infectious dose along with an efficient transmission tactic is a potent strategy for a microorganism to sustain its progeny. Genomic analyses indicate that many molecular factors (toxins, islands of pathogenicity, type III secretion systems, ureases, etc.) are involved both in pathogenic and mutualistic relationships. Toxins are virulence factors expressed by some pathogens to establish in a particular host and transmit to new susceptible hosts by causing direct harm to the so-called AG resines, (e.g. ADP-ribosylation toxins, neurotoxins, Rnase, cytotoxins, etc.) or indirect harm, the so-called modulines (e.g. lipopolysaccharide, superantigens, etc.) to the hosts. Examples of toxins include the diphtheria toxin, the botulimun toxin, the tetanus toxin or the cholera toxin. However, toxins may also play a role in mutualistic associations. For instance, facultative symbionts produce antifungal compounds and toxins that are known or suspected to be involved in protecting the host against natural enemies. For example, symbionts produce antifungal metabolites that protect their crustacean hosts *Palaemon macrodactylus* and *Homarus americanus* from the pathogenic fungus *Lagenidium callinectes* (Gil-Turnes et al. 1989; Gil-Turnes and Fenical 1992). Similarly, possession of the facultative symbiont "*Ca.* Regiella insecticola" is associated with host resistance to a fungal pathogen *Pandora neoaphidis* in aphids (Ferrari et al. 2004; Scarborough et al. 2005) and "*Ca.* Serratia symbiotica" and "*Ca.* Hamiltonella defensa" confer protection to their aphid host from the parasitoid wasps *Aphidius ervi* and *Aphidius eadyi* (Oliver et al. 2003, 2005, 2009; Ferrari et al. 2004). Another element is that of genomic islands which are movable genetic elements located in the chromosome, flanked by specific sequences (direct repeats), associated with tRNA loci, and often possess genes encoding for genetic mobility (phages, IS elements, integrases, transposases, origins of replication). There are pathogenicity islands (e.g. PaiI and PaiII of E. coli, linked to resistance to blood serum and union to urinary epithelium, respectively), but more recently symbiosis islands (e.g. *Mesorhizobium loti* R7A symbiotic island, Sullivan et al. 2002) have also been reported.

Even endosymbiotic bacteria with a long-time established relationship with their hosts and dramatic genome size reduction maintain genes that encode essential endosymbiotic factors that are proposed to be virulence associated in bacterial pathogens, such as type III secretion systems and urease (Gil et al. 2003; Goebel and Gross 2001; Shigenobu et al. 2000). In many free-living bacteria, genes encoding the type III secretion system are located within pathogenicity islands that have been acquired by horizontal gene transfer. This system is present in many insect endosymbiotic bacteria where it has been proposed to be essential to invade the host cells, thus playing an essential role in the establishment of the symbiosis (Dale et al. 2001, 2002). In some pathogenic microorganisms, ureases have been identified as virulence factors (e.g. Rokita et al. 1998) whereas in the P-endosymbiont of carpenter ants, "*Ca.* Blochmannia", it has become beneficial in the recycling of nitrogen in phases such as the metamorphosis (Gil et al. 2003; Degnan et al. 2005).

Little is known about the molecular basis of interactions of mutualists-hosts in comparison to that of pathogens-hosts. The coexistence of these bacteria with the host without triggering major immune responses suggests no recognition of the bacteria or active evasion the immune system of the host. The genome sequences of several primary endosymbionts of insects show the loss of enzymes required for the biosynthesis of peptidoglycan and lipopolysaccharide (Zientz et al. 2004). This may indicate a reduction in that the visibility of the bacteria by the host immune system. In addition, their intracellular location may protect them from immune recognition and defense mechanisms. The recent characterization of the bacteriocyte transcriptome of the aphid *A. pisum* harboring the primary endosymbiont *B. aphidicola* revealed a specific up-regulation of genes involved in defense responses including those encoding putative lysozymes and may possibly function to control the endosymbiont population (Nakabachi et al. 2005). Some intracellular endosymbionts may also be found extracellularly at least during some phases of the host's life cycle. For example, during metamorphosis the primary endosymbiont of the weevil *Sitophilus zeamais* (SZPE) migrates from the larval to the adult bacteriocytes situated around the foregut and close to the hindgut, respectively. As a reaction the host increases the expression of a peptidoglycan recognition protein (PGRP), possibly to avoid bacterial invasion into insect tissues other than the bacteriome (Anselme et al. 2006). In the case of the secondary endosymbiont of tsetse flies, *Sodalis glossinidius* found in the hemolymph and bacteriocytes, the host constitutively expresses high levels of the antimicrobial peptide Diptericin to which *S. glossinidius* is about ten times more resistant than *E. coli*. Thus, the endosymbiont may have adapted to the antimicrobial activity and/or the host attempts to control symbiont number by the constitutive expression of high levels of Diptericin (Hao et al. 2001).

3 What Do Pathogens and Symbionts Have in Common and What Separates Them?

Approaches from the Ecology, Immunology, Comparative Genomics and Biochemistry can provide us with some clues to understand similarities and differences between mutualists and pathogens. Thus, some traits shared by all symbionts, no matter the outcome of the symbiosis, illustrate the existence of convergences despite their apparently different lifestyles.

Mutualists and parasites share some common genetic and molecular features (Goebel and Gross 2001), to the point that they can be considered as two different results of a process that would begin similarly: with the encounter between host and symbiont, the infection and colonization of the host cells and tissues, the avoidance of the defense mechanisms and the persistence and multiplication within the host. As shown in Sect. 2, molecular factors (toxins, islands of pathogenicity, type III secretion systems, ureases ...) are involved both in pathogenic and mutualistic relationships. In fact, the definition of pathogen given by Fallow 1997 as "any microorganism whose survival is dependent upon its capacity to replicate and persist on or within another species by actively breaching (...) a cellular or humoral host barrier that ordinarily restricts or inhibits other microorganisms" may apply with some nuances to mutualists, because it does not imply the final outcome for the host fitness, which is regarded as the main line that separates pathogens and mutualists. In this sense, the cardinal difference can be established between a commensal species and a pathogen or mutualistic one is that the latter has a capacity to break the host cell barriers to gain access to niches that commensals cannot. However, commensals may cause opportunistic infections if the host defense is breached.

Comparative genomic analysis of obligate intracellular bacteria has revealed a series of features shared by intracellular parasites and mutualistic endosymbionts, when compared with their free-living relatives. These include genome size reduction, almost total absence of recombination, increased rate of nucleotide substitution, high A + T content, accumulation of deleterious mutations by random genetic drift, loss of codon bias toward A or T and accelerated sequence evolution (Moran 1996; Andersson and Kurland 1998; Clark et al. 1999; Moya et al. 2002). These genome features of obligate symbionts including elevated sequence evolution, gene loss and shift toward higher AT content (Dale and Maudlin 1999; McCutcheon and Moran 2007, Moya et al. 2002) are observed at less extreme levels in facultative symbionts.

Interestingly, there is a remarkable convergence in which genes have been lost from entirely free-living ancestors in the current obligate pathogens and mutualists. Both retain genes underlying the basic functions of cell growth and division, such as replication, transcription, translation and energy metabolism. On the other hand, lost loci include in both groups a large proportion of genes encoding DNA repair and recombination functions, although particular repair genes are retained and lost vary. Also, many genes required for production of cell envelope components are missing. Nevertheless, similarities end when we observe the metabolic capabilities retained by each type of symbiont. Small genome pathogenic bacteria have also lost genes for intermediate metabolism and biosynthetic pathways (Andersson and Kurland 1998; Fraser et al. 1995), observations congruent with the fact that many of the required metabolites are available in the nutrient-rich environment of the host cell. On the other hand, small genome obligate mutualists have retained many metabolic genes required for the biosynthesis of metabolites required by the host. For example genes required for the synthesis of essential amino acids (e.g. in *B. aphidicola*, "*Ca.* Sulcia muelleri", and SOPE) or vitamins and cofactors (e.g. in "*Ca.* Baumannia cicadellinicola" and "*Ca.* Hodgkinia cicadicola", SOPE and *Wigglesworthia glossinidia*) (Gosalbes et al. 2010).

Symbionts that form chronic infections share a common syndrome of genomic reduction and accelerated sequence evolution because they also share some of the population dynamics. They both reside in sequestered habitats where their reduced effective population size and reduced opportunities for recombination (by physical barriers and loss of recombinase genes) imply that the effectiveness of selection is decreased whereas the levels of genetic drift increase, resulting in the accumulation of mildly deleterious mutations, which unlike free-living bacteria, are not effectively eliminated by selection and thus they might fix by genetic drift. Bottlenecks are inherent to the mode of transmission of mutualists and pathogens, either due to the small number of bacteria transmitted maternally, where the host progeny receives only a limited diversity of bacteria present in their mother or, in horizontally transmitted pathogens (also in facultative symbionts), where each infection event implies a small number of bacteria passing to the new host. As a result of the accumulation of mildly deleterious mutations in this context of relaxed purifying selection plus reduced mechanisms of repair and recombination, gene inactivation is easily followed by gene erosion and loss, which is not balanced by gene uptake, unlike free-living bacteria. Unlike chronic pathogens and mutualists, which spend their life cycles closely associated with host cells, many other symbionts establish more labile interactions with eukaryotes. These so-called facultative or opportunistic symbionts may shift easily the type of association through gene loss or acquisition via frequent horizontal transfer, and as long as they can grow as free-living bacteria, they can also grow to achieve large effective population sizes and retain metabolic capabilities necessary to grow as free-living bacteria. This results in a population dynamics notably different from obligate symbionts, and more similar to that of entirely free-living bacteria. Not surprisingly, these bacteria do not show the strong effect of drift or the genome reduction that characterizes obligate mutualists and pathogens.

Another hypothetically major difference between mutualists and pathogens that is blurred when we consider the whole spectrum of interactions is their mode of transmission. Obligate heritable mutualists such as the bacteriome-associated symbionts of insects display a mode of transmission strictly vertical, with a process starting at early stages of oogenesis or embryogenesis. This maternal transmission is controlled by the host, since the bacteria cannot invade naïve hosts and have evolved to be dependent on host-based mechanisms for transmission. This is demonstrated repeatedly by the codiversification of obligate mutualists and their hosts since the ancient acquisition in each case. On the contrary, pathogens can be more or less host-specific, but, unlike mutualists, they do not show a pattern of codiversification with their hosts since the mode of transmission is intrinsically horizontal, imposed by the own nature of this kind of association. Since there is a trade-off between fitness gained through increased among-host transmission (infectivity) and fitness lost through increased virulence, virulence of pathogens will increase with horizontal transmission, whereas with vertical transmission, virulence will decrease because a parasite's reproductive potential will be maximized only by decreasing harm to the host, allowing parasite transmission to more host offspring (Bull et al. 1991; Agnew and Koella 1997; Stewart et al. 2005). In other words, as pathogens become more harmful and decrease the reproduction rate of their host, they are less likely to be passed on to the hosts' offspring since there will be fewer offspring. Actually, as shown in Sect. 2, the mode of transmission plays a key role in determining the virulence of a symbiont (Ewald 1995; Day 2001; Ferdy and Godelle 2005). To complicate things further, facultative symbionts resemble obligate and opportunistic pathogens in that their successful persistence in their hosts rely on their own mechanisms and capabilities for invading new hosts, entering cells and tissues and countering host immune responses. There are observations, such as variation in their presence within or among individuals or species, coexistence of several lineages within a host or the presence of the same facultative symbiont in non-related hosts, which support the presence of horizontal transfer in these bacteria within and between species (O'Neill et al. 1992; Sandström et al. 2001). Also, phylogenetic evidence suggests that secondary mutualistic symbionts have undergone horizontal transfers. This occurs when the phylogenies of endosymbionts and their hosts are not congruent (Russell et al. 2003; Sandström et al. 2001). For example, fungus-growing ants have a mutualism with bacteria that produce antibiotics with activity against *Escovopsis*, a specialized microparasite capable of rapidly devastating the fungus garden. These mutualistic bacteria from the genus *Pseudonocardia* are primarily vertically transmitted by colony-founding queens. However, strict cocladogenesis between the ants and the bacteria is disrupted, especially at the finer phylogenetic levels, by *Pseudonocardi*a strains switching between ant species, both within and between genera. Poulsen et al. (2005) found evidence for frequent *Pseudonocardia* switches between sympatric ant species within the genus *Acromyrmex*, indicating that horizontal symbiont transmission occurs and may be common. However, this transfer is not often observed in laboratory experiments. On the other hand, *Wolbachia* is a maternally inherited intracellular reproductive parasite of arthropods that distorts the sex ratio in host's progeny. It imposes fitness costs on its host, such as physiological impairment (Fleury et al. 2000), decreased sperm quality (Champion de Crespigny and Wedell 2006), reduced immune response (Fytrou et al. 2006), mortality of male embryos (Hurst et al. 1999a; Zeh and Zeh 2006) and reproductive failure resulting from cytoplasmic incompatibility (Perrot-Minnot et al. 2002). Thus, it is a parasite vertically transmitted. Furthermore, it has also developed long-term associations with nematodes (Langworthy et al. 2000; Werren et al. 2008) and even with other arthropods (Hosokawa et al. 2010). Also, Rickettsia comprises pathogens of vertebrates, reproductive manipulators of arthropods and bacteriocyte-associated mutualists, having been suggested that *Rickettsia* species have exchanged a putative maternal inheritance in arthropods for mammal that serve as new "vectors" of transmission between arthropod hosts (Darby et al. 2007). Thus, exceptions to the rule of mutualists vertically transmitted and parasites horizontally transmitted are found in both sides.

Phylogenetic placement of bacterial pathogens and mutualists suggests that bacterial symbionts that form chronic infections in animals belong to deeply branching clades that are strictly parasitic or strictly mutualistic (Moran and Wernegreen 2000). However, the more genomes are available and more the effect of the symbionts on their hosts are known, the tree becomes more intricate as different categories of symbionts according to their effects fall into the same or closely related clades (Ecker et al. 2005; Moran et al. 2008). An increasing number of exceptions are challenging the classical trees as the isolated exceptions become more abundant. For instance, *Wolbachia* is a reproductive manipulator of arthropods (see above) but apparently a mutualist in filarial worms; the Flavobacteria contain obligate mutualists of cockroaches (*Blattabacterium*) as well as obligate pathogens of trouts and salmons (*Flavobacterium psychrophilum*); or Spirochaetes, which contain chronic pathogens (*Borrelia* spp.) but also mutualists of oligochaete worms, *Nautilus* and protozoa from termites. Moreover, the fact that certain groups concentrate more symbionts of either type (e.g. Gammaproteobacteria especially Enterobacteriaceae, and the Bacteroidetes represent an overwhelming majority within the obligate insect symbionts) could result simply from our limited knowledge and far-away from exhaustive sampling rather than a representative picture of the actual diversity of those symbionts. For example, newly characterized insect endosymbionts other than Gammaproteobacteria are being added to the existing ones (for example the Alphaproteobacterium "*Ca.* Midichloria mitochondrii" (Sassera et al. 2006), "*Ca.* Hodgkinia cicadicola" (Wu et al. 2006; McCutcheon et al. 2009), or the Betaproteobacterium "*Ca.* Zinderia insecticola" (McCutcheon and Moran 2010).

4 Classification of Symbioses: Different Strokes for Different Folks

In view of the fact that so many factors intermingle, when trying to define what pathogens and mutualists are, the criterion of choice determines different ways to classify the symbionts. On the one hand, there are "static" classifications based upon ecological principles. On the other hand, symbionts can be classified from a genomic point of view, with more dynamic evolutionary and temporal implications. In the first classification, factors such as the contribution to the host fitness, degree of dependence, location or specificity are contemplated. Thus, symbiotic associations have been generally divided into three categories based on whether the symbiont has beneficial, harmful or no effect on the host. When both partners reciprocally benefit from the relationship, the association represents mutualism, whereas if the symbiont uses the host as a resource and causes it harm as a result, it is qualified as a parasite. Finally, if the symbiont utilizes the host without benefiting or harming it, it is considered as a commensal, which might be regarded as a hodgepodge that reflects our ignorance of the effects of the symbiont on the host. Also, according to the degree of dependence of the symbiont (in the case of parasites) or both partners (in the case of mutualists), symbionts can be classified as obligate, primary or P-symbionts which only can be isolated from their host even if sporadically and during short periods some of them can be found in other environments. Their long-term survival absolutely depends upon its ability to replicate and to be transmitted in a particular host. In the case of mutualists of many invertebrates and in particular of insects, bacteria are typically restricted to a specialized organ called bacteriome, which varies depending on the host group, and are vertically or maternally inherited by the offspring. Primary pathogens cause disease in healthy individuals, with immune defenses unaltered. For example, *Bordetella pertussis* the etiological agent of the whooping cough. A special case is that of the co-primary symbionts (Wu et al. 2006; McCutcheon and Moran 2007). In this association or consortium, the symbionts coexist in the same host, being both essential for its survival. In contrast, the so-called opportunistic, facultative, secondary or S-symbionts show the capacity to survive outside of their natural hosts for long periods of time, and sometimes they have a natural habitat other than the host. Their long-term survival does not depend upon a particular host. They are usually more erratically distributed, with differences in their presence among populations and species, instances of co-infections with two or more of them in the same host, etc. In the case of mutualists of insects, these bacteria may invade various cell types and also reside extracellularly, and although vertical transmission is also common, they can also colonize novel hosts through horizontal transmission among host individual belonging to the same or different species (Dale and Moran 2006). Opportunistic pathogens cause disease only in individuals who are compromised in their immune defenses. For example, *Clostridium difficile* is a commensal bacterium of the intestine in a 2–5% of the human population which normally does not result in significant disease, but the introduction of broad-spectrum antibiotics causes disruption of normal intestinal flora, leading to an overgrowth of *C. difficile* and diarrhea and severe infection of the colon leading to a disease called pseudomembranous colitis. This is an emergent nosocomial infection especially in hospitalized patients, who have a higher risk of contracting it. A third factor considered in this ecological classification is the location of the symbiont on or within the host. Those who live on the surface of the host, including the skin, gut, respiratory or genitourinary tract, or in the blood or hemolymph are called ecto-symbionts or extracellular symbionts. They can adhere and multiply on the cells but they do not invade epithelial barriers. Others have developed mechanisms of invasion and can be internalized by the eukaryotic cells, where they can remain transitorily in transit to their final niche or where they become residents. These are the so-called facultative or obligate endosymbionts or intra-cellular symbionts, respectively. In this case, the bacteria can be more or less integrated within the cell (in vacuoles or phagosomes, in the cytoplasm or even in the nucleus). For example, the pathogens *Legionella pneumophila*, *Chlamydia* spp., and *Mycobaterium* spp, or the mutualists *B. aphidicola*, and *Blattabacterium* spp. are harbored in vacuoles, whereas the pathogens *Listeria monocytogenes* or *Shigella* spp. and the mutualists "*Ca*. Blochmannia floridanus" and *Wigglesworthia glossinidia* live free in the cytosol. Finally the pathogen Rickettsia rickettsii has the ability to grow within the cytoplasm and the nuclei of their host cells (Silverman and Bound 1984). Finally, the symbiosis can be more host-specific or restricted to a single strain, species or genus, or alternatively, the symbiont can infect different non-related hosts. For example, *B. aphidicola* has been found exclusively in aphids, whereas "*Ca*. Sulcia muellerii" is found in sharpshooters, cicadas and spittlebugs. Another example, *Salmonella enterica* serovar Typhi can affect humans only, whereas serovar Thyphimurium has a broad range of hosts that include mice (Edwards et al. 2002).

The second perspective conceives the symbiosis from an evolutionary and temporal point of view, focusing on the transformations that affect the bacterial genomes from the establishment and the early stages of the symbiotic relationships to the final stages. A series of changes arise in both symbionts and hosts as a result of an adaptive process. This classification has been developed particularly well in the case of heritable bacterial symbionts of arthropods, where a quite complete picture of the process is being made available. The path from facultative symbiosis to early obligate endosymbiosis first, and toward long-established primary endosymbioses and the final stages in endosymbiotic relationships have been extensively reviewed in mutualistic symbioses between prokaryote and arthropods (Gil et al. 2010), but some of their principles may also be valid for other mutualistic and even parasitic symbioses. At the very beginning, it is not possible to determine if the relationship that would be established will be parasitic or mutualistic, since this distinction is based on the effect of the bacterium in the eukaryotic host but, from the bacterial point of view, the biological processes needed to successfully infect hosts are largely the same for both types of microorganisms (Gil et al. 2004).

Facultative or S-symbionts have been described in several lineages of aphids ("*Ca*. Hamiltonella defensa", "*Ca*. Regiella insecticola", and "*Ca*. Serratia symbiotica"), psyllids ("*Ca*. Arshenophonus"), whiteflies ("*Ca*. Hamiltonella", "*Ca*. Arsenophonus", "*Ca*. Cardinium", *Wolbachia*, *Rickettsia*, and *Fritschea bemisiae*), leafhoppers (*Wolbachia*), scale insects (*Fritschea*, "*Ca.* Cardinium"), tsetse flies (*S. glossinidius*), fruit flies (*Spiroplasma*, *Wolbachia pipientis* wRi and wMel), hippoboscid flies (*Sodalis*), parasitic wasps ("*Ca.* Cardinium"), butterflies (*Spiroplasma*) and mosquitoes (*Wolbachia pipientis* wPip). (Moran et al. 2005a; Hansen et al. 2007; Gottlieb et al. 2008; Mitsuhashi et al. 2002; Everett et al. 2005; Zchori-Fein and Perlman 2004; Toh et al. 2006; Williamson and Poulson 1979; Stouthamer et al. 1999; Nováková and Hypsa 2007; Klasson et al. 2009; Jiggins et al. 2000; Klasson et al. 2008). Although they are generally maternally transmitted through host generations, their distribution patterns suggest sporadic horizontal transmission events among host individuals and species (Russell et al. 2003; Russell and Moran 2005), and, therefore they do not share long evolutionary histories with their hosts. In addition, their phylogeny points to relatively recent associations with their hosts (Dale and Moran 2006), unlike primary symbionts. Their genome traits may resemble those in the early stages of a transition from a free-living lifestyle to an obligate mutualism. S-symbionts show a wide distribution in tissues and cells or in the body cavity (hemolymph), sheath cells or specialized cells and organs called secondary bacteriocytes and bacteriomes, respectively (Fukatsu et al. 2000; Moran et al. 2005b), that are infected through their own mechanisms, rather than host adaptations. Their irregular distribution among individual hosts and species suggests that they are not essential for host survival, but their effects on host fitness have been studied and increasingly, we are gaining insight into the benefits and costs of their associations. Although some nutritional benefits attributed to their bacterial biosynthetic and metabolic capabilities are possible, experiments have focused mostly on their environmental effects. Those include detrimental consequences on growth and reproduction to the host (Koga et al. 2003; Sakurai et al. 2005). The reported benefits that facultative mutualists confer include protection against natural enemies (Ferrari et al. 2004; Guay et al. 2009; Oliver et al. 2005; Scarborough et al. 2005), heat damage (Chen et al. 2000; Montllor et al. 2002) or stress (Russell and Moran 2006). They are involved in host plant specialization and reproduction (Simon et al. 2003; Ferrari et al. 2004; Tsuchida et al. 2004), and even compensate the loss of the essential endosymbiont, as it was experimentally proven (Koga et al. 2003). Some facultative symbionts are reproductive manipulators, parasites that manipulate host reproduction to spread by increasing matrilineal transmission through reproductive incompatibility, son killing, feminization of genetic males and parthenogenesis. This is the case of *Wolbachia pipientis*, "*Ca.* Cardinium hertigii", *Arsenophus nasoniae*, *Spiroplasma*, some *Rickettsia* and ladybird male killers. Regarding the possible routes of transmission, it has been suggested that in some cases parasitic wasps can act as vectors of transmission (Vavre et al. 1999). Moreover, given the similarity between some symbionts found in aphids and ladybirds, such as *Rickettsia* (Werren et al. 1994), *Wolbachia* (Hurst et al. 1999b) and *Spiroplasma* (Hurst et al. 1999a), it has been postulated that predator–prey interactions also play a role in the transmission of these bacteria. Since many of these bacteria have an affinity for bacteria found in the intestines of these insects, has also suggested an oral route of transmission through food by plants whose surface has molasses, remains of other individuals or other infected phloem insects, something possible in crowded colonies. All of these bacteria are accessory for their hosts since the infected insects can survive, and often even increase their fitness, in the absence of them. However that does not imply that the opposite is true. Thus, despite the fact that some of these S-symbionts having closely related free-living bacteria, as for example "*Ca.* Serratia symbiotica", they have never been found in the environment and are not cultivable, relying on the aphid and probably the P-symbiont *B. aphidicola* for its own survival.

A second step toward integration into a more close relationship is constituted by obligate mutualists, which become essential for the survival and reproduction of the insect. Their role becomes less environmentally dependent than in facultative symbionts and more nutritional and/or metabolic. Within the obligate symbionts, several stages of integration, revealed by their genome features can be distinguished. The most recent stages of this association are represented by clades of bacteria that have recently established such associations, such as SOPE and SZPE, the P-endosymbiont of the rice weevil *Sitophilus oryzae* and the maize weevil *Sitophilus zeamais*, respectively. These obligate mutualistic endosymbionts are closely related with the S-symbiont of tsetse flies, *S. glossinidius* (Dale and Welburn 2001; Heddi et al. 1998), but unlike it they cannot be cultured outside the host. SOPE and SZPE provide their hosts with amino acids and vitamins, having effects on fertility, development and the flying ability of adult insects (Heddi et al. 1999). The bacteria live inside bacteriocytes organized in an organ called bacteriome surrounding the midgut of the insect and near the female ovaries. With an estimated 3 Mb genome (Charles et al. 1997), within the range of many free-living bacteria, the association of this gamma-proteobacteria is not antique. Actually, some data indicate a recent endosymbiont replacement of an ancestral endosymbiont in the family Dryophtoridae to which the rice and maize weevils belong (Lefevre et al. 2004). Big amounts of transposable elements mainly IS elements have been identified in the SOPE and SZPE genomes (Gil et al. 2008; Plague et al. 2008). This abundance of repetitive DNA was unexpected in obligate mutualistic endosymbionts, because bacterial endosymbionts with a long-established obligatory relationship with their hosts show total absence of phages or transposable elements, whereas repetitive DNA is common in free-living bacteria, especially in recent specialized pathogens (e.g. the enteric bacteria *Shigella* and *Salmonella enterica* Typhi) (Jin et al. 2002; Wei et al. 2003), intracellular parasites (e.g. *W. pipientis* strains) (Klasson et al. 2008, 2009; Wu et al. 2004), or facultative insect symbionts (e.g. "*Ca.* Hamiltonella defensa", "*Ca.* Arsenophonus arthropodicus" and "*Ca.* Regiella insecticola") (Dale and Moran 2006; Degnan et al. 2009, 2010). The IS are able to move within genomes and also between genomes of different organisms by horizontal gene transfer. In free-living bacteria the IS transposition is tightly controlled, so that only a few copies of a limited number of categories appear in each genome. Their dramatic increase in intracellular bacteria must reflect the uncontrolled proliferation of elements that were already present at the beginning of symbiosis. This proliferation could be involved in the inactivation of non-essential genes and can also serve as a substrate for unequal recombination, thus promoting genome size reduction and genome rearrangements in early stages.

After the establishment of an obligate endosymbiont lifestyle, repetitive DNA tends to diminish until its total disappearance as observed in old endosymbionts, where the current lack of repetitive sequences, appears to be in the origin of their high genomic architecture stability levels (Silva et al. 2003). However, the early presence of such repetitive elements in the past might explain the genome reorganizations observer when compared to their free-living relatives. This is an indication that most of genomic modeling, including chromosomal rearrangements and the loss of many functionally dispensable genes, must take place at an early stage of the process of genomic adaptation to intracellular life (Dougherty and Plague 2008; Touchon and Rocha 2007). In addition to their genomic architecture stability, other genomic features of the long-established P-endosymbionts, many other characteristics indicate a convergent evolution among groups of bacteria non-related belonging to the Gammaproteobacteria and Bacteroidetes (López-Sánchez et al. 2009). These mature endosymbionts constitute the first and most studied ones so far (Buchner 1965), and have been considered the archetype of the mutualistic symbioses in arthropods. In general, these endosymbionts have highly compact genomes eight to ten times smaller than those of their free-living relatives. The reduced genome size of their genomes correlates quite well with the presence of a smaller number of genes compared to their free-living relatives. Additional characteristic genome features include almost total absence of recombination, increased rate of nucleotide substitution, high A + T content (although this no longer can be considered a general trait), accumulation of deleterious mutations by random genetic drift, loss of codon bias toward A or T, and accelerated sequence evolution (Andersson and Kurland 1998; Clark et al. 1999; Moya et al. 2002; Wernegreen 2005). The striking similarities in independently evolved obligate symbiont genomes indicate that common forces are at work. The aforementioned population dynamics of these microorganisms, including elevated genetic drift resulting from relaxed selection in populations with small effective size, frequent bottlenecks and horizontal transmission among hosts eliminated has as result the elevated fixation of mildly deleterious mutations, lower stability of proteins and inactivation and loss of non-essential genes. Nonetheless, variations according to the age of the association or the host lifestyle exist among these bacteria. Some adaptive changes include the plasmid-associated amplification of genes for biosynthesis of leucine and tryptophan (Latorre et al. 2005) in *B.aphidicola*, with different arrangements in divergent lineages, or the highly constitutive expression of the chaperone GroEL in *B.aphidicola* (Fares et al. 2002, 2004).

The fact that even long-established symbionts are still suffering a reductive process is illustrated by *B.aphidicola* where different strains from several aphid subfamilies show differences up to 200 Kb (Gil et al. 2002), and present pseudogenes and differential gene losses in the *B. aphidicola* genomes that have been sequenced (Perez-Brocal et al. 2006; Shigenobu et al. 2000; Tamas et al. 2002; van Ham et al. 2003). Those losses are randomly affecting different genes in each genome in a way that correlates with their hosts since the different strains of the bacteria diverged from their last common symbiont ancestor (LCSA).

In addition to a gradual reduction in the genome size, P-endosymbionts with an old association with their hosts have in general an increase in the A + T content compared with P-endosymbionts with a younger association and with S-symbionts, which in turn have intermediate values between the former and free-living bacteria. This enrichment in A + T has been related to the loss of DNA repair enzymes. However, several cases that do not follow this nucleotide composition rule, such as "*Ca.* Tremblaya princeps", the P-endosymbiont of the mealybug *Planococcus citri*, and "*Ca.* Hogkinia cicadicola", P-endosymbiont of the cicada *Dieroprocta semicincta* with a 57.1 and a 58.4% G + C content, respectively (Baumann et al. 2002; McCutcheon et al. 2009), which suggest that mutational pressure favoring A + T is not a critical step in endosymbiont genome evolution.

The evolution of the endosymbiotic integration through a random process of gradual pseudogenization and gene loss scattered throughout the genome (Gomez-Valero et al. 2004; Silva et al. 2001) might, theoretically, lead to the loss of all genes except those involved in the symbiotic relationship, as well as a reduced repertoire of genes necessary to maintain the essential functions. However, there are several cases described of extremely reduced genomes that have exceeded this threshold and have lost part of such essential functions. The most striking cases is that of "*Ca.* Carsonella ruddii", initially considered the P-endosymbiont of the psylid *Pachpsylla venusta*, with a 160 Kb genome, and only 182 predicted open reading frames, many of which overlapping (Nakabachi et al. 2006). The analysis of its coding capacities revealed that the extensive degradation of the genome is affecting vital and symbiotic functions (Tamames et al. 2007). "Thus, "*Ca.* Carsonella rudii" is not able to sustain its own essential life-related functions since most genes for DNA replication, transcription and translation are absent, as well as essential domains of many proteins due to a gene length shortening. In addition, "*Ca.* Carsonella ruddii" is not able to sustain the requirements of its host since the pathways for the synthesis of several essential amino acids are missing. Since it appears to be the sole symbiont in its host, one explanation is that somehow this insect can enhance the supply of nutrients provisioned by plants, as other sap-feeders do (Larson and Whitham 1991) in which case the former P-endosymbiont might become expendable and therefore nearing extinction. Alternatively it might be possible that some "*Ca.* Carsonella ruddii" genes have been transferred to the host nuclear DNA, as proved for present organelles. In this case "*Ca.* Carsonella ruddii" could be considered as a subcellular new entity between living cells and organelles (Tamames et al. 2007). The status as a cell or organelle and the fate of "*Ca.* Carsonella ruddii" represent an enigma that has yet to be clarified.

In other cases of extreme genomes, which are also reduced beyond the theoretical capacity of maintenance of vital and symbiotic functions, we can distinguish two scenarios offering more well-supported explanations. Both imply the eventual addition of a second bacterium in an existing symbiotic association between a bacterium and an animal host. Although at first this novel association can be facultative, the second bacterium can become essential for host fitness if it provides benefits to the organization. As a result, the evolutionary process of genome shrinkage will affect both bacteria of the association, as new genes become redundant. However, which one of the two bacterial genomes loses them will be a matter of chance. Depending on which genome is affected by the loss of genes needed for the synthesis of essential molecules, either one bacterium can enter an extreme degenerative process, which may end with its extinction (replacement), and the retained bacteria will continue the degenerative process alone (Moya et al. 2009) or both of them will become indispensable to keep a healthy consortium (complementation). Replacement has already been reported, for example, in the Family Dryophthoridae, where a former endosymbiont "*Ca.* Nardonella" was replaced by the ancestor of the Sitophilus P-endosymbionts (Lefevre et al. 2004). However, there are many more described cases of consortia reported. One case of metabolic complementation involves *B. aphidicola* and "*Ca.* Serratia symbiotica" in the aphid *Cinara cedri*. "*Ca.* Serratia symbiotica" appears as a facultative symbiont in many aphid species. However, it is reported to have become an obligate symbiont in the cedar aphid (Lamelas et al. 2008), where it coexists with *B. aphidicola* BCc in the insect bacteriome. Perez-Brocal et al. (2006) proposed that "*Ca.* Serratia symbiotica" SCc might be replacing B *aphidicola* BCc because, unlike other sequenced *B.aphidicola* strains, it has partially lost its symbiotic role, as it cannot synthesize the essential amino acid tryptophan. However, using a metagenomic approach Gosalbes et al. 2008 discovered a plasmid in *B.aphidicola* BCc containing the *trpEG* genes, coding for anthranilate synthase, the first enzyme of the tryptophan biosynthesis pathway. The remaining genes for the pathway (*trpDCBA*) are located in the chromosome of "*Ca.* Serratia symbiotica". These data show that both endosymbionts are involved in the tryptophan biosynthesis that supplies both their host and themselves. *B.aphidicola* produces a metabolic intermediate that is then provided to "*Ca.* Serratia symbiotica" to synthesize the final product. Therefore, both bacteria coexist in intracellular obligatory mutualistic association with their host. The establishment of an endosymbiotic bacterium consortium can be in the origin of big evolutionary changes in host lifestyle. This could be the case of the following example. At least three different consortia involving the Bacteroidetes member "*Ca.* Sulcia muelleri" and three non-related co-primary symbionts have been characterized. The first one in the xylem-feeding sharpshooter *Homalodisca vitripennis*, with the Gammaproteobacterium "*Ca.* Baumannia cicadellinicola" (Wu et al. 2006), which is one with largest genome of the three co-resident bacteria (686 Kb) and with a 33% GC. The second one is found in the cicada *Diceroprocta semicincta*, with the Alphaproteobacterium "*Ca.* Hodgkinia cicadicola" (McCutcheon et al. 2009), the one with the most reduced genome size of any know bacterium (144 kb) but surprisingly high G + C content (58.4%). More recently the consortium formed in the spittlebug *Clastoptera arizonana* has been reported with the Betaproteobacterium "*Ca.* Zinderia insecticola" (McCutcheon and Moran, 2010) with a tiny genome of 208 kb and the most extreme nucleotide composition (13.5% G + C) reported to date The three strains of "*Ca.* Sulcia muelleri" differ in their genome size (from 245 to 276 kb), G + C content (21.1–22.6%) and gene content (e.g. lacks the *trp* operon in the spittlebug, which is encoded by the two other strains). In all cases their whole genome analysis revealed that they have complementary sets of biosynthetic capabilities needed to provide to their host the nutrients that are lacking in the xylem sap. "*Ca.* Sulcia muelleri" encodes the enzymes involved in the biosynthesis of eight of the ten essential amino acids, whereas co-resident symbionts synthesize the remaining two. Interestingly, in the most recently described system, that of *C.arizonana*, "*Ca.* Sulcia muelleri" has completely lost the pathway for the biosynthesis of tryptophan and, therefore, retains the ability to make only seven of the ten essential amino acids, being the remaining three encoded by "*Ca.* Zinderia insecticola". "*Ca.* Baumannia cicadellinicola" contains a large number of pathways for biosynthesis of vitamins and cofactors, a role less evident in "*Ca.* Hodgkinia cicadicola", since it has lost all vitamin and cofactor biosynthetic abilities, implying that the cicada and its symbionts must access to external sources of these compounds, possibly from plant root xylem.

Another symbiotic consortium is that established between the sap-sucking mealybugs and the nested endosymbiotic system that comprises two bacteria: the P-endosymbiont "*Ca.* Tremblaya princeps", a Betaproteobacterium, which contains inside a Gammaproteobacterium (von Dohlen et al. 2001), considered as an S-symbiont based on its polyphyletic origin (Thao et al. 2002). The S-endosymbionts are distinct in each mealybug species and also differ from other insect-associated bacteria. "*Ca.* Tremblaya princeps" was the first endosymbiotic genome in which a high G + C content was found (57.1% for a fragment of 64 kb of its DNA) (Baumann et al. 2002). This is one of the very few described cases of a double endosymbiosis, [another one is the nested endosymbiosis of molgulid tunicates with the protist *Nephromyces* which is itself chronically infected with Gram-negative, intracellular bacteria, (Saffo, 1990)] and the first involving arthropods, although the symbiotic relationship between the two bacteria has not been elucidated (Kono et al. 2008).

In an example of convergent intracellular lifestyle, the genome reduction and the general (although not universal) trend toward lower GC content are also two characteristics of obligate intracellular pathogens, although the extremes reached by the insect endosymbionts have not been matched by any parasite. The most extreme cases are found within the Mollicutes. Thus, genera *Mycoplasma* (0.58–1.36 Mb, 23.8–40% GC), *Phytoplasma.* (0.60–0.88 Mb, 21.4–27.8% GC), *Ureaplasma.* (0.75–0.87 Mb, 25.5–25.8% GC) and *Mesoplasma* (0.79 Mb, 27.0% GC) account for the smallest genomes among parasitic bacteria. The intracellular lifestyle also imposed reduced genomes in other non-related bacteria, such as genera *Borrelia* (0.92–1.57 Mb, 27.5–29.8% GC), *Chlamydia* (1.00–1.08 Mb, 40.3–41.3% GC), *Chlamydophila* (1.14–1.23 Mb, 39.2–40.6% GC) or *Rickettsia* (1.10–1.58 Mb, 28.9–32.6% GC). This trend is evident even for the most extremely reduced genome described in archaea so far, that of *Nanoarchaeum equitans* (0.49 Mb, 31.6% GC), only known archaeal parasite, in this case of the Crenarcheote *Ignococcus hospitalis* (Waters et al. 2003). Therefore, there are universal principles that affect host-associated bacteria that transcends the character of the relationship with the host and that shape the genome of the bacterial partners, making our efforts ineffective to classify the whole diversity of the symbionts in a single and unambiguous way.

5 Concluding Remarks

The traditional textbook division of symbionts as parasites, commensals and mutualists no longer prevails as a set of discrete categories. Although intracellular obligate symbionts can usually be ascribed to either of them, virtually any criterion employed can bump into exceptions to the rules when it comes to the enormous quantity of newly identified and sequenced genomes of more labile symbionts, be they opportunistic pathogens or facultative mutualists. As shown in this review, many beneficial or detrimental effects can be defined only for a particular environment and under particular conditions, which are under constant transformation, imposing changing evolutionary pressures on the partners. We have shown how typical characteristics that traditionally have been ascribed to mutualists and parasites have, in fact examples from the each side that breach the norm, such as the mode of transmission, the mechanisms of invasion, the phylogeny. Also, transitions between mutualism and parasitism and vice versa are possible, even in evolutionary short periods of time. Even a priori considered essential long-established mutualists can become non-essential, losing the mutualistic role and therefore opening the door to extreme degradation and extinction and/or replacement. Due to the novel methodologies the study of symbiotic relationships is entering a new age beyond the study of relatively simple systems composed by two species. As more multiple interactions involving several bacteria, which interact the ones with the others in addition to the host, are reported the more this knowledge may help us in understanding relationships in more complex microbial communities, such as the human, animal or environmental microbiomes.

Acknowledgments Financial support was provided by grants BFU2009-12895-C02-01 (Ministerio de Ciencia e Innovación, Spain) to A. Latorre, and SAF2009-13302-C02-01 (Ministerio de Ciencia e Innovación, Spain) and PROMETEO/2009/092 (Conselleria d'Educació, Generalitat Valenciana, Spain) to A. Moya

References

Agnew P, Koella JC (1997) Virulence, parasite mode of transmission, and host fluctuating asymmetry. Proc Biol Sci 264:9–15

Andersson SG, Kurland CG (1998) Reductive evolution of resident genomes. Trends Microbiol 6:263–268

Anselme C, Vallier A, Balmand S, Fauvarque MO, Heddi A (2006) Host PGRP gene expression and bacterial release in endosymbiosis of the weevil *Sitophilus zeamais*. Appl Environ Microbiol 72:6766–6772

Baumann L, Thao ML, Hess JM, Johnson MW, Baumann P (2002) The genetic properties of the primary endosymbionts of mealybugs differ from those of other endosymbionts of plant sapsucking insects. Appl Environ Microbiol 68:3198–3205

Buchner P (1965) Endosymbiosis of animals with plant microorganisms. Interscience Publishers, New York

Bull JJ, Molineux IJ, Rice WR (1991) Selection of benevolence in a host–parasite system. Evolution 45:875–882

Champion de Crespigny FE, Wedell N (2006) *Wolbachia* infection reduces sperm competitive ability in an insect. P Roy Soc B-Biol Sci 273:1455–1458
Charles H, Heddi A, Guillaud J, Nardon C, Nardon P (1997) A molecular aspect of symbiotic interactions between the weevil *Sitophilus oryzae* and its endosymbiotic bacteria: over-expression of a chaperonin. Biochem Biophys Res Commun 239:769–774
Chen D-Q, Montllor CB, Purcell AH (2000) Fitness effects of two facultative endosymbiotic bacteria on the pea aphid, *Acyrthosiphon pisum*, and the blue alfalfa aphid *A. kondoi*. Entomol Exp Appl 95:315–323
Clark MA, Moran NA, Baumann P (1999) Sequence evolution in bacterial endosymbionts having extreme base compositions. Mol Biol Evol 16:1586–1598
Cushman JH, Beattie AJ (1991) Mutualisms: assessing the benefits to hosts and visitors. Trends Ecol Evol 6:193–195
Dale C, Maudlin I (1999) Sodalis gen. nov. and *Sodalis glossinidius* sp. nov., a microaerophilic secondary endosymbiont of the tsetse fly *Glossina morsitans morsitans*. Int J Syst Bacteriol 49:267–275
Dale C, Moran NA (2006) Molecular interactions between bacterial symbionts and their hosts. Cell 126:453–465
Dale C, Welburn SC (2001) The endosymbionts of tsetse flies: manipulating host-parasite interactions. Int J Parasitol 31:628–631
Dale C, Young SA, Haydon DT, Welburn SC (2001) The insect endosymbiont *Sodalis glossinidius* utilizes a type III secretion system for cell invasion. In: Proceedings of National Academy of Science, USA 98:1883–1888
Dale C, Plague GR, Wang B, Ochman H, Moran NA (2002) Type III secretion systems and the evolution of mutualistic endosymbiosis. In: Proceedings of National Academy of Science 99:12397–12402
Darby AC, Cho NH, Fuxellus HH, Westberg J, Andersson SGE (2007) Intracellular pathogens go extreme: genome evolution in the Rickettsiales. Trends Genet 23:511–520
Davis MM, Bjorkman PJ (1988) T-cell antigen receptor genes and T-cell recognition. Nature 334:395–401
Day T (2001) Parasite transmission modes and the evolution of virulence. Evolution 55:2389–2400
Degnan H, Lazarus AB, Wernegreen JJ (2005) Genome sequence of *Blochmannia pennsylvanicus* indicates parallel evolutionary trends among bacterial mutualists of insects. Ge-nome Res 15:1023–1033
Degnan PH, Leonardo TE, Cass BN, Hurwitz B, Stern D, Gibbs RA, Richards S, Moran NA (2010) Dynamics of genome evolution in facultative symbionts of aphids. Environ Microbiol 12:2060–2069
Degnan PH, Yu Y, Sisneros N, Wing RA, Moran NA (2009) *Hamiltonella defensa*, genome evolution of protective bacterial endosymbiont from pathogenic ancestors. Proc Natl Acad Sci USA 106:9063–9068
Dionne MS, Schneider DS (2008) Models of infectious diseases in the fruit fly *Drosophila* melanogaster. Dis Model Mech 1:43–49
Dougherty KM, Plague GR (2008) Transposable element loads in a bacterial symbiont of weevils are extremely variable. Appl Environ Microbiol 74:7832–7834
Ecker DJ, Sampath R, Willett P, Wyatt JR, Samant V, Massire C, Hall TH, Hari K, McNeil JA, Büchen-Osmond C, Budowle B (2005) The microbial rosetta stone database: a compilation of global and emerging infectious microorganisms and bioterrorist threat agents. BMC Microbiol 5:19
Edwards RA, Olsen GJ, Maloy SR (2002) Comparative genomics of closely related Salmonellae. Trends Microbiol 10:94–99
Everett KDE, Thao M, Horn M, Dyszynski GE, Baumann P (2005) Novel chlamydiae in whiteflies and scale insects: endosymbionts '*Candidatus* Fritschea bemisiae' strain Falk and '*Candidatus* Fritschea eriococci' strain Elm. Int J Syst Evol Microbiol 55:1581–1587

Ewald PW (1995) The evolution of virulence: a unifying link between parasitology and ecology. J Parasitol 81:659–669
Falkow S (1997) What is a pathogen? ASM News 7:359–365
Fares MA, Ruiz-González MX, Moya A, Elena SF, Barrio E (2002) Endosymbiotic bacteria: GroEL buffers against deleterious mutations. Nature 417:398
Fares MA, Moya A, Barrio E (2004) GroEL and the maintenance of bacterial endosymbiosis. Trends Genet 20:413–416
Feldhaar H, Gross R (2008) Immune reactions of insects on bacterial pathogens and mutualists. Microb Infect 10:1082–1088
Ferdy JB, Godelle B (2005) Diversification of transmission modes and the evolution of the mutualism. Am Nat 166:613–627
Ferrari J, Darby AC, Daniell HCJG, Douglas AE (2004) Linking the bacterial community in pea aphids with host-plant use and natural enemy resistance. Ecol Entomol 29:60–65
Ferrière R, Gauduchon M, Bronstein JL (2007) Evolution and persistence of obligate mutualists and exploiters: competition for partners and evolutionary immunization. Ecol Lett 10:115–126
Fleury F, Vavre F, Ris N, Fouillet P, Boulétreau M (2000) Physiological cost induced by the maternally-transmitted endosymbiont *Wolbachia* in *Drosophilla* parasitoid *Leptopilina heterotoma*. Parasitology 121:493–500
Fraser CM, Gocayne JD, White O, Adams MD, Clayton RA, Fleischmann RD, Bult CJ, Kerlavage AR, Sutton G, Kelley JM, Fritchman RD, Weidman JF, Small KV, Sandusky M, Fuhrmann J, Nguyen D, Utterback TR, Saudek DM, Phillips CA, Merrick JM, Tomb JF, Dougherty BA, Bott KF, Hu PC, Lucier TS, Peterson SN, Smith HO, Hutchison CA, Venter JC (1995) The minimal gene complement of *Mycoplasma genitalium*. Science 270:397–403
Fukatsu T, Nikoh N, Kawai R, Koga R (2000) The secondary endosymbiotic bacterium of the pea aphid *Acyrthosiphon pisum* (Insecta: Homoptera). Appl Environ Microbiol 66:2748–2758
Fytrou A, Schofield PG, Kraaijeveld AR, Hubbard SF (2006) *Wolbachia* infection sup-presses both host defence and parasitoid counter-defence. P Roy Soc B-Biol Sci 273:791–796
Gerardo NM, Altincicek B, Anselme C, Atamian H, Barribeau SM, de Vos M, Duncan EJ, Evans JD, Gabaldón T, Ghanim M, Heddi A, Kaloshian I, Latorre A, Moya A, Nakabachi A, Parker BJ, Pérez-Brocal V, Pignatelli M, Rahbé Y, Ramsey JS, Spragg CJ, Tamames J, Tamarit D, Tamborindeguy C, Vincent-Monegat C, Vilcinskas A (2010) Immunity and other defenses in pea aphids, *Acyrthosiphon pisum*. Genome Biol. doi: 10.1186/gb-2010-11-2-r21
Gil R, Silva FJ, Zientz E, Delmotte F, Gonzalez-Candelas F, Latorre A, Rausell C, Kamerbeek J, Gadau J, Holldobler B, van Ham RCHJ, Gross R, Moya A (2003) The genome sequence of *Blochmannia floridanus*: comparative analysis of reduced genomes. In: Proceedings of National Academy of Science, USA 100:9388–9393
Gil R, Latorre A, Moya A (2004) Bacterial endosymbionts of insects: insights from comparative genomics. Environ Microbiol 6:1109–1122
Gil R, Belda E, Gosalbes MJ, Delaye L, Vallier A, Vincent-Monegat C, Heddi A, Silva FJ, Moya A, Latorre A (2008) Massive presence of insertion sequences in the genome of SOPE, the primary endosymbiont of the rice weevil *Sitophilus oryzae*. Int Microbiol 11:41–48
Gil R, Latorre A, Moya A (2010) Evolution of prokaryote-animal symbiosis from a genomics perspective. Microbiol Monographs 19:207–233
Gil-Turnes MS, Hay ME, Fenical W (1989) Symbiotic marine bacteria chemically defend crustacean embryos from a pathogenic fungus. Science 246:116–118
Gil-Turnes MS, Fenical W (1992) Embryos of *Homarus americanus* are protected by epibiotic bacteria. Biol Bull 182:105–108
Goebel W, Gross R (2001) Intracellular survival strategies of mutualistic and parasitic prokaryotes. Trends Microbiol 9:267–273
Gomez-Valero L, Latorre A, Silva FJ (2004) The evolutionary fate of nonfunctional DNA in the bacterial endosymbiont *Buchnera aphidicola*. Mol Biol Evol 21:2172–2181
Gosalbes MJ, Lamelas A, Moya A, Latorre A (2008) The striking case of tryptophan provision in the cedar aphid *Cinara cedri*. J Bacteriol 190:6026–6029

Gosalbes MJ, Latorre A, Lamelas A, Moya A (2010) Genomics of intracellular symbionts in insects. Int J Med Microbiol 300:271–278

Gottlieb Y, Ghanim M, Gueguen G, Kontsedalov S, Vavre F, Fleury F, Zchori-Fein E (2008) Inherited intracellular ecosystem: symbiotic bacteria share bacteriocytes in whiteflies. FASEB J 22:2591–2599

Govind S (2008) Innate immunity in *Drosophila*: pathogens and pathways. Insect Sci 8(15):29–43

Guay J-F, Boudreault S, Michaud D, Cloutier C (2009) Impact of environmental stress on aphid clonal resistance to parasitoids: role of *Hamiltonella defensa* bacterial symbiosis in association with a new facultative symbiont of the pea aphid. J Insect Physiol 55:919–926

Hansen AK, Jeong G, Paine TD, Stouthamer R (2007) Frequency of secondary symbiont infection in an invasive psyllid relates to parasitism pressure on a geographic scale in California. Appl Environ Microbiol 73:7531–7535

Hao Z, Kasumba I, Lehane MJ, Gibson WC, Kwon J, Aksoy S (2001) Tsetse immune responses and trypanosome transmission: implications for the development of tsetse-based strategies to reduce trypanosomiasis. In: Proceedings of National Acadmy of Science, USA 98:12648–12653

Haynes S, Darby AC, Daniell TJ, Webster G, van Veen FJF, Godfray HCJ, Prosser JI, Douglas AE (2003) Diversity of bacteria associated with natural aphid populations. Appl Envir Microbiol 69:7216–7223

Heddi A, Charles H, Khatchadourian C, Bonnot G, Nardon P (1998) Molecular characterization of the principal symbiotic bacteria of the weevil *Sitophilus oryzae*: a peculiar G + C content of an endocytobiotic DNA. J Mol Evol 47:52–61

Heddi A, Grenier AM, Khatchadourian C, Charles H, Nardon P (1999) Four intracellular genomes direct weevil biology: nuclear, mitochondrial, principal endosymbiont, and *Wolbachia*. In: Proceedings of National Academy of Science 96:6814–6819

Hosokawa T, Koga R, Kikuchi Y, Meng XY, Fukatsu T (2010) *Wolbachia* as a bacteriocyte-associated nutritional mutualist. In: Proceedings of National Academy of Science 107:769–774

Hurst GD, Graf von der Schulenburg JH, Majerus TM, Bertrand D, Zakharov IA, Baungaard J, Völkl W, Stouthamer R, Majerus ME (1999a) Invasion of one insect species, *Adalia bipunctata*, by two different male-killing bacteria. Insect Mol Biol 8:133–139

Hurst GDD, Jiggins FM, von der Schulenberg JHG, Bertrand D, West SA, Goriacheva II, Zakharov MEN, Werren JH, Stouthamer R, Majerus EN (1999b) Male-killing *Wolbachia* in *Trichogramma* wasps. P Roy Soc B Biol Sci 266:735–740

Jiggins FM, Hurst GDD, Jiggins CD, von der Schulenburg JHG, Majerus MEN (2000) The butterfly *Danaus chrysippus* is infected by a male-killing *Spiroplasma* bacterium. Parasitology 120:439–446

Jin Q, Yuan Z, Xu J, Wang Y, Shen Y, Lu W, Wang J, Liu H, Yang J, Yang F, Zhang X, Zhang J, Yang G, Wu H, Qu D, Dong J, Sun L, Xue Y, Zhao A, Gao Y, Zhu J, Kan B, Ding K, Chen S, Cheng H, Yao Z, He B, Chen R, Ma D, Qiang B, Wen Y, Hou Y, Yu J (2002) Genome sequence of *Shigella flexneri* 2a: insights into pathogenicity through comparison with genomes of *Escherichia coli* K12 and O157. Nucleic Acids Res 30:4432–4441

Klasson L, Walker T, Sebaihia M, Sanders MJ, Quail MA, Lord A, Sanders S, Earl J, O'Neill SL, Thomson N, Sinkins SP, Parkhill J (2008) Genome evolution of *Wolbachia* strain wPip from the *Culex pipiens* group. Mol Biol Evol 25:1877–1887

Klasson L, Westberg J, Sapountzis P, Naslund K, Lutnaes Y, Darby AC, Veneti Z, Chen L, Braig HR, Garrett R, Bourtzis K, Andersson SG (2009) The mosaic genome structure of the *Wolbachia* wRi strain infecting *Drosophila simulans*. In: Proceedings of National Academy of Science 106:5725–5730

Koga R, Tsuchida T, Fukatsu T (2003) Changing partners in an obligate symbiosis: a facultative endosymbiont can compensate for loss of the essential endosymbiont *Buchnera* in an aphid. Proc Biol Sci 270:2543–2550

Kono M, Koga R, Shimada M, Fukatsu T (2008) Infection dynamics of coexisting beta- and gammaproteobacteria in the nested endosymbiotic system of mealybugs. Appl Environ Microbiol 74:4175–4184

Kurtz J, Armitage SAO (2006) Alternative adaptive immunity in invertebrates. Trends Immunol 27:493–496

Lamelas A, Pérez-Brocal V, Gómez-Valero L, Gosalbes MJ, Moya A, Latorre A (2008) Evolution of the secondary symbiont ''*Candidatus* Serratia symbiotica'' in aphid species of the subfamily Lachninae. Appl Environ Microbiol 74:4236–4240

Langworthy NG, Renz A, Mackenstedt U, Henkle-Duhrsen K, De Bronsvoort MB, Tanya VN, Donnelly MJ, Trees AJ (2000) Macrofilaricidal activity of tetracycline against the filarial nematode *Onchocerca ochengi*: elimination of *Wolbachia* precedes worm death and suggests a dependent relationship. Proc R Soc Lond B Biol Sci 267:1063–1069

Larson KC, Whitham TG (1991) Manipulation of food resources by a gall-forming aphid: the physiology of sink-source interactions. Oecologia 88:15–21

Latorre A, Gil R, Silva FJ, Moya A (2005) Chromosomal stasis versus plasmid plasticity in aphid endosymbiont *Buchnera aphidicola*. Heredity 95:339–347

Lebecque SG, Bearhart PJ (1990) Boundaries of somatic mutation in rearranged immunoglobulin genes: 5b boundary is near the promoter and 3b boundary is 1 Kb from V(D)J gene. J Exp Med 172:1717–1727

Lefevre C, Charles H, Vallier A, Delobel B, Farrell B, Heddi A (2004) Endosymbiont phylogenesis in the Dryophthoridae weevils: evidence for bacterial replacement. Mol Biol Evol 21:965–973

Leung TLF, Poulin R (2008) Parasitism, commensalism, and mutualism: exploring the many shades of symbioses. Vie Milieu 58:107–115

López-Sánchez MJ, Neef A, Peretó J, Patiño-Navarrete R, Pignatelli M, Latorre A, Moya A (2009) Evolutionary convergence and nitrogen metabolism in *Blattabacterium* strain Bge, primary endosymbiont of the cockroach *Blattella germanica*. PLoS Genet 5:e1000721

Male D (2004) Immunology. Elsevier, London

Margulis L (1981) Symbiosis in cell evolution, 1st edn. Freeman, New York

Margulis L (1993) Symbiosis in Cell Evolution, 2nd edn. Freeman, New York

McCutcheon JP, Moran NA (2007) Parallel genomic evolution and metabolic interdependence in an ancient symbiosis. In: Proceedings of National Academy of Science 104:19392–19397

McCutcheon JP, Moran NA (2010) Functional convergence in reduced genomes of bacterial symbionts spanning 200 My of evolution. Genome Biol Evol 2:708–718

McCutcheon JP, McDonald BR, Moran NA (2009) Origin of an alternative genetic code in the extremely small and GC-rich genome of a bacterial symbiont. PLoS Genet 5:e1000565

Meyers RA (2007) Immunology: from cell biology to disease. Wiley, Weinheim

Miller MR, White A, Boots M (2006) The evolution of parasites in response to tolerance in their host: the good, the bad, and apparent commensalism. Evolution 60:945–956

Mitsuhashi W, Saiki T, Wei W, Kawakita H, Sato M (2002) Two novel strains of Wolbachia coexist in both species of mulberry leafhoppers, *Hishimonoides sellatiformis* and *Hishimonus sell*atus which are vectors of mulberry dwarf phytoplasma. Insect Mol Biol 11:577–584

Montllor CB, Maxmen A, Purcell AH (2002) Facultative bacterial endosymbionts benefit pea aphids *Acyrthosiphon pisum* under heat stress. Ecol Entomol 27:189–195

Moran NA (1996) Accelerated evolution and Muller's rachet in endosymbiotic bacteria. In: Proceedings of National Academy of Science USA 93:2873–2878

Moran NA (2007) Symbiosis as an adaptive process and source of phenotypic complexity. In: Proceedings of National Academy of Science 104:8627–8633

Moran NA, Wernegreen JJ (2000) Lifestyle evolution in symbiotic bacteria: insights from genomics. Trends Ecol Evol 15:321–326

Moran NA, Russell JA, Koga R, Fukatsu T (2005a) Evolutionary relationships of three new species of Enterobacteriaceae living as symbionts of aphids and other insects. Appl Environ Microbiol 71:3302–3310

Moran NA, Tran P, Gerardo NM (2005b) Symbiosis and insect diversification: an ancient symbiont of sapfeeding insects from the bacterial phylum Bacteroidetes. App Environ Microbiol 71:8802–8810

Moran NA, McCutcheon JP, Nakabachi A (2008) Genomics and evolution of heritable bacterial symbionts. Annu Rev Genet 42:165–190

Moya A, Latorre A, Sabater-Munoz B, Silva FJ (2002) Comparative molecular evolution of primary (*Buchnera*) and secondary symbionts of aphids based on two protein-coding genes. J Mol Evol 55:127–137

Moya A, Gil R, Latorre A (2009) The evolutionary history of symbiotic associations among bacteria and their animal hosts: a model. Clin Microbiol Infect 15:11–13

Müller WEG, Müller I (2003) Analysis of the sponge (Porifera) gene repertoire: implication for the evolution of the Metazoan body plan. In: Müller WEG (ed) Marine Molecular Biotechnology. Springer-Verlag, Berlin

Nakabachi A, Shigenobu S, Sakazume N, Shiraki T, Hayashizaki Y, Carninci P, Ishikawa H, Kudo T, Fukatsu T (2005) Transcriptome analysis of the aphid bacteriocyte, the symbiotic host cell that harbors an endocellular mutualistic bacterium, *Buchnera*. In: Proceedings of National Academy of Science 102:5477–5482

Nakabachi A, Yamashita A, Toh H, Ishikawa H, Dunbar HE, Moran NA, Hattori M (2006) The 160-kilobase genome of the bacterial endosymbiont Carsonella. Science 314:267

Nováková E, Hypsa V (2007) A new *Sodalis* lineage from bloodsucking fly *Craterina melbae* (Diptera, Hippoboscoidea) originated independently of the tsetse flies symbiont *Sodalis glossinidius*. FEMS Microbiol Lett 269:131–135

Oliver KM, Russell JA, Moran NA, Hunter MS (2003) Facultative bacterial symbionts in aphids confer resistance to parasitic wasps. Proc Natl Acad Sci USA 100:1803–1807

Oliver KM, Moran NA, Hunter MS (2005) Variation in resistance to parasitism in aphids is due to symbionts not host genotype. In: Proceedings of National Academy of Science 102:12795–12800

Oliver KM, Moran NA, Hunter MS (2006) Costs and benefits of a superinfection of facultative symbionts in aphids. P Roy Soc B Biol Sci 273:1273–1280

Oliver KM, Degnan PH, Hunter MS, Moran NA (2009) Bacteriophages encode factors required for protection in a symbiotic mutualism. Science 325:992–994

O'Neill SL, Giordano R, Colbert AME, Karr TL, Robertson HM (1992) 16S ribosomal RNA phylogenetic analysis of the bacterial endosymbionts associated with cytoplasmic incompatibility in insects. In: Proceedings of National Academy of Science 89:2699–2702

Perez-Brocal V, Gil R, Ramos S, Lamelas A, Postigo M, Michelena JM, Silva FJ, Moya A, Latorre A (2006) A small microbial genome: the end of a long symbiotic relationship? Science 314:312–313

Perrot-Minnot MJ, Cheval B, Migeon A, Navajas M (2002) Contrasting effects of Wolbachia on cytoplasmic incompatibility and fecundity in the haplodiploid mite *Tetranychus urticae*. J Evol Biol 15:808–817

Pier GB, Lyczak JB, Wetzler LM (2004) Immunology, infection, and immunity. ASM Press, Washington DC

Plague GR, Dunbar HE, Tran PL, Moran NA (2008) Extensive proliferation of transposable elements in heritable bacterial symbionts. J Bacteriol 190:777–779

Poulsen M, Cafaro M, Boosma JJ, Currie CR (2005) Specificity of the mutualistic associ-ation between actinomycete bacteria and two sympatric species of Acromyrmex leaf-cutting ants. Mol Ecol 14:3597–3604

Rokita E, Makristathis A, Presterl E, Rotter ML, Hirschl AM (1998) Helicobacter pylori urease significantly reduces opsonization by human complement. J Infect Dis 178:1521–1525

Russell JA, Moran NA (2005) Horizontal transfer of bacterial symbionts: heritability and fitness effects in a novel aphid host. Appl Environ Microbiol 71:7987–7994

Russell JA, Moran NA (2006) Costs and benefits of symbiont infection in aphids: variation among symbionts and across temperatures. Proc Biol Sci 273:603–610

Russell JA, Latorre A, Sabater-Munoz B, Moya A, Moran NA (2003) Side-stepping secondary symbionts: widespread horizontal transfer across and beyond the Aphidoidea. Mol Ecol 12:1061–1075
Saffo MB (1990) Symbiosis within a symbiosis: Intracellular bacteria within the endosymbiotic protist *Nephromyces*. Marine Biol 107:291–296
Sakurai M, Koga R, Tsuchida T, Meng X-Y, Fukatsu T (2005) *Rickettsia* symbiont in the pea aphid Acyrthosiphon pisum: novel cellular tropism, effect on host fitness, and interaction with the essential symbiont *Buchnera*. Appl Environ Microbiol 71:4069–4075
Sandström JP, Russell JA, White JP, Moran NA (2001) Independent origins and horizontal transfer of bacterial symbionts of aphids. Mol Ecol 10:217–228
Sassera D, Beninati T, Bandi C, Bouman EAP, Sacchi L, Fabby M, Lo N (2006) '*Candidatus* Midichloria mitochondrii', an endosymbiont of the tick *Ixodes ricinus* with a unique intramitochondrial lifestyle. Int J Syst Evol Microbiol 56:2535–2540
Scarborough CL, Ferrari J, Godfray HCJ (2005) Bacterial endosymbiont increases aphid inclusive fitness after pathogen attack. Science 310:1781
Schulenburg H, Boehnisch C, Michiels NK (2007) How do invertebrates generate a highly specific innate immune response? Mol Immunol 44:3338–3344
Shigenobu S, Watanabe H, Hattori M, Sakaki Y, Ishikawa H (2000) Genome sequence of the endocellular bacterial symbiont of aphids *Buchnera* sp. APS. Nature 407:81–86
Silva FJ, Latorre A, Moya A (2001) Genome size reduction through multiple events of gene disintegration in *Buchnera* APS. Trends Genet 17:615–618
Silva FJ, Latorre A, Moya A (2003) Why are the genomes of endosymbiotic bacteria so stable? Trends Genet 19:176–180
Silverman DJ, Bound SB (1984) Infection of human vascular endothelial cells by *Rickettsia rickettsii*. J Infect Dis 149:201–206
Simon JC, Carre S, Boutin M, Prunier-Leterme N, Sabater-Mun B, Latorre A, Bournoville R (2003) Host-based divergence in populations of the pea aphid: insights from nuclear markers and the prevalence of facultative symbionts. Proc Biol Sci 270:1703–1712
Steinert M, Hentschel U, Hacker J (2000) Symbiosis and pathogenesis: evolution of the microbe-host interaction. Naturwissenschaften 87:1–11
Stewart AD, Logsdon JM Jr, Kelley SE (2005) An empirical study of the evolution of virulence under both horizontal and vertical transmission. Evolution 59:730–739
Stouthamer R, Breeuwer JAJ, Hurst GDD (1999) *Wolbachia pipientis*: Microbial manipulator of arthropod reproduction. Annu Rev Microbiol 53:71–102
Strand MR (2008) The insect cellular immune response. Insect Sci 15:1–14
Sullivan JT, Trzebiatowski JR, Cruickshank RW, Gouzy J, Brown SD, Elliot RM, Fleet wood DJ, McCallum NG, Rossbach U, Stuart GS, Weaver JE, Webby RJ, De Bruijn FJ, Ronson CW (2002) Comparative sequence analysis of the symbiosis island of *Mesorhizobium loti* strain R7A. J Bacteriol 184:3086–3095
Tamames J, Gil R, Latorre A, Pereto J, Silva FJ, Moya A (2007) The frontier between cell and organelle: genome analysis of Candidatus Carsonella ruddii. BMC Evol Biol 7:181
Tamas I, Klasson L, Canback B, Naslund AK, Eriksson AS, Wernegreen JJ, Sandstrom JP, Moran NA, Andersson SG (2002) 50 million years of genomic stasis in endosymbiotic bacteria. Science 296:2376–2379
Thao ML, Gullan PJ, Baumann P (2002) Secondary (gamma-Proteobacteria) endosymbionts infect the primary (beta-Proteobacteria) endosymbionts of mealybugs multiple times and coevolve with their hosts. Appl Environ Microbiol 68:3190–3197
Toft C, Andersson SGE (2010) Evolutionary microbial genomics: insights into bacterial host adaptation. Nature Rev Genet 11:465–475
Toh H, Weiss BL, Perkin SA, Yamashita A, Oshima K, Hattori M, Aksoy S (2006) Massive genome erosion and functional adaptations provide insights into the symbiotic lifestyle of *Sodalis glossinidius* in the tsetse host. Genome Res 16:149–156
Tonegawa S (1983) Somatic generation of antibody diversity. Nature 302:575–581

Touchon M, Rocha EP (2007) Causes of insertion sequences abundance in prokaryotic genomes. Mol Biol Evol 24:969–981
Tsuchida TR, Koga R, Fukatsu T (2004) Host plant specialization governed by facultative symbiont. Science 303:1989
van Ham RC, Kamerbeek J, Palacios C, Rausell C, Abascal F, Bastolla U, Fernandez JM, Jimenez L, Postigo M, Silva FJ, Tamames J, Viguera E, Latorre A, Valencia A, Moran F, Moya A (2003) Reductive genome evolution in *Buchnera aphidicola*. In: Proceedings of National Academy of Science 100:581–586
Vavre F, Fleury F, Lepetit D, Fouillet P, Bouletreau M (1999) Phylogenetic evidence for horizontal transmission of *Wolbachia* in host-parasitoid associations. Mol Biol Evol 16:1711–1723
von Dohlen CD, Kohler S, Alsop ST, McManus WR (2001) Mealybug beta-proteobacterial endosymbionts contain gamma-proteobacterial symbionts. Nature 412:433–436
Waters E, Hohn MJ, Ahel I, Graham DE, Adams MD, Barnstead M, Beeson KY, Bibbs L, Bolanos R, Keller M, Kretz K, Lin X, Mathur E, Ni J, Podar M, Richardson T, Sutton GG, Simon M, Soll D, Stetter KO, Short JM, Noordewier M (2003) The genome of *Nanoarchaeum equitans*: insights into early archaeal evolution and derived parasitism. In: Proceedings of National Academy of Science 100:12984–12988
Weeks AR, Turelli M, Harcombe WR, Reynolds KT, Hoffmann AA (2007) From parasite to mutualist: Rapid evolution of *Wolbachia* in natural populations of *Drosophila*. PLoS Biol 5:e114
Wei J, Goldberg MB, Burland V, Venkatesan MM, Deng W, Fournier G, Mayhew GF, Plunkett G 3rd, Rose DJ, Darling A, Mau B, Perna NT, Payne SM, Runyen-Janecky LJ, Zhou S, Schwartz DC, Blattner FR (2003) Complete genome sequence and comparative genomics of *Shigella flexneri* serotype 2a strain 2457T. Infect Immun 71:2775–2786
Wernegreen JJ (2005) For better or worse: genomic consequences of intracellular mutualism and parasitism. Curr Opin Genet Dev 15:572–583
Werren JH, Hurst GDD, Zhang W, Breeuwer JAJ, Stouthamer R, Majerus MEN (1994) Rickettsial relative associated with male-killing in the ladybird beetle (*Adalia bipunctata*). J Bacteriol 176:388–394
Werren JH, Baldo L, Clark ME (2008) *Wolbachia*: master manipulators of invertebrate biology. Nat Rev Microbiol 6:741–751
Williamson DL, Poulson DF (1979) Sex ratio organisms (Spiroplasmas) of *Drosophila*. In: Whitcomb RF, Tully JG (eds) The Mycoplasmas. Academic Press, New York
Wu M, Sun LV, Vamathevan J, Riegler M, Deboy R, Brownlie JC, McGraw EA, Martin W, Esser C, Ahmadinejad N, Wiegand C, Madupu R, Beanan MJ, Brinkac LM, Daugherty SC, Durkin AS, Kolonay JF, Nelson WC, Mohamoud Y, Lee P, Berry K, Young MB, Utterback T, Weidman J, Nierman WC, Paulsen IT, Nelson KE, Tettelin H, O'Neill SL, Eisen JA (2004) Phylogenomics of the reproductive parasite *Wolbachia pipientis* wMel: a streamlined genome overrun by mobile genetic elements. PLoS Biol 2:E69
Wu D, Daugherty SC, Van Aken SE, Pai GH, Watkins KL, Khouri H, Tallon LJ, Zaborsky JM, Dunbar HE, Tran PL, Moran NA, Eisen JA (2006) Metabolic complementarity and genomics of the dual bacterial symbiosis of sharpshooters. PLoS Biol 4:e188
Zchori-Fein E, Perlman SJ (2004) Distribution of the bacterial symbiont *Cardinium* in arthropods. Mol Ecol 13:2009–2016
Zeh JA, Zeh DW (2006) Male-killing *Wolbachia* in a live-bearing arthropod: brood abortion as a constraint on the spread of a selfish microbe. J Invertebr Pathol 92:33–38
Zientz E, Dandekar T, Gross R (2004) Metabolic interdependence of obligate intracellular bacteria and their insect hosts. Microbiol Mol Biol Rev 68:745–770

Part II
Host-microbe Interaction in the Intestinal Tract

Ecology and Physiology of the Intestinal Tract

Michael Blaut

Abstract The number of microorganisms inhabiting the human digestive tract exceeds the number of body cells by a factor of ten. This microbial community affects host physiology and host health. The metabolic potential of the gut microbiota is immense affording the extraction of energy from otherwise indigestible carbohydrates (dietary fiber) and the conversion of host-derived substances, non-nutritive dietary components and drugs. Recognized functions of the gut microbiota include provision of colonization resistance against pathogens and priming of both the innate and the acquired immune systems. However, the intestinal microbiota may also contribute to the development of diseases such as ulcerative colitis and colorectal cancer. Culture-dependent studies provided basic knowledge on the gut microbiota, but only the advent of culture-independent molecular methods led to a better understanding of host-microbe interactions. The application of metagenomics to the gut microbial ecosystem revealed truly remarkable correlations between certain diseases and the gut microbiome. It also led to the suggestion of the existence of a 'core microbiome' that encompasses key functions shared by each individual. However, the mechanisms underlying host-microbe interactions have not yet been unraveled.

M. Blaut (✉)
Department of Gastrointestinal Microbiology,
German Institute of Human Nutrition Potsdam-Rehbruecke (DIfE),
Arthur-Scheunert-Allee 114-116,
14558 Nuthetal, Germany
e-mail: blaut@dife.de

Current Topics in Microbiology and Immunology (2013) 358: 247–272
DOI: 10.1007/82_2011_192

Published Online: 26 November 2011

Contents

1 Introduction

The human digestive tract is populated by a large number of microbes, whose cell number (10^{14}) has been estimated to exceed the number of body cells (10^{13}) by a factor of ten. The vast majority of these microorganisms are bacteria, while small proportions are archaea and yeasts. Interest in this microbial community has considerably increased owing to the realization that gut microbes profoundly influence host functions and thereby contribute to health maintenance or disease development. Recognized functions of the gut microbiota include provision of a barrier against pathogens, also referred to as colonization resistance, priming of both the innate and the acquired immune systems and extraction of energy from otherwise indigestible carbohydrates (dietary fiber). However, the intestinal microbiota may also contribute to the development of diseases such as ulcerative colitis and colorectal cancer. The application of metagenomics to the gut microbial ecosystem revealed truly remarkable correlations between certain diseases and the gut microbiome. It also led to the suggestion of the existence of a 'core microbiome' that encompasses key functions shared by each individual. However, the mechanisms underlying host-microbe interactions have not yet been unraveled.

2 Culture-Dependent Studies of the Gut Microbial Ecoystem

Until the advent of culture-independent approaches, the exploration of the gut microbial ecosystem has been hampered by the necessity to culture the community members in order to gain information about their features and possible roles in the intestine. Finegold as well as Holdeman and Moore were the first researchers in gut microbiology to take advantage of techniques for growing gut bacteria in the absence of oxygen. Holdeman and Moore (1972) used oxygen-free roll tubes whose inner surface was covered with a thin layer of solidified medium. This method had originally been developed by Hungate (1950) for the growth of rumen bacteria. Finegold (1974) introduced the use of anaerobic chambers for the handling and growing of oxygen-sensitive intestinal bacteria. This pioneering work led to the isolation and valid description of a large number of new species whose identification was entirely based on their phenotypic features. Even though the proportion of validly described intestinal bacteria is small compared to the proportion of intestinal bacteria whose presence is deduced from their 16S ribosomal RNA gene sequences found in fecal samples, much of our knowledge on their metabolic features is based on the phenotypic characterization of these isolates.

3 Culture-Independent Exploration of the Human Gut Ecosystem

Molecular biology has revolutionized the characterization and identification of microorganisms. In particular the 16S ribosomal RNA, which is also called the small subunit ribosomal RNA, was identified as a molecule that reflects the phylogenetic interrelationships of microorganisms (Woese 1987). By using PCR technologies, even microorganisms that are difficult to culture became accessible (Amann et al. 1991). When this technique was applied to the gut microbiota it became clear that the diversity of this ecosystem had previously been underestimated and that validly described isolates from this habitat only comprised a small proportion of the total community (Suau et al. 1999). Taking advantage of the steadily increasing database of 16S rRNA gene sequences, oligonucleotide probes targeting diagnostic stretches within the rRNA were designed. These probes were fluorescently labeled and used for the detection and enumeration of intestinal bacteria at various levels of the phylogenetic hierarchy using whole-cell fluorescence in situ hybridization combined with fluorescence microscopy or flow cytometry (Franks et al. 1998; Doré et al. 1998; Blaut et al. 2002). Alternative 16S rRNA-targeted approaches applied include phylogenetic microarrays (Suau et al. 2001; Rajilic-Stojanovic et al. 2009) and quantitative real-time PCR. For the latter approach other conserved genes such as those encoding heat-shock proteins have also been used (Chiang et al. 2008).

Technological progress and the drastic fall in DNA sequencing costs facilitated the in-depth analysis of the gut microbial ecosystem. Initial studies concentrated on creating inventories of 16S rRNA gene sequences (Eckburg et al. 2005). More recently, sequencing efforts have been extended to the gut microbiome, which encompasses all genes of the microbial community in the intestine (Gill et al. 2006; Turnbaugh et al. 2009, 2010; Qin et al. 2010). These metagenomic studies revealed that the gut microbiota displays low diversity at the phylum level but high diversity at the species and subspecies level. Prokaryotic members of the microbial

Table 1 Relevant genera of the human gut microbiota and their taxonomic affiliation

Domain	Phylum	Order	Genus
Archaea	Euryarchaeota	Methanobacteriales	Methanobrevibacter
Bacteria	Actinobacteria	Bifidobacteriales	Bifidobacterium
		Coriobacteriales	Adlercreutzia
			Atopobium
			Collinsella
			Coriobacterium
			Eggerthella
			Slackia
	Bacteroidetes	Bacteroidales	Alistipes
			Bacteroides
			Prevotella
			Porphyromonas
	Firmicutes	Bacillales	Staphylococcus
		Clostridiales	Anaerostipes
			Blautia
			Butyrivibrio
			Clostridium
			Coprococcus
			Dorea
			Eubacterium
			Faecalibacterium
			Finegoldia
			Lachnospira
			Lactonifactor
			Roseburia
			Ruminococcus
			Subdoligranulum
		Erysipelotrichales	Coprobacillus
			Holdemania
			Catenibacterium
		Lactobacillales	Enterococcus
			Lactobacillus
			Lactococcus
			Streptococcus
	Fusobacteria	Fusobacteriales	Fusobacterium
	Proteobacteria	Enterobacteriales	Escherichia
			Enterobacter
	Verrucomicrobia	Verrrucomicrobiales	Akkermannsia
Eukarya	Ascomycota	Saccharomycetales	Candida

community belong to only one archaeal phylum and six bacterial phyla (Table 1). Five major phyla, the Firmicutes, Bacteroidetes, Actinobacteria, Proteobacteria, and Verrucomicrobia account for more than 95% of all bacterial cells in the human intestinal tract. Other microbial groups in the human intestine include Fusobacteria, eukaryotic yeasts, and the archaeon *Methanobrevibacter smithii*. The number of bacterial species prevalent in the human gut has been estimated to be 1,000–1,150 for a cohort of 124 European individuals and at least 160 for each of these subjects (Qin et al. 2010).

These metagenomic studies not only revealed the presence of previously undetected species but also identified gene functions that are part of everybody's gut microbiome and that are found in different gut bacteria. However, it should be kept in mind that almost all gene assignments are based on the sequence similarity to known genes. It is by no means sure that all assignments are correct. A high proportion of genes are less common or even rare and the functions of a considerable proportion of genes in the gut microbiome are still unknown (Qin et al. 2010). The elucidation of their function is of importance for a better understanding of the gut ecosystem.

4 Physicochemical Conditions in the Human Intestinal Tract

The human digestive tract is not a uniform habitat but differs in the various sections. The stomach of the healthy adult is acidic (pH 2) and therefore contains only 10^3 colony forming units (cfu)/ml (Finegold et al. 1983). Viable organisms in the stomach include primarily acid-tolerant bacteria such as *Lactobacillus* spp., *Streptococcus* spp., *Staphylococcus* spp. and Enterobacteriaceae. Interestingly, recent 16S rRNA gene sequence analysis revealed a considerable microbial diversity in the gastric mucosa encompassing 128 phylotypes, the majority of which were assigned to the Proteobacteria, Firmicutes, Actinobacteria, Bacteroidetes, and Fusobacteria (Bik et al. 2006). The transit from the stomach to the small intestine is characterized by an increase in pH which lies between 6 and 6.5 in the duodenum and between 7 and 8 in jejunum and ileum. The concentration of bacteria in stomach juice and duodenal contents is low ($<10^4$ cfu/ml) but increases steadily along the small intestine and reaches 10^9 cfu/ml in the cecum. The relatively low bacterial concentration in the duodenum is due to the rather sudden change in pH from the stomach to the duodenum and to the secretion of digestive enzymes and bile acids into the intestinal tract which impede the proliferation of many bacteria. Bacteria cultured from the small intestine include species that are also found in the stomach and in addition *Bifidobacterium* spp., *Bacteroides* spp., and *Clostridium* spp. (Finegold et al. 1983). The colonic pH, which ranges from 5.5 to 7, is lower in the proximal colon than in the distal colon. In the colon, intestinal bacteria reach their highest density (10^{12} cfu/g) and also the highest diversity. Cultured organisms include *Bacteroides, Eubacterium, Clostridium, Peptostreptococcus, Streptococcus, Bifidobacterium, Ruminococcus, Fusobacterium*, Enterobacteriaceae, *Lactobacillus*, and *Staphylococcus* (Finegold et al. 1983).

Other factors influencing the gut microbiota composition include the redox potential (E_h) and the oxygen concentration. The E_h in the distal regions of the human intestinal tract is expected to be in the same range as that reported for the cecum of piglets (−215 mV) (Vervaeke et al. 1973). In accordance with this reduced environment the vast majority of human fecal bacteria (>99%) are strict anaerobes, i.e. they only grow under reducing conditions in the absence of oxygen. Less than 1% of the bacterial cells in the human intestine is facultatively aerobic or aerotolerant. Facultative aerobes such as Enterobacteriaceae may take advantage of oxygen as terminal electron acceptor if available but they are also capable of growing anaerobically. Aerotolerant bacteria, such as certain lactic acid bacteria, are not capable of gaining energy by respiration but they are not killed by oxygen because they are devoid of cellular components that lead to the generation of reactive oxygen species. Even though the E_h in distal ileum and colon is low, measurements in the porcine intestinal tract indicate that the gut lumen contains some dissolved oxygen which decreases from the stomach to the colon (Hillman et al. 1993). It may be surmised that oxygen stems from both swallowed air and the mucosal surface where the oxygen partial pressure is elevated due to blood circulation. Some gut bacteria that have previously been considered as strict anaerobes turned out to benefit from nanomolar concentrations of dissolved oxygen (Baughn and Malamy 2004). *Bacteroides fragilis* was demonstrated to reduce oxygen by taking advantage of a high-affinity cytochrome *bd* oxidase allowing ATP generation by aerobic respiration. The corresponding genes were also detected in other *Bacteroides* species except *Bacteroides distasonis*. This indicates that most *Bacteroides* species are capable of aerobic respiration provided the concentration of dissolved oxygen is <1 μM (Baughn and Malamy 2004).

5 Metabolic Activities of the Gut Microbiota

The metabolic capacity of the gut microbiota has been compared to that of the liver. Intestinal bacteria ferment a large range of substrates including carbohydrates, proteins, secondary plant metabolites, and xenobiotics. The underlying metabolic pathways provide energy for the bacteria and fulfill their anabolic requirements. However, bacterial fermentation products may also affect the host in various ways.

5.1 Dietary Carbohydrates Fermented by Intestinal Bacteria

One of the main functions of the gut microbiota is the extraction of energy from otherwise non-digestible food constituents. In particular carbohydrates, which in healthy humans escape digestion in the small intestine, undergo fermentation in the large intestine. These carbohydrates include non-digestible sugar alcohols like sorbitol, synthetic carbohydrates like lactulose, and polymeric carbohydrates such as resistant starch and non-starch polysaccharides (Table 2). The latter are referred to as dietary fibers.

Resistant starch is the primary carbon and energy source of intestinal bacteria (Cummings 1994). Four types of resistant starch (RS) can be distinguished

Table 2 Major substrates utilized by the gut microbiota

Substrates Amount	(g/day)
Carbohydrates	
Resistant starches	8–40
Non-starch polysaccharides	8–18
Cellulose	
Hemicellulose	
Pectin	
Inulin	
Unabsorbed sugars and sugar alcohols	2–10
Chitin and amino sugars	1–2
Synthetic Carbohydrates	*
Lactulose	
Lactitol	
Polydextrose	
Protein	
Dietary protein	3–12
Pancreatic enzymes and intestinal secretions	4–6
Others	
Mucus	2–3
Bacterial recycling	?
Sloughed epithelial cells	?

* small on average but variable in individuals; ?: unknown
Modified after (Cummings 1994)

(Sajilata et al. 2006): RS1 is physically inaccessible to human amylases. It is found in unprocessed seeds, legumes, and whole grains. RS2 is a granular form of starch, as found in green bananas, uncooked potatoes, and corn. RS3 is formed during cooking and cooling of starch-containing foods by retrogradation of amylose. RS4 refers to chemically modified starch that resists digestion and is not found in nature.

Non-starch polysaccharides include cellulose, hemicellulose and pectin, all of which are structural components of the cell wall of green plants. In contrast to cellulose, a polysaccharide consisting of a linear chain of up to 10,000 D-glucose units linked by $\beta(1 \rightarrow 4)$ bonds, hemicelluloses are composed of a range of different monomers such as xylose, arabinose, galactose, mannose, and rhamnose, which are linked in various ways. Hemicelluloses include xylan, arabinoxylan, glucuronoxylan, glucomannan, and xyloglucan. The degradation of cellulose by colonic bacteria depends on its degree of polymerization and crystallinity. Less crystalline forms of cellulose, as found in cabbage or salad, are more easily degraded than more crystalline forms such as bran or wood cellulose (Slavin et al. 1981; Van Soest 1984). Approximately one-third of the dietary cellulose and one half of the hemicellulose contained in a typical Indian diet (30–40 g of dietary fiber per day) was degraded during passage through the gastrointestinal tract of Indian vegetarians indicating the capacity of the gut microbiota to degrade these dietary fibers (Joshi and Agte 1995). However, it was also noted that the study participants displayed a high inter-individual variability in the degree of degradation.

Other non-starch polysaccharides include fructans such as inulin, which are polymers of $\alpha(1 \rightarrow 2)$-linked fructose and serve as reserve material in plants. They occur in the tubers of food plants such as Jerusalem artichoke and chicory, and are also found in onions, garlic, and wheat. Fructans such as inulin and oligofructose selectively stimulate the growth of intestinal bifidobacteria (Gibson et al. 1995). Food ingredients capable of selectively stimulating beneficial intestinal bacteria and thereby supporting host health have been termed prebiotics (Gibson 1999).

The range of dietary non-digestible polysaccharides is considerable. This is due to the large variety of carbohydrate monomers that can be linear or branched, linked in various ways, and can differ in their degree of polymerization. For example, galactomannans consist of an α (1–4)-linked backbone of mannose residues with α (1–6)-linked galactose side groups and occur in guar beans and carob. They are widely used in foods for increasing the viscosity of the liquid phase. Interestingly, in conjunction with arabinogalactans they are also found in coffee brew (Gniechwitz et al. 2007). Gut bacteria are capable of utilizing these carbohydrates as carbon and energy source (Reichardt et al. 2009).

5.2 *Enzymes and Pathways Involved in Bacterial Carbohydrate Breakdown in the Colon*

The above examples highlight the fact that a large variety of non-digestible polysaccharides present in various foods and beverages reach the colon, where they undergo fermentation by the gut microbiota. Recent metagenomic studies revealed that gut microbes are equipped with an arsenal of enzymes that catalyze the breakdown of dietary polysaccharides (Turnbaugh et al. 2006). Compared to all sequenced microbial genomes, the metagenome of the human colon is enriched with genes involved in carbohydrate breakdown while genes encoding other functions such as energy production and lipid metabolism are underrepresented (Gill et al. 2006). The distal human gut metagenome contains genes that represent more than 80 different glycoside hydrolase families, many of which are absent from the human genome (Gill et al. 2006). By providing these enzymes, the gut microbial community enables the host to take advantage of carbohydrates that would otherwise be excreted. In this way the gut microbiota helps the host in extracting as much nutrients from the diet as possible. Intestinal bacteria convert these carbohydrates mainly to short chain fatty acids (SCFA) such as acetate, propionate, and butyrate as well as to the gases CO_2, H_2 and CH_4. SCFA are absorbed from the colon and subsequently serve either as energy source in various tissues (acetate, butyrate) or are used for gluconeogenesis in the liver (propionate) (Cummings 1995).

Breakdown of plant polysaccarides by colonic bacteria occurs in several steps and involves the formation of a number of intermediates such as lactate, succinate, ethanol, and formate (Fig. 1). The first step in the bacterial breakdown of polysaccharides results in the formation of oligomeric and monomeric carbohydrates.

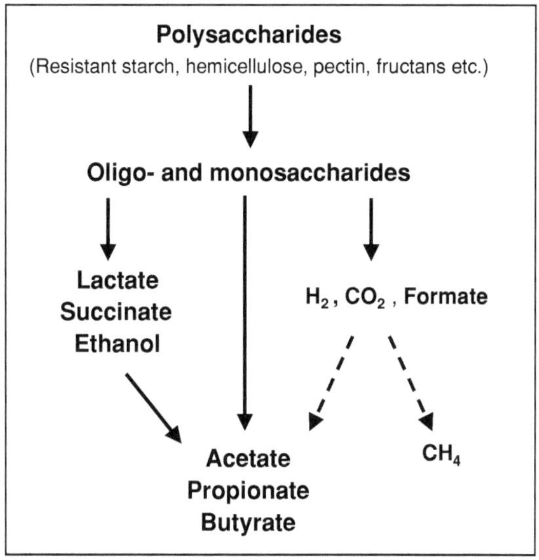

Fig. 1 Steps in the breakdown of polysaccharides by intestinal bacteria

Genera playing a major role in colonic polysaccharide depolymerization include *Bacteroides, Ruminococcus, Butyrivibrio, Eubacterium, Clostridium*, and *Bifidobacterium* (Salyers et al. 1977). However, this list is far from being complete. As mentioned, resistant starch is the most important energy source for the gut microbiota in people consuming a Western diet (Cummings 1994). The conversion of starch by *Bacteroides thetaiotaomicron* has been characterized in detail (Shipman et al. 2000); (Cho et al. 2001) and may serve as an example for the utilization of polysaccharides by *Bacteroides* spp. Uptake and degradation of starch by *B. thetaiotaomicron* involves two sets of proteins. The starch utilization system (Sus) encompasses eight proteins (SusA, B, C, D, E, F, G) including one regulatory protein (SusR). The other set of proteins belongs to the Mal system, which consists of the regulatory protein MalR and unknown proteins, such as an α-glucosidase, whose expression is regulated by MalR (Cho et al. 2001). Deletion of MalR reduces the expression of the *sus* genes, indicating that MalR together with SusR regulates the expression of *sus* genes. The exact mechanism how MalR affects *sus* gene expression is still unknown. All *sus* genes reside in one locus of the *B. thetaiotaomicron* chromosome (Reeves et al. 1997). Transcription of *susBCDEFG* to be consistent: 'susB-G' or susB, C, D, E, F, G' is controlled by SusR, which upon binding of maltose, maltotriose or longer oligmers acts as a transcriptional activator by binding to the promoter region located upstream of *susB* while *susA*, located upstream of *susB*, has its own promoter (Fig. 2). SusA, located in the periplasm, is a pullulanase that cleaves α-1,6 linkages as present in pullulan and amylopectin. SusB, located in the cytoplasm, is an α-glucosidase that cleaves α-1,4 linkages. SusC–F form a starch binding complex in the outer membrane of *B. thetaiotaomicron* (Shipman et al. 2000). SusG is essential for starch utilization as it catalyzes the hydrolysis of starch on the cell surface (Shipman et al. 1999). It is conceivable that this activity leads to

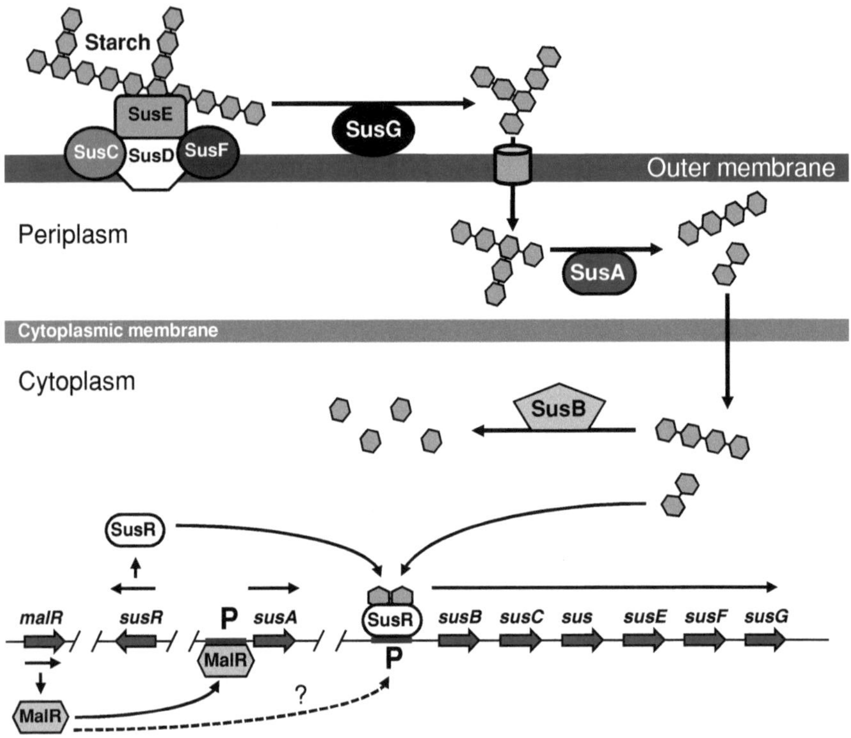

Fig. 2 Genes and enzymes involved in the degradation of starch by *Bacteroides thetaiotaomicron.* P, promoter

the formation of oligomeric degradation products that are capable of passing into the periplasm through porins in the outer membrane.

It may appear that starch is the preferred substrate of *B. thetaiotaomicron* in the gut environment, but experiments in monoassociated mice indicated that this organism does not use one preferred substrate but rather small amounts of various polysaccharides (Salyers and Pajeau 1989). This finding is in agreement with observations that have been made for *Escherichia coli*: based on a proteomic analysis, this organism utilizes a larger variety of substrates, including galactose, ribose, maltose, glucuronate, galacturonate, and gluconate in the intestinal environment as compared to in vitro conditions (Alpert et al. 2009).

B. thetaiotaomicron's ability to utilize a large variety of carbohydrates is also reflected by its genome. Genomic sequence analysis of this organism revealed genes predicted to encode a wide range of glycosylhydrolases including amylases, α- and β-glucosidases, α- and β-galactosidases, α-mannosidases, β-glucuronidases, β-fructofuranosidases, and endo-1,2-β-xylanases (Xu et al. 2003). The authors pointed out that the presence of such enzymes in *B. thetaiotaomicron* "exceeds that in any other sequenced bacteria, including other human gut commensals and symbionts" (Xu et al. 2003).

5.3 Conversion of Host-Derived Substrates

Saliva, gastric juice, and mucus are constantly secreted into the gastrointestinal tract. They all contain glycoproteins. Members of the gut microbial community are capable of utilizing glycoproteins as growth substrates. Glycoproteins are composed of a polypeptide backbone with oligosaccharide chains covalently attached to it via serine and threonine residues. The oligosaccharide side chains, which make up 70–80% (w/w) of the glycoprotein, typically contain glucose, galactose, mannose, fucose, *N*-acetylgalactosamine, *N*-acetylglucosamine, *N*-acetylneuraminic acid (sialic acid), and xylose. The carbohydrate moieties of certain glycoproteins in the intestinal tract can be acetylated or sulfated (Filipe 1979). One major intestinal mucin is MUC2, which is formed by goblet cells and plays a key role in the formation of two discernible mucus layers in the colon (Hansson and Johansson 2010). A stratified dense and cross-linked layer is firmly attached to the epithelium; a thicker mucus layer on top of it is less dense and nonattached. The lower mucus layer is impermeable to bacteria, while the upper layer is occupied by bacteria (Hansson and Johansson 2010), which use it for attachment (Kline et al. 2009) and as energy source. 23 human fecal isolates grown on gastric mucin produced various extracellular and cell-bound enzymes required for the degradation of mucin, including sialidase, sialate-*O*-acetylesterase, and *N*-acetylneuraminate lyase (Corfield et al. 1992). These isolates were identified as strains of *Bacteroides, Ruminococcus, Bifidobacterium,* and *Enterococcus faecalis*. Acetylated and sulfated mucins (natural form) were more slowly degraded than the corresponding deacetylated and desulfated forms (Corfield et al. 1992). The mucin protein is degraded to a lesser extent than the carbohydrate moieties (Variyam and Hoskins 1981). *Akkermansia muciniphila*, a member of the phylum Verrucomicrobia, has been identified as an important mucin-degrading member of the gut microbial community (Derrien et al. 2004). The organism colonizes the intestinal tract early in life and reaches concentrations of 10^7–10^8 cells/g feces in healthy adults (Collado et al. 2007).

More detailed data on the utilization of mucins are available for *Bacteroides thetaiotaomicron*: Transcriptional analysis revealed that this organism preferentially utilizes mucin *O*-glycans in the intestinal tract (Martens et al. 2008). Deletion of transcription factors that activate the expression of mucin *O*-glycan utilization genes reduced the persistence of this organism in the gut and its transmission from the mother to the offspring. Eighteen percent of all genes in the genome of *B. thetaiotaomicron* encode enzymes involved in the utilization of host and dietary glycans indicating the importance of these substrates for the establishment of this organism in the gut environment.

B. thetaiotaomicon even triggers the host to decorate the epithelium with fucose residues that it utilizes as a growth substrate (Bry et al. 1996). This fucosylation program starts in germfree mice on day 17 after birth but gradually ends unless the animals are colonized with a complete gut microbiota or with *B. thetaiotaomicron*. Induction of fucosyltransferases by transcriptional activation of the corresponding

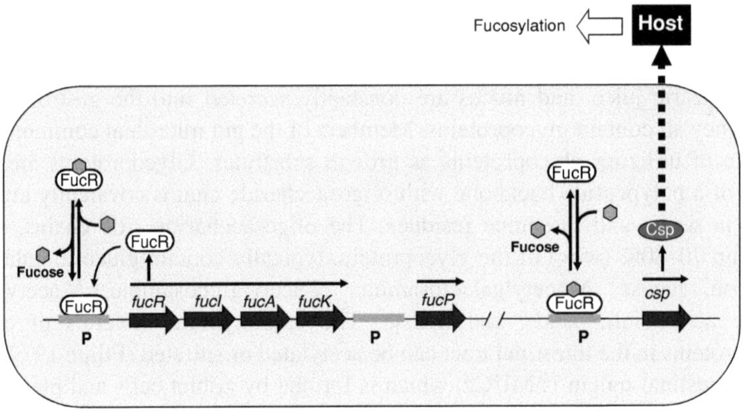

Fig. 3 Induction of host fucosyltransferases and utilization of fucose by *Bacteroides thetaiotaomicron*

genes affords fucosylation of the epithelial surface. *B. thetaiotaomicron* mutants devoid of their fucose-utilization genes are no longer capable of inducing epithelial fucosytransferases in the mouse ileum (Hooper et al. 1999). The authors proposed a model according to which *B. thetaiotaomicron* produces a transcriptional regulator (FucR) that is capable of sensing fucose (Fig. 3). In the absence of fucose, FucR binds to the promoter region of the fucose-utilization genes and thereby represses their transcription. If fucose is present it binds to FucR. As a result, FucR no longer blocks transcription of the *fuc* genes. In this case, fucose acts as an inducer (Hooper et al. 1999). The authors postulated the presence of a second locus on the *B. thetaiotaomicron* chromosome, called control of signal production (*csp*), whose expression is also regulated by FucR. Its gene product Csp signals to the host that fucosyltransferases are to be expressed. Fucose acts as a corepressor in *csp* transcription and in conjunction with FucR blocks *csp* transcription as long as fucose is available. In the absence of fucose, *csp* is transcribed because FucR alone does not bind to the *csp* promoter region and therefore no longer blocks transcription (Hooper et al. 1999).

Degradation of the host-derived glycoproteins chondroitin sulfate and hyaluronate by *B. thetaiotaomicron* has been studied in detail. As a major component of intestinal epithelium, chondroitin sulfate and hyaluronate contribute to maintaining tissue integrity. These polysaccharides are released into the intestinal lumen along with sloughed epithelial cells (Linn et al. 1983). In humans, intestinal epithelial cells are replaced every 4–5 days. Stem cells in the crypts differentiate to various cell types including epithelial cells. The latter migrate to the villus tips, from where they are sloughed off, and undergo bacterial degradation. Chondroitin sulfate has a molecular weight of approximately 10 kDa and consists of a chain of alternating *N*-acetylgalactosamine and glucuronic acid, both of which can be sulfated in various positions (Fig. 4). *N*-acetylgalactosamine may be sulfated in carbons 4, 6, or 4 and 6, while glucuronic acid may be sulfated in carbon 2.

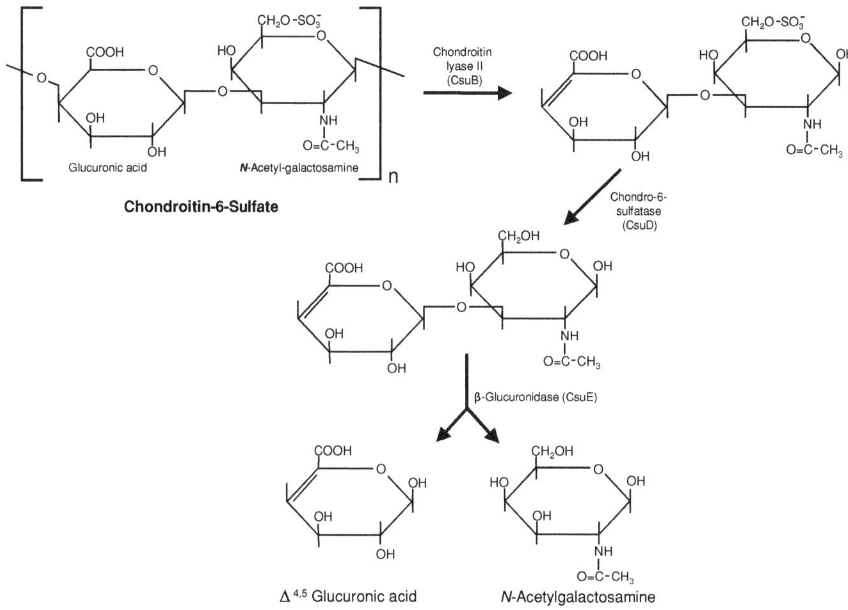

Fig. 4 Breakdown of chondroitin sulfate by *Bacteroides thetaiotaomicron*

The structure of hyaluronic acid is similar to that of chondroitin sulfate, except that it contains no sulfate groups and the *N*-acetylgalactosamine residues are replaced by *N*-acetylglucosamine. *B. thetaiotaomicron* produces two enzymes for the degradation of chondroitin sulfate: chondroitin lyase I and II (Linn et al. 1983). The two enzymes are located in the periplasm (Salyers and O'Brien 1980), are induced by chondroitin sulfate (Salyers and Kotarski 1980), have similar biochemical properties (Linn et al. 1983) and are encoded by the genes *csuA* and *csuB* (csu = chondroitin sulfate utilization) (Hwa and Salyers 1992). Other *csu* genes that are induced by chondroitin sulfate include *csuC*, *csuD*, and *csuE,* which encode a chondro-4-sulfatase, a chondro-6-sulfatase, and a *β*-glucuronidase respectively. All three enzymes have a cytoplasmic localization. Chondroitin sulfate not only induces the expression of *csuA*, *csuB*, *csuD* and *csuE* but also of *csuF*, which encodes an outer membrane protein (Salyers and O'Brien 1980) that is essential for growth on this sulfated polysaccharide and on hyaluronic acid (Cheng et al. 1995). Mutants defective in *csuF* cannot grow on intact chondroitin sulfate or hyaluronic sulfate, but still grow on their disaccharide components. Analogous to the roles of SusC-F in the starch degradation by *B. thetaiotaomicron,* CsuF has been proposed to bind chondroitin sulfate and hyaluronic acid to the cell surface and to deliver the two polysaccharides to the periplasmic chondroitin lyases (CsuA and CsuB) (Kotarski et al. 1985). The latter cleave chondroitin sulfate and hyaluronic acid to the corresponding disaccharides. These are subsequently taken up into the cytoplasm where the disaccharide originating from chondroitin sulfate undergoes desulfation by CsuC and CsuD. The resulting

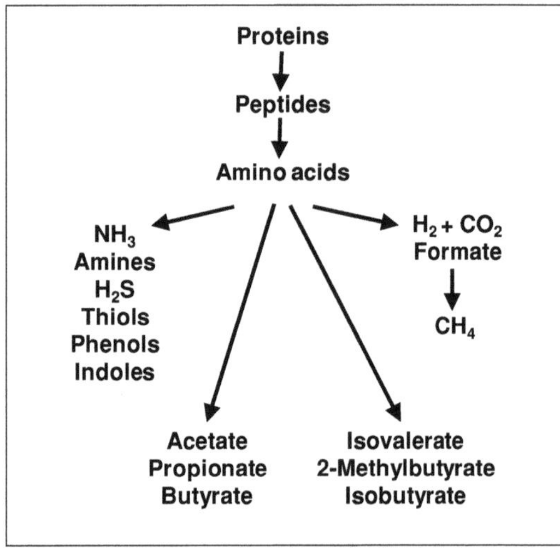

Fig. 5 Steps in the breakdown of proteins by intestinal bacteria

disaccharides are cleaved by CsuE to the corresponding monomers (Fig. 4). The expression of enzymes involved in chondroitin sulfate utilization are controlled by the regulatory protein ChuR except the β-glucuronidase CsuE (Cheng et al. 1992).

5.4 Utilization of Proteins by the Gut Microbiota

Even though the digestion of proteins in the human intestinal tract is highly efficient, it has been estimated that 3–12 g of dietary protein, in particular myofibrillar and sarcoplasmatic muscle protein, daily escapes digestion in the small intestine (Cummings 1994). Moreover, 4–6 g of digestive host enzymes, such as trypsin, chymotrypsin, elastase, lipase, amylase, and nucleic acid hydrolases, pass into the colon, where they undergo fermentation by intestinal bacteria. Proteolytic activity decreases along the intestinal tract because digestive enzymes including proteases undergo digestion during intestinal passage (Gibson et al. 1989). Even though the majority of gut bacteria prefer carbohydrates as an energy source, protein is nevertheless an important bacterial substrate, in particular in the distal colon. There, carbohydrates are largely exhausted, as evident from the analysis of intestinal contents of sudden death victims (Macfarlane et al. 1986). Protein degradation in the colon occurs in several steps and involves different bacteria. In the first step protein is hydrolyzed to oligopeptides and amino acids (Fig. 5). Strains of *Bacteroides* spp. including *B. fragilis* and *B. vulgatus* produce cell-bound proteases while strains of the genera *Clostridium*, *Propionibacterium*, and *Bacillus* produce extracellular proteases. Fecal *Streptococcus* and *Staphylococcus* isolates have both forms of proteases (Macfarlane et al. 1986). The peptides and amino acids formed in this process may serve as carbon, nitrogen, and energy source. If peptides serve as nitrogen source, intestinal bacteria such as

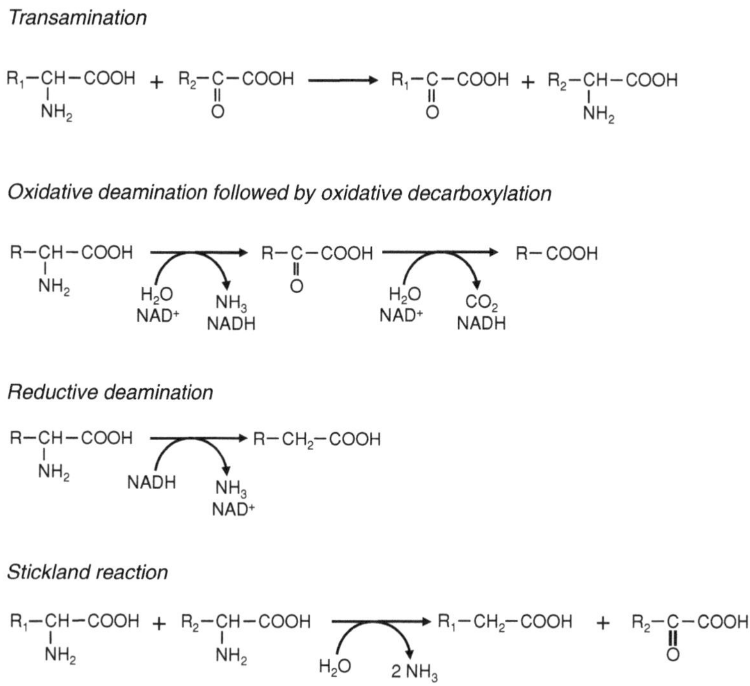

Fig. 6 Principal reactions used by gut bacteria for the degradation of amino acids

bifidobacteria and clostridia often utilize only certain amino acids of a peptide and excrete the remaining amino acids (Hespell and Smith 1983). Degradation of amino acids by gut bacteria involves various types of reactions: transamination, oxidative deamination, oxidative decarboxylation, reductive deamination, and Stickland reaction (Fig. 6). Conversion of amino acids leads to the formation of short chain fatty acids and a number of products characteristic of bacterial protein utilization in the colon, including ammonia, the branched-chain fatty acids isovalerate and isobutyrate, amines, polyamines, hydrogen sulfide, thiols, phenols, and indoles (Fig. 5). Ammonia is formed by oxidative or reductive deamination, while amines and polyamines arise from the decarboxylation of amino acids. Deamination and decarboxylation of valine, isoleucine, and leucine lead to the formation of isovalerate (3-methylbutyrate), isobutyrate (2-methylpropionate), and 2-methylbutyrate. Branched-chain fatty acids have been taken as indicators of bacterial proteolysis in the colon (Macfarlane et al. 1986).

Phenols and indoles arise from the conversion of the aromatic amino acids tyrosin, phenylalanine, and tryptophan. Hydrogen sulfide and thiols originate from cysteine and methionine. Many intestinal bacteria including *Bacteroides* spp. and *Escherichia coli* are capable of fermenting amino acids in addition to carbohydrates (Hespell and Smith 1983). Some members of the genera *Fusobacterium, Clostridium, Acidaminococcus*, and *Veillonella* even prefer amino acids over carbohydrates for energy generation. The degradation pathways for a given amino acid may differ from species to species. In general, knowledge about the utilization of proteins, peptides, and amino acids is limited compared to that of carbohydrates.

5.5 *Conversion of Non-Nutritive Substances by Intestinal Bacteria*

The intestinal microbiota is also confronted with a whole range of non-nutritive substances ingested with the diet or produced by the host. Non-nutritive substances include dietary secondary plant metabolites. Bile acids are an example for host-derived non-nutritive substances.

5.5.1 Bile Acids

Bile acids are synthesized in the liver involving the cytochrome P450-dependent oxidation of cholesterol and the conjugation of the resulting hydroxylated bile acids acids with glycine or taurine. These primary bile acids, taurocholic acid and glycocholic acid, are stored in the gall bladder, from where they are excreted into the gut lumen to solubilize dietary fat and support the formation of micelles. Many intestinal bacteria are capable of deconjugating the primary bile acids to the corresponding unconjugated forms. Genes encoding bile salt hydrolases have been detected in *Clostridium perfringens, Lactobacillus plantarum, Lactobacillus johnsonii, Bifidobacterium longum, Bifidobacterium bifidum*, and *Bifidobacterium adolescentis* (Ridlon et al. 2006) The unconjugated bile acids can subsequently be dehydroxylated by bacterial dehydroxylases. Bacterial bile acid dehydroxylase were first detected in *Eggerthella lenta* (Bokkenheuser et al. 1979; MacDonald et al. 1979). However, the list of intestinal bacteria possessing enzymes involved in bile acid dehydroxylation now includes *C. perfringens, Blautia producta* (previously *Peptostreptococcus productus*), *B. fragilis, B. thetaiotaomicron, E. coli, Clostridium absonum, Clostridium sordellii, Clostridium innoccuum, Clostridium scincens, Clostridium bifermentans, Clostridium limosum, Clostridium leptum,* and *Clostridium paraputrificum* (Ridlon et al. 2006). The respective enzymes are stereo-selective and specific for the position of the hydroxyl group. Dehydroxylation of bile acids involves several steps that have been studied in *Clostridium scindens* in detail. The genes encoding the proteins required for bile acid dehydroxylation are organized in the bile acid-inducible (*bai*) operon, which encompasses eight genes (*baiB*, *CD*, *E*, *A*, *F*, *G*, *H*, *I*). The functions of the gene products are as follows: BaiA is a 27 kDa 3α-hydroxysteroid dehydrogenase. BaiB is a 58 kDa bile acid CoA ligase. BaiCD is a 70 kDa 3-dehydro-4-chenodeoxycholic acid/cholic acid steroid oxidoreductase. BaiH is a 72 kDa 3-dehydro-4-ursodeoxycholic acid/7-epi cholic acid steroid oxidoreductase. BaiE is a 19.5 kDa 7α-dehydratase. BaiI is a hypothetical 22 kDa 7β-dehydratase, BaiF is a 47.5 kDa bile acid CoA hydrolase and a hypothetical bile acid CoA transferase. BaiG is a

50 kDa transmembrane protein catalyzing H^+-dependent bile acid transport (Ridlon et al. 2006).

A variable proportion of the resulting secondary bile acids, deoxycholic acid, and lithocholic acid, is re-absorbed, conjugated in the liver, and secreted again into the intestine. This process is referred to as enterohepatic circulation. Whether the bacteria benefit from the conversion of primary to secondary bile acids is still a matter of speculation. Proposed benefits include the utilization of glycine from glycocholic acid as an energy, carbon and/or nitrogen source and of taurine from taurodeoxycholate as a sulfur source (Ridlon et al. 2006). A role of taurine as a nitrogen source is supported by the observation that the bile salt hydrolase gene from *B. longum* is co-transcribed with the glutamine synthetase adenylyltransferase gene (*glnE*), which is part of the nitrogen regulation cascade (Tanaka et al. 2000). Deconjugation followed by dehydroxylation has also been proposed to be a mechanism of bile acid detoxification. Conjugated bile acids are less cytotoxic than the free bile acids. However, the latter undergoes a more efficient fecal excretion because following their dehydroxylation they precipitate more easily and bind to dietary fiber. The formation of secondary bile acids may also be a mechanism to eliminate bacteria that are susceptible to these detergents (Ridlon et al. 2006).

5.5.2 Secondary Plant Metabolites

A large proportion of the human diet stems from plants, which in addition to nutrients contain secondary plant metabolites. The latter are thought to play a role in the defense of the plant against herbivores. Examples of such metabolites found in plants of nutritional relevance include glucosinolates in cruciferous vegetables and polyphenols in grains, fruit, and vegetables. Fate and bioavailability of some of these substances have been investigated in detail because of their purported beneficial health effects. In the case of lignans and isoflavones (a subgroup of the flavonoids) these effects have in part been connected to their antioxidant activity and their estrogen-like structure. Owing to the latter feature they have collectively been called phytoestrogens (Adlercreutz and Mazur 1997). Purported beneficial effects include the protection against sex-hormone-dependent cancers and cardiovascular diseases (Arts and Hollman 2005).

Polyphenols comprise several hundred different substances. The fate of these compounds in the gastrointestinal tract depends on their structure and chemical features. Hardly any of them is completely absorbed in the small intestine. In nature, polyphenols occur in their glycosylated form and usually have to be deglycosylated before being absorbed. Dependent on their structure, a considerable proportion of polyphenolic substances escape digestion and become available for intestinal bacteria. Bacterial conversion of some polyphenolic model compounds has been studied in detail. For example, the lignan secoisolariciresinol diglucoside (SDG), which occurs in flaxseed, rye, blackberry, broccoli, and garlic, is converted to enterolactone (Axelson et al. 1982). A number of fecal bacteria

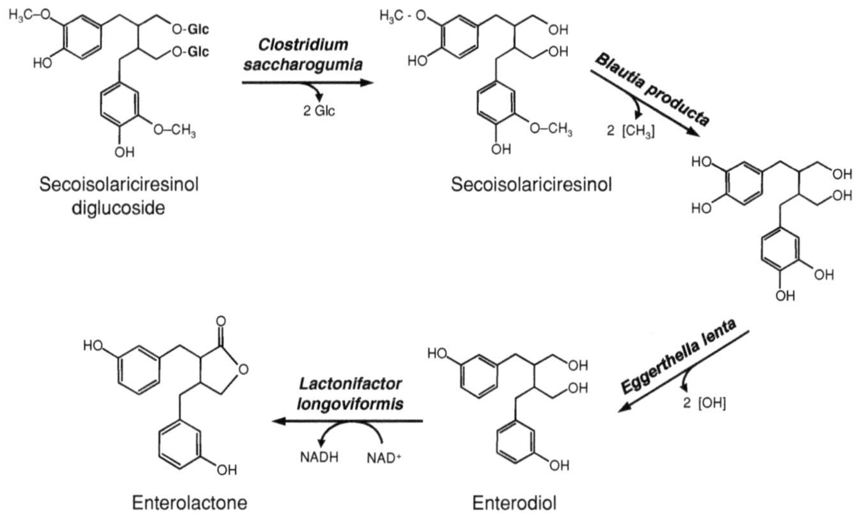

Fig. 7 Reactions involved in the bioactivation of the plant lignan secoisolariciresinol to the mammalian lignan enterolactone and intestinal bacterial species catalyzing the various steps

involved in the various steps of this transformation have been isolated (Clavel et al. 2006a, 2007). The first step, the deglycosylation of SDG, is catalyzed by various gut bacteria, including several *Bacteroides* and *Clostridium* species, with *Clostridium saccharogumia* being particularly effective (Clavel et al. 2006b, 2007). The second step in this transformation, the *O*-demethylation of matairesinol, is catalyzed by isolates identified as *Butyribacterium methylotrophicum, Eubacterium calendari, Eubacterium limosum*, and *Blautia producta*. Step three, the dehydroxlyation of the *O*-demethylated matairesinol to enterodiol, is catalyzed by *Clostridium scindens* and *Eggerthella lenta*, organisms that are capable of dehydroxylating bile acids (see above). Only one bacterial strain that catalyzes the last step of this conversion, the dehydrogenation of enterodiol to enterolactone, has so far been isolated (Fig. 7). This isolate was subsequently described as *Lactonifactor longoviformis* (Clavel et al. 2007). Gnotobiotic rats associated with a defined consortium of four species, each of which catalyzes one of the four steps, excrete enterodiol and enterolactone in their urine and feces when fed a flaxseed diet, indicating their ability to transform lignans in vivo. In contrast, germfree rats fed the same diet do not do so (Woting et al. 2010). Conversion of SDG to the estrogen-like metabolites ED and EL was found to be widely distributed among humans. Interestingly, women tended to harbor higher concentrations of both ED- and EL-producing organisms (Clavel et al. 2005). The bacterial conversion of SDG to enterodiol and enterolactone is of relevance because the bacterial transformation products have a higher biological activity than the corresponding plant lignans.

Daidzein, genistein, and their glycosides are the most common isoflavones present in soy and they may therefore be consumed in considerable amounts. Similar to the lignans, they have been implicated in the prevention of hormone-related cancers and cardiovascular diseases as well as the alleviation of menopausal symptoms. In the case of daidzein, these effects have mainly been attributed to one of its bacterial transformation products, equol, which undergoes urinary excretion (Setchell and Clerici 2010). Equol formation was observed in 33–50% of healthy adults challenged with soy, while *O*-desmethylangolensin was produced by 80–90% of the tested subjects (Atkinson et al. 2004). The underlying inter-individual differences may be due to a competition between equol- and *O*-desmethylangolensin-producing intestinal bacteria. *Eubacterium ramulus* converts daidzein to the biologically inactive *O*-desmethylangolensin (Schoefer et al. 2002). In contrast, *Adlercreutzia equolifaciens* (Maruo et al. 2008) and *Slackia isoflavoniconvertens* (Matthies et al. 2009) convert daidzein to the biologically active equol. The latter moreover converts genistein to 5-hydroxy-equol. Daidzein and genistein are first reduced to dihydrodaidzein and dihydrogenistein, respectively, and subsequently to equol and 5-hydroxy-equol, respectively.

Plants belonging to the Ericaceae, Asteraceae, and Rosaceae may contain considerable levels of arbutin (hydroquinone-β-D-glucopyranoside) in their leaves and fruits (Deisinger et al. 1996). For example, pears may contain up to 4.8 mg per portion. Ingested arbutin passes into the colon where intestinal bacteria such as *Eubacterium ramulus, Enterococcus casseliflavus*, *Bacteroides distasonis*, and *Bifidobacterium adolescentis*, deglycosylate arbutin to hydroquinone (Blaut et al. 2006). Hydroquinone is mutagenic to eukaryotic cells, indicating the potential genotoxicity of dietary arbutin and the ability of intestinal bacteria to activate it.

These examples show that the conversion of bioactive food constituents by intestinal bacteria may lead to the formation of potentially beneficial or adverse substances. It should be kept in mind that the examples given here cover only a small proportion of substances that undergo bacterial conversion in the intestine. The structural diversity of non-nutritive food components is immense. In contrast, the intestinal bacterial enzymes acting on these substances catalyze only a limited number of reactions including deglycosylation, deconjugation, demethylation, dehydroxylation, dehydrogenation, reduction, and demethylation.

5.5.3 Drugs

Drugs prescribed to patients to alleviate disease symptoms or to cure diseases are xenobiotics that undergo conversion by both host enzymes and the gut microbiota. Hydrophobic drugs absorbed in the small intestine are subsequently oxidized in epithelial cells or in the liver by microsomal Cytochrome P450-dependent monooxygenases and the resulting hydroxylated products are subsequently conjugated with glucuronic acid, sulfate, or glutathione. Depending on their structure and size, the resulting molecules undergo urinary or biliary excretion. Conjugated xenobiotics secreted with the bile may be deconjugated to the corresponding unconjugated compounds. The latter may be reabsorbed (enterohepatic circulation), as described above for the bile acids. In this way the intestinal microbiota influences the retention period of drugs in the human body.

5.6 Core Activities Versus Specialized Activities

While many metabolic activities of the gut microbiota are found in every subject, others are only present in some individuals. The degradation of non-digestible carbohydrates and proteins is a general feature of the human gut microbiota. Conversely, metabolic activities such as methanogenesis, equol formation, or oxalate degradation are found in some individuals but not at all or at only very low levels in others (Atkinson et al. 2004; Wolin and Miller 1983; Kumar et al. 2002). Metagenomic analysis revealed that human individuals share a wide range of intestinal microbial genes, referred to as core microbiome; it was concluded that microbial communities differing in species composition may nevertheless share a core microbiome at a functional level (Turnbaugh et al. 2009). More recent work supports the concept of a functional core metagenome which is shared by essentially every individual and which is required for the gut microbial community to function properly (Qin et al. 2010). However, it is also important to note that the function of a large proportion of the detected genes is still unknown.

6 Ecological Relationships

Intestinal bacteria may live in a fierce competition for substrates available in the gut ecosystem but they may also have a cooperative relationship. For example, different bacteria are capable of utilizing the same carbohydrate as a substrate. In this situation they are competitors. In contrast, bacteria that utilize lactate, succinate, or ethanol profit from those bacteria that produce these fermentation products in the colon. While most lactate and succinate-utilizing bacteria are capable of utilizing alternative substrates, methanogenic archaea in the colon (*Methanobrevibacter smithii*) are absolutely dependent on the formation of H_2 or formate by other organisms in the ecosystem.

Not only bacterial population groups may profit from each other but also the host and his/her intestinal microbes. Such a relationship is referred to as mutualistic. A commensal relationship exists if only one of the partners benefits while the other one is not affected. The character of host-microbe interactions may differ in that certain aspects of their relationship may be considered mutualistic while others may be considered commensal.

The gut is a very dynamic microbial ecosystem (Ley et al. 2006a). Autochthonous bacteria are permanent members of the community. They are optimally adapted to the habitat and prevail in a given niche. In contrast, bacteria ingested with food and drinks usually do not establish in the ecosystem permanently. These allochthonous bacteria are transient members of the community. Since they are not optimally adapted to the environment they sooner or later are out-competed by other community members and eventually washed out. Microbial diversity and the functional redundancy may contribute to the stability of the gut microbial ecosystem as proposed for the earth ecosystem (McCann 2000).

7 Microbial Activities Related to Host Health and Disease

Owing to their immense catalytic potential and their multifaceted interactions with the host, intestinal bacteria have been implicated in disease prevention but also in disease development. Positive aspects of this double-edged role include the fortification of the barrier function against pathogens (Wells et al. 2011), salvage of energy by converting dietary fiber to short chain fatty acids (Cummings et al. 1987), activation of health-promoting food constituents (Clavel et al. 2005), contribution to the maturation and maintenance of the immune system (Bry et al. 1996; Round and Mazmanian 2009) and regulation of the differentiation and growth of gut epithelial cells (Falk et al. 1998). Adverse effects of gut bacteria include a role in the onset of inflammatory bowel diseases (ulcerative colitis and Crohn's disease) (Sartor 2006), formation of reactive metabolites that may lead to colorectal cancer (Hughes et al. 2000), and obesity (Ley et al. 2006b). Final experimental proof for a preventive or causative role of the gut microbiota in any of these diseases is not available but circumstantial evidence is in support of this notion. For example, *Faecalibacterium prausnitzii* has been implicated in the prevention of inflammatory bowel diseases (Sokol et al. 2008, 2009). However, the underlying mechanism is still obscure.

An interesting case is the proposed role of *Oxalobacter formigenes* in the formation of renal stones. *O. formigenes*, which has been isolated from the human gut and the bovine rumen (Allison et al. 1985) gains energy by cleavage of oxalate to formate and CO_2. Human studies suggest that the absence of *O. formigenes* may be an important risk factor for recurrence of renal stones composed of calcium oxalate (Mittal et al. 2003). Patients at high risk for calcium oxalate kidney stones have low concentrations of *O. formigenes* in their gut and they display higher serum oxalate concentrations, indicating that this organism plays a role in oxalate homeostasis (Sidhu et al. 1999; Troxel et al. 2003).

8 Concluding Remarks

It is increasingly recognized that the microbial community inhabiting the human intestinal tract profoundly affects host physiology. Metagenomic analyses not only led to the definition of a core microbiome but also to the discovery that the microbiome of humans belongs to one of three enterotypes (Arumugam et al. 2011). However, there is still a gap in our understanding of the molecular mechanisms underlying host-microbe interactions. Neither is the exact role of most members of the gut microbiota in the gastrointestinal tract known nor has the basis of their competitive or cooperative interactions and the consequences been identified. Moreover, bacterial activities relevant for host health and disease have not fully been recognized. Recent studies found correlations between the gut microbiome and various diseases, but in most cases it has remained unclear whether the observed differences between diseased and healthy subjects are the cause or rather the consequence of the disease. It therefore remains necessary to continue intense research in both humans and animal models. Especially animals with a defined microbiota (gnotobiotic animals) offer the opportunity to test hypotheses that emerge from human studies and to identify bacterial factors underlying host-microbe interactions. Since the functions of a large proportion of genes in the human metagenome are still unknown, it also is mandatory to unravel their function. All of these investigations will contribute to a better understanding of host-microbe interactions and will help in developing strategies for manipulating the gut microbiota in a targeted way.

References

Adlercreutz H, Mazur W (1997) Phyto-oestrogens and Western diseases. Ann. Med. 29:95–120

Allison MJ, Dawson KA, Mayberry WR et al (1985) *Oxalobacter formigenes* gen. nov., sp. nov.: oxalate-degrading anaerobes that inhabit the gastrointestinal tract. Arch Microbiol 141:1–7

Alpert C, Scheel J, Engst W et al (2009) Adaptation of protein expression by *Escherichia coli* in the gastrointestinal tract of gnotobiotic mice. Environ Microbiol 11:751–761

Amann R, Springer N, Ludwig W et al (1991) Identification in situ and phylogeny of uncultured bacterial endosymbionts. Nature 351:161–164

Arts IC, Hollman PC (2005) Polyphenols and disease risk in epidemiologic studies. Am J Clin Nutr 81:317S–325S

Arumugam M, Raes J, Pelletier E et al (2011) Enterotypes of the human gut microbiome. Nature 473:174–180

Atkinson C, Berman S, Humbert O et al (2004) In vitro incubation of human feces with daidzein and antibiotics suggests interindividual differences in the bacteria responsible for equol production. J Nutr 134:596–599

Axelson M, Sjövall J, Gustafsson BE et al (1982) Origin of lignans in mammals and identification of a precursor from plants. Nature 298:659–660

Baughn AD, Malamy MH (2004) The strict anaerobe Bacteroides fragilis grows in and benefits from nanomolar concentrations of oxygen. Nature 427:441–444

Bik EM, Eckburg PB, Gill SR et al. (2006) Molecular analysis of the bacterial microbiota in the human stomach. In: Proceedings of National Academy of Science U S A 103:732–737

Blaut M, Collins MD, Welling GW et al (2002) Molecular biological methods for studying the gut microbiota: the EU human gut flora project. Br J Nutr 2:S203–S211

Blaut M, Braune A, Wunderlich S et al (2006) Mutagenicity of arbutin in mammalian cells after activation by human intestinal bacteria. Food Chem Toxicol 44:1940–1947

Bokkenheuser VD, Winter J, Finegold SM et al (1979) New markers for Eubacterium lentum. Appl Environ Microbiol 37:1001–1006

Bry L, Falk PG, Midtvedt T et al (1996) A model of host-microbial interactions in an open mammalian ecosystem. Science 273:1380–1383

Cheng Q, Hwa V, Salyers AA (1992) A locus that contributes to colonization of the intestinal tract by Bacteroides thetaiotaomicron contains a single regulatory gene (chuR) that links two polysaccharide utilization pathways. J Bacteriol 174:7185–7193

Cheng Q, Yu MC, Reeves AR et al (1995) Identification and characterization of a Bacteroides gene, csuF, which encodes an outer membrane protein that is essential for growth on chondroitin sulfate. J Bacteriol 177:3721–3727

Chiang YC, Pai WY, Chen CY et al (2008) Use of primers based on the heat shock protein genes hsp70, hsp40, and hsp10, for the detection of bovine mastitis pathogens *Streptococcus agalactiae*, *Streptococcus uberis* and *Streptococcus bovis*. Mol Cell Probes 22:262–266

Cho KH, Cho D, Wang GR et al (2001) New regulatory gene that contributes to control of *Bacteroides thetaiotaomicron* starch utilization genes. J Bacteriol 183:7198–7205
Clavel T, Henderson G, Alpert CA et al (2005) Intestinal bacterial communities that produce active estrogen-like compounds enterodiol and enterolactone in humans. Appl Environ Microbiol 71:6077–6085
Clavel T, Henderson G, Engst W et al (2006a) Phylogeny of human intestinal bacteria that activate the dietary lignan secoisolariciresinol diglucoside. FEMS Microbiol Ecol 55:471–478
Clavel T, Borrmann D, Braune A et al (2006b) Occurrence and activity of human intestinal bacteria involved in the conversion of dietary lignans. Anaerobe 12:140–147
Clavel T, Lippman R, Gavini F et al (2007) *Clostridium saccharogumia* sp. nov. and *Lactonifactor longoviformis* gen. nov., sp. nov., two novel human faecal bacteria involved in the conversion of the dietary phytoestrogen secoisolariciresinol diglucoside. Syst Appl Microbiol 30:16–26
Collado MC, Derrien M, Isolauri E et al (2007) Intestinal integrity and *Akkermansia muciniphila*, a mucin-degrading member of the intestinal microbiota present in infants, adults, and the elderly. Appl Environ Microbiol 73:7767–7770
Corfield AP, Wagner SA, Clamp JR et al (1992) Mucin degradation in the human colon: production of sialidase, sialate O-acetylesterase, N-acetylneuraminate lyase, arylesterase, and glycosulfatase activities by strains of fecal bacteria. Infect Immun 60:3971–3978
Cummings JH (1994) Anatomy and physiology of the human colon. In: ILSI Workshop on colonic microflora: nutrition and health, Barcelona, Spain
Cummings JH (1995) Short chain fatty acids. In: Gibson GR, Macfarlane GT (eds) Human colonic bacteria: role in nutrition physiology and pathology. CRC Press, Boca Raton, pp 101–130
Cummings JH, Pomare EW, Branch WJ et al (1987) Short chain fatty acids in human large intestine, portal, hepatic and venous blood. Gut 28:1221–1227
Deisinger PJ, Hill TS, English JC (1996) Human exposure to naturally occurring hydroquinone. J Toxicol Environ Health 47:31–46
Derrien M, Vaughan EE, Plugge CM et al (2004) *Akkermansia muciniphila* gen. nov., sp. nov., a human intestinal mucin-degrading bacterium. Int J Syst Evol Microbiol 54:1469–1476
Doré J, Sghir A, Hannequart-Gramet G et al (1998) Design and evaluation of a 16S rRNA-targeted oligonucleotide probe for specific detection and quantitation of human faecal Bacteroides populations. Syst Appl Microbiol 21:65–71
Eckburg PB, Bik EM, Bernstein CN et al (2005) Diversity of the human intestinal microbial flora. Science 308:1635–1638
Falk PG, Hooper LV, Midtvedt T et al (1998) Creating and maintaining the gastrointestinal ecosystem: what we know and need to know from gnotobiology. Microbiol Mol Biol Rev 62:1157–1170
Filipe MI (1979) Mucins in the human gastrointestinal epithelium: a review. Invest Cell Pathol 2:195–216
Finegold SM, Attebery HR, Sutter VL (1974) Effect of diet on human fecal flora: comparison of Japanese and American diets. Am J Clin Nutr 27:1456–1469
Finegold SM, Sutter VL, Mathisen GE (1983) Normal indigenous intestinal flora. In: Hentges DJ (ed) Human intestinal microflora in health and disease. Academic Press, New York, pp 3–31
Franks AH, Harmsen HJM, Raangs GC et al (1998) Variations of bacterial populations in human feces measured by fluorescent in situ hybridization with group-specific 16S rRNA-targeted oligonucleotide probes. Appl Environ Microbiol 64:3336–3345
Gibson GR (1999) Dietary modulation of the human gut microflora using the prebiotics oligofructose and inulin. J Nutr 129:1438S–1441S
Gibson SA, McFarlan C, Hay S et al (1989) Significance of microflora in proteolysis in the colon. Appl Environ Microbiol 55:679–683
Gibson GR, Beatty ER, Wang X et al (1995) Selective stimulation of *bifidobacteria* in the human colon by oligofructose and inulin. Gastroenterology 108:975–982
Gill SR, Pop M, Deboy RT et al (2006) Metagenomic analysis of the human distal gut microbiome. Science 312:1355–1359

Gniechwitz D, Brueckel B, Reichardt N et al (2007) Coffee dietary fiber contents and structural characteristics as influenced by coffee type and technological and brewing procedures. J Agric Food Chem 55:11027–11034
Hansson GC, Johansson ME (2010) The inner of the two Muc2 mucin-dependent mucus layers in colon is devoid of bacteria. Gut Microbes 1:51–54
Hespell RB, Smith CJ (1983) Utilization of nitrogen sources by gastrointestinal tract bacteria. In: Hentges DJ (ed) Human intestinal microflora in health and disease. Academic Press, New York, p 21
Hillman K, Whyte AL, Stewart CS (1993) Dissolved oxygen in the porcine gastrointestinal tract. Lett. Appl. Microbiol. 16:299–302
Holdeman LV, Moore WE (1972) Roll-tube techniques for anaerobic bacteria. Am J Clin Nutr 25:1314–1317
Hooper LV, Xu J, Falk PG et al. (1999) A molecular sensor that allows a gut commensal to control its nutrient foundation in a competitive ecosystem. In: In: Proceedings of National Academy of Science U S A 96:9833–9838
Hughes R, Magee EA, Bingham S (2000) Protein degradation in the large intestine: relevance to colorectal cancer. Curr Issues Intest Microbiol 1:51–58
Hungate RE (1950) The anaerobic mesophilic cellulolytic bacteria. Bacteriol Rev 14:1–49
Hwa V, Salyers AA (1992) Evidence for differential regulation of genes in the chondroitin sulfate utilization pathway of *Bacteroides thetaiotaomicron*. J Bacteriol 174:342–344
Joshi S, Agte V (1995) Digestibility of dietary fiber components in vegetarian men. Plant Foods Hum Nutr 48:39–44
Kline KA, Falker S, Dahlberg S et al (2009) Bacterial adhesins in host-microbe interactions. Cell Host Microbe 5:580–592
Kotarski SF, Linz J, Braun DM et al (1985) Analysis of outer membrane proteins which are associated with growth of *Bacteroides thetaiotaomicron* on chondroitin sulfate. J Bacteriol 163:1080–1086
Kumar R, Mukherjee M, Bhandari M et al (2002) Role of *Oxalobacter formigenes* in calcium oxalate stone disease: a study from North India. Eur Urol 41:318–322
Ley RE, Peterson DA, Gordon JI (2006a) Ecological and evolutionary forces shaping microbial diversity in the human intestine. Cell 124:837–848
Ley RE, Turnbaugh PJ, Klein S et al (2006b) Human gut microbes associated with obesity. Nature 444:1022–1023
Linn S, Chan T, Lipeski L et al (1983) Isolation and characterization of two chondroitin lyases from *Bacteroides thetaiotaomicron*. J Bacteriol 156:859–866
MacDonald IA, Jellett JF, Mahony DE et al (1979) Bile salt 3 alpha- and 12 alpha-hydroxysteroid dehydrogenases from *Eubacterium lentum* and related organisms. Appl Environ Microbiol 37:992–1000
Macfarlane GT, Cummings JH, Allison C (1986) Protein degradation by human intestinal bacteria. J Gen Microbiol 132:1647–1656
Martens EC, Chiang HC, Gordon JI (2008) Mucosal glycan foraging enhances fitness and transmission of a saccharolytic human gut bacterial symbiont. Cell Host Microbe 4:447–457
Maruo T, Sakamoto M, Ito C et al (2008) *Adlercreutzia equolifaciens* gen. nov., sp. nov., an equol-producing bacterium isolated from human faeces, and emended description of the genus *Eggerthella*. Int J Syst Evol Microbiol 58:1221–1227
Matthies A, Blaut M, Braune A (2009) Isolation of a human intestinal bacterium capable of daidzein and genistein conversion. Appl Environ Microbiol 75:1740–1744
McCann KS (2000) The diversity-stability debate. Nature 405:228–233
Mittal RD, Kumar R, Mittal B et al (2003) Stone composition, metabolic profile and the presence of the gut-inhabiting bacterium *Oxalobacter formigenes* as risk factors for renal stone formation. Med Princ Pract 12:208–213
Qin J, Li R, Raes J et al (2010) A human gut microbial gene catalogue established by metagenomic sequencing. Nature 464:59–65

Rajilic-Stojanovic M, Heilig HG, Molenaar D et al (2009) Development and application of the human intestinal tract chip, a phylogenetic microarray: analysis of universally conserved phylotypes in the abundant microbiota of young and elderly adults. Environ Microbiol 11:1736–1751

Reeves AR, Wang GR, Salyers AA (1997) Characterization of four outer membrane proteins that play a role in utilization of starch by *Bacteroides thetaiotaomicron*. J Bacteriol 179:643–649

Reichardt N, Gniechwitz D, Steinhart H et al (2009) Characterization of high molecular weight coffee fractions and their fermentation by human intestinal microbiota. Mol Nutr Food Res 53:287–299

Ridlon JM, Kang DJ, Hylemon PB (2006) Bile salt biotransformations by human intestinal bacteria. J Lipid Res 47:241–259

Round JL, Mazmanian SK (2009) The gut microbiota shapes intestinal immune responses during health and disease. Nat Rev Immunol 9:313–323

Sajilata MG, Singhal RS, Kulkarni PR (2006) Resistant starch - a review. Compr Rev Food Sci F 5:1–17

Salyers AA, Kotarski SF (1980) Induction of chondroitin sulfate lyase activity in *Bacteroides thetaiotaomicron*. J Bacteriol 143:781–788

Salyers AA, O'Brien M (1980) Cellular location of enzymes involved in chondroitin sulfate breakdown by Bacteroides thetaiotaomicron. J Bacteriol 143:772–780

Salyers AA, Pajeau M (1989) Competitiveness of different polysaccharide utilization mutants of Bacteroides thetaiotaomicron in the intestinal tracts of germfree mice. Appl Environ Microbiol 55:2572–2578

Salyers AA, West SE, Vercellotti JR et al (1977) Fermentation of mucins and plant polysaccharides by anaerobic bacteria from the human colon. Appl Environ Microbiol 34:529–533

Sartor RB (2006) Mechanisms of disease: pathogenesis of Crohn's disease and ulcerative colitis. Nat Clin Pract Gastroenterol Hepatol 3:390–407

Schoefer L, Mohan R, Braune A et al (2002) Anaerobic C-ring cleavage of genistein and daidzein by *Eubacterium ramulus*. FEMS Microbiol. Lett. 208:197–202

Setchell KD, Clerici C (2010) Equol: history, chemistry, and formation. J Nutr 140:1355S–1362S

Shipman JA, Cho KH, Siegel HA et al (1999) Physiological characterization of SusG, an outer membrane protein essential for starch utilization by *Bacteroides thetaiotaomicron*. J Bacteriol 181:7206–7211

Shipman JA, Berleman JE, Salyers AA (2000) Characterization of four outer membrane proteins involved in binding starch to the cell surface of *Bacteroides thetaiotaomicron*. J Bacteriol 182:5365–5372

Sidhu H, Schmidt ME, Cornelius JG et al (1999) Direct correlation between hyperoxaluria/oxalate stone disease and the absence of the gastrointestinal tract-dwelling bacterium *Oxalobacter formigenes*: possible prevention by gut recolonization or enzyme replacement therapy. J Am Soc Nephrol 10:S334–S340

Slavin JL, Brauer PM, Marlett JA (1981) Neutral detergent fiber, hemicellulose and cellulose digestibility in human subjects. J Nutr 111:287–297

Sokol H, Pigneur B, Watterlot L et al. (2008) *Faecalibacterium prausnitzii* is an anti-inflammatory commensal bacterium identified by gut microbiota analysis of Crohn disease patients. In: Proceedings of National Academy of Science, U S A 105:16731–16736

Sokol H, Seksik P, Furet JP et al (2009) Low counts of *Faecalibacterium prausnitzii* in colitis microbiota. Inflamm Bowel Dis 15:1183–1189

Suau A, Bonnet R, Sutren M et al (1999) Direct analysis of genes encoding 16S rRNA from complex communities reveals many novel molecular species within the human gut. Appl Environ Microbiol 65:4799–4807

Suau A, Rochet V, Sghir A et al (2001) *Fusobacterium prausnitzii* and related species represent a dominant group within the human fecal flora. Syst Appl Microbiol 24:139–145

Tanaka H, Hashiba H, Kok J et al (2000) Bile salt hydrolase of *Bifidobacterium longum*-biochemical and genetic characterization. Appl Environ Microbiol 66:2502–2512

Troxel SA, Sidhu H, Kaul P et al (2003) Intestinal *Oxalobacter formigenes* colonization in calcium oxalate stone formers and its relation to urinary oxalate. J Endourol 17:173–176
Turnbaugh PJ, Ley RE, Mahowald MA et al (2006) An obesity-associated gut microbiome with increased capacity for energy harvest. Nature 444:1027–1031
Turnbaugh PJ, Hamady M, Yatsunenko T et al (2009) A core gut microbiome in obese and lean twins. Nature 457:480–484
Turnbaugh PJ, Quince C, Faith JJ et al. (2010) Organismal, genetic, and transcriptional variation in the deeply sequenced gut microbiomes of identical twins. In: Proceedings of National Academy of Science, U S A 107:7503-7508
Van Soest PJ (1984) Some physical characteristics of dietary fibres and their influence on the microbial ecology of the human colon. Proc Nutr Soc 43:25–33
Variyam EP, Hoskins LC (1981) Mucin degradation in human colon ecosystems: degradation of hog gastric mucin by fecal extracts and fecal cultures. Gastroenterology 81:751–758
Vervaeke IJ, Van Nevel CJ, Decuypere JA et al (1973) A comparison of two methods for obtaining anaerobic counts in different segments of the gastro-intestinal tract of piglets. J Appl Bacteriol 36:397–405
Wells JM, Rossi O, Meijerink M et al. (2011) Epithelial crosstalk at the microbiota-mucosal interface. In: Proceedings of National Academy of Science, U S A 1:4607–4614 (108 Supplementary)
Woese CR (1987) Bacterial evolution. Microbiol. Rev. 51:221–271
Wolin MJ, Miller TL (1983) Carbohydrate fermentation. In: Hentges DJ (ed) Human intestinal microflora in health and disease. Academic Press, New York, p 19
Woting A, Clavel T, Loh G et al (2010) Bacterial transformation of dietary lignans in gnotobiotic rats. FEMS Microbiol Ecol 72:507–514
Xu J, Bjursell MK, Himrod J et al (2003) A genomic view of the human *Bacteroides thetaiotaomicron* symbiosis. Science 299:2074–2076

The Gut Microflora and Its Variety of Roles in Health and Disease

Julia-Stefanie Frick and Ingo B. Autenrieth

Abstract The intestinal microbiota is a complex community of microorganisms that colonizes the gastrointestinal tract. The composition of the intestinal microbiota and the number of microorganisms differ in dependency of the local environmental conditions. The intestinal microbiota has an important impact on the development of the intestinal architecture and function, it influences the development of the gut-associated immune system, and epithelial cell functions. One of the most important functions of the intestinal microbiota is the prevention of bacterial overgrowth and susceptibility to infection with enteropathogenic organisms. Additionally, the intestinale microbiota plays a crucial role in the development of the systemic immunity and has an important influence on the host nutrition and metabolism. However, in genetically predisposed hosts, the intestinal microbiota is involved in the pathophysiology of inflammatory bowel diseases and pouchitis. Additionally, recent studies suggest that there might be an inflammation triggering effect of the intestinal microbiota in necrotizing enterocolitis. Here, we give an overview of the intestinal microbiota and its variety of roles in health and disease.

J.-S. Frick (✉) · I. B. Autenrieth
Institute for Medical Microbiology and Hygiene,
University Hospital of Tübingen, Elfriede-Aulhorn-Str. 6,
72076 Tübingen, Germany
e-mail: julia-stefanie.frick@med.uni-tuebingen.de

I. B. Autenrieth
e-mail: ingo.autenrieth@med.uni-tuebingen.de

Current Topics in Microbiology and Immunology (2013) 358: 273–289
DOI: 10.1007/82_2012_217

Published Online: 5 April 2012

Contents

1 The Gut Microflora and Its Variety of Roles in Health

1.1 Composition of the Intestinal Microbiota

It is estimated that the human microbiota contain as many as 10^{14} (Ley et al. 2006a; Whitman et al. 1998) bacterial cells and that more than 70% are located in the colon (Ley et al. 2006a; Whitman et al. 1998) which reach an impressive density of 10^{12} bacteria per gram of luminal content. Estimates of the number of bacterial species present in the intestinale microbiota vary between 500 and 1,150 species (Xu and Gordon 2003; Qin et al. 2010). The intestinale microbiota influence strongly the architecture and function of the intestine and play an important role in the balance between health and disease.

1.2 Influence of the Intestinal Microbiota on the Intestinal Architecture and Function

Studies with germfree (sterile) animals revealed that the intestinal microbiota contribute essentially to the development of the intestinal architecture. Germfree animals show enlarged cecae, often leading to gastrointestinal and reproductive disorders (Wostmann and Bruckner-Kardoss 1959), and a reduced intestinal surface (Gordon and Bruckner-Kardoss 1961). Among others, the intestinal surface reduction is based on smaller villus thickness due to reduced cell regeneration

(Banasaz et al. 2002) and reduced leukocyte infiltration might also account for a reduced gut surface (Sekirov et al. 2010). Additionally germfree animals show a reduced villus capillary network (Stappenbeck et al. 2002) that might also account to less effective nutrient absorption (Sekirov et al. 2010). Different commensal bacteria of the intestinal microbiota have been shown to induce Angiogenin-3 transcription which results in a protein with angiogenic activity (Hooper et al. 2001).

The intestinal microbiota is also important for the physiologic intestinal peristalsis (Husebye et al. 1994) which is impaired in germfree animals. Furthermore, germfree animals show abnormal cholesterol and bile acid metabolism (Gustafsson et al. 1975; Madsen et al. 1976; Wostmann et al. 1966) indicating the important contribution of the microbiota to these pathways.

1.3 Influence of the Intestinal Microbiota on Epithelial Cell Functions

Several studies show that the gene expression elicited by pattern recognition receptor (PRR) signaling plays an essential role in maintaining gut homeostasis (Neish 2009). It was shown that mice deficient in luminal bacteria were significantly more susceptible to dextran-sodium-sulfate (DSS) colitis. This effect could be abrogated by oral administration of isolated microbe- associated molecular patterns (MAMP) e.g., lipopolysaccharide (LPS) or lipoteichoic acid (LTA) (Rakoff-Nahoum et al. 2004). This protective effects were lost in TLR2- and TLR4 null mice, indicating that TLR signaling mediates the cytoprotective effect (Neish 2009; Rakoff-Nahoum et al. 2004). It is known that in germfree mice the regenerative responses to colonic injury are markedly attenuated, indicating a discernable role of the intestinal microbiota in induction of epithelial proliferation and response to injury (Pull et al. 2005). These and related observations in mice with defects in epithelial NFκB pathway components suggest that PRR signaling is necessary for gut homeostasis (Ben-Neriah and Schmidt-Supprian 2007; Nenci et al. 2007; Zaph et al. 2007; Chen et al. 2003), presumably because of the tonic upregulation of cytoprotective genes like e.g., gene products with antiapoptotic, stress response, and antioxidant effects (Zaph et al. 2007).

1.4 Influence of the Intestinal Microbiota on the Development of the Gut-Associated Immune System

Germfree mice represent a powerful tool to investigate the contribution of intestinal microbiota on the development of the gut-associated lymphoid tissue (GALT). Germfree mice have fewer and smaller Peyer's Patches, spleens, and mesenteric lymph nodes (Macpherson and Harris 2004; Falk et al. 1998; Pollard and Sharon 1970; Sekirov et al. 2010) as compared to mice with complex physiological intestinal microbiota (e.g., SPF). Additionally, the development and maturation of isolated lymphoid follicles (ILFs) is impaired (Bouskra et al. 2008) and the mice show greatly reduced numbers of IgA producing plasma cells, lamina propria T-cells (Macpherson and Harris 2004; Macpherson et al. 2001, 2002) and intraepithelial lymphocytes which additionally are compromised in their cytotoxicity (Imaoka et al. 1996; Umesaki et al. 1993). They also exhibit irregularities in cytokine levels and profiles (O'Hara and Shanahan 2006) and are impaired in the generation of oral tolerance (Ishikawa et al. 2008; Sekirov et al. 2010). Germfree mice also show defects in the innate immune system. Studies report a reduced expression of TLR and class II major histocompatibility complex on intestinal epithelial cells which form a physical barrier between luminal contents and the host immune system (Lundin et al. 2008; Matsumoto et al. 1992). Therefore germfree mice show extensive defects in the development of the GALT.

1.5 Influence of the Intestinal Microbiota on the Development of the Systemic Immunity

Intestinal microbiota is not only essential for the development of the GALT. They play additionally a crucial role in systemic immunity. It was e.g., shown that germfree animals show significantly reduced numbers of $CD4^+$ T-cells in spleen. The architecture of the spleen is also altered, spleens of germfree animals show fewer germinal centers; additionally germfree animals exhibit significantly reduced systemic antibody levels (Noverr and Huffnagle 2004; Mazmanian et al. 2005; Bauer et al. 1963; Lee and Mazmanian 2010).

1.6 Influence of the Intestinal Microbiota on Nutrition and Metabolism

Recently, it was shown that the intestinal microbiota has also a very important impact on the host metabolism. In contrast to mice with physiologic intestinal microbiota, germfree mice are protected against the obesity that develops after feeding with a western-style high fat high sugar diet (Backhed et al. 2007). In humans a shift in the ratio between Firmicutes and Bacteroidetes in favor of the Firmicutes was shown to be associated with obesity (Ley et al. 2006b). Functional studies using *ob/ob* mice, a preclinical model of obesity revealed that in this animal model obesity is also related to changes in the relative abundance of the two dominant bacterial divisions, the Firmicutes and the Bacteroidetes. It was shown that the intestinal microbiota of obesed mice has an increased capacity to harvest energy from diet. Additionally, the intestinal microbiota promotes the absorption of monosaccharide from the gut with resulting induction of de novo hepatic lipogenesis (Backhed et al. 2004). Studies of germfree mice and mice harboring a physiologic intestinal microbiota revealed that the fasting-induced adipocyte factor (FIAF), a member of the angiopoietin-like family of proteins, and circulating lipoprotein-lipase inhibitor is selectively suppressed by the intestinal microbiota and that this suppression is essential for the deposition of triglycerides in adipocytes (Backhed et al. 2004). Therefore, the gut microbiota directly contributes to the regulation of fat storage (Backhed et al. 2004).

2 The Gut Microflora and Its Variety of Roles in Disease

2.1 Inflammatory Bowel Diseases

The development of suitable animal models was crucial for the investigation of the immunological mechanism underlying inflammatory bowel diseases. Different knockout and transgenic mice serve as models for intestinal inflammatory disorders like e.g., *IL-10*$^{-/-}$ mice (Davidson et al. 1996), *IL-2*$^{-/-}$ mice (Sadlack et al. 1993) or models of adoptive T-cell transfer like e.g., *Rag1*$^{-/-}$ mice (Trobonjaca et al. 2001) or SCID mice (Claesson et al. 1996). Beside mice, rat models of colitis are known e.g., the HLA-B27 transgenic rat (Hammer et al. 1995; Aiko and Grisham 1995). Most of these animal models are T-cell driven models with a predominant polarization of T-cells in Th1 direction (Bertrand et al. 1998; Davidson et al. 1996; Ma et al. 1995; Muller et al. 2008; Trobonjaca et al. 2001; Claesson et al. 1999, 1996).

These animal models suggest that a dysregulation of the mucosal and the systemic immune system is associated with chronic inflammatory disorders of the intestine. On the other hand, it is known that environmental factors play an important role in the maintenance of intestinal homeostasis and the intestinal microbiota are one of the most important environmental factors associated with inflammatory bowel disease. However, the intestinal microbiota accounts for both induction as well as prevention of IBD in genetically predisposed hosts (Atarashi et al. 2011; Dieleman et al. 2000; Niess et al. 2008; Rath et al. 1996, 1999a; Sellon et al. 1998; Waidmann et al. 2003; Yoshida et al. 2002).

Commensal bacteria in these genetically predisposed hosts can induce colitis, can be inert, or can even prevent colitis. However, the immunological effects commensal bacteria exert on the host immune system might be both, host as well as strain specific.

Different commensal bacteria have been tested in different genetically predisposed hosts for their ability to induce or prevent IBD (Table 1). It was shown that bacteria of the same species trigger opposed host immune responses, depending on the host genotype and species. In HLA-B27/Human β_2 Microglobulin transgenic rats it was shown that *Bacteroides vulgatus* triggers the induction of an intestinal inflammation whereas *E. coli* is able to prevent inflammation (Rath et al. 1999b, 1996). In contrast, we reported that in *IL-2*$^{-/-}$ mice *E. coli* induces severe colitis, whereas *B. vulgatus* does not and even prevents the *E. coli*-induced colitis (Waidmann et al. 2003). These divergent results may indicate that the effects of commensal bacteria on the

Table 1 Overview of microbiota-dependent animal models of IBD

Bacterium	Model	Outcome	Reference
Bacteroides vulgatus	*IL-2*$^{-/-}$ mice	Prevention of inflammation	Waidmann et al. 2003 225/id; Muller et al. 2008 1255/id
	HLA-B27 transgenic rats	Inflammation	Rath et al. 1999a, b 159/id; Rath et al. 1996 112/id
Bacteroides fragilis	*H. hepaticus* positive *Rag1*$^{-/-}$ mice	Prevention of inflammation	Mazmanian et al. 2008 1277/id
Escherichia coli	*IL-2*$^{-/-}$ mice	Inflammation	Waidmann et al. 2003 225/id; Muller, et al. 2008 1255/id
	HLA-B27 transgenic rats	Prevention of inflammation	Rath et al. 1999a, b 159/id; Rath et al. 1996 112/id
Citrobacter rodentium	Wild type	Inflammation	Borenshtein 2008 1792/id
Enterococcus faecalis	*IL-10*$^{-/-}$	Inflammation	Balish 2002 179/id
F. prausnitzii	TNBS colitis	Antiinflammatoy	Sokol et al. 2008 1747/id
SFB (segmented filamentous bacteria)	SCID mice	Inflammation	Stepankova 2007 1791/id

immunological host responses feature both host specificity as well as strain specificity. This hypothesis is supported by the finding that *E. coli* strain Nissle 1917 does not induce colitis in *IL-2*$^{-/-}$ mice (Waidmann et al. 2003). Additionally, there might be a disease specificity of commensal bacteria. This is indicated by a clinical study showing that children suffering from acute diarrhea, benefit from the application of *Lactobacillus rhamnosus* strain GG or a mix of *Lactobacillus delbrueckii var bulgaricus*, *Streptococcus thermophilus*, *Lactobacillus acidophilus*, and *Bifidobacterium bifidum* but not from application of preparations containing *Saccharomyces boulardi*, *Bacillus clausii*, or *Enterococcus faecium* SF68 (Canani et al. 2007).

Although till now in humans no single bacterial species of the intestinal microbiota responsible for the induction of maintenance of IBD in otherwise unpredisposed hosts has been identified, several studies report on a high incidence of pathogenic *E. coli* in ileal biopsies of patients suffering from Crohn's disease (Packey and Sartor 2009; Rolhion and Rfeuille-Michaud 2007; Sekirov et al. 2010; Sartor 2008; Srikanth and McCormick 2008) This *E. coli* stain was described as an adherent/invasive *E. coli* strain (AIEC) and has the ability to bind to CECAM6 which is expressed on the apical side of ileal epithelial cells, and, interestingly, in Crohn's disease the expression of CEAM6 is upregulated (Darfeuille-Michaud et al. 1998, 2004; Rfeuille-Michaud 2002; Barnich et al. 2007; Barnich and Rfeuille-Michaud 2010).

Several studies in human and mice report on a significant overgrowth of *Enterobacteriaceae* on the inflamed ileal or colonic mucosa. The *E. coli* strains associated to the ileal mucosa adhere to and invade epithelial cell lines although these strains lack conventional pathogenicity genes. Therefore, mechanisms of *E. coli* that account for the epithelial adherence remain unclear (Martin et al. 2004; Sekirov et al. 2010). The presence of increased numbers of adherent hemaglutinin expressing *E. coli* on the mucosa of resected colon cancer specimens indicates a role of commensal bacteria in colon cancer. It was shown that the isolated *E. coli* strains induce release of IL-8 from intestinal epithelial cells. This inflammatory host response might account for a failure of apoptosis and therefore for an enhanced risk of colon cancer (Martin et al. 2004). Recently, a relative decrease in the number of *Faecalibacterium prausnitzii* was shown to be associated with and highly indicative of Crohn's disease localized to the ileum (Sekirov et al. 2010; Willing et al. 2010; Sokol et al. 2008). However, till now it is still unclear whether the IBD-associated changes in the composition of the intestinal microbiota are a trigger or a result of the intestinal inflammation (Sekirov et al. 2010).

Mycobacterium avium subspecies *paratuberculosis* (MAP) causes a granulomatous enterocolitis (Johne's disease) in ruminants. This disease is characterized by diarrhea and wasting of the host and makes this pathogen a credible aetiologic agent of Crohn's disease. However, the link between MAP and Crohn's disease remains neither substantiated nor invalidated (Packey and Sartor 2009). There are studies reporting on no detectable MAP DNA in tissue samples from Crohn's disease (Baumgart et al. 2007; Frank et al. 2007; Packey and Sartor 2009) or ulcerative colitis patients as well as contradictious results detecting MPP DNA in 87% of Crohn's disease tissues and 15% of controls (Scanu et al. 2007; Packey and Sartor 2009). Despite continued suggestions of a link between MAP and IBD, it remains doubtful that MAP is a causative agent of most Crohn's disease patients (Packey and Sartor 2009; Sartor 2005).

The aetiology of IBD is thought to result from dysregulated interactions between the commensal microbiota and the host immune system (Rakoff-Nahoum et al. 2006; Strober et al. 2002). Toll-like receptors play an essential role in the recognition of the highly conserved MAMP like e.g., LPS, Flagellin, etc. The microbial recognition by TLR is essential for the induction of inflammation and plays an important role in the development of the adaptive immune response,in particular, the TH1 response (Iwasaki and Medzhitov 2004; Rakoff-Nahoum et al. 2006). However, the impact of commensal-induced activation of TLR in induction of inflammation differs between the preclinical models of IBD. It was shown that the development of colitis in $IL\text{-}10^{-/-}$ mice was completely dependent on TLR-MyD88 signaling whereas $IL\text{-}2^{-/-}$ mice developed colitis even in the absence of TLR signaling pathways (Rakoff-Nahoum et al. 2006). Genetic association studies have linked PRR like e.g., NOD2, NLRP3 as well as multiple TLR genes with the development of IBD (Hugot et al. 2001; Ogura et al. 2001; Villani 2009). However, the functional characterization of PRRs in IBD is just starting to be unraveled (Asquith and Powrie 2010).

It is thought that in intestinal homeostasis the intestine has a limited and controlled uptake of microbial antigens by e.g., m cells and dendritic cells and that it is able to exclude viable commensal organisms by e.g., the production of defensins or IgA (Koslowski et al. 2010; Macpherson and Uhr 2004; Sartor 2008). Furthermore, it is assumed that in intestinal homeostasis epithelial cells, cells of the innate and adaptive immune system are in a state of non responsiveness toward microbial antigens in terms of secretion of proinflammatory cytokines, and that the secretion of cytokines maintaining homeostasis like e.g., IL-10 or TGF-β by regulatory T-cells dominates (Izcue et al. 2006; Sartor 2008). In contrast, in IBD functionally abnormal bacteria can attach to and invade epithelial cells due to reduced expression of antimicrobial peptides (Sartor 2008; Koslowski et al. 2010). These bacteria can persist within epithelial cells and phagocytic cells if the intracellular bacterial killing is defective (Sartor 2008; Korzenik 2007). The persistent secretion of proinflammatory cytokines like e.g., TNF-α, IL-6, IL-1ß, IL-12, and IL-23 leads to an inflammatory host reaction that results in a Th1/Th17 response (Sartor 2008; Shen and Durum 2010). The stimulation of these Th1/Th17 cells by a persistent secretion of TNF-α, IL-12, IL23, and IL-6 leads to a chronic tissue injury and epithelial damage (Sartor 2008; Shen and Durum 2010). The ongoing translocation of bacteria, microbial antigens, and TLR ligands via the disturbed epithelial barrier leads to a perpetuation of these pathophysiological processes and to a chronification of the inflammatory disorder (Sartor 2008).

2.2 Pouchitis

Restorative proctocolectomy with ileal pouch-anal anastomosis is the procedure of choice in patients with ulcerative colitis. An inflammation of the ileoanal reservoir (pouchitis) is the most common long-term complication (McLaughlin et al. 2010a). Pouchitis is a form of IBD and occurs predominantly in patients operated on for ulcerative colitis (McLaughlin et al. 2010b). The aetiology of pouchitis is still not completely understood; however, there is incidence for at least a triggering role of the intestinal microbiota. The mucosal inflammation that occurs in pouchitis is localized to areas with the highest concentration of bacteria (McLaughlin et al. 2010a) and antibiotic treatment has been reported to be effective in reducing the inflammation on pouchitis (McLaughlin et al. 2010b; Mimura et al. 2002). Additionally, probiotics have been shown to reduce both the risk of disease onset and disease relapse (Gionchetti et al. 2000; Mimura et al. 2004; McLaughlin et al. 2010b). However, the microbial changes leading to or induced by pouchitis are still poorly understood. Results from early studies using culture techniques are highly diverse and do not show a strong evidence for dysbiosis as the case of pouchitis (McLaughlin et al. 2010b; Lim et al. 2006). However, studies using molecular techniques suggest that a reduction in the bacterial diversity but not a dysbiosis occurs in pouchitis and that UC pouchitis patients have a different less-diverse microbiota with increased *Proteobacteria* and *reduced* Bacteroidetes and *F. prausnitzii* (McLaughlin et al. 2010a, b).

2.3 Necrotizing Enterocolitis

Necrotizing enterocolitis (NEC) is an acute inflammation that affects the intestinal tract of neonates (Lin and Stoll 2006). Neonates who are of low birth weight and under 28-week gestation are the most susceptible (Lee and Polin 2003). Before birth the gut is sterile. The intestine is generally initially colonized with a complex microbiota, which reflects maternal microbiota (Gronlund et al. 1999; Harmsen et al. 2000). Studies have shown that the delivery method influences the composition of the intestinal microbiota. The fecal colonization of infants born by cesarean delivery is described to be delayed, especially Bifidobacterium-like bacteria, and Lactobacillus-like bacteria colonization rates are affected (Gronlund et al. 1999). Gestation age also affects the colonization of the intestine. Premature children in the neonatal intensive care units undergo a delayed colonization with a limited number of bacterial species that tend to be virulent (Orrhage and Nord 1999; Claud and Walker 2008; Kosloske 1994). However, although specific pathogens have been cultured in outbreaks of NEC in single institutions, no organism has consistently been implicated (Neu 2011). Investigations support a role for different bacteria e.g., for clostridial-like species including *Clostridium perfringens*, *Clostridium butyricum*, and *Clostridium neonatale* (Sturm et al. 1980; Warren et al. 1984; Alfa et al. 2002). Nevertheless, other studies have clearly demonstrated that the prevalence of colonization with clostridial species does not differ between NEC affected children and healthy controls (Westra-Meijer et al. 1983; Kelsey and Vince 1979; El-Mohandes et al. 1993; Mshvildadze et al. 2010). A recent study working with preclinical models of NEC and using 16S rRNA pyrosequencing techniques, reports on an increased relative abundance of Gram-negative bacteria like e.g., *Citrobacter, Klebsiella*, or *Tatumella* in the colonic flora of animals with disease as compared to healthy control mice (Carlisle et al. 2011). Overgrowth of Gram-negative bacteria has been previously described in clinical studies using culturing techniques (Bell et al. 1979; Carlisle et al. 2011; Cordero et al. 2004). Additionally a recent study describes that the overall microbiotic profiles in babies with NEC had increased levels of *Citrobacter* and *Enterococcus* sp. as compared to babies without NEC (Mshvildadze et al. 2010). Conversely, other studies using 16S rRNA sequencing of the fecal microbiome of infants did not find differences between healthy and diseased children (Millar et al. 1996; Carlisle et al. 2011). The inconsistency of these findings strongly suggests that no single causative pathogen seems to be responsible for the NEC pathogenesis (Morowitz et al. 2010). Additionally, the role of viruses in the pathogenesis of NEC remains unclear and further studies are required to elucidate this topic (Morowitz et al. 2010).

2.4 Bacterial Overgrowth and Susceptibility to Infection

Several studies show that the host intestinal microbiota is significantly influenced by antibiotic treatment, in both, human subjects (De La Cochetiere et al. 2008; Dethlefsen et al. 2008; Jernberg et al. 2007; Lofmark et al. 2006) and mice (Antonopoulos et al. 2009; Yap et al. 2008). The alterations in the intestinal microbiota persist for a long time period after discontinuation of the administration of antibiotics (Dethlefsen et al. 2008; Lofmark et al. 2006; Jernberg et al. 2007; Lindgren et al. 2009).

These changes in the intestinal microbiota might result in bacterial overgrowth of certain species or to an increased susceptibility toward infections with enteropathogenic bacteria (Sekirov et al. 2010).

The overgrowth of e.g., *Clostridium difficile* in the intestine, due to antibiotic treatment, is a well-known complication following antibiotic treatment and a frequent cause of antibiotic-associated diarrhea (McFarland and Dublin 2008; Sekirov et al. 2010).

The presence of Vancomycin resistant *Enterococci* (VRE) in the intestine is also described to be correlated with antibiotic treatment. Recently, it was reported that VRE are able to exploit an immune deficit, due to downregulation of RegIIIγ, at the intestinal mucosa. This immune deficit seems to be caused by antibiotic-mediated disruption of the gut microbiota and helps VRE to prolong their colonization of the mouse intestine (Brandl et al. 2008).

Additionally, an antibiotic-induced disruption of the intestinal microbiota enhances the susceptibility of the human host to an infection with nontyphoidal *Salmonellae* (Gradel et al. 2009), and is a prerequisite for infection of mice with *Salmonella Typhimurium* (Barthel et al. 2003).

An intricate symbiotic relationship between humans and microbes has evolved (Neish 2009). However, in genetic predisposed hosts the intestinal microbiota seem to be essential for the pathogenesis of inflammatory disorders. We need to more fully understand the relationship and the host microbe interactions in physiology and pathophysiology to be able to manipulate this relationship therapeutically (Neish 2009). Perhaps there are therapeutic or preventive strategies to suppress the pathogenic side and maintain or restore the beneficial side of this relationship.

References

Aiko S, Grisham MB (1995) Spontaneous intestinal inflammation and nitric oxide metabolism in HLA-B27 transgenic rats. Gastroenterology 109:142–150

Alfa MJ, Robson D, Davi M, Bernard K, Van CP, Harding GK (2002) An outbreak of necrotizing enterocolitis associated with a novel *clostridium* species in a neonatal intensive care unit. Clin Infect Dis 35:S101–S105

Antonopoulos DA, Huse SM, Morrison HG, Schmidt TM, Sogin ML, Young VB (2009) Reproducible community dynamics of the gastrointestinal microbiota following antibiotic perturbation. Infect Immun 77:2367–2375

Asquith M, Powrie F (2010) An innately dangerous balancing act: intestinal homeostasis, inflammation, and colitis-associated cancer. J Exp Med 207:1573–1577

Atarashi K, Tanoue T, Shima T, Imaoka A, Kuwahara T, Momose Y, Cheng G, Yamasaki S, Saito T, Ohba Y, Taniguchi T, Takeda K, Hori S, Ivanov II, Umesaki Y, Itoh K, Honda K (2011) Induction of colonic regulatory T cells by indigenous *Clostridium* species. Science 331:337–341

Backhed F, Ding H, Wang T, Hooper LV, Koh GY, Nagy A, Semenkovich CF, Gordon JI (2004) The gut microbiota as an environmental factor that regulates fat storage. Proc Natl Acad Sci U S A 101:15718–15723

Backhed F, Manchester JK, Semenkovich CF, Gordon JI (2007) Mechanisms underlying the resistance to diet-induced obesity in germ-free mice. Proc Natl Acad Sci U S A 104:979–984

Balish E, Warner T (2002) Enterococcus faecalis induces inflammatory bowel disease in interleukin-10 knockout mice. Am J Pathol 160(6):2253–2257

Banasaz M, Norin E, Holma R, Midtvedt T (2002) Increased enterocyte production in gnotobiotic rats mono-associated with *Lactobacillus rhamnosus* GG. Appl Environ Microbiol 68: 3031–3034

Barnich N, Carvalho FA, Glasser AL, Darcha C, Jantscheff P, Allez M, Peeters H, Bommelaer G, Desreumaux P, Colombel JF, Rfeuille-Michaud A (2007) CEACAM6 acts as a receptor for adherent-invasive *E. coli*, supporting ileal mucosa colonization in Crohn disease. J Clin Invest 117:1566–1574

Barnich N, Rfeuille-Michaud A (2010) Abnormal CEACAM6 expression in Crohn disease patients favors gut colonization and inflammation by adherent-invasive *E. coli*. Virulence 1:281–282

Barthel M, Hapfelmeier S, Quintanilla-Martinez L, Kremer M, Rohde M, Hogardt M, Pfeffer K, Russmann H, Hardt WD (2003) Pretreatment of mice with streptomycin provides a Salmonella enterica serovar Typhimurium colitis model that allows analysis of both pathogen and host. Infect Immun 71:2839–2858

Bauer H, Horowitz RE, Levenson SM, Popper H (1963) The response of the lymphatic tissue to the microbial flora. Studies on germfree mice. Am J Pathol 42:471–483

Baumgart M, Dogan B, Rishniw M, Weitzman G, Bosworth B, Yantiss R, Orsi RH, Wiedmann M, McDonough P, Kim SG, Berg D, Schukken Y, Scherl E, Simpson KW (2007) Culture independent analysis of ileal mucosa reveals a selective increase in invasive *Escherichia coli* of novel phylogeny relative to depletion of Clostridiales in Crohn's disease involving the ileum. J ISME 1:403–418

Bell MJ, Feigin RD, Ternberg JL (1979) Changes in the incidence of necrotizing enterocolitis associated with variation of the gastrointestinal microflora in neonates. Am J Surg 138: 629–631

Ben-Neriah Y, Schmidt-Supprian M (2007) Epithelial NF-kappaB maintains host gut microflora homeostasis. Nat Immunol 8:479–481

Bertrand V, Quere S, Guimbaud R, Sogni P, Chauvelot-Moachon L, Tulliez M, Lamarque D, Charreire J, Giroud JP, Couturier D, Chaussade S, Breban M (1998) Effects of murine recombinant interleukin-10 on the inflammatory disease of rats transgenic for HLA-B27 and human beta 2-microglobulin. Eur Cytokine Netw 9:161–170

Borenshtein D, McBee ME, Schauer DB (2008) Utility of the Citrobacter rodentium infection model in laboratory mice. Curr Opin Gastroenterol 4(1):32–37

Bouskra D, Brezillon C, Berard M, Werts C, Varona R, Boneca IG, Eberl G (2008) Lymphoid tissue genesis induced by commensals through NOD1 regulates intestinal homeostasis. Nature 456:507–510

Brandl K, Plitas G, Mihu CN, Ubeda C, Jia T, Fleisher M, Schnabl B, DeMatteo RP, Pamer EG (2008) Vancomycin-resistant enterococci exploit antibiotic-induced innate immune deficits. Nature 455:804–807

Canani RB, Cirillo P, Terrin G, Cesarano L, Spagnuolo MI, De VA, Albano F, Passariello A, De MG, Manguso F, Guarino A (2007) Probiotics for treatment of acute diarrhoea in children: randomised clinical trial of five different preparations. BMJ 335(7615):340

Carlisle EM, Poroyko V, Caplan MS, Alverdy JA, Liu D (2011) Gram negative bacteria are associated with the early stages of necrotizing enterocolitis. PLoS One 6:e18084

Chen LW, Egan L, Li ZW, Greten FR, Kagnoff MF, Karin M (2003) The two faces of IKK and NF-kappaB inhibition: prevention of systemic inflammation but increased local injury following intestinal ischemia-reperfusion. Nat Med 9:575–581

Claesson MH, Bregenholt S, Bonhagen K, Thoma S, Moller P, Grusby MJ, Leithauser F, Nissen MH, Reimann J (1999) Colitis-inducing potency of CD4 + T cells in immunodeficient, adoptive hosts depends on their state of activation, IL-12 responsiveness, and CD45RB surface phenotype. J Immunol 162:3702–3710

Claesson MH, Rudolphi A, Kofoed S, Poulsen SS, Reimann J (1996) CD4 + T lymphocytes injected into severe combined immunodeficient (SCID) mice lead to an inflammatory and lethal bowel disease. Clin Exp Immunol 104:491–500

Claud EC, Walker WA (2008) Bacterial colonization, probiotics, and necrotizing enterocolitis. J Clin Gastroenterol 42(Suppl 2):S46–S52

Cordero L, Rau R, Taylor D, Ayers LW (2004) Enteric gram-negative bacilli bloodstream infections: 17 years' experience in a neonatal intensive care unit. Am J Infect Control 32:189–195

Darfeuille-Michaud A, Boudeau J, Bulois P, Neut C, Glasser AL, Barnich N, Bringer MA, Swidsinski A, Beaugerie L, Colombel JF (2004) High prevalence of adherent-invasive *Escherichia coli* associated with ileal mucosa in Crohn's disease. Gastroenterology 127: 412–421

Darfeuille-Michaud A, Neut C, Barnich N, Lederman E, Di Martino P, Desreumaux P, Gambiez L, Joly B, Cortot A, Colombel JF (1998) Presence of adherent *Escherichia coli* strains in ileal mucosa of patients with Crohn's disease. Gastroenterology 115:1405–1413

Davidson NJ, Leach MW, Fort MM, Thompson-Snipes L, Kuhn R, Muller W, Berg DJ, Rennick DM (1996) T helper cell 1-type CD4 + T cells, but not B cells, mediate colitis in interleukin 10-deficient mice. J Exp Med 184:241–251

De La Cochetiere MF, Durand T, Lalande V, Petit JC, Potel G, Beaugerie L (2008) Effect of antibiotic therapy on human fecal microbiota and the relation to the development of Clostridium difficile. Microb Ecol 56:395–402

Dethlefsen L, Huse S, Sogin ML, Relman DA (2008) The pervasive effects of an antibiotic on the human gut microbiota, as revealed by deep 16S rRNA sequencing. PLoS Biol 6:e280

Dieleman LA, Arends A, Tonkonogy SL, Goerres MS, Craft DW, Grenther W, Sellon RK, Balish E, Sartor RB (2000) Helicobacter hepaticus does not induce or potentiate colitis in interleukin-10-deficient mice. Infect Immun 68:5107–5113

El-Mohandes AE, Keiser JF, Refat M, Jackson BJ (1993) Prevalence and toxigenicity of Clostridium difficile isolates in fecal microflora of preterm infants in the intensive care nursery. Biol Neonate 63:225–229

Falk PG, Hooper LV, Midtvedt T, Gordon JI (1998) Creating and maintaining the gastrointestinal ecosystem: what we know and need to know from gnotobiology. Microbiol Mol Biol Rev 62:1157–1170

Frank DN, St Amand AL, Feldman RA, Boedeker EC, Harpaz N, Pace NR (2007) Molecular-phylogenetic characterization of microbial community imbalances in human inflammatory bowel diseases. Proc Natl Acad Sci U S A 104:13780–13785

Gionchetti P, Rizzello F, Venturi A, Brigidi P, Matteuzzi D, Bazzocchi G, Poggioli G, Miglioli M, Campieri M (2000) Oral bacteriotherapy as maintenance treatment in patients with chronic pouchitis: a double-blind, placebo-controlled trial. Gastroenterology 119:305–309

Gordon HA, Bruckner-Kardoss E (1961) Effect of normal microbial flora on intestinal surface area. Am J Physiol 201:175–178

Gradel KO, Schonheyder HC, Kristensen B, Dethlefsen C, Ejlertsen T, Nielsen H (2009) Are host characteristics or exposure factors mainly involved in the acquisition of zoonotic Salmonella and Campylobacter coinfection in humans? Foodborne Pathog Dis 6:251–255

Gronlund MM, Lehtonen OP, Eerola E, Kero P (1999) Fecal microflora in healthy infants born by different methods of delivery: permanent changes in intestinal flora after cesarean delivery. J Pediatr Gastroenterol Nutr 28:19–25

Gustafsson BE, Einarsson K, Gustafsson J (1975) Influence of cholesterol feeding on liver microsomal metabolism of steroids and bile acids in conventional and germ-free rats. J Biol Chem 250:8496–8502

Hammer RE, Richardson JA, Simmons WA, White AL, Breban M, Taurog JD (1995) High prevalence of colorectal cancer in HLA-B27 transgenic F344 rats with chronic inflammatory bowel disease. J Investig Med 43:262–268
Harmsen HJ, Wildeboer-Veloo AC, Raangs GC, Wagendorp AA, Klijn N, Bindels JG, Welling GW (2000) Analysis of intestinal flora development in breast-fed and formula-fed infants by using molecular identification and detection methods. J Pediatr Gastroenterol Nutr 30:61–67
Hooper LV, Wong MH, Thelin A, Hansson L, Falk PG, Gordon JI (2001) Molecular analysis of commensal host-microbial relationships in the intestine. Science 291:881–884
Hugot JP, Chamaillard M, Zouali H, Lesage S, Cezard JP, Belaiche J, Almer S, Tysk C, O'Morain CA, Gassull M, Binder V, Finkel Y, Cortot A, Modigliani R, Laurent-Puig P, Gower-Rousseau C, Macry J, Colombel JF, Sahbatou M, Thomas G (2001) Association of NOD2 leucine-rich repeat variants with susceptibility to Crohn's disease. Nature 411:599–603
Husebye E, Hellstrom PM, Midtvedt T (1994) Intestinal microflora stimulates myoelectric activity of rat small intestine by promoting cyclic initiation and aboral propagation of migrating myoelectric complex. Dig Dis Sci 39:946–956
Imaoka A, Matsumoto S, Setoyama H, Okada Y, Umesaki Y (1996) Proliferative recruitment of intestinal intraepithelial lymphocytes after microbial colonization of germ-free mice. Eur J Immunol 26:945–948
Ishikawa H, Tanaka K, Maeda Y, Aiba Y, Hata A, Tsuji NM, Koga Y, Matsumoto T (2008) Effect of intestinal microbiota on the induction of regulatory CD25+ CD4+ T cells. Clin Exp Immunol 153:127–135
Iwasaki A, Medzhitov R (2004) Toll-like receptor control of the adaptive immune responses. Nat Immunol 5:987–995
Izcue A, Coombes JL, Powrie F (2006) Regulatory T cells suppress systemic and mucosal immune activation to control intestinal inflammation. Immunol Rev 212:256–271
Jernberg C, Lofmark S, Edlund C, Jansson JK (2007) Long-term ecological impacts of antibiotic administration on the human intestinal microbiota. J ISME 1:56–66
Kelsey MC, Vince AJ (1979) Clostridia in neonatal faeces. Lancet 2:100
Korzenik JR (2007) Is Crohn's disease due to defective immunity? Gut 56(1):2–5
Kosloske AM (1994) Epidemiology of necrotizing enterocolitis. Acta Paediatr 396(Suppl):2–7
Koslowski MJ, Beisner J, Stange EF, Wehkamp J (2010) Innate antimicrobial host defense in small intestinal Crohn's disease. Int J Med Microbiol 300:34–40
Lee JS, Polin RA (2003) Treatment and prevention of necrotizing enterocolitis. Semin Neonatol 8:449–459
Lee YK, Mazmanian SK (2010) Has the microbiota played a critical role in the evolution of the adaptive immune system? Science 330:1768–1773
Ley RE, Peterson DA, Gordon JI (2006a) Ecological and evolutionary forces shaping microbial diversity in the human intestine. Cell 124:837–848
Ley RE, Turnbaugh PJ, Klein S, Gordon JI (2006b) Microbial ecology: human gut microbes associated with obesity. Nature 444:1022–1023
Lim M, Sagar P, Finan P, Burke D, Schuster H (2006) Dysbiosis and pouchitis. Br J Surg 93:1325–1334
Lin PW, Stoll BJ (2006) Necrotising enterocolitis. Lancet 368:1271–1283
Lindgren M, Lofmark S, Edlund C, Huovinen P, Jalava J (2009) Prolonged impact of a one-week course of clindamycin on Enterococcus spp. in human normal microbiota. Scand J Infect Dis 41:215–219
Lofmark S, Jernberg C, Jansson JK, Edlund C (2006) Clindamycin-induced enrichment and long-term persistence of resistant Bacteroides spp. and resistance genes. J Antimicrob Chemother 58:1160–1167
Lundin A, Bok CM, Aronsson L, Bjorkholm B, Gustafsson JA, Pott S, Arulampalam V, Hibberd M, Rafter J, Pettersson S (2008) Gut flora, Toll-like receptors and nuclear receptors: a tripartite communication that tunes innate immunity in large intestine. Cell Microbiol 10:1093–1103

Ma A, Datta M, Margosian E, Chen J, Horak I (1995) T cells, but not B cells, are required for bowel inflammation in interleukin 2-deficient mice. J Exp Med 182:1567–1572
Macpherson AJ, Harris NL (2004) Interactions between commensal intestinal bacteria and the immune system. Nat Rev Immunol 4:478–485
Macpherson AJ, Uhr T (2004) Induction of protective IgA by intestinal dendritic cells carrying commensal bacteria. Science 303:1662–1665
Macpherson AJ, Hunziker L, McCoy K, Lamarre A (2001) IgA responses in the intestinal mucosa against pathogenic and non-pathogenic microorganisms. Microbes Infect 3:1021–1035
Macpherson AJ, Martinic MM, Harris N (2002) The functions of mucosal T cells in containing the indigenous commensal flora of the intestine. Cell Mol Life Sci 59:2088–2096
Madsen D, Beaver M, Chang L, Bruckner-Kardoss E, Wostmann B (1976) Analysis of bile acids in conventional and germfree rats. J Lipid Res 17:107–111
Martin HM, Campbell BJ, Hart CA, Mpofu C, Nayar M, Singh R, Englyst H, Williams HF, Rhodes JM (2004) Enhanced Escherichia coli adherence and invasion in Crohn's disease and colon cancer. Gastroenterology 127:80–93
Matsumoto S, Setoyama H, Umesaki Y (1992) Differential induction of major histocompatibility complex molecules on mouse intestine by bacterial colonization. Gastroenterology 103: 1777–1782
Mazmanian SK, Liu CH, Tzianabos AO, Kasper DL (2005) An immunomodulatory molecule of symbiotic bacteria directs maturation of the host immune system. Cell 122:107–118
Mazmanian SK, Round JL, Kasper DL (2008) A microbial symbiosis factor prevents intestinal inflammatory disease. Nature 453(7195):620–625
McFarland LV, Dublin S (2008) Meta-analysis of probiotics for the treatment of irritable bowel syndrome. World J Gastroenterol 14:2650–2661
McLaughlin SD, Clark SK, Tekkis PP, Nicholls RJ, Ciclitira PJ (2010a) The bacterial pathogenesis and treatment of pouchitis. Therap Adv Gastroenterol 3:335–348
McLaughlin SD, Walker AW, Churcher C, Clark SK, Tekkis PP, Johnson MW, Parkhill J, Ciclitira PJ, Dougan G, Nicholls RJ, Petrovska L (2010b) The bacteriology of pouchitis: a molecular phylogenetic analysis using 16S rRNA gene cloning and sequencing. Ann Surg 252:90–98
Millar MR, Linton CJ, Cade A, Glancy D, Hall M, Jalal H (1996) Application of 16S rRNA gene PCR to study bowel flora of preterm infants with and without necrotizing enterocolitis. J Clin Microbiol 34:2506–2510
Mimura T, Rizzello F, Helwig U, Poggioli G, Schreiber S, Talbot IC, Nicholls RJ, Gionchetti P, Campieri M, Kamm MA (2002) Four-week open-label trial of metronidazole and ciprofloxacin for the treatment of recurrent or refractory pouchitis. Aliment Pharmacol Ther 16:909–917
Mimura T, Rizzello F, Helwig U, Poggioli G, Schreiber S, Talbot IC, Nicholls RJ, Gionchetti P, Campieri M, Kamm MA (2004) Once daily high dose probiotic therapy (VSL#3) for maintaining remission in recurrent or refractory pouchitis. Gut 53:108–114
Morowitz MJ, Poroyko V, Caplan M, Alverdy J, Liu DC (2010) Redefining the role of intestinal microbes in the pathogenesis of necrotizing enterocolitis. Pediatrics 125:777–785
Mshvildadze M, Neu J, Shuster J, Theriaque D, Li N, Mai V (2010) Intestinal microbial ecology in premature infants assessed with non-culture-based techniques. J Pediatr 156:20–25
Muller M, Fink K, Geisel J, Kahl F, Jilge B, Reimann J, Mach N, Autenrieth IB, Frick JS (2008) Intestinal colonization of IL-2 deficient mice with non-colitogenic B. vulgatus prevents DC maturation and T-cell polarization. PLoS One 3:e2376
Neish AS (2009) Microbes in gastrointestinal health and disease. gastroenterology 136:65–80
Nenci A, Becker C, Wullaert A, Gareus R, Van LG, Danese S, Huth M, Nikolaev A, Neufert C, Madison B, Gumucio D, Neurath MF, Pasparakis M (2007) Epithelial NEMO links innate immunity to chronic intestinal inflammation. Nature 446:557–561
Neu J, Walker WA (2011) Necrotizing enterocolitis. N Engl J Med 364:255–264

Niess JH, Leithauser F, Adler G, Reimann J (2008) Commensal gut flora drives the expansion of proinflammatory CD4 T cells in the colonic lamina propria under normal and inflammatory conditions. J Immunol 180:559–568
Noverr MC, Huffnagle GB (2004) Does the microbiota regulate immune responses outside the gut? Trends Microbiol 12:562–568
O'Hara AM, Shanahan F (2006) The gut flora as a forgotten organ. EMBO Report 7:688–693
Ogura Y, Bonen DK, Inohara N, Nicolae DL, Chen FF, Ramos R, Britton H, Moran T, Karaliuskas R, Duerr RH, Achkar JP, Brant SR, Bayless TM, Kirschner BS, Hanauer SB, Nunez G, Cho JH (2001) A frameshift mutation in NOD2 associated with susceptibility to Crohn's disease. Nature 411:603–606
Orrhage K, Nord CE (1999) Factors controlling the bacterial colonization of the intestine in breastfed infants. Acta Paediatr 88(Suppl):47–57
Packey CD, Sartor RB (2009) Commensal bacteria, traditional and opportunistic pathogens, dysbiosis and bacterial killing in inflammatory bowel diseases. Curr Opin Infect Dis 22: 292–301
Pollard M, Sharon N (1970) Responses of the Peyer's patches in germ-free mice to antigenic stimulation. Infect Immun 2:96–100
Pull SL, Doherty JM, Mills JC, Gordon JI, Stappenbeck TS (2005) Activated macrophages are an adaptive element of the colonic epithelial progenitor niche necessary for regenerative responses to injury. Proc Natl Acad Sci U S A 102:99–104
Qin J, Li R, Raes J, Arumugam M, Burgdorf KS, Manichanh C, Nielsen T, Pons N, Levenez F, Yamada T, Mende DR, Li J, Xu J, Li S, Li D, Cao J, Wang B, Liang H, Zheng H, Xie Y, Tap J, Lepage P, Bertalan M, Batto JM, Hansen T, Le PD, Linneberg A, Nielsen HB, Pelletier E, Renault P, Sicheritz-Ponten T, Turner K, Zhu H, Yu C, Li S, Jian M, Zhou Y, Li Y, Zhang X, Li S, Qin N, Yang H, Wang J, Brunak S, Dore J, Guarner F, Kristiansen K, Pedersen O, Parkhill J, Weissenbach J, Bork P, Ehrlich SD, Wang J (2010) A human gut microbial gene catalogue established by metagenomic sequencing. Nature 464:59–65
Rakoff-Nahoum S, Hao L, Medzhitov R (2006) Role of toll-like receptors in spontaneous commensal-dependent colitis. Immunity 25:319–329
Rakoff-Nahoum S, Paglino J, Eslami-Varzaneh F, Edberg S, Medzhitov R (2004) Recognition of commensal microflora by toll-like receptors is required for intestinal homeostasis. Cell 118: 229–241
Rath HC, Herfarth HH, Ikeda JS, Grenther WB, Hamm-TE J, Balish E, Taurog JD, Hammer RE, Wilson KH, Sartor RB (1996) Normal luminal bacteria, especially Bacteroides species, mediate chronic colitis, gastritis, and arthritis in HLA-B27/human beta2 microglobulin transgenic rats. J Clin Invest 98:945–953
Rath HC, Ikeda JS, Linde HJ, Scholmerich J, Wilson KH, Sartor RB (1999a) Varying cecal bacterial loads influences colitis and gastritis in HLA-B27 transgenic rats. Gastroenterology 116:310–319
Rath HC, Wilson KH, Sartor RB (1999b) Differential induction of colitis and gastritis in HLA-B27 transgenic rats selectively colonized with Bacteroides vulgatus or Escherichia coli. Infect Immunol 67:2969–2974
Rfeuille-Michaud A (2002) Adherent-invasive *Escherichia coli*: a putative new *E. coli* pathotype associated with Crohn's disease. Int J Med Microbiol 292:185–193
Rolhion N, Rfeuille-Michaud A (2007) Adherent-invasive *Escherichia coli* in inflammatory bowel disease. Inflamm Bowel Dis 13:1277–1283
Sadlack B, Merz H, Schorle H, Schimpl A, Feller AC, Horak I (1993) Ulcerative colitis-like disease in mice with a disrupted interleukin-2 gene. Cell 75:253–261
Sartor RB (2005) Does Mycobacterium avium subspecies paratuberculosis cause Crohn's disease? Gut 54:896–898
Sartor RB (2008) Microbial influences in inflammatory bowel diseases. Gastroenterology 134: 577–594
Scanu AM, Bull TJ, Cannas S, Sanderson JD, Sechi LA, Dettori G, Zanetti S, Hermon-Taylor J (2007) *Mycobacterium avium* subspecies paratuberculosis infection in cases of irritable bowel

syndrome and comparison with Crohn's disease and Johne's disease: common neural and immune pathogenicities. J Clin Microbiol 45:3883–3890
Sekirov I, Russell SL, Antunes LC, Finlay BB (2010) Gut microbiota in health and disease. Physiol Rev 90:859–904
Sellon RK, Tonkonogy S, Schultz M, Dieleman LA, Grenther W, Balish E, Rennick DM, Sartor RB (1998) Resident enteric bacteria are necessary for development of spontaneous colitis and immune system activation in interleukin-10-deficient mice. Infect Immun 66: 5224–5231
Shen W, Durum SK (2010) Synergy of IL-23 and Th17 cytokines: new light on inflammatory bowel disease. Neurochem Res 35:940–946
Sokol H, Pigneur B, Watterlot L, Lakhdari O, Bermudez-Humaran LG, Gratadoux JJ, Blugeon S, Bridonneau C, Furet JP, Corthier G, Grangette C, Vasquez N, Pochart P, Trugnan G, Thomas G, Blottiere HM, Dore J, Marteau P, Seksik P, Langella P (2008) Faecalibacterium prausnitzii is an anti-inflammatory commensal bacterium identified by gut microbiota analysis of Crohn disease patients. Proc Natl Acad Sci U S A 105:16731–16736
Srikanth CV, McCormick BA (2008) Interactions of the intestinal epithelium with the pathogen and the indigenous microbiota: a three-way crosstalk. Interdiscip Perspect Infect Dis 2008: 626827
Stappenbeck TS, Hooper LV, Gordon JI (2002) Developmental regulation of intestinal angiogenesis by indigenous microbes via Paneth cells. Proc Natl Acad Sci U S A 99: 15451–15455
Stepankova R, Powrie F, Kofronova O, Kozakova H, Hudcovic T, Hrncir T, Uhlig H, Read S, Rehakova Z, Benada O, Heczko P, Strus M, Bland P, Tlaskalova-Hogenova H (2007) Segmented filamentous bacteria in a defined bacterial cocktail induce intestinal inflammation in SCID mice reconstituted with CD45RBhigh CD4+ T cells. Inflamm Bowel Dis 13(10): 1202–1211
Strober W, Fuss IJ, Blumberg RS (2002) The immunology of mucosal models of inflammation. Annu Rev Immunol 20:495–549
Sturm R, Staneck JL, Stauffer LR, Neblett WW III (1980) Neonatal necrotizing enterocolitis associated with penicillin-resistant, toxigenic *Clostridium butyricum*. Pediatrics 66:928–931
Trobonjaca Z, Leithauser F, Moller P, Bluethmann H, Koezuka Y, MacDonald HR, Reimann J (2001) MHC-II-independent CD4+ T cells induce colitis in immunodeficient RAG-/- hosts. J Immunol 166:3804–3812
Umesaki Y, Setoyama H, Matsumoto S, Okada Y (1993) Expansion of alpha beta T-cell receptor-bearing intestinal intraepithelial lymphocytes after microbial colonization in germ-free mice and its independence from thymus. Immunology 79:32–37
Villani AC, Lemire M, Louis E, Silverberg MS, Collette C, Fortin G, Nimmo ER, Renaud Y, Brunet S, Libioulle C, Belaiche J, Bitton A, Gaudet D, Cohen A, Langelier D, Rioux JD, Arnott ID, Wild GE, Rutgeerts P, Satsangi J, Vermeire S, Hudson TJ, Franchimont D (2009) Genetic variation in the familial Mediterranean fever gene (MEFV) and risk for Crohn's disease and ulcerative colitis. PLoS One 4:e7154
Waidmann M, Bechtold O, Frick JS, Lehr HA, Schubert S, Dobrindt U, Loeffler J, Bohn E, Autenrieth IB (2003) Bacteroides vulgatus protects against Escherichia coli-induced colitis in gnotobiotic interleukin-2-deficient mice. Gastroenterology 125:162–177
Warren S, Schreiber JR, Epstein MF (1984) Necrotizing enterocolitis and hemolysis associated with Clostridium perfringens. Am J Dis Child 138:686–688
Westra-Meijer CM, Degener JE, Dzoljic-Danilovic G, Michel MF, Mettau JW (1983) Quantitative study of the aerobic and anaerobic faecal flora in neonatal necrotising enterocolitis. Arch Dis Child 58:523–528
Whitman WB, Coleman DC, Wiebe WJ (1998) Prokaryotes: the unseen majority. Proc Natl Acad Sci U S A 95:6578–6583
Willing BP, Dicksved J, Halfvarson J, Andersson AF, Lucio M, Zheng Z, Jarnerot G, Tysk C, Jansson JK, Engstrand L (2010) A pyrosequencing study in twins shows that gastrointestinal

microbial profiles vary with inflammatory bowel disease phenotypes. Gastroenterology 139(6):1844–1854
Wostmann B, Bruckner-Kardoss E (1959) Development of cecal distention in germ-free baby rats. Am J Physiol 197:1345–1346
Wostmann BS, Wiech NL, Kung E (1966) Catabolism and elimination of cholesterol in germfree rats. J Lipid Res 7:77–82
Xu J, Gordon JI (2003) Inaugural article: honor thy symbionts. Proc Natl Acad Sci U S A 100:10452–10459
Yap IK, Li JV, Saric J, Martin FP, Davies H, Wang Y, Wilson ID, Nicholson JK, Utzinger J, Marchesi JR, Holmes E (2008) Metabonomic and microbiological analysis of the dynamic effect of vancomycin-induced gut microbiota modification in the mouse. J Proteome Res 7:3718–3728
Yoshida M, Shirai Y, Watanabe T, Yamori M, Iwakura Y, Chiba T, Kita T, Wakatsuki Y (2002) Differential localization of colitogenic Th1 and Th2 cells monospecific to a microflora-associated antigen in mice. Gastroenterology 123:1949–1961
Zaph C, Troy AE, Taylor BC, Berman-Booty LD, Guild KJ, Du Y, Yost EA, Gruber AD, May MJ, Greten FR, Eckmann L, Karin M, Artis D (2007) Epithelial-cell-intrinsic IKK-beta expression regulates intestinal immune homeostasis. Nature 446:552–556

Host-Recognition of Pathogens and Commensals in the Mammalian Intestine

Oriana Rossi, Peter van Baarlen and Jerry M. Wells

Abstract To peacefully coexist with the microbial inhabitants of the intestine, mammals have evolved elaborate and interconnected regulatory mechanisms to maintain immune homeostasis in the face of potential infection and tissue damage by pathogenic microorganisms. Physical barriers, antimicrobial factors and secretory antibodies act in concert to keep microbes at a distance from the epithelium and initiate repair mechanisms in the event of damage. Commensal bacteria are not ignored but dynamically controlled via many complex overlapping and intertwined mechanisms involving intestinal epithelial cells (IECs) and signals from the microbiota. Polarized IECs play a decisive role in homeostasis by regulating the expression and activity of the pattern-recognition receptors (PRRs), in different compartments of the intestine. The differential signaling and expression of receptors on apical and basal membranes of the epithelium also plays its part in distinguishing commensals from harmful invaders. In steady state conditions macrophages and dendritic cells (DCs) in the lamina propria (LP) are conditioned by environmental factors to induce immune tolerance. The distinction between pathogen and non-pathogen is linked to the ability of pathogens to invade and cause damage to the host cells and tissues. This induces local inflammatory responses and the attraction of capillary leukocytes by chemokines released from colonized and invaded epithelial cells. This bypasses the tolerogenic mechanisms controlling the responses of resident DCs and macrophages leading to pathogen killing and adaptive immune responses. Research on this topic has important implications for the development of novel therapeutic approaches to treat or prevent inflammatory bowel disease (IBD), inflammation-related cancer and other gut-related diseases and disorders.

O. Rossi · P. van Baarlen · J. M. Wells (✉)
Host-Microbe Interactomics Group, Wageningen University and Research Centre, 338, 6700 AH Wageningen, The Netherlands
e-mail: jerry.wells@wur.nl

Current Topics in Microbiology and Immunology (2013) 358: 291–321

DOI: 10.1007/82_2011_191

Published Online: 17 December 2011

Abbreviation	
GIT	Gastrointestinal tract
SI	Small intestine
LP	Lamina propria
M cell	Microfold cell
GALT	Gut-associated lymphoid tissue
DSS	Dextran sodium sulphate
TFF3	Trefoil factor 3
IECs	Intestinal epithelial cells
APC	Antigen presenting cell
TJ	Tight junction
EPEC	Enteropathogenic *Escherichia coli*
PRR	Pattern recognition receptor
HD	Human α-defensin
MMP	Matrix metalloprotease
HBD-1	Beta-defensin-1
MBD-2	Murine β-defensin 2
DC	Dendritic cell
sIgA	Secretory immunoglobulin A
MLNs	Mesenteric lymph nodes
TGF-β	Transforming growth factor-β
RA	Retinoic acid
J chain	'joining' chain
pIg	Polymeric-immunoglobulin
V_H	Variable region
BAFF	B cell-activating factor
APRIL	A B cell proliferation-inducing ligand
LPS	Lipopolysaccharide
MAMPs	Microbe-associated molecular patterns
TLRs	Toll-like receptors
NLRs	Nod-like receptors
LRR	Leucine-rich repeat
TIR	Toll-interleukin 1 receptor
CARD	Caspase recruitment domain
NOD	Nucleotide-binding oligomerization domain
PG	Peptidoglycan
MyD88	Myeloid differentiation primary response gene 88
MAPK	Mitogen-activated protein kinase
NF-κB	Nuclear factor κB
IKK	IκB kinase
IL	Interleukin
IFN	Interferon
TNF	Tumor necrosis factor
MCP-1 or CCL-2	Monocyte chemotactic protein-1
PPARγ	Peroxisome proliferator-activated receptor-γ

TOLLIP	Toll-interacting protein
TSLP	Thymic stromal lymphopoietin
IDO	Indoleamine 2,3 dioxygenase
IBD	Inflammatory bowel disease
AMPs	Antimicrobial peptides
PP	Peyer's patch
FAE	Follicle-associated epithelium

Contents

1 A Delicate Balancing Act

The mammalian gastrointestinal tract (GI-tract) is specialized in the digestion and absorption of nutrients and in multicellular animals, it has become the dominant arena for host-microbial interactions. The concentration of bacteria forms a gradient from the stomach, and small intestine (SI) to the colon where densities reach around 10^{12} bacteria per gram of luminal content. Here mammals exploit the genetic and metabolic attributes of the resident bacteria to derive energy from the fermentation of otherwise indigestible carbohydrates. In humans, this accounts for 6–9% of the total energy requirement whereas this can be as high as 44% for ruminants (McNeil 1984; Hume et al. 1997). Additional symbiotic functions include the competitive exclusion of pathogens and the production of essential vitamins and amino acids.

The coevolution of mammals with intestinal bacteria had a strong impact on the host immune system, which needed to develop the ability to avoid excessive inflammatory responses to antigens and commensal bacteria while retaining the capacity to defend the body against infections of pathogenic microorganisms. Inflammation, the acute response to harmful stimuli including microorganisms, is a crucial component of this host defense. In part, the protective consequences of triggering inflammation result from increased vascular permeability and attraction of phagocytes to engulf and kill invading microorganisms. The interaction of antigen-presenting phagocytes with the adaptive immune system then leads to protracted cellular and humoral responses and immunological memory against "non-self" antigens. However, if the host is not immunologically tolerant to commensal microbes it can cause considerable damage to its own tissues as seen in IBD. The host must also elicit a localized inflammatory response to an invading pathogen while, at the same time, preserving homeostasis in the uninfected tissues. It is this delicate act of balancing opposing functions that we focus on in this review.

In the intestine, homeostasis of inflammation is maintained by a complex network of microbe and host cell interactions involving numerous negative feedback and feed-forward controls. The interacting cell types include epithelial cells, connective tissue cells and both innate and adaptive immune cells (macrophages, DCs and lymphocytes) residing in the LP (Fig. 1). The LP is a layer of lose connective tissue between the epithelium and the muscularis mucosa, containing lymphatic ducts and blood vessels. The main homeostatic mechanisms involve the compartmentalization of bacteria by physical and chemical barriers, the continuous surveillance of the microbial content by innate receptors, the induction of secretory antibody responses and a state of immunological tolerance to commensals and food antigens. In this review, we discuss in more detail how the interplay between these complex systems maintains homeostasis of inflammation in the SI and colon, which are permanently colonized by commensal bacteria and sometimes opportunistic or obligate pathogens.

2 Keeping Microbes at a Distance

At the tissue level, antimicrobial factors, secretory antibodies and physical barriers limit exposure of the mucosal tissue to microorganisms. The mucus layer covering the epithelium and the epithelium itself constitute the two major physical barriers (Fig. 1).

2.1 *Physical Barriers, the Mucus Layer*

The first physical barrier encountered by the luminal microbes is the mucus, a thick layer of mucin glycoprotein secreted by specialized goblet cells in the

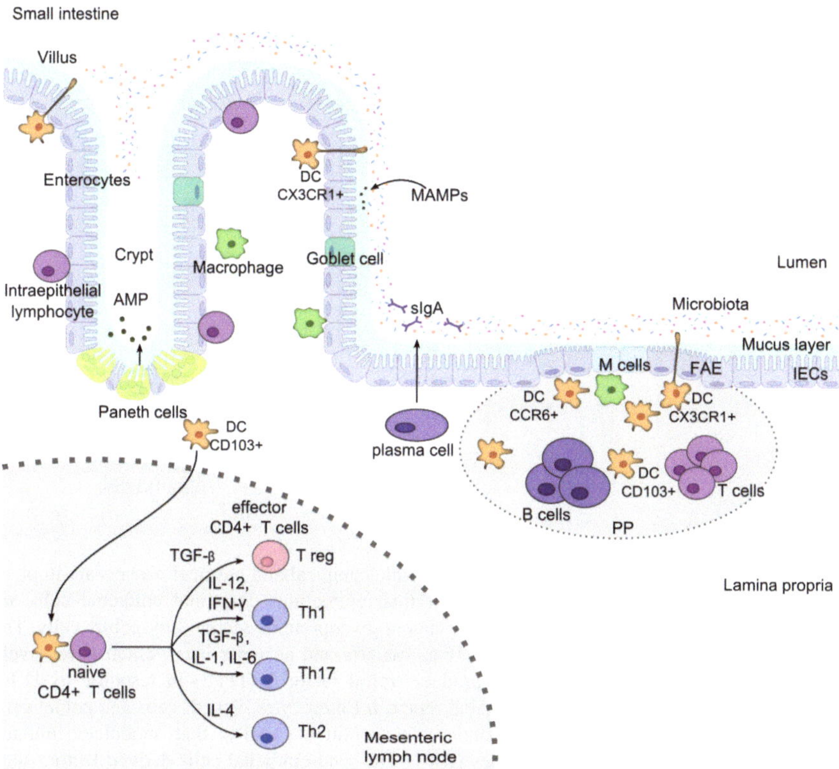

Fig. 1 The mucosal epithelium, lamina propria cells and gut-associated lymphoid tissue in the small intestine. Intestinal homeostasis is maintained by several mechanisms involving epithelial cells, connective tissue and immune cells. A single layer of intestinal epithelial cells (IECs) separates the luminal contents from the lamina propria (LP). In the basolateral space between epithelial cells are positioned intraepithelial lymphocytes. The epithelium is covered by a thick layer of mucus produced by goblet cells. Antimicrobial peptides (AMPs) and secretory antibodies provide additional protection mechanisms. Enterocytes and specialized Paneth cells in the crypts are able to secrete AMPs preventing bacterial growth in proximity to the epithelium. Plasma cells in the LP produce secretory immunoglobulin A (sIgA) that mediate bacterial exclusion. The luminal content is continuously sampled by specialized epithelial cells and immune cells. In the follicle-associated epithelium (FAE) of the Peyer's patch (PP) or isolated lymphoid follicles microfold (M) cells transcytose bacteria and antigens. In the LP and PPs, $CX3CR1^+$ macrophages and dendritic cells (DCs) sample microbes and antigens through the epithelium to support inflammatory responses. DCs, macrophages and M cells deliver antigens and bacteria to antigen-presenting cells (APCs), such as $CCR6^+$ DCs. Activated APCs then move to the follicular area of the PP or they migrate to the mesenteric lymph nodes (MLNs) through the lymphatic system and present the antigen to naïve lymphocytes. In the MLNs or in the PPs, mature DCs drive the differentiation of naive $CD4^+$ T lymphocytes into effector T cells, helper T cells (Th1, Th2, Th17) or regulatory T cells (Tregs). The $CD103^+$ population of DCs in the LP plays a key role in generating inducible Tregs

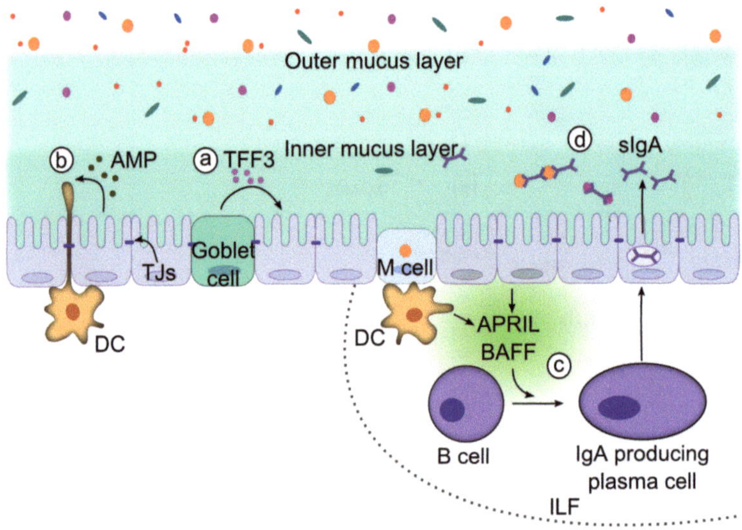

Fig. 2 The physical and chemical defenses. Multiple chemical and physical barriers are in place to separate the luminal content from the intestinal epithelium. Intestinal epithelial cells are covered by a layer of mucus that consists of mucin glycoproteins secreted by goblet cells. The mucus layer forms an outer layer permeable to bacteria and an inner layer which is relatively devoid of bacteria. **a** Goblet cells also produce trefoil factor 3 (TFF3) in response to TLR2 signaling which mediates intestinal epithelial repair. **b** Enterocytes, Paneth cells and goblet cells secrete constitutive and inducible antimicrobial peptides (AMPs) that modulate immune functions. **c** In isolated lymphoid follicles (ILFs), DC- and epithelial cell- derived factors such as the B cell-activating factor (BAFF) and a proliferation-inducing ligand (APRIL), drive B cell differentiation into plasma cells that produce secretory immunoglobulin A (sIgA). **d** sIgA is transported through IEC to the lumen where it can bind to bacteria to prevent their attachment or invasion and neutralize toxic molecules

epithelium. The mucus is constantly produced and removed by peristaltic flushing. In the colon, the mucus layer extends about 150 μm from the epithelial cell surface and can be divided into two structurally different layers, an outer mucus layer containing bacteria and a firmly attached inner layer that is almost devoid of bacteria (Johansson et al. 2008) (Fig. 2). However, enteric pathogens such as *Campylobacter jejuni, Samonella enterica* and *Vibrio cholerae* are able to penetrate the mucus using flagella, which allow movement toward or away from a chemical stimulus. The spiral-shape and secreted proteases of these pathogens are also thought to facilitate movement through viscous mucus (Lee et al. 1986; Guerry 2007; Hugdahl et al. 1988; Takata et al. 1992). Flagella are not only a feature of pathogenic bacteria and some prominent members of the microbiota of animals and humans such as *Roseburia* sp. and *Wolinella succinogenes*, a close relative of *C. jejuni*, are flagellated. It is currently not known whether the mucus layer in the SI is also stratified as described for the colon and if some epithelial surfaces, such as the tips of villi, are more accessible to the contents of the lumen.

Mice lacking MUC2, the gene encoding the major mucin in the human intestine, spontaneously develop colonic inflammation, highlighting the protective role of the mucus in limiting direct contact between bacteria and epithelial cells (Johansson et al. 2008; Van der Sluis et al. 2006). Furthermore, it was shown recently that one of the initial events in the development of dextran sodium sulfate (DSS)-induced colitis is the alteration of the inner mucus layer. DSS-induced alterations in the inner mucus layer make it permeable to bacteria, which, therefore, can reach and contact epithelial cells triggering an inflammatory reaction (Johansson et al. 2010).

The outer layer of the mucus is home to many commensal mucin-degrading bacteria that derive nutrients from enzymatic breakdown of the glycan structures. Nevertheless very few mucin dwelling bacteria have been isolated and characterized except for *Akkermansia muciniphila* (Derrien et al. 2004). Interestingly, short-chain fatty acids, one major product of bacterial fermentation, have been shown to modulate the expression of MUC2 in a human goblet cell-like cell line. This highlights a potentially important regulatory link between metabolites produced by fermentative colonic bacteria and the host defenses (Burger-van Paassen et al. 2009).

Goblet cells also facilitate mucosal protection and epithelial repair through the production of trefoil factor 3 (TFF3) which induces migration and anti-apoptotic pathways in IECs (Fig. 2) (Mashimo et al. 1996). Goblet cells appear to regulate TFF3 via a Toll-like receptor 2 (TLR2) dependent pathway (described in Sect. 3) (Podolsky et al. 2009). Thus, TLR2 knockout mice are deficient in TFF3 and have an increased susceptibility to DSS-induced colitis. This provides another example of how the sensing of bacteria by the host intestinal epithelium regulates the expression of homeostatic factors.

It is not known why commensal bacteria can colonize the outer mucus layer but not the inner layer of the colonic mucus. We also need more knowledge about the diffusion of bacterial TLR ligands through the mucus, mucus degrading bacteria and the mucus structure and function in the SI.

2.2 Physical Barriers, the Intestinal Epithelium

The most important physical barrier is the intestinal epithelium, a 30 μm single layer of IECs which separates the lumen from the underlying LP and the inside of the body. IECs are polarized having an apical and a basolateral pole each with a distinct composition of membrane proteins, lipids and receptors. The apical pole of mature enterocytes contains microvilli, finger-like membrane protrusions that increase the surface area for exchange of nutrients and water. The basolateral pole is attached to a matrix of loose connective in LP. The intestinal epithelium is renewed approximately every 5 days in humans due to proliferation and differentiation of multipotential Lgr5-expressing stem cells located in the crypts of Lieberkühn (Sato et al. 2009). Terminally differentiated cells undergo apoptosis and exfoliation at the tips of small intestinal villi or the equivalent epithelial cuff in the colon. This dynamic renewal process also serves as a protective mechanism to remove cells that may have been damaged or invaded by pathogens. As the cells migrate upwards from the crypt, they differentiate into three different cell lineages; enterocytes, enteroendocrine cells and mucus producing goblet cells (Yen and Wright 2006). In the SI, the stem cells migrating to the bottom of the crypt differentiate into Paneth cells which produce a range of antimicrobial factors to protect the crypt cells from infection with microorganisms (Garabedian et al. 1997; Salzman et al. 2007). Remarkably, the process of continuous epithelial replacement occurs without compromising the integrity of cell-to-cell junctions and intestinal permeability. The importance of maintaining junctional integrity is evident from studies in mice which have molecular defects in tight junction (TJ) proteins, such as JAM-A (Laukoetter et al. 2007), or regulatory mechanisms affecting TJ properties (Su et al. 2009). TJs are the most apical structures of cell-to-cell junctions and are composed of transmembrane proteins (such as occludins and claudins) and intracellular scaffold proteins (such as ZO-1, -2 and -3). TJs seal the intercellular space between adjacent cells thus creating a selective barrier keeping out microorganisms and macromolecules while allowing diffusion of fluids, electrolytes and small molecules through the paracellular space. TJs are crucial for the development and maintenance of epithelial polarity as they also prevent diffusion of plasma membrane components from the apical to the basolateral pole and vice versa. In mouse models, increased intestinal epithelial permeability may precede the development of colitis (Olson et al. 2006; John et al. 2011) and is characteristic of chronic intestinal inflammation in humans and animals due to the effect of inflammatory cytokines on TJ regulation (Ye et al. 2006).

Many enteric pathogens have evolved ways to disrupt the TJs of epithelial cells as a means to gain nutrients or to cross the epithelial barrier. For example, this can involve proteolytic degradation of specific TJs, e.g. with enterotoxin/fragilysin producing strains of *Bacteroides fragilis.* Disruption of TJs may also occur by phosphorylation or dephosphorylation of TJ proteins as a consequence of the injection of specific effectors into the host cell by a type III secretion system. For example, during infection with Enteropathogenic *Escherichia coli* (EPEC) a type III secretion system delivers several effectors into the host cytoplasm (McDaniel et al. 1995). One of the effectors, the intimin receptor Tir, induces the accumulation of actin and cytoskeletal proteins beneath the attached microcolonies of EPEC. This leads to effacement of the microvilli and formation of a cell protrusion known as a pedestal. Injection of the EspF effector contributes to the disruption of TJs. In addition to these morphological and physiological changes, EPEC also causes the dephosphorylation of occludin (Simonovic et al. 2000) and altered distribution of ZO-1 (Philpott et al. 1996). Similarly, several viral pathogens breach the epithelial barrier by entering through junctions (O'Hara and Buret 2008; Bergelson 2009) and exploiting the binding to TJ proteins (Raschperger et al. 2006; Antar et al. 2009). Rotaviruses alter occludin localization to TJs by changing the levels of non-phosphorylated occludin (Beau et al. 2007).

In contrast to the enteric pathogens, different species of commensal lactobacilli have been reported to enhance barrier function and/or protect against barrier disruption by pathogens in vitro (Wells et al. 2008). Recently, *Lactobacillus plantarum* was shown to modulate TJ composition in human volunteers. In this landmark study, administration of *L. plantarum* to human volunteers increased the localization (immunofluorescent staining) of occludin and ZO-1 in the TJs in duodenal tissue biopsies. In vitro, TJ modification was mediated by TLR2 ligands and conferred protection against disruption of the TJs by phorbol ester (Karczewski et al. 2010). These results are in agreement with recent studies in mice showing protection from colitis by administration of TLR2 ligands and growing evidence that TLR signaling in the intestinal epithelium is crucial for homeostasis (Cario et al. 2007). Apart from serving as a semi-permeable physical barrier, the epithelium plays a key role in secreting antimicrobial factors to kill pathogens and microbes that come into close proximity to the apical surface which is discussed in the following section.

2.3 Homeostatic Effects of Epithelial Antimicrobial Peptides and Polypeptides

Antimicrobial peptides (AMPs) and polypeptides provide the first line of chemical protection against pathogenic or commensal organisms that penetrate the mucosal barrier. In this section, we specifically review the contribution of the enterocyte and Paneth cell produced antimicrobial factors to homeostasis. There have been excellent reviews covering other aspects such as structure and mechanism of action, including also immune cell defensins (Klotman and Chang 2006; Lai and Gallo 2009). Paneth cells are localized in the crypts of Lieberkühn throughout the SI of most mammals. A characteristic feature of Paneth cells is the extensive Golgi network that directs large electron dense secretory granules to the apical membrane for release into the crypt. The granules contain several antimicrobial factors including α-defensins, angiogenin-4, secretory phospholipases, lysozyme, Reg3-γ as well as pro-inflammatory cytokines and PRR co-receptors (Stappenbeck 2009). In the mouse, six α-defensins have been characterized but genetic evidence suggests there may be as many as 17 defensin encoding mRNA expressed in the crypt. In humans, Paneth cells express α-defensins 5 and 6 (human α-defensin, HD). The α-defensins are secreted as a precursor form which is activated by trypsin (matrix metalloprotease (MMP) 7 in the mouse) to confer antimicrobial activity.

Evidence for the role of Paneth cell defensins in protection of the epithelial barrier was demonstrated by expression of human HD-5 in transgenic mice. As a result, these mice were much more resistant to challenge with *Salmonella typhimurium* (Salzman et al. 2003) than the wild-type mice strain. Mice deficient for MMP7 were used to show that the activity of the α-defensins against the normal microbiota is important in shaping the composition of the intestinal microbiota (Salzman et al. 2010).This may be due to differences in the sensitivity of different microorganisms to α-defensins, their location and inter-dependencies in the intestinal environment. This has important implications for intestinal inflammatory diseases because changes in defensin production may contribute to pathophysiology of a disease by altering the composition and thus the symbiotic or antagonistic activities of the microbiota. Notably, NOD2 polymorphisms in Crohn's disease patients are associated with reduced expression of AMPs and loss of immune tolerance to the microbiota (Hugot et al. 2001; Wehkamp et al. 2005; Maeda et al. 2005). Whereas some components of Paneth cell granules are expressed at equal level in germ-free and colonized mice, others are induced by microbial exposure (e.g. Reg3-γ and angiogenin-4) (Cash et al. 2006; Hooper et al. 2003). Using a transgenic mouse model in which MyD88 was expressed in Paneth cells in a MyD88-/- background, it was shown that TLR signaling by commensal bacteria induces expression of several antimicrobial factors including Reg3-γ and Reg3-β that inhibit growth of bacterial pathogens and commensals in the crypts of the SI (Vaishnava et al. 2008). Loss of MyD88-dependent TLR signaling did not influence the density of the luminal bacteria emphasizing the key role for Paneth cells in protecting the crypt from colonization with commensal or pathogenic bacteria.

Intestinal enterocytes have also been shown to express antimicrobial factors including β-defensins and Reg3 proteins, which are evolutionarily conserved in mammals. In humans, beta-defensin-1 (HBD-1) is constitutively expressed, while HBD-2, HBD-3 and HBD-4 are inducible (Pazgier et al. 2006). Although HBD-1 is highly expressed in all human epithelia it was suggested to have only minor antimicrobial activity compared to other defensins. However, recently it was shown that reducing the disulfide bonds in HBD-1, mimicking the reducing potential found in the anaerobic environment of the intestine, unmasks a potent antimicrobial activity (Schroeder et al. 2011). HBD-2 and HBD-3 are induced by microbial products and inflammatory cytokines (Harder et al. 2001; Liu et al. 2002; Garcia et al. 2001; Sorensen et al. 2005). Inducible expression of HBD-2 and Reg3 proteins in enterocytes depends on TLR and MyD88-dependent signaling (Vaishnava et al. 2008; Vora et al. 2004).

In addition to antimicrobial activities, HBD-2 and HBD-3 have been proposed to serve as a chemo-attractants for a broad spectrum of leukocytes in a CCR6- and CCR2-dependent manner (Rohrl et al. 2010; Biragyn et al. 2001). Murine β-defensin 2 (MBD-2) and HBD-3 have also been shown to have immunomodulatory effects on DC function by acting as an endogenous ligand for TLR4 and TLR1/2 heterodimers, respectively (Fig. 2) (Biragyn et al. 2002; Funderburg et al. 2007). Additionally, HBD-2 has been shown to stimulate epithelial restitution and barrier repair in an autocrine fashion by binding to CCR6 expressed on the apical membrane of epithelial cells (Fig. 2) (Vongsa et al. 2009). Taken together, these findings suggest a dual or even multifunctional role for certain intestinal defensins in intestinal homeostasis. The homeostatic mechanisms of AMPs include epithelial restitution, immunomodulation and the innate killing of microorganisms.

2.4 Secretory Antibodies and the "Mucosal Firewall"

Mammals produce much larger amounts of secretory immunoglobulin A (sIgA) in the mucosa than any other class of immunoglobulin in the body, reflecting its central role in immunity and protection against microorganisms in the intestine. B cell activation and proliferation leading to sIgA production occurs in the organized follicular lymphoid structures of the GALT. The lymphoid structures are present as isolated follicles intestinal cell wall or as large aggregates in the Peyer's patches (PPs) of the SI. The follicle-associated epithelium (FAE) overlaying the PP contains specialized M cells, which transport antigens and microorganisms to APC in close proximity to the basal membrane (Fig. 1). Upon activation, the APCs move from the sub-epithelial dome to the follicular area and present the antigen to naïve lymphocytes; alternatively, they migrate to the mesenteric lymph nodes (MLNs) through the efferent lymphatic ducts (Fig. 1). The engagement of B cells with antigens in the GALT causes them to migrate to the interface between the B and T cell zones where they become fully activated as a result of interaction with $CD4^+$ T cells and DCs. In the GALT and MLNs, class switching of B cells from IgM to IgA is driven by a subset of DCs which release interleukin-10 (IL-10), transforming growth factor-β (TGF-β), IL-6 and retinoic acid (RA), a hormone-like growth factor for epithelial and other cell types (Iwasaki and Kelsall 2001; Mora et al. 2006; Sato et al. 2003). Activated B cells then migrate to the LP from the lymph or blood, or directly from isolated lymphoid follicles. In the LP, the IgA^+ B cells differentiate into plasma cells which secrete dimeric IgA coupled to the antibody 'joining' (J) chain. Secretory IgA or IgM is transported across the epithelium by the polymeric-immunoglobulin (pIg) receptor expressed on the basolateral membrane of epithelial enterocytes. The pIg receptor interacts with antibodies carrying the J chain, especially pentameric IgM or dimeric IgA. The conventional T cell-dependent pathway for high affinity IgA antibody production described above, is associated with B-2 cells and takes around 5–7 days to develop.

In addition to the T cell-dependent pathway for IgA production described above there is a faster mechanism for generating IgA responses to highly conserved antigenic determinants on commensal bacteria and pathogens (Macpherson et al. 2008; Suzuki and Fagarasan 2008). In mice, this involves a specialized subset of B-1 cells that can rapidly produce secretory IgA in the absence of help from $CD4^+$ T cells (Fagarasan and Honjo 2000). Much of this secretory antibody appears to be polyreactive, binding several antigens and is commonly known as "natural" antibody. The repertoire of 'natural IgA' is restricted and affinity maturation is limited, although the heavy-chain variable region (V_H) genes used by IgA-producing plasma cells in the gut are somatically hyper-mutated, to diversify antibody specificity (Macpherson et al. 2008). Humans lack canonical B-1 cells but it has been suggested that they may have functionally equivalent cells that can colonize the lymphoid tissues (Weller et al. 2004). T cell-independent secretory antibody responses are generated in the isolated lymphoid follicles by the direct presentation of antigen to B cells by DCs and M cells (Fig. 2) (Batista and Harwood 2009; Batista et al. 2001; Bergtold et al. 2005; Tsuji et al. 2008). IECs also participate in B cell responses by releasing B cell-activating factor (BAFF) and a B cell proliferation-inducing ligand (APRIL) which play roles in IgA class-switching and B cell survival (reviewed in Wells et al. 2010).

B cells and T cells activated in the GALT and MLN acquire homing receptors that target them back to mucosal tissues and this ensures that the immune responses occur selectively and locally in the mucosa. Furthermore, DCs activated by bacterial interaction in the PP are retained in the MLNs and PP (Macpherson and Uhr 2004). This has led to the "mucosal firewall" concept that proposes that productive mucosal sIgA responses to intestinal bacteria can develop without inducing unnecessary systemic immunity (Macpherson and Uhr 2004).

Much of the IgA produced in the gut appears not to be reactive to the commensal microbiota. Nevertheless, in germ-free mice IgA is present at very low levels in the gut and soon after colonization with bacteria the number of IgA positive B cells increases in the lymphoid tissue followed by the appearance of sIgA secreting plasma cells in the intestinal LP (Shroff et al. 1995). At the epithelial surface, sIgA can bind to viruses and bacteria to prevent or inhibit their attachment to and/or invasion of epithelial cells, a process known as immune exclusion (Macpherson et al. 2001). Just as important may be the fact that sIgA can also interact with the antigens presented by intracellular pathogens in endosomes during the pIgR–mediated transport through epithelial cells. These interactions have been shown to inhibit key steps in viral assembly and play an important role in immunity to viral infection (Mazanec et al. 1995). Recent studies suggest that immune exclusion also helps shape the bacterial community through immunological selection (Suzuki and Fagarasan 2008; Sonnenburg et al. 2004; Peterson et al. 2007). Furthermore, when mice lacking a functional adaptive immune system, are colonized with bacteria, they display a more robust innate immune response than their immunocompetent wild-type counterparts (Cash et al. 2006; Keilbaugh et al. 2005). This indicates that in normal mice the adaptive immune response plays a critical role in minimizing activation of the innate immune system by the gut microbiota.

Other homeostatic mechanisms for sIgA include the removal of antigen complexes formed in the LP thereby reducing the likelihood of inflammatory reactions and systemic responses (Mestecky et al. 1999). This concept is supported by the finding that individuals carrying secretory antibody deficiencies are more prone to autoimmune diseases, such as celiac disease. Secretory IgA also helps to prevent inflammatory reactions by neutralizing intracellular pro-inflammatory components such as lipopolysaccharide (LPS) (Fernandez et al. 2003). Moreover, in contrast to IgG and IgM, IgA does not activate the classical complement cascade leading to the production of C5a which is chemotactic for phagocytes.

To avoid immune exclusion, many pathogens that colonize the mucosal surface produce IgA proteases that cleave IgA1 in the hinge region. Bacteria producing IgA proteases include the major bacterial pathogens that instigate infection at mucosal surfaces of the respiratory tract, such as *Streptococcus pneumoniae*, *Haemophilus influenzae* and *Neisseria meningitidis*, or the genital tract such as *Neisseria gonorrhoeae*. As the IgA1 proteases of human pathogens only cleave the IgA1 subclass found in humans and closely related primates it has not been easy to prove their role in virulence. However, IgA1 protease has been shown to compromise IgA-mediated killing of *S. pneumoniae* both in vitro and in vivo (Kadioglu et al. 2008).

In summary, it is clear that IgAs are critical to mucosal protection against pathogens but it is only recently that we began to appreciate its homeostatic roles in shaping the intestinal microbiota, preventing mucosal inflammation by immune exclusion, removal of antigen–antibody complexes in the LP and the neutralization of inflammatory mediators.

3 Detecting Bacterial Incursions

Apart from its critical role as a physical barrier, the intestinal epithelium provides an early warning system for sensing microbes. Through the expression of pattern-recognition receptors (PRR), the epithelium senses the microbial content of the lumen and participates in the innate immune response by secreting a potent array of cytokines and chemokines. PRR recognize microbe-associated molecular patterns (MAMPs), formerly known as pathogen-associated molecular patterns, consisting of conserved structures found both on pathogenic and non-pathogenic microorganisms. PRR are expressed along the entire GI-tract by enterocytes (Cario et al. 2000), Paneth cells (Rumio et al. 2004) and immune cells such as DCs (Hart et al. 2005). The innate sensing mediated by PRR is essential to detect pathogens and initiate immune responses, but needs to be tightly regulated to maintain immune homeostasis toward the trillions of microbes that inhabit the GI-tract.

3.1 Gut Innate Sensors and Microbial Interactions

IECs have the capacity to express a range of PRR including members of the well-characterized families of Toll-like receptors (TLRs), and intracellular Nod-like receptors (NLRs). Several TLRs are expressed in IECs in human and mice, while among the NLRs only NOD1 and NOD2 have been well characterized in the intestine. Ten TLRs have been identified in humans and 12 in mice. TLRs are type I transmembrane glycoproteins, that share a conserved structure, with an extracellular domain that contains multiple leucine-rich repeats (LRR), a short trans-membrane domain and a highly conserved intracellular domain, the

Table 1 PRR, ligands and subcellular localization

Receptor	Subcellular localization	Ligand	Origin of the ligand
TLR2	Cell surface	Lipopeptide	Bacteria
		PG	G+ bacteria
		LTA	G+ bacteria
TLR2/1	Cell surface	Triacylated lipopeptide	G− bacteria, mycoplasma
TLR2/6	Cell surface	Diacylated lipopeptide	G+ bacteria, mycoplasma
TLR3	Cellular compartment	dsRNA	Viruses, virus infected cells
TLR4/ MD2	Cell surface	LPS	G− bacteria
TLR5	Cell surface	Flagellin protein	Bacteria
TLR7	Cellular compartment	ssRNA	ssRNA viruses
TLR9	Cellular compartment, cell surface	DNA	DNA viruses, bacteria
TLR11	Cell surface	Uropathogenic bacterial components	Uropathogenic bacteria
NOD1	Cell cytoplasm	Meso-DAP	PG from G−, some G+ , mycobacterium
NOD2	Cell cytoplasm	MDP	PG from G−, G+ bacteria, mycobacterium

PG peptidoglycan, *LTA* lipoteichoic acid, *LPS* lipopolysaccharide, *DAP* diaminopimelic acid, *MDP* muramyl dipeptide, *G+* Gram-positive, *G−* Gram-negative, *ssRNA* single-stranded RNA, *dsRNA* double-stranded RNA

Toll-interleukin 1 receptor (TIR) domain. The extracellular LRR domains are responsible for the binding to MAMPs and contribute to the homo- or hetero-dimerization of the receptors. Dimerization of the intracellular TIR domains leads to the recruitment of adaptor molecules (e.g. MyD88, TIRAP and TRIF) to initiate the signaling cascade. Each PRR recognizes a specific molecular pattern and can be expressed on the cell surface, in intracellular compartments or in the cytosol (Table 1). TLR1, 2, 4, 5, 6 and 11 recognize mainly microbial membrane components and are expressed on the cell surface, TLR3, 7, 8 and 9 recognize microbial nucleic acids and are expressed in intracellular compartments such as the endoplasmic reticulum, endosomes, lysosomes and endolysosome (Table 1). NOD1 and NOD2 are intracellular proteins that contain an N-terminal caspase recruitment domain (CARD), a nucleotide-binding oligomerization domain (NOD) and a C-terminal regulatory domain with multiple LRRs that bind to peptidoglycan (PG) fragments in the cytosol. NOD1 senses Meso-diaminopimelic acid from the PG of Gram-negative bacteria, while NOD2 recognizes muramyl dipeptides that are common to the PG of both Gram-negative and Gram-positive bacteria (Girardin et al. 2003a, b; Chamaillard et al. 2003) (Table 1). TLR2 can form heterodimers with TLR1 or TLR6 leading to the recognition of distinct ligands. The heterodimers TLR2/1 and TLR2/6, bind triacylated and diacylated lipoproteins from Gram-positive and Gram-negative bacteria, respectively. Additionally, TLR2/6 can bind lipoteichoic acid, a lipidated glycan polymer present in the cell wall of Gram-positive bacteria. In all cases, two lipid chains present on each ligand insert into a hydrophobic pocket in TLR2 whereas the dimerization with TLR1 or 6 is determined by specific interactions with other parts of the ligands. The highly conserved lipid A of LPS, a component of the outer membrane of Gram-negative bacteria, is bound to MD-2 in a complex with TLR4. Presentation of LPS to MD-2 is facilitated by CD14 and LPS binding protein. Another important bacterial MAMP is flagellin, a protein subunit of bacterial flagella that is typically recognized by TLR5.

TLR signaling pathways have been reviewed in detail recently (Kawai and Akira 2010). Essentially, after ligand binding, TLRs undergo dimerization and conformational changes that result in the recruitment of TIR-containing adaptor proteins. All TLRs, except TLR3, recruit the myeloid differentiation primary response gene 88 (MyD88), which in turn activates the mitogen-activated protein kinase (MAPK) pathway and the nuclear factor κB (NF-κB) pathway signaling cascades (Janssens and Beyaert 2002). TLR3 and TLR4 can signal also through a different pathway via the recruitment of the adaptor protein TRIF, leading to the expression of type 1 interferons. Similarly, NLRs activate the MAPK pathway and the canonical NF-κB pathway signaling cascades. These signaling cascades lead to the activation of transcription factors, such as NF-κB, that, after translocation to the nucleus, stimulate the transcription of genes encoding AMPs, pro-inflammatory cytokines and chemokines, ultimately initiate the inflammatory response. In an inactivated state, NF-κB is located in the cytosol as a protein complex with the inhibitory protein IκBα. Activation of NF-κB is triggered by different stimuli (e.g. TLR, cytokine, NLR, stress inducers) that lead to the phosphorylation of IκBα by the enzyme IκB kinase (IKK). Subsequently, IκBα is ubiquitinated and degraded by the cell proteasome, leading to its dissociation from NF-κB. Activated NF-κB then translocates into the nucleus and induces the transcription of specific genes. In IECs, activated NF-κB can induce the production of a broad range of chemokines and cytokines including interleukins (ILs), interferons (IFNs), tumor necrosis factors (TNFs) and growth factors. In case of microbial colonization or invasion, the most abundant cytokine secreted by human IECs is IL-8. IL-8 forms a gradient that, together with others factors, acts as chemoattractant for neutrophils that are recruited at the site of infection. IECs can also produce the DC chemokine CCL20 and the monocyte chemotactic protein-1 (MCP-1 or CCL-2), inducing recruitment of monocytes and basophils.

The importance of PRR-mediated recognition in the induction of inflammatory responses to pathogens is evident from infection studies in PRR-deficient mice. Mice lacking TLR4, NOD1 or the common adaptor protein MyD88 have impaired immunity to infectious agents (Schilling et al. 2003; Brandl et al. 2007; Lebeis et al. 2007; Viala et al. 2004). It is not surprising, therefore, that pathogens have evolved different strategies to evade the innate recognition. For example, they can target TLR signaling in order to suppress the host immune responses to benefit their survival and persistence. *Brucella* species are highly infectious intracellular pathogens, this bacteria encode a TIR domain-containing protein that, mimicks the human adaptor protein TIRAP and blocks TLR2- and TLR4- mediated NF-κB activation (Radhakrishnan et al. 2009). The NF-κB pathway can also be a target for pathogens. For example, the pathogen *S. typhimurium* can inject virulence effectors into the host cells through a type III secretion system. One of these effector proteins, AvrA, inhibits the NF-κB pathway and induces apoptosis in IECs (Collier-Hyams et al. 2002). Some pathogens express MAMPs that are not recognized by the PRRs, for instance *Campylobacter jejuni* and *Helicobacter pylori* produce subclasses of flagellin that do not activate TLR5 (Andersen-Nissen et al. 2005). Interestingly LPS from a commensal *Bacteroidetes* species has also been found to have a low endotoxicity (Weintraub et al. 1989). The lipid A part of LPS is pentacylated and consequently a weak activator of TLR4 compared with the hexacylated lipid A of other commensals and pathogens. This indicates that commensals may have also evolved to avoid detection by innate receptors as a means to favor their persistence in the intestinal tract.

3.2 Regulation of Innate Sensing

The section above clearly illustrates the requirement for TLR signaling and inflammatory responses to protect the host from infection with pathogens. However, to peacefully coexist with the commensal microbiota and avoid perpetual inflammation, PRR signaling needs to be carefully regulated. Mice lacking TLR9, 4, 2 and MyD88 become more susceptible to DSS-induced colitis (Lee et al. 2006; Rakoff-Nahoum et al. 2004). In TLR2 and TLR4 deficient mice or mice depleted of bacteria with multiple antibiotics, the severity of colitis and mortality levels are worse than in wild-type mice, suggesting an additional role for NF-κB in homeostasis, epithelial repair and renewal processes (Rakoff-Nahoum et al. 2004). This concept is also supported by the finding that mice lacking TLR5 have the tendency to develop spontaneous colitis (Vijay-Kumar et al. 2007). Further evidence for the crucial role of the NF-κB regulator in intestinal homeostasis came from studies in mice with a conditional knockout of the intestinal epithelial IKKγ, or both IKKα and IKKβ. IKK knockout mice lose the capacity to activate NF-κB signaling and develop severe chronic intestinal inflammation (Nenci et al. 2007).

The chemical and physical barriers described in the preceding sections play a key role in keeping the microbes at a distance but in addition several other sophisticated regulatory mechanisms have evolved to avoid excessive immune responses to non-pathogens. These mechanisms include (i) the regulation of TLR expression, (ii) TLR localization, (iii) differential apical and basolateral TLR signaling, (iv) negative feedback regulation of the NF-kB pathway and (v) the attenuation of NF-κB activation by commensal bacteria.

In the absence of inflammation, TLRs are expressed at low levels on the epithelial surface and their expression is highest in the intestinal crypts to guarantee the protection of stem cells. For example, TLR2 and TLR4 are expressed at low levels in human colonic epithelium and mainly in crypt cells (Abreu et al. 2001; Cario and Podolsky 2000; Furrie et al. 2005), TLR5 is expressed mainly in the colon (Cario and Podolsky 2000). In the mouse, TLR2, 4, 5 and 9 are expressed in

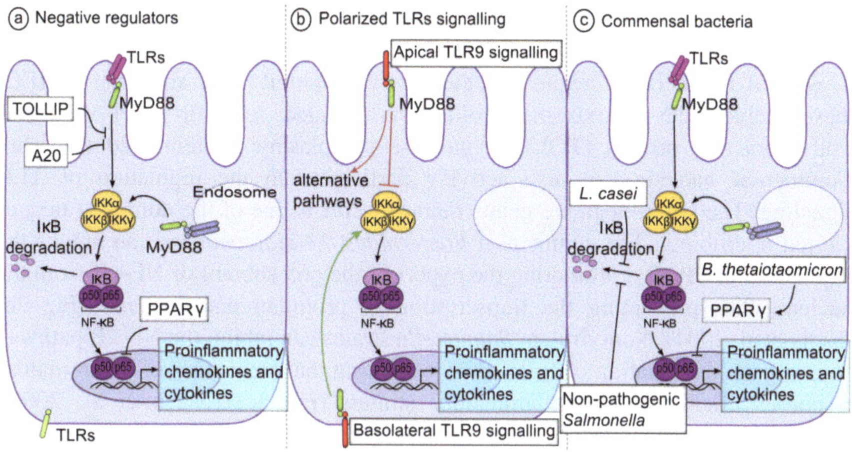

Fig. 3 Homeostasis of innate sensing in IEC. Toll-like receptor (TLR) sensing and signaling is regulated at multiple levels via interacting pathways that contribute to the maintenance of intestinal homeostasis. **a** Negative regulators of pro-inflammatory nuclear factor κB (NF-κB) signaling pathway: NF-κB activation pathway can be inhibited at different levels by NF-κB inhibitory regulators such as the peroxisome proliferator-activated receptor-γ (PPARγ), the Toll-interacting protein (TOLLIP) and the cytoplasmic protein A20. PPARγ competes with the p65 subunit of NF-kB and induces its nuclear export, thus limiting the activation of NF-kB. TOLLIP and A20 interfere with adaptor proteins essential for the signal transduction. **b** Polarized TLR signaling: Basolateral TLR9 stimulation (in *green*) induces NF-κB activation and transcription of proinflammatory genes. Whereas apical stimulation (in *red*), can suppress the inflammatory response and induces alternative signaling pathways. **c** TLR signaling can be modulated by commensal bacteria. *Bacteroides thetaiotaomicron* stimulates the expression of PPARγ, promoting the export of the p65 subunit of NF-κB from the nucleus. *Lactobacillus casei* and non-pathogenic strains of *Salmonella* inhibit the degradation of IκB

villus and crypt epithelial cells and in the FAE. Expression of TLRs may also be polarized in IECs and located on either the apical or basolateral membranes. For example in villi and crypts TLR2 is located on the apical side of epithelial cells. In the FAE, TLR4 and TLR5 are located at the apical pole, while TLR2 and TLR9 are located on both the apical and the basolateral side (Chabot et al. 2006). In polarized human colonic epithelial cell lines, TLR9 has been shown to be expressed on both the apical and the basolateral poles but activation leads to distinct responses depending on the location of the stimulus (Fig. 3b) (Lee et al. 2006). Basolateral TLR9 stimulation activates the NF-κB pathway while apical stimulation prevents NF-κB activation conferring tolerance to the chronic TLR challenges represented by the commensal microbiota (Fig. 3b) (Lee et al. 2006). TLR5 has been reported to be expressed on the basolateral surface in the human colonic epithelial cell line T84, such that it would only respond to flagella on pathogens that have crossed the epithelium (Gewirtz et al. 2001). However, his appears not to be the case for human intestinal Caco2 and HT29 cells (unpublished data O. Rossi) and in the mouse, TLR5 is present on both apical and basal membranes of the villus epithelium (Chabot et al. 2006) bringing into question the validity of this model in humans and mice in vivo.

Several intracellular negative regulators can inhibit TLR signaling in IECs these include the peroxisome proliferator-activated receptor-γ (PPARγ), the Toll-interacting protein (TOLLIP) and the cytoplasmic protein A20 (Fig. 3a). Commensal bacteria can also actively participate in the regulation of TLR signaling. The Gram-negative genus *Bacteroidetes* is one of the dominant taxa of the gut microbiota. The commensal *Bacteroides thetaiotaomicron* can induce the expression of PPARγ, promoting the export of the p65 subunit of NF-κB from the nucleus, thus preventing the transcription of proinflammatory genes (Fig. 3c) (Kelly et al. 2004). Non-virulent *Salmonella* strains can inhibit the NF-κB pathway blocking the degradation of IκBα, thus promoting the attenuation of inflammatory responses elicited by proinflammatory stimuli (Fig. 3c) (Neish et al. 2000). A similar mechanism is used by *Lactobacillus casei*, which was reported to down-regulate inflammation induced by an entero-invasive pathogen, protecting IκB from proteolytic degradation and blocking NF-κB translocation to the nucleus (Fig. 3c) (Tien et al. 2006).

4 Immune Cell Responses to Microbes

Besides the recognition by PRR, the microbial content of the lumen is constantly sampled by specialized cells and lymphoid structures along the gut. DCs and macrophages can directly sample the luminal antigens across the mucosa or encounter bacteria/antigens sampled by M cells in the GALT. Once activated, a subset of LP DCs migrates to the MLNs where they induce T cell responses. PPs APCs also interact with T cells in the inter-follicular regions driving their clonal expansion and differentiation. Macrophages, DCs and other immune cells are also responsible for the detection and phagocytic killing of invading bacteria. The crosstalk between IECs and immune cells is essential in driving the immune response against pathogens and in preserving the intestinal homeostasis (Wells et al. 2010).

4.1 Intestinal Dendritic Cells

DCs are specialized APCs, that regulate both innate and adaptive immunity (recently reviewed (Rescigno 2010)). They are found in the LP of both the small and large intestine and in the GALT, which includes the PP and draining lymph nodes. Typically, DCs migrate from the peripheral tissues to the lymphoid tissue where they present antigens to T cells. Most tissue resident DCs are immature and poorly immunogenic due to the low expression of MHC molecules and co-stimulatory molecules. Immature DCs undergo maturation when they sense

MAMPs or "danger"-associated molecular patterns released by damaged or stressed cells. Mature DCs express high levels of MHC, co-stimulatory molecules and cytokines for antigen presentation and immune signaling. In the lymphoid tissues, mature DCs interact with naive $CD4^+$ T lymphocytes inducing their activation, clonal expansion and differentiation. Depending on the phenotype of the DC and the cytokine environment, T cells will differentiate into different subsets of effector and regulatory T cells. $CD4^+$ T cell subsets include T helper cells, Th1, Th2 and Th17, and regulatory T cells, Tregs (Fig. 1). Each subset is characterized by a specific cytokine secretion profile to facilitate a different type of immune response. Th1 cells are important in promoting cell-mediated immunity against intracellular pathogens and secrete mainly IL-12, IFN-γ and TNF-α. Th2 cells produce IL-4, IL-5 and IL-13 to promote B cell development and antibody production (Mosmann and Coffman 1989). Th17 respond to extracellular bacteria and fungi and secrete mainly IL-17, IL-22 and IL-23 (Bettelli et al. 2007). The pro-inflammatory responses mediated by effector T cells are regulated by inducible Tregs. Tregs are characterized by the expression of the transcription factor Foxp3 and secrete mainly IL-10 and TGF-β, which suppress activation of the immune system and thereby maintain immune system homeostasis and tolerance to self-antigens (Powrie et al. 2003). The intestinal mucosa contains numerous DCs that are specialized in different functions depending on their location and subtype. In the following sections, we describe their main characteristics and roles in regulating immunity and tolerance.

4.2 Lamina Propria Dendritic Cells

In the LP, $CD11c^+$ DCs are divided in two major subsets depending on the expression of the CD103 receptor. These two subsets are derived from two different blood cell precursors (Bogunovic et al. 2009; Varol et al. 2009) and have different functions. The $CD103^-$ DCs differentiate from blood monocytes and are typically in close contact with the epithelium. They can also be distinguished from the $CD103^+$ DC subset for the expression of CX3CR1, the receptor for fractalkine, which is expressed on the basal membrane of polarized epithelial cells. The $CX3CR1^+$ DCs can sample luminal antigens and microorganisms directly by extending dendrites through the epithelial cell junctions to gain access to the luminal content (Niess et al. 2005; Rescigno et al. 2001). Mice deficient for CX3CR1 are more susceptible to the entero-invasive pathogen *S. typhimurium* (Niess et al. 2005). The formation of trans-epithelial dendrites is dependent on the presence of the microbiota and on TLR signaling in epithelial cells (Niess et al. 2005; Chieppa et al. 2006). The $CX3CR1^+$ DCs also express CD70 and can promote Th17 differentiation from naive T cells in response to bacterial ATP or TLR5 signaling triggered by flagellin (Atarashi et al. 2008; Uematsu et al. 2006). This is in agreement with the finding that germ-free mice lack Th17 cells and that colonization by segmented filamentous bacteria or administration of bacterial ATP induces Th17 cells (Atarashi et al. 2008; Gaboriau-Routhiau et al. 2009).

The other major subset of LP DCs is the CD103$^+$ CX3CR1$^-$ cells that promote the development of inducible Foxp3$^+$ T cells (Coombes et al. 2007). The migratory CD103$^+$ DCs found in the MLNs express the integrin CCR7 while those in the LP are CCR7 negative. The Tregs generated in the MLNs upregulate the expression of the homing receptor CCR9 an alpha4 beta7 which is required for homing of these cells back to the mucosal tissues. The tolerogenic property of CD103$^+$ LP DCs is determined by their local tissue environment and in particular by epithelial derived TGF-β, RA and in humans, also thymic stromal lymphopoietin (TSLP) (Rimoldi et al. 2005; Iliev et al. 2009). Interestingly, these factors are induced in IECs after apical interaction with bacteria in vitro. This suggests that that luminal contact with bacteria or bacterial factors can potentially modulate DC function in the LP. The CD103$^+$ DC also express the enzyme indoleamine 2,3 dioxygenase (IDO) which is required for their tolerogenic function (Matteoli et al. 2010). The immunosuppressive effects of IDO are linked to its catabolism of tryptophan and and/or production of immunomodulatory metabolites (Munn and Mellor 2007). Although these models are based on mechanistic studies in mice, the CD103$^+$ DC subset is also found in the LP and MLNs of humans. Furthermore, these DCs also produce IDO suggesting that this tolerogenic mechanism is conserved in humans and mice (Matteoli et al. 2010).

4.3 Peyer's Patch Dendritic Cells

Several DC subsets have been described in the mouse PP that perform different functions. These subsets are distinguished by the expression of CX3CR1 and the chemokine receptors CCR6 and CCR7, which direct their migration into the lymphoid tissues. In the dome area of the PP, CX3CR1$^+$ DCs are found associated with or in close proximity to the follicular epithelium. These cells can transcytose antigen but do not seem to induce T cell responses (Salazar-Gonzalez et al. 2006) suggesting that they may play a role in the maintenance of tolerance under steady state conditions.

DCs expressing CCR6, the receptor for CCL20, are found only in the PP. They are required for localized CD4$^+$ T cell activation and defense against intestinal pathogens. This was elegantly shown using CCR6 knockout mice, GFP-CCR6 reporter mice and *S. typhimurium* flagellin-specific T cell receptor transgenic mice to identify the DC subset responsible for T cell activation in the PP (Salazar-Gonzalez et al. 2006). The results were consistent with a model where CCR6$^+$ DCs migrate toward the FAE during infection to engulf transcytosed bacteria. The mature DCs then activate specific effector T cells to mediate defense against entero-invasive pathogens. The CCR6$^+$ DC subset is unique to the PP, despite the fact that IECs also produce CCL20, the chemokine for this receptor.

The PP has an important role in IgA isotype switching (discussed above) and this is mediated by a CD11b$^+$ subset of DCs which release IL-10, TGF-β, IL-6 and RA (Iwasaki and Kelsall 2001; Mora et al. 2006; Sato et al. 2003). PP DCs are also involved in the T cell-dependent IgA class switching through TNF-α and inducible nitric oxide synthase (Tezuka et al. 2007).

4.4 Gut Macrophages

The mucosal tissues contain the largest pool of tissue macrophages of the human body. They derive from blood monocytes that migrate into the tissues from the blood capillaries. Despite their potent capacity for phagocytosis and pathogen killing, the mucosal macrophages are relatively attenuated for proliferation and chemotactic activity compared to those found in other tissues (Smith et al. 2005). Additionally, in mucosal macrophages innate signaling via TLRs or NLRs does not induce potent inflammatory responses as in the case of other tissue macrophages. The intestinal macrophages are also involved in the maintenance of immunological tolerance via the induction of Tregs (Coombes et al. 2007). This is attributed to a population of $CD11b^{+}$ $F4/80^{+}$ $CD11c^{-}$ macrophages that constitutively produce high levels of IL-10 and express retinol dehydrogenases that convert retinol into RA. This population may also negatively regulate the ability of $CX3CR1^{+}$ DCs to drive inflammatory Th17 responses (Denning et al. 2007).

5 Conclusions

To peacefully coexist with the microbial inhabitants of the intestine, mammals have evolved elaborate and interconnected regulatory mechanisms. Mucus secreted by goblet cells plays a key role in keeping microbes at a distance from the epithelium where they could damage or invade the tissues, triggering inflammatory responses. Mucus gene expression is regulated by short-chain fatty acids which are bacterial metabolites produced mainly in the colon, coupling mucus production to the activities of the colonic microbes. In steady state the mucus layer is permeated by a gradient of antimicrobial factors including defensins and degradative enzymes that are constitutively produced by epithelial enterocytes or/and Paneth cells (Fig. 4). These factors can be supplemented by the inducible AMPs and other antimicrobial proteins following recognition of MAMPs by PRR expressed in epithelial cells. These mechanisms have a key role in preventing infections and in shaping the microbiota. Pathogens have evolved mechanisms to avoid killing by host defensins but it is not yet clear whether this is also the case for commensal species.

Recently, the epithelial produced defensins were shown to have a dual role in homeostasis by signaling through host receptors on epithelial cells, DCs and macrophages. In the epithelium HBD-2 stimulates restitution and barrier repair in an autocrine fashion. In addition, HBD-2 and HBD-3 have been proposed to serve

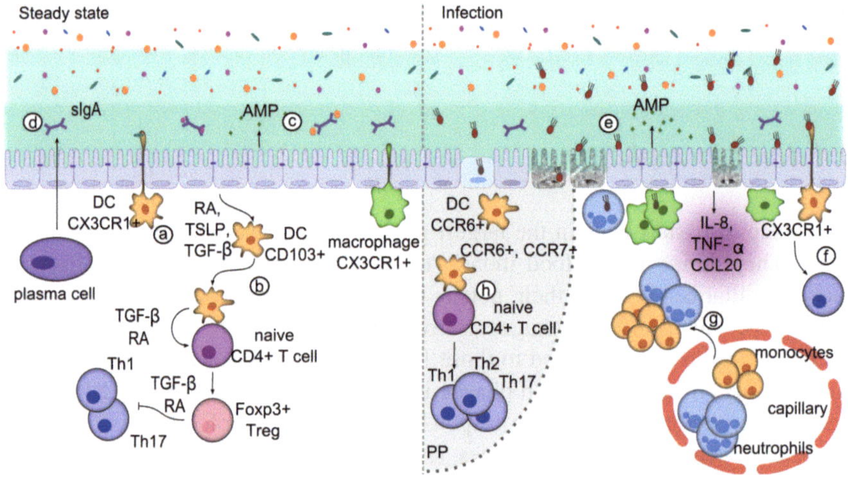

Fig. 4 Immune and epithelial mechanisms involved in the maintenance of homeostasis in steady state conditions (*left panel*) and in the response to infection (*right panel*). In steady state: **a** $CX3CR1^+$ dendritic cells (DCs) and macrophages extend their dendrites in the lumen to directly sample microbes or antigens. **b** $CD103^+$ DCs in the lamina propria (LP) acquire a tolerogenic phenotype by imprinting from tissue factors, in particular transforming growth factor (TGF)-β, retinoic acid (RA) and thymic stromal lymphopoietin (TSLP). In the draining lymph nodes, tolerogenic DCs induce the development of regulatory T cells (Tregs). **c** Intestinal epithelial cells constitutively produce antimicrobial peptides (AMPs), in particular β-defensins. **d** Plasma cells secrete secretory immunoglobulin A (sIgA) which is transported to the lumen. Together these mechanisms limit contact with microbes and create an anti-inflammatory environment. During infection by pathogenic microbes. **e** Toll-like receptor (TLR) signaling induces increased production of AMPs which also contribute to epithelial repair. **f** Pathogens colonizing the epithelium drive inflammatory responses via their interaction with $CX3CR1^+$ antigen-presenting cells. **g** TLR signaling induces chemokine secretion (interleukin-8 (IL-8) and CCL20) which induce the migration of monocytes and neutrophils to the site of infection. **h** In the Peyer's patch (PP), $CCR6^+$ DCs induce Th1 and Th17 cell polarization and adaptive T cell responses to the pathogen

as chemo-attractants for a broad spectrum of leukocytes. MBD-2 and HBD-3 have also been shown to have immunomodulatory effects on DC function. Further studies on the host effects of other defensins are therefore warranted as they may reveal novel mechanisms of homeostasis.

Epithelial cells play a key role in directing the responses in the gut and maintaining homeostasis via the sensing of microbes or MAMPs (reviewed in Wells et al. 2010). In most parts of the body, TLR signaling leads to inflammatory responses but in the polarized epithelium of the intestine several adaptations have developed to prevent uncontrolled inflammatory responses to luminal MAMPs. For example, basolateral TLR9 stimulation activates the NF-κB pathway while apical stimulation prevents NF-κB activation and confers a protective response (Lee et al. 2006). Several studies in transgenic mice have demonstrated the importance of TLR signaling and NF-kB activation to homeostasis.

In germ-free mice sIgA is produced in response to colonization by bacteria and plays a role in shaping the composition of the microbiota (Peterson et al. 2007). Nevertheless much of the luminal sIgA appears not to be reactive to commensal bacteria. Mucosal IgA production is regulated by the B cell cytokines APRIL and BAFF that are produced by IECs and resident DCs in the LP which is in turn influenced by the microbiota. In steady state sIgA prevents intimate contact of bacteria with the epithelial cells through binding to bacteria and mucus, a process known as immune exclusion (Fig. 4). Apart from its barrier properties sIgA can help maintain homeostasis by binding to antigens in the LP which are then transported to the lumen via pIgR-mediated transport through epithelial cells.

The LP is populated by $CD11c^+$ DCs from two different lineages distinguished by differential expression of CD103 and CX3CR1 (Fig. 4). At the epithelial surface $CX3CR1^+$ DCs and macrophages associated with the epithelium participate in innate immune clearance and support inflammatory Th17 T cell responses. In steady-state, the $CD103^+$ $CX3CR1^-$ DC subset play a tolerogenic role in the gut. These resident DCs are conditioned by tissue factors such as TGF-β and RA and traffic to the draining lymphoid tissue where they can induce Tregs (Fig. 4). DCs from both PPs and MLNs have the ability to imprint gut-homing properties to T cells and B cells (Mora et al. 2006; Johansson-Lindbom et al. 2003). This ensures that activated lymphocytes migrate back to mucosal sites to combat invading pathogens. Thus it appears that in steady state innate clearance of commensals is maintained in the absence of inflammatory responses due to the tolerizing properties of the resident DCs and macrophages.

Clearly, commensal bacteria are not ignored but actively controlled in an anti-inflammatory environment via many complex overlapping and intertwined mechanisms involving IECs and signals from the microbiota. The distinction between pathogen and non-pathogen is linked to the ability of pathogens to invade and cause damage to the host cells and tissues. In the case of infection, local inflammatory responses are initiated due to the recruitment of blood leukocytes by chemokines (e.g. CCL20 and IL-8) released by colonized and invaded epithelial cells (Fig. 4). Additionally, TLR signaling increases production of antimicrobial factors helping to kill invading microbes and mediate repair mechanisms through autocrine pathways (Fig. 4). The production of chemokines in the LP is essential to bypass the tolerogenic mechanisms modulating the responses of resident DCs and macrophages and induce adaptive immune responses. In the PP invading microbes also trigger the engagement of distinct DC subsets (e.g. the $CCR6^+$ DCs in the PP) which are critical to the initiation of pathogen-specific T cell responses and protection (Fig. 4). Failure of this finely tuned network of mechanisms is clearly implicated in IBD where several immunity gene defects interact with environmental factors including the microbiota. A better understanding of these mechanisms may provide new strategies to treat or prevent IBD and other gut-related diseases and disorders.

References

Abreu MT, Vora P, Faure E, Thomas LS, Arnold ET, Arditi M (2001) Decreased expression of Toll-like receptor-4 and MD-2 correlates with intestinal epithelial cell protection against dysregulated proinflammatory gene expression in response to bacterial lipopolysaccharide. J Immunol 167(3):1609–1616

Andersen-Nissen E, Smith KD, Strobe KL, Barrett SL, Cookson BT, Logan SM, Aderem A (2005) Evasion of Toll-like receptor 5 by flagellated bacteria. Proc Natl Acad Sci U S A 102(26):9247–9252. doi:0502040102[pii]10.1073/pnas.0502040102

Antar AA, Konopka JL, Campbell JA, Henry RA, Perdigoto AL, Carter BD, Pozzi A, Abel TW, Dermody TS (2009) Junctional adhesion molecule-A is required for hematogenous dissemination of reovirus. Cell Host Microbe 5(1):59–71. doi:S1931-3128(08)00398-3[pii]10.1016/j.chom.2008.12.001

Atarashi K, Nishimura J, Shima T, Umesaki Y, Yamamoto M, Onoue M, Yagita H, Ishii N, Evans R, Honda K, Takeda K (2008) ATP drives lamina propria T(H)17 cell differentiation. Nature 455(7214):808–812. doi:nature07240[pii]10.1038/nature07240

Batista FD, Harwood NE (2009) The who, how and where of antigen presentation to B cells. Nat Rev Immunol 9(1):15–27. doi:nri2454[pii]10.1038/nri2454

Batista FD, Iber D, Neuberger MS (2001) B cells acquire antigen from target cells after synapse formation. Nature 411(6836):489–494. doi:10.1038/3507809935078099[pii]

Beau I, Cotte-Laffitte J, Amsellem R, Servin AL (2007) A protein kinase A-dependent mechanism by which rotavirus affects the distribution and mRNA level of the functional tight junction-associated protein, occludin, in human differentiated intestinal Caco-2 cells. J Virol 81(16):8579–8586. doi:JVI.00263-07[pii]10.1128/JVI.00263-07

Bergelson JM (2009) Intercellular junctional proteins as receptors and barriers to virus infection and spread. Cell Host Microbe 5(6):517–521. doi:S1931-3128(09)00176-0[pii]10.1016/j.chom.2009.05.009

Bergtold A, Desai DD, Gavhane A, Clynes R (2005) Cell surface recycling of internalized antigen permits dendritic cell priming of B cells. Immunity 23(5):503–514. doi:S1074-7613(05)00309-2[pii]10.1016/j.immuni.2005.09.013

Bettelli E, Korn T, Kuchroo VK (2007) Th17: the third member of the effector T cell trilogy. Curr Opin Immunol 19(6):652–657. doi:S0952-7915(07)00133-1[pii]10.1016/j.coi.2007.07.020

Biragyn A, Surenhu M, Yang D, Ruffini PA, Haines BA, Klyushnenkova E, Oppenheim JJ, Kwak LW (2001) Mediators of innate immunity that target immature, but not mature, dendritic cells induce antitumor immunity when genetically fused with nonimmunogenic tumor antigens. J Immunol 167(11):6644–6653

Biragyn A, Belyakov IM, Chow YH, Dimitrov DS, Berzofsky JA, Kwak LW (2002) DNA vaccines encoding human immunodeficiency virus-1 glycoprotein 120 fusions with proinflammatory chemoattractants induce systemic and mucosal immune responses. Blood 100(4):1153–1159. doi:10.1182/blood-2002-01-0086

Bogunovic M, Ginhoux F, Helft J, Shang L, Hashimoto D, Greter M, Liu K, Jakubzick C, Ingersoll MA, Leboeuf M, Stanley ER, Nussenzweig M, Lira SA, Randolph GJ, Merad M (2009) Origin of the lamina propria dendritic cell network. Immunity 31(3):513–525. doi:S1074-7613(09)00368-9[pii]10.1016/j.immuni.2009.08.010

Brandl K, Plitas G, Schnabl B, DeMatteo RP, Pamer EG (2007) MyD88-mediated signals induce the bactericidal lectin RegIII gamma and protect mice against intestinal Listeria monocytogenes infection. J Exp Med 204(8):1891–1900. doi:jem.20070563[pii]10.1084/jem.20070563

Burger-van Paassen N, Vincent A, Puiman PJ, van der Sluis M, Bouma J, Boehm G, van Goudoever JB, van Seuningen I, Renes IB (2009) The regulation of intestinal mucin MUC2 expression by short-chain fatty acids: implications for epithelial protection. Biochem J 420(2):211–219. doi:BJ20082222[pii]10.1042/BJ20082222

Cario E, Podolsky DK (2000) Differential alteration in intestinal epithelial cell expression of toll-like receptor 3 (TLR3) and TLR4 in inflammatory bowel disease. Infect Immun 68(12):7010–7017

Cario E, Rosenberg IM, Brandwein SL, Beck PL, Reinecker HC, Podolsky DK (2000) Lipopolysaccharide activates distinct signaling pathways in intestinal epithelial cell lines expressing toll-like receptors. J Immunol 164(2):966–972. doi:ji_v164n2p966[pii]

Cario E, Gerken G, Podolsky DK (2007) Toll-like receptor 2 controls mucosal inflammation by regulating epithelial barrier function. Gastroenterology 132(4):1359–1374. doi:S0016-5085(07)00415-5[pii]10.1053/j.gastro.2007.02.056

Cash HL, Whitham CV, Behrendt CL, Hooper LV (2006) Symbiotic bacteria direct expression of an intestinal bactericidal lectin. Science 313(5790):1126–1130. doi:313/5790/1126[pii]10.1126/science.1127119

Chabot S, Wagner JS, Farrant S, Neutra MR (2006) TLRs regulate the gatekeeping functions of the intestinal follicle-associated epithelium. J Immunol 176(7):4275–4283. doi:176/7/4275[pii]

Chamaillard M, Hashimoto M, Horie Y, Masumoto J, Qiu S, Saab L, Ogura Y, Kawasaki A, Fukase K, Kusumoto S, Valvano MA, Foster SJ, Mak TW, Nunez G, Inohara N (2003) An essential role for NOD1 in host recognition of bacterial peptidoglycan containing diaminopimelic acid. Nat Immunol 4(7):702–707. doi:10.1038/ni945ni945[pii]

Chieppa M, Rescigno M, Huang AY, Germain RN (2006) Dynamic imaging of dendritic cell extension into the small bowel lumen in response to epithelial cell TLR engagement. J Exp Med 203(13):2841–2852. doi:jem.20061884[pii]10.1084/jem.20061884

Collier-Hyams LS, Zeng H, Sun J, Tomlinson AD, Bao ZQ, Chen H, Madara JL, Orth K, Neish AS (2002) Cutting edge: Salmonella AvrA effector inhibits the key proinflammatory, anti-apoptotic NF-kappa B pathway. J Immunol 169(6):2846–2850

Coombes JL, Siddiqui KR, Arancibia-Carcamo CV, Hall J, Sun CM, Belkaid Y, Powrie F (2007) A functionally specialized population of mucosal CD103+ DCs induces Foxp3+ regulatory T cells via a TGF-beta and retinoic acid-dependent mechanism. J Exp Med 204(8):1757–1764. doi:jem.20070590[pii]10.1084/jem.20070590

Denning TL, Wang YC, Patel SR, Williams IR, Pulendran B (2007) Lamina propria macrophages and dendritic cells differentially induce regulatory and interleukin 17-producing T cell responses. Nat Immunol 8(10):1086–1094. doi:ni1511[pii]10.1038/ni1511

Derrien M, Vaughan EE, Plugge CM, de Vos WM (2004) Akkermansia muciniphila gen. nov., sp. nov., a human intestinal mucin-degrading bacterium. Int J Syst Evol Microbiol 54 (Pt 5): 1469–1476. doi:10.1099/ijs.0.02873-054/5/1469 [pii]

Fagarasan S, Honjo T (2000) T-Independent immune response: new aspects of B cell biology. Science 290(5489):89–92. doi:8882[pii]

Fernandez MI, Pedron T, Tournebize R, Olivo-Marin JC, Sansonetti PJ, Phalipon A (2003) Anti-inflammatory role for intracellular dimeric immunoglobulin a by neutralization of lipopolysaccharide in epithelial cells. Immunity 18(6):739–749. doi:S1074761303001225[pii]

Funderburg N, Lederman MM, Feng Z, Drage MG, Jadlowsky J, Harding CV, Weinberg A, Sieg SF (2007) Human -defensin-3 activates professional antigen-presenting cells via Toll-like receptors 1 and 2. Proc Natl Acad Sci U S A 104(47):18631–18635. doi:0702130104[pii]10.1073/pnas.0702130104

Furrie E, Macfarlane S, Thomson G, Macfarlane GT (2005) Toll-like receptors-2, -3 and -4 expression patterns on human colon and their regulation by mucosal-associated bacteria. Immunology 115(4):565–574. doi:IMM2200[pii]10.1111/j.1365-2567.2005.02200.x

Gaboriau-Routhiau V, Rakotobe S, Lecuyer E, Mulder I, Lan A, Bridonneau C, Rochet V, Pisi A, De Paepe M, Brandi G, Eberl G, Snel J, Kelly D, Cerf-Bensussan N (2009) The key role of segmented filamentous bacteria in the coordinated maturation of gut helper T cell responses. Immunity 31(4):677–689. doi:S1074-7613(09)00404-X[pii]10.1016/j.immuni.2009.08.020

Garabedian EM, Roberts LJ, McNevin MS, Gordon JI (1997) Examining the role of Paneth cells in the small intestine by lineage ablation in transgenic mice. J Biol Chem 272(38):23729–23740

Garcia JR, Jaumann F, Schulz S, Krause A, Rodriguez-Jimenez J, Forssmann U, Adermann K, Kluver E, Vogelmeier C, Becker D, Hedrich R, Forssmann WG, Bals R (2001) Identification of a novel, multifunctional beta-defensin (human beta-defensin 3) with specific antimicrobial activity. Its interaction with plasma membranes of Xenopus oocytes and the induction of macrophage chemoattraction. Cell Tissue Res 306 (2):257–264. doi:10.1007/s004410100433

Gewirtz AT, Navas TA, Lyons S, Godowski PJ, Madara JL (2001) Cutting edge: bacterial flagellin activates basolaterally expressed TLR5 to induce epithelial proinflammatory gene expression. J Immunol 167(4):1882–1885

Girardin SE, Boneca IG, Carneiro LA, Antignac A, Jehanno M, Viala J, Tedin K, Taha MK, Labigne A, Zahringer U, Coyle AJ, DiStefano PS, Bertin J, Sansonetti PJ, Philpott DJ (2003a) Nod1 detects a unique muropeptide from gram-negative bacterial peptidoglycan. Science 300 (5625):1584–1587. doi:10.1126/science.1084677300/5625/1584 [pii]

Girardin SE, Boneca IG, Viala J, Chamaillard M, Labigne A, Thomas G, Philpott DJ, Sansonetti PJ (2003b) Nod2 is a general sensor of peptidoglycan through muramyl dipeptide (MDP) detection. J Biol Chem 278 (11):8869–8872. doi:10.1074/jbc.C200651200

Guerry P (2007) Campylobacter flagella: not just for motility. Trends Microbiol 15(10):456–461. doi:S0966-842X(07)00174-6[pii]10.1016/j.tim.2007.09.006

Harder J, Bartels J, Christophers E, Schroder JM (2001) Isolation and characterization of human beta -defensin-3, a novel human inducible peptide antibiotic. J Biol Chem 276(8):5707–5713. doi:10.1074/jbc.M008557200M008557200[pii]

Hart AL, Al-Hassi HO, Rigby RJ, Bell SJ, Emmanuel AV, Knight SC, Kamm MA, Stagg AJ (2005) Characteristics of intestinal dendritic cells in inflammatory bowel diseases. Gastroenterology 129(1):50–65. doi:S0016508505008838[pii]

Hooper LV, Stappenbeck TS, Hong CV, Gordon JI (2003) Angiogenins: a new class of microbicidal proteins involved in innate immunity. Nat Immunol 4(3):269–273. doi:10.1038/ni888[pii]

Hugdahl MB, Beery JT, Doyle MP (1988) Chemotactic behavior of Campylobacter jejuni. Infect Immun 56(6):1560–1566

Hugot JP, Chamaillard M, Zouali H, Lesage S, Cezard JP, Belaiche J, Almer S, Tysk C, O'Morain CA, Gassull M, Binder V, Finkel Y, Cortot A, Modigliani R, Laurent-Puig P, Gower-Rousseau C, Macry J, Colombel JF, Sahbatou M, Thomas G (2001) Association of NOD2 leucine-rich repeat variants with susceptibility to Crohn's disease. Nature 411(6837):599–603. doi:10.1038/3507910735079107[pii]

Hume ME, Nisbet DJ, DeLoach JR (1997) In vitro 14C-amino acid fermentation by CF3, a characterized continuous-flow competitive exclusion culture of caecal bacteria. J Appl Microbiol 83(2):236–242

Iliev ID, Spadoni I, Mileti E, Matteoli G, Sonzogni A, Sampietro GM, Foschi D, Caprioli F, Viale G, Rescigno M (2009) Human intestinal epithelial cells promote the differentiation of tolerogenic dendritic cells. Gut 58(11):1481–1489. doi:gut.2008.175166[pii]10.1136/gut.2008.175166

Iwasaki A, Kelsall BL (2001) Unique functions of CD11b+ , CD8 alpha+ , and double-negative Peyer's patch dendritic cells. J Immunol 166(8):4884–4890

Janssens S, Beyaert R (2002) A universal role for MyD88 in TLR/IL-1R-mediated signaling. Trends Biochem Sci 27(9):474–482. doi:S096800040202145X[pii]

Johansson ME, Phillipson M, Petersson J, Velcich A, Holm L, Hansson GC (2008) The inner of the two Muc2 mucin-dependent mucus layers in colon is devoid of bacteria. Proc Natl Acad Sci U S A 105(39):15064–15069. doi:0803124105[pii]10.1073/pnas0803124105

Johansson ME, Gustafsson JK, Sjoberg KE, Petersson J, Holm L, Sjovall H, Hansson GC (2010) Bacteria penetrate the inner mucus layer before inflammation in the dextran sulfate colitis model. PLoS One 5 (8):e12238. doi:e12238 [pii] 10.1371/journal.pone.0012238

Johansson-Lindbom B, Svensson M, Wurbel MA, Malissen B, Marquez G, Agace W (2003) Selective generation of gut tropic T cells in gut-associated lymphoid tissue (GALT): requirement for GALT dendritic cells and adjuvant. J Exp Med 198(6):963–969. doi:10.1084/jem.20031244jem.20031244[pii]

John LJ, Fromm M, Schulzke JD (2011) Epithelial barriers in intestinal inflammation. antioxid redox signal. doi:10.1089/ars.2011.3892

Kadioglu A, Weiser JN, Paton JC, Andrew PW (2008) The role of *Streptococcus pneumoniae* virulence factors in host respiratory colonization and disease. Nat Rev Microbiol 6(4):288–301. doi:nrmicro1871[pii]10.1038/nrmicro1871

Karczewski J, Troost FJ, Konings I, Dekker J, Kleerebezem M, Brummer RJ, Wells JM (2010) Regulation of human epithelial tight junction proteins by *Lactobacillus plantarum* in vivo and protective effects on the epithelial barrier. Am J Physiol Gastrointest Liver Physiol 298(6):G851–G859. doi:ajpgi.00327.2009[pii]10.1152/ajpgi.00327.2009

Kawai T, Akira S (2010) The role of pattern-recognition receptors in innate immunity: update on Toll-like receptors. Nat Immunol 11(5):373–384. doi:ni.1863[pii]10.1038/ni.1863

Keilbaugh SA, Shin ME, Banchereau RF, McVay LD, Boyko N, Artis D, Cebra JJ, Wu GD (2005) Activation of RegIIIbeta/gamma and interferon gamma expression in the intestinal tract of SCID mice: an innate response to bacterial colonisation of the gut. Gut 54(5):623–629. doi:54/5/623[pii]10.1136/gut.2004.056028

Kelly D, Campbell JI, King TP, Grant G, Jansson EA, Coutts AG, Pettersson S, Conway S (2004) Commensal anaerobic gut bacteria attenuate inflammation by regulating nuclear-cytoplasmic shuttling of PPAR-gamma and RelA. Nat Immunol 5(1):104–112. doi:10.1038/ni1018ni1018[pii]

Klotman ME, Chang TL (2006) Defensins in innate antiviral immunity. Nat Rev Immunol 6(6):447–456. doi:nri1860[pii]10.1038/nri1860

Lai Y, Gallo RL (2009) AMPed up immunity: how antimicrobial peptides have multiple roles in immune defense. Trends Immunol 30(3):131–141. doi:S1471-4906(09)00005-2[pii]10.1016/j.it.2008.12.003

Laukoetter MG, Nava P, Lee WY, Severson EA, Capaldo CT, Babbin BA, Williams IR, Koval M, Peatman E, Campbell JA, Dermody TS, Nusrat A, Parkos CA (2007) JAM-A regulates permeability and inflammation in the intestine in vivo. J Exp Med 204(13):3067–3076. doi:jem.20071416[pii]10.1084/jem.20071416

Lebeis SL, Bommarius B, Parkos CA, Sherman MA, Kalman D (2007) TLR signaling mediated by MyD88 is required for a protective innate immune response by neutrophils to *Citrobacter rodentium*. J Immunol 179(1):566–577. doi:179/1/566[pii]

Lee A, O'Rourke JL, Barrington PJ, Trust TJ (1986) Mucus colonization as a determinant of pathogenicity in intestinal infection by *Campylobacter jejuni*: a mouse cecal model. Infect Immun 51(2):536–546

Lee J, Mo JH, Katakura K, Alkalay I, Rucker AN, Liu YT, Lee HK, Shen C, Cojocaru G, Shenouda S, Kagnoff M, Eckmann L, Ben-Neriah Y, Raz E (2006) Maintenance of colonic homeostasis by distinctive apical TLR9 signaling in intestinal epithelial cells. Nat Cell Biol 8(12):1327–1336. doi:ncb1500[pii]10.1038/ncb1500

Liu AY, Destoumieux D, Wong AV, Park CH, Valore EV, Liu L, Ganz T (2002) Human beta-defensin-2 production in keratinocytes is regulated by interleukin-1, bacteria, and the state of differentiation. J Invest Dermatol 118(2):275–281. doi:1651[pii]10.1046/j.0022-202x.2001.01651.x

Macpherson AJ, Uhr T (2004) Induction of protective IgA by intestinal dendritic cells carrying commensal bacteria. Science 303(5664):1662–1665. doi:10.1126/science.1091334303/5664/1662[pii]

Macpherson AJ, Hunziker L, McCoy K, Lamarre A (2001) IgA responses in the intestinal mucosa against pathogenic and non-pathogenic microorganisms. Microbes Infect 3(12):1021–1035. doi:S1286457901014605[pii]

Macpherson AJ, McCoy KD, Johansen FE, Brandtzaeg P (2008) The immune geography of IgA induction and function. Mucosal Immunol 1(1):11–22. doi:mi20076[pii]10.1038/mi.2007.6

Maeda S, Hsu LC, Liu H, Bankston LA, Iimura M, Kagnoff MF, Eckmann L, Karin M (2005) Nod2 mutation in Crohn's disease potentiates NF-kappaB activity and IL-1beta processing. Science 307(5710):734–738. doi:307/5710/734[pii]0.1126/science.1103685

Mashimo H, Wu DC, Podolsky DK, Fishman MC (1996) Impaired defense of intestinal mucosa in mice lacking intestinal trefoil factor. Science 274(5285):262–265

Matteoli G, Mazzini E, Iliev ID, Mileti E, Fallarino F, Puccetti P, Chieppa M, Rescigno M (2010) Gut CD103 + dendritic cells express indoleamine 2, 3-dioxygenase which influences T regulatory/T effector cell balance and oral tolerance induction. Gut 59(5):595–604. doi:59/5/595[pii]10.1136/gut.2009.185108

Mazanec MB, Coudret CL, Fletcher DR (1995) Intracellular neutralization of influenza virus by immunoglobulin A anti-hemagglutinin monoclonal antibodies. J Virol 69(2):1339–1343

McDaniel TK, Jarvis KG, Donnenberg MS, Kaper JB (1995) A genetic locus of enterocyte effacement conserved among diverse enterobacterial pathogens. Proc Natl Acad Sci U S A 92(5):1664–1668
McNeil NI (1984) The contribution of the large intestine to energy supplies in man. Am J Clin Nutr 39(2):338–342
Mestecky J, Russell MW, Elson CO (1999) Intestinal IgA: novel views on its function in the defence of the largest mucosal surface. Gut 44(1):2–5
Mora JR, Iwata M, Eksteen B, Song SY, Junt T, Senman B, Otipoby KL, Yokota A, Takeuchi H, Ricciardi-Castagnoli P, Rajewsky K, Adams DH, von Andrian UH (2006) Generation of gut-homing IgA-secreting B cells by intestinal dendritic cells. Science 314(5802):1157–1160. doi:314/5802/1157[pii]10.1126/science.1132742
Mosmann TR, Coffman RL (1989) TH1 and TH2 cells: different patterns of lymphokine secretion lead to different functional properties. Annu Rev Immunol 7:145–173. doi:10.1146/annurev.iy.07.040189.001045
Munn DH, Mellor AL (2007) Indoleamine 2, 3-dioxygenase and tumor-induced tolerance. J Clin Invest 117(5):1147–1154. doi:10.1172/JCI31178
Neish AS, Gewirtz AT, Zeng H, Young AN, Hobert ME, Karmali V, Rao AS, Madara JL (2000) Prokaryotic regulation of epithelial responses by inhibition of IkappaB-alpha ubiquitination. Science 289(5484):1560–1563. doi:8795[pii]
Nenci A, Becker C, Wullaert A, Gareus R, van Loo G, Danese S, Huth M, Nikolaev A, Neufert C, Madison B, Gumucio D, Neurath MF, Pasparakis M (2007) Epithelial NEMO links innate immunity to chronic intestinal inflammation. Nature 446(7135):557–561. doi:nature05698[pii]10.1038/nature05698
Niess JH, Brand S, Gu X, Landsman L, Jung S, McCormick BA, Vyas JM, Boes M, Ploegh HL, Fox JG, Littman DR, Reinecker HC (2005) CX3CR1-mediated dendritic cell access to the intestinal lumen and bacterial clearance. Science 307(5707):254–258. doi:307/5707/254[pii]10.1126/science.1102901
O'Hara JR, Buret AG (2008) Mechanisms of intestinal tight junctional disruption during infection. Front Biosci 13:7008–7021. doi:3206[pii]
Olson TS, Reuter BK, Scott KG, Morris MA, Wang XM, Hancock LN, Burcin TL, Cohn SM, Ernst PB, Cominelli F, Meddings JB, Ley K, Pizarro TT (2006) The primary defect in experimental ileitis originates from a nonhematopoietic source. J Exp Med 203(3):541–552. doi:jem.20050407[pii]10.1084/jem.20050407
Pazgier M, Hoover DM, Yang D, Lu W, Lubkowski J (2006) Human beta-defensins. Cell Mol Life Sci 63(11):1294–1313. doi:10.1007/s00018-005-5540-2
Peterson DA, McNulty NP, Guruge JL, Gordon JI (2007) IgA response to symbiotic bacteria as a mediator of gut homeostasis. Cell Host Microbe 2(5):328–339. doi:S1931-3128(07)00246-6[pii]10.1016/j.chom.2007.09.013
Philpott DJ, McKay DM, Sherman PM, Perdue MH (1996) Infection of T84 cells with enteropathogenic Escherichia coli alters barrier and transport functions. Am J Physiol 270(4 Pt 1),: G634–G645
Podolsky DK, Gerken G, Eyking A, Cario E (2009) Colitis-associated variant of TLR2 causes impaired mucosal repair because of TFF3 deficiency. Gastroenterology 137(1):209–220. doi:S0016-5085(09)00378-3[pii]10.1053/j.gastro.2009.03.007
Powrie F, Read S, Mottet C, Uhlig H, Maloy K (2003) Control of immune pathology by regulatory T cells. Novartis Found Symp 252:92–98 discussion 98–105, 106–114
Radhakrishnan GK, Yu Q, Harms JS, Splitter GA (2009) Brucella TIR domain-containing protein mimics properties of the toll-like receptor adaptor protein TIRAP. J Biol Chem 284(15): 9892–9898. doi:M805458200[pii]10.1074/jbc.M805458200
Rakoff-Nahoum S, Paglino J, Eslami-Varzaneh F, Edberg S, Medzhitov R (2004) Recognition of commensal microflora by toll-like receptors is required for intestinal homeostasis. Cell 118(2):229–241. doi:10.1016/j.cell.2004.07.002S0092867404006610[pii]
Raschperger E, Thyberg J, Pettersson S, Philipson L, Fuxe J, Pettersson RF (2006) The coxsackie- and adenovirus receptor (CAR) is an in vivo marker for epithelial tight junctions,

with a potential role in regulating permeability and tissue homeostasis. Exp Cell Res 312(9):1566–1580. doi:S0014-4827(06)00023-1[pii]10.1016/j.yexcr.2006.01.025
Rescigno M (2010) Intestinal dendritic cells. Adv Immunol 107:109–138. doi:B978-0-12-381300-8.00004-6[pii]10.1016/B978-0-12-381300-8.00004-6
Rescigno M, Rotta G, Valzasina B, Ricciardi-Castagnoli P (2001) Dendritic cells shuttle microbes across gut epithelial monolayers. Immunobiology 204(5):572–581
Rimoldi M, Chieppa M, Salucci V, Avogadri F, Sonzogni A, Sampietro GM, Nespoli A, Viale G, Allavena P, Rescigno M (2005) Intestinal immune homeostasis is regulated by the crosstalk between epithelial cells and dendritic cells. Nat Immunol 6(5):507–514. doi:ni1192[pii]10.1038/ni1192
Rohrl J, Yang D, Oppenheim JJ, Hehlgans T (2010) Human beta-defensin 2 and 3 and their mouse orthologs induce chemotaxis through interaction with CCR2. J Immunol 184(12):6688–6694. doi:jimmunol.0903984[pii]10.4049/jimmunol.0903984
Rumio C, Besusso D, Palazzo M, Selleri S, Sfondrini L, Dubini F, Menard S, Balsari A (2004) Degranulation of paneth cells via toll-like receptor 9. Am J Pathol 165(2):373–381. doi:S0002-9440(10)63304-4[pii]10.1016/S0002-9440(10)63304-4
Salazar-Gonzalez RM, Niess JH, Zammit DJ, Ravindran R, Srinivasan A, Maxwell JR, Stoklasek T, Yadav R, Williams IR, Gu X, McCormick BA, Pazos MA, Vella AT, Lefrancois L, Reinecker HC, McSorley SJ (2006) CCR6-mediated dendritic cell activation of pathogen-specific T cells in Peyer's patches. Immunity 24(5):623–632. doi:S1074-7613(06)00215-9[pii]10.1016/j.immuni.2006.02.015
Salzman NH, Ghosh D, Huttner KM, Paterson Y, Bevins CL (2003) Protection against enteric salmonellosis in transgenic mice expressing a human intestinal defensin. Nature 422(6931):522–526. doi:10.1038/nature01520nature01520[pii]
Salzman NH, Underwood MA, Bevins CL (2007) Paneth cells, defensins, and the commensal microbiota: a hypothesis on intimate interplay at the intestinal mucosa. Semin Immunol 19(2):70–83. doi:S1044-5323(07)00032-2[pii]10.1016/j.smim.2007.04.002
Salzman NH, Hung K, Haribhai D, Chu H, Karlsson-Sjoberg J, Amir E, Teggatz P, Barman M, Hayward M, Eastwood D, Stoel M, Zhou Y, Sodergren E, Weinstock GM, Bevins CL, Williams CB, Bos NA (2010) Enteric defensins are essential regulators of intestinal microbial ecology. Nat Immunol 11(1):76–83. doi:ni.1825[pii]10.1038/ni.1825
Sato A, Hashiguchi M, Toda E, Iwasaki A, Hachimura S, Kaminogawa S (2003) CD11b + Peyer's patch dendritic cells secrete IL-6 and induce IgA secretion from naive B cells. J Immunol 171(7):3684–3690
Sato T, Vries RG, Snippert HJ, van de Wetering M, Barker N, Stange DE, van Es JH, Abo A, Kujala P, Peters PJ, Clevers H (2009) Single Lgr5 stem cells build crypt-villus structures in vitro without a mesenchymal niche. Nature 459(7244):262–265. doi:nature07935[pii]10.1038/nature07935
Schilling JD, Martin SM, Hung CS, Lorenz RG, Hultgren SJ (2003) Toll-like receptor 4 on stromal and hematopoietic cells mediates innate resistance to uropathogenic *Escherichia coli*. Proc Natl Acad Sci U S A 100(7):4203–4208. doi:10.1073/pnas.07364731000736473100[pii]
Schroeder BO, Wu Z, Nuding S, Groscurth S, Marcinowski M, Beisner J, Buchner J, Schaller M, Stange EF, Wehkamp J (2011) Reduction of disulphide bonds unmasks potent antimicrobial activity of human beta-defensin 1. Nature 469(7330):419–423. doi:nature09674[pii]10.1038/nature09674
Shroff KE, Meslin K, Cebra JJ (1995) Commensal enteric bacteria engender a self-limiting humoral mucosal immune response while permanently colonizing the gut. Infect Immun 63(10):3904–3913
Simonovic I, Rosenberg J, Koutsouris A, Hecht G (2000) Enteropathogenic *Escherichia coli* dephosphorylates and dissociates occludin from intestinal epithelial tight junctions. Cell Microbiol 2(4):305–315. doi:cmi55[pii]
Smith PD, Ochsenbauer-Jambor C, Smythies LE (2005) Intestinal macrophages: unique effector cells of the innate immune system. Immunol Rev 206:149–159. doi:IMR288[pii]10.1111/j.0105-2896.2005.00288.x

Sonnenburg JL, Angenent LT, Gordon JI (2004) Getting a grip on things: how do communities of bacterial symbionts become established in our intestine? Nat Immunol 5(6):569–573. doi:10.1038/ni1079ni1079[pii]

Sorensen OE, Thapa DR, Rosenthal A, Liu L, Roberts AA, Ganz T (2005) Differential regulation of beta-defensin expression in human skin by microbial stimuli. J Immunol 174(8): 4870–4879. doi:174/8/4870[pii]

Stappenbeck TS (2009) Paneth cell development, differentiation, and function: new molecular cues. Gastroenterology 137(1):30–33. doi:S0016-5085(09)00778-1[pii]10.1053/j.gastro.2009.05.013

Su L, Shen L, Clayburgh DR, Nalle SC, Sullivan EA, Meddings JB, Abraham C, Turner JR (2009) Targeted epithelial tight junction dysfunction causes immune activation and contributes to development of experimental colitis. Gastroenterology 136(2):551–563. doi:S0016-5085(08)01979-3[pii]10.1053/j.gastro.2008.10.081

Suzuki K, Fagarasan S (2008) How host-bacterial interactions lead to IgA synthesis in the gut. Trends Immunol 29(11):523–531. doi:S1471-4906(08)00221-4[pii]10.1016/j.it.2008.08.001

Takata T, Fujimoto S, Amako K (1992) Isolation of nonchemotactic mutants of *Campylobacter jejuni* and their colonization of the mouse intestinal tract. Infect Immun 60(9):3596–3600

Tezuka H, Abe Y, Iwata M, Takeuchi H, Ishikawa H, Matsushita M, Shiohara T, Akira S, Ohteki T (2007) Regulation of IgA production by naturally occurring TNF/iNOS-producing dendritic cells. Nature 448(7156):929–933. doi:nature06033[pii]10.1038/nature06033

Tien MT, Girardin SE, Regnault B, Le Bourhis L, Dillies MA, Coppee JY, Bourdet-Sicard R, Sansonetti PJ, Pedron T (2006) Anti-inflammatory effect of *Lactobacillus casei* on Shigella-infected human intestinal epithelial cells. J Immunol 176(2):1228–1237. doi:176/2/1228[pii]

Tsuji M, Suzuki K, Kitamura H, Maruya M, Kinoshita K, Ivanov II, Itoh K, Littman DR, Fagarasan S (2008) Requirement for lymphoid tissue-inducer cells in isolated follicle formation and T cell-independent immunoglobulin A generation in the gut. Immunity 29(2):261–271. doi:S1074-7613(08)00303-8[pii]10.1016/j.immuni.2008.05.014

Uematsu S, Jang MH, Chevrier N, Guo Z, Kumagai Y, Yamamoto M, Kato H, Sougawa N, Matsui H, Kuwata H, Hemmi H, Coban C, Kawai T, Ishii KJ, Takeuchi O, Miyasaka M, Takeda K, Akira S (2006) Detection of pathogenic intestinal bacteria by Toll-like receptor 5 on intestinal CD11c+ lamina propria cells. Nat Immunol 7(8):868–874. doi:ni1362[pii]10.1038/ni1362

Vaishnava S, Behrendt CL, Ismail AS, Eckmann L, Hooper LV (2008) Paneth cells directly sense gut commensals and maintain homeostasis at the intestinal host-microbial interface. Proc Natl Acad Sci U S A 105(52):20858–20863. doi:0808723105[pii]10.1073/pnas.0808723105

Van der Sluis M, De Koning BA, De Bruijn AC, Velcich A, Meijerink JP, Van Goudoever JB, Buller HA, Dekker J, Van Seuningen I, Renes IB, Einerhand AW (2006) Muc2-deficient mice spontaneously develop colitis, indicating that MUC2 is critical for colonic protection. Gastroenterology 131(1):117–129. doi:S0016-5085(06)00762-1[pii]10.1053/j.gastro.2006.04.020

Varol C, Vallon-Eberhard A, Elinav E, Aychek T, Shapira Y, Luche H, Fehling HJ, Hardt WD, Shakhar G, Jung S (2009) Intestinal lamina propria dendritic cell subsets have different origin and functions. Immunity 31(3):502–512. doi:S1074-7613(09)00362-8[pii]10.1016/j.immuni.2009.06.025

Viala J, Chaput C, Boneca IG, Cardona A, Girardin SE, Moran AP, Athman R, Memet S, Huerre MR, Coyle AJ, DiStefano PS, Sansonetti PJ, Labigne A, Bertin J, Philpott DJ, Ferrero RL (2004) Nod1 responds to peptidoglycan delivered by the *Helicobacter pylori* cag pathogenicity island. Nat Immunol 5(11):1166–1174. doi:ni1131[pii]10.1038/ni1131

Vijay-Kumar M, Sanders CJ, Taylor RT, Kumar A, Aitken JD, Sitaraman SV, Neish AS, Uematsu S, Akira S, Williams IR, Gewirtz AT (2007) Deletion of TLR5 results in spontaneous colitis in mice. J Clin Invest 117(12):3909–3921. doi:10.1172/JCI33084

Vongsa RA, Zimmerman NP, Dwinell MB (2009) CCR6 regulation of the actin cytoskeleton orchestrates human beta defensin-2- and CCL20-mediated restitution of colonic epithelial cells. J Biol Chem 284(15):10034–10045. doi:M805289200[pii]10.1074/jbc.M805289200

Vora P, Youdim A, Thomas LS, Fukata M, Tesfay SY, Lukasek K, Michelsen KS, Wada A, Hirayama T, Arditi M, Abreu MT (2004) Beta-defensin-2 expression is regulated by TLR signaling in intestinal epithelial cells. J Immunol 173(9):5398–5405. doi:173/9/5398[pii]

Wehkamp J, Salzman NH, Porter E, Nuding S, Weichenthal M, Petras RE, Shen B, Schaeffeler E, Schwab M, Linzmeier R, Feathers RW, Chu H, Lima H Jr, Fellermann K, Ganz T, Stange EF, Bevins CL (2005) Reduced Paneth cell alpha-defensins in ileal Crohn's disease. Proc Natl Acad Sci USA 102(50):18129–18134. doi:0505256102[pii]10.1073/pnas.0505256102

Weintraub A, Zahringer U, Wollenweber HW, Seydel U, Rietschel ET (1989) Structural characterization of the lipid A component of Bacteroides fragilis strain NCTC 9343 lipopolysaccharide. Eur J Biochem 183(2):425–431

Weller S, Braun MC, Tan BK, Rosenwald A, Cordier C, Conley ME, Plebani A, Kumararatne DS, Bonnet D, Tournilhac O, Tchernia G, Steiniger B, Staudt LM, Casanova JL, Reynaud CA, Weill JC (2004) Human blood IgM "memory" B cells are circulating splenic marginal zone B cells harboring a prediversified immunoglobulin repertoire. Blood 104(12):3647–3654. doi:10.1182/blood-2004-01-03462004-01-0346[pii]

Wells JM, Kostantinov S, Konings I, Karczewski J (2008) Effects of probiotic and commensals on epithelial barrier function. Int J Probiotics Prebiotics 3(3):127–132

Wells JM, Rossi O, Meijerink M, van Baarlen P (2010) Microbes and health sackler colloquium: epithelial crosstalk at the microbiota-mucosal interface. Proc Natl Acad Sci U S A. doi:1000092107 [pii] 10.1073/pnas.1000092107

Ye D, Ma I, Ma TY (2006) Molecular mechanism of tumor necrosis factor-alpha modulation of intestinal epithelial tight junction barrier. Am J Physiol Gastrointest Liver Physiol 290(3): G496–G504. doi:290/3/G496[pii]10.1152/ajpgi.00318.2005

Yen TH, Wright NA (2006) The gastrointestinal tract stem cell niche. Stem Cell Rev 2(3):203–212. doi:SCR:2:3:203[pii]10.1007/s12015-006-0048-1

Contribution of the Intestinal Microbiota to Human Health: From Birth to 100 Years of Age

Jing Cheng, Airi M. Palva, Willem M. de Vos and Reetta Satokari

Abstract Our intestinal tract is colonized since birth by multiple microbial species that show a characteristic succession in time. Notably the establishment of the microbiota in early life is important as it appears to impact later health. While apparently stable in healthy adults, the intestinal microbiota is changing significantly during aging. After 100 years of symbiosis marked changes have been observed that may relate to an increased level of intestinal inflammation. There is considerable interest in the microbiota in health and disease as it may provide functional biomarkers, the possibility to differentiate subjects, and avenues for interventions. This chapter reviews the present state of the art on the research to investigate the contribution of the intestinal microbiota to human health. Specific

J. Cheng · A. M. Palva · W. M. de Vos · R. Satokari (✉)
Department of Veterinary Biosciences, University of Helsinki,
P.O. Box 66 FIN-00014, Helsinki, Finland
e-mail: reetta.satokari@helsinki.fi

J. Cheng
e-mail: jing.cheng@helsinki.fi

A. M. Palva
e-mail: airi.palva@helsinki.fi

W. M. de Vos
e-mail: willem.devos@helsinki.fi

W. M. de Vos
Laboratory, Department of Bacteriology and Immunology,
Haartman Institute, University of Helsinki, Helsinki, Finland

W. M. de Vos
Laboratory of Microbiology, Wageningen University,
6703 HB, Wageningen, Netherlands

Current Topics in Microbiology and Immunology (2013) 358: 323–346
DOI: 10.1007/82_2011_189

Published Online: 19 November 2011

attention will be given to the healthy microbiota and aberrations due to disturbances such as celiac disease, irritable bowel syndrome, inflammatory bowel disease, obesity and diabetes, and non-alcoholic fatty liver disease.

Contents

Abbreviations

BF	Breast feeding
FF	Formula feeding
CeD	Celiac disease
TLRs	Toll-like receptors
NOD	Nucleotide-binding oligomerization domain containing
IBS	Irritable bowel syndrome
HCs	Healthy controls
IBD	Inflammatory bowel diease
UC	Ulcerative colitis
CD	Crohn's disease
TNBS	2,4,6-trinitrobenzenesulphonic acid
SCFA	Short chain fatty acid
BMI	Body mass index
FIAF	Fasting-induced adipose factor
LPS	Lipopolysaccharide
T1D	Type 1 diabetes
T2D	Type 2 diabetes
DGGE	Denaturing gradient gel electrophoresis
NAFLD	Non-alcoholic fatty liver disease
NASH	Non-alcoholic steatohepatitis
SIBO	Small intestinal bacterial overgrowth
LGG	*Lactobacillus rhamnosus GG*

1 Microbiota Succession From Birth to Hundred Years of Age

1.1 Establishment of Microbiota in Early Life

The prevailing assumption is that the human fetus is microbiologically sterile and the bacterial colonization starts during and after birth when the newborn comes into contact with the microbes in the birth canal and the surrounding environment. The detection of bacteria in the amniotic fluid bacterial is generally linked to the pathogenesis of preterm birth (DiGiulio et al. 2008, 2010; Zhou et al. 2010). However, the recent discoveries of bacterial DNA signatures from placenta and live bacteria in the umbilical cord blood and meconium of healthy neonates born by cesarean section suggest that exposure to low levels of microbes *in utero* may also occur during normal course of pregnancy and without pathologic consequences (Jiménez et al. 2005, 2008; Satokari et al. 2009). While *in utero* exposure to microbes may prime the infant's immune system already during fetal life, the major microbial colonization of the infant gut starts after delivery.

The bacterial colonization of the infant gut after birth is a gradual process. Typically the first colonizers are facultative anaerobic bacteria followed by strictly anaerobic genera such as *Bifidobacterium*, *Bacteroides*, *Clostridium* and *Eubacterium* (Favier et al. 2002; Palmer et al. 2007). A number of factors including the mode of delivery, gestational age, infant hospitalization, antibiotic therapy, mother's microbiota and mode of feeding during early life have an impact on the development of infant microbiota (Penders et al. 2006; Palmer et al. 2007; Collado et al. 2010; Dominguez-Bello et al. 2010). Also a significant impact of geographic location (country) on the microbiota of infants was revealed in a recent cross-European study (Fallani et al. 2010). It seems that vaginally born full term infants who are consequently exclusively breast-fed during the first months of life have the most favorable microbiota. Cesarean delivered infants may have generally delayed colonization and lower counts of *Bacteroides* and bifidobacteria and be more frequently colonized and have higher counts of *Clostridium* sp. as compared to vaginally delivered infants (Grönlund et al. 1999; Adlerberth et al. 2007; Palmer et al. 2007; Kuitunen et al. 2009). The differences in microbiota between these two groups of infants may persist up to one year of age (Grönlund et al. 1999; Adlerberth et al. 2006; Penders et al. 2006).

The mode of feeding, either breast-feeding (BF) or formula-feeding (FF) has a profound effect on the infant microbiota. Human breast milk contains 3–16 g/l of complex oligosaccharides, whereas cow's milk contains only 0.03–0.06 g/l (Kunz and Rudloff 1993). These oligosaccharides stimulate the growth of bifidobacteria and recent genomic analysis of a number of *B. longum* strains has revealed their specialization in the utilization of this source on nutrients and high level of adaptation to the infant gut (Schell et al. 2002; Sela et al. 2008; Zivkovic et al. 2011). The traditional view that the microbiota of breast-fed infants is predominated by bifidobactera still seems valid. Although some studies have not found significant differences in the bifidobacterial counts between BF and FF infants, the latter seem to be more frequently colonized also with other bacterial groups and harbor them in higher numbers (Adlerberth and Wold 2009; Fallani et al. 2010). Thus, the FF infants generally have a more mixed-type microbiota. The product development of infant formulas has improved their composition and subsequently narrowed the microbiota differences between BF and FF infants (Rinne et al. 2005; Penders et al. 2006; van der Aa et al. 2010). However, a recent study showed that BF infants have a more complex *Bifidobacterium* microbiota in terms of species and strain diversity as compared to FF infants (Roger et al. 2010). While breast milk stimulates the growth of bifidobacteria it also serves as a source of living bacteria, including bifidobacteria, to the infant gut (Martín et al. 2003, 2009; Grönlund et al. 2007; Perez et al. 2007). The common infant species of bifidobacteria include *B. breve, B. bifidum, B. longum* and *B. longum* subsp. *infantis* (Satokari et al. 2002). Further, BF is recognized to reduce the risk of not only GI tract infections (diarrhea, necrotizing enterocolitis) but also other infections such as otitis media and respiratory and urinary tract infections in infants (Hanson et al. 2002). The action of human breast milk is mediated via several mechanisms, which include providing support to the infant's immune system, stimulation of potentially protective gut microbes (bifidobacteria) and binding of milk oligosaccharides to pathogens and thus preventing infections i.e. acting as receptor analogs (Hanson et al. 2002).

During the first months and year of life individual-specific temporal patterns of bacterial colonization can be seen (Palmer et al. 2007). After weaning and introduction of solid foods the infantile microbial population gradually starts to diversify and convert to an adult-type microbiota (Palmer et al. 2007). It has been generally considered that by the age of 1–2 years of age the microbiota starts to resemble that of an adult (Mackie et al. 1999), but the actual age of microbiota stabilization has not been addressed adequately in long-term studies. Shkoporov et al. investigated fecal bifidobacteria in eight children when they were 1 and 6 years old and demonstrated a shift in the *Bifidobacterium* population from an infantile species profile to an adult-type species profile, but the actual turning point remained unspecified (Shkoporov et al. 2008). Our high-throughput microbiota profiling studies of infant microbiota by using the bacterial phylogenetic microarray HITChip (Rajilić-Stojanović et al. 2009) have shown that at the age of 1.5 years the microbiota diversity is still much lower than in adults, although all major bacterial groups are already present (Nylund et al. in preparation).

1.2 The Normal Microbiota in Adults

The complex microbial community of an adult individual typically consists of hundreds of species, but in some individuals the species richness may account in thousands (Eckburg et al. 2005; Tap et al. 2009). The estimates of the full microbial richness in the human population vary from several thousands up to 40,000 species or phylotypes (Eckburg et al. 2005; Frank et al. 2007; Tap et al. 2009).

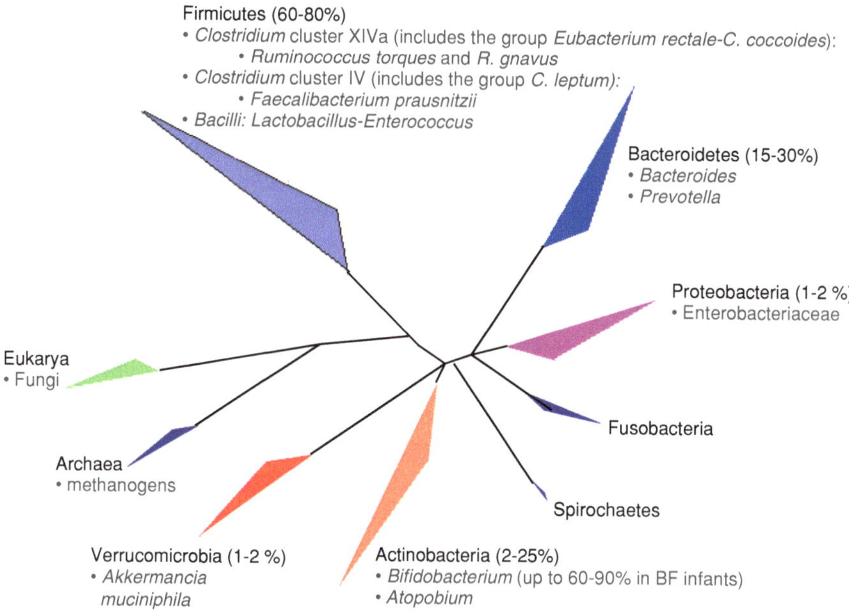

Fig. 1 The schematic presentation of a phylogenetic tree of bacteria showing the relative abundance of major phyla of human intestinal microbiota. In parenthesis, the proportion of the total microbiota in healthy adults. The major groups and genera and some species presently known to be relevant to human health are indicated—see the text. Abbreviations: BF—breast-fed

However, only approximately 1,000 can be considered as abundantly present species (Qin et al. 2010). In contrast to the enormous species diversity, the microbiota is dominated by very few phyla, the major ones being Firmicutes and Bacteroidetes, which typically constitute 60–80 and 15–30% of the total bacteria, respectively (Eckburg et al. 2005; Ley et al. 2006; Frank et al. 2007; Andersson et al. 2008; Tap et al. 2009). The phylum Actinobacteria have a share of 2–10% or even as high as 25%, while Proteobacteria and Verrumicrobia typically represent only 1–2% or less of the total microbiota (Andersson et al. 2008; Krogius-Kurikka et al. 2009; Tap et al. 2009; Jalanka-Tuovinen et al. 2011). According to a cross-European cohort study of 91 individuals the major bacterial groups or genera in the gut are *Eubacterium rectale-Clostridium coccoides* (28%), *Clostridium leptum* (25.2%), *Bacteroides* (8.5%), *Bifidobacterium* (4.4%), *Atopobium* (3.1%) and *Lactobacillus-Enterococcus* (1.3%) (Lay et al. 2005). The measurement was based on fluorescent in situ hybridization with specific 16S rRNA probes and flow cytometric counting of cells and thus, the quantification can be considered highly reliable. In this study, only a limited impact of the geographic location (country) on the microbiota composition was observed (Lay et al. 2005). The phyla to which these groups belong to are presented in Fig. 1.

There have been attempts to define a common core microbiota among people i.e. microbial phylotypes that we all share. In a recent metagenomic microbiota analysis of 142 individuals from Europe it was constituted that each individual carries at least 160 phylotypes (species), which are largely shared with other people and that 57 phylotypes can be found in more than 90% of individuals (Qin et al. 2010). The number of detected phylotypes is strongly dependent on the analysis depth and therefore the estimates on the phylogenic core still vary considerably (Tap et al. 2009; Turnbaugh et al. 2009; Qin et al. 2010; Jalanka-Tuovinen et al. 2011). Thus, the more detailed composition of the normal microbiota still remains undefined, but nevertheless several important features of the so-called normal microbiota have been established. In healthy adults the bacterial profiles of the total gut microbiota are characterized by individual-specificity and relative temporal stability (Zoetendal et al. 1998; Rajilić-Stojanović et al. 2009; Jalanka-Tuovinen et al. 2011). Further, different bacterial groups such as Actinobacteria, including bifidobacteria and *Clostridium* cluster XIVa, and the *Eubacterium rectale-Clostridium coccoides* group have also shown remarkable stability in healthy adults (Satokari et al. 2001a, b; Maukonen et al. 2006; Rajilić-Stojanović et al. 2009), but there may be large variation in the temporal behavior of different bacterial groups between individuals (Jalanka-Tuovinen et al. 2011).

1.3 Microbiota in the Old Age

When we age, our microbiota also "ages" i.e. age-related changes of microbiota occur. Naturally, the basis for the microbiota in old age is the individual-specific microbiota during adulthood and it has been demonstrated that among the elderly each individual has a unique microbial profile (Claesson et al. 2010). Despite the interindividual variation in microbiota composition, some general trends of the age-related changes of microbiota can be seen, although the age-related changes may be partly country-and population-specific. Recent high-throughput microbiota analysis studies have shown an increase in the ratio of Bacteroidetes to Firmicutes related to aging. However, while Biagi et al. (2010) reported the relative decrease in Firmicutes and no change in the Bacteroidetes, Claesson et al. (2010) noticed also a significant increase in Bacteroidetes. Earlier culture-based and lower-resolution molecular studies have shown discrepancy in the results concerning the genus *Bacteroides* (Woodmansey et al. 2004; Zwielehner et al. 2009). Within the Firmicutes phylum *Clostridium* cluster XIVa has been found to decrease in the elderly (Biagi et al. 2010; Claesson et al. 2010) and the proportion of *Clostridium* cluster IV of the total microbiota either to increase, decrease or remain unaltered (Zwielehner et al. 2009; Biagi et al. 2010; Claesson et al. 2010). Interestingly, Biagi et al. (2010) found that *Clostridium* cluster IV was subject to compositional rearrangement in the centenarians, although no quantitative cluster-level changes were observed. Within this group of bacteria a significant reduction was observed in *Faecalibacterium prausnitzii*, a species with anti-inflammatory properties in the centenarians (Biagi et al. 2010). On the other hand, the centenarians had a more than tenfold increase in *Eubacterium limosum* (*Clostridium* cluster XV), also a species with anti-inflammatory activity, which may have contributed as a balancing factor in the aged intestine (Biagi et al. 2010).

A constant finding in the elderly is that the proportion of facultative anaerobic bacteria increases (Woodmansey et al. 2004; Tiihonen et al. 2008; Biagi et al. 2010; Claesson et al. 2010). Particularly the increase in Proteobacteria, a group containing many opportunistic pathogens, may affect the health significantly. In a recent study, the increased proportion of Proteobacteria was positively correlated with the increased inflammatory status in centenarians (Biagi et al. 2010). In another study, numbers of *Enterobacteriaceae*, a family which belongs to Proteobacteria, were found to be higher in elderly with high frailty scores as compared to less frail elderly (van Tongeren et al. 2005). The age-related microbiota alterations may either contribute to the inflammatory status or be a consequence of the compromised immunity in the old age. Another group of bacteria that seems to be consistently altered in the aged people in most human populations studied is bifidobacteria. Decreased counts of bifidobacteria in the elderly have been reported frequently from both cultivation-based and molecular studies (Woodmansey et al. 2004; Mueller et al. 2006; Zwielehner et al. 2009; Biagi et al. 2010). Reduced diversity and compromised stability of the total bacterial population have been observed in the elderly and extremely old people (Rajilić-Stojanović et al. 2009; Zwielehner et al. 2009; Biagi et al. 2010). Also the compositional profiles of bifidobacteria appeared to be less stable in elderly subjects as compared to healthy adults (Rajilić-Stojanović et al. 2009; Claesson et al. 2010). The reduced stability and decrease in bifidobacteria together with the use of antibiotics are considered as main factors of the increased susceptibility of elderly to GI tract infections.

Taken together, while age-related microbiota changes are generally seen when elderly people are compared to young adults from the same country or region, there are significant country-specific differences in the bacterial groups that have found to differ and also at what age the microbiota changes start to take place (Mueller et al. 2006; Biagi et al. 2010). For example, in the recent study by Biagi et al. (2010) no microbiota changes were observed in an Italian population when young adults and old people with mean age of 73 years and no recent use of antibiotics were compared, but significant changes were found between centenarians and these two groups. In contrast, in the Claesson et al. (2010) study the Irish elderly subjects who had an average age of 77 years and no recent use of antibiotics showed significant differences in their microbiota composition as compared to young controls. This emphasizes the need of baseline studies on the effect of aging on microbiota in defined populations and community settings.

2 Microbial Disbalance and Health

There is considerable interest in the intestinal microbiota in health and disease as it may provide functional biomarkers, the possibility to differentiate subjects and avenues for interventions. The used approaches build on high-throughput and other

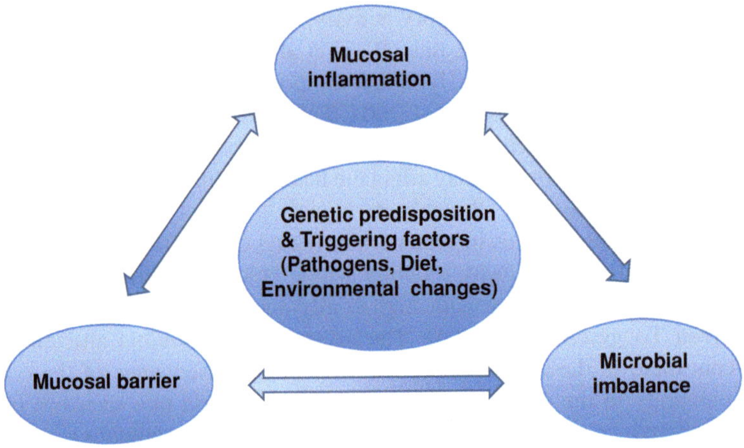

Fig. 2 Model for the relationship between microbial imbalance, mucosal barrier and inflammation. For explanation see text

molecular approaches to determine the microbiota and its function that are used to compare and contrast intestinal samples from healthy and compromised subjects (Zoetendal et al. 2008). One of the strongest disturbing factors for intestinal microbiota is the use of antibiotics. In healthy adults the microbiota composition may be restored in a relatively short period of time after the antibiotic treatment has been stopped (Dethlefsen et al. 2008). This points to a certain resilience of the microbial ecosystem as it returns to its original composition (De La Cochetière et al. 2005). As a consequence, this does not lead to permanent changes, such as in dysbiosis or imbalance (see below). However, repeated antibiotics treatments may result in incomplete recovery of the microbiota and subsequently permanent changes in its composition (Dethlefsen and Relman 2010). In infants and elderly with a less stable microbiota, the effects can be even more long-term, although they have not been studied systematically. In the elderly, antibiotics seem to fortify the age-related microbiota changes (Bartosch et al. 2004; Woodmansey et al. 2004; Claesson et al. 2010) and increase the risk of GI tract infections including *Clostridium difficile* infection.

When comparing the microbiota in healthy and compromised subjects, the concept of microbial dysbiosis or imbalance is often applied. This relates to the absence of resilience in the microbial ecosystem and results in permanent disturbances in the microbiota that contrast with the stability observed in healthy subjects (Zoetendal et al. 2008; Jalanka-Tuovinen et al. 2011). As the intestinal microbiota is highly subject-specific and complex, our databases are still limited and it is not yet possible to define the microbial imbalance in molecular terms. However, it is expected that the mining of large datasets will be instrumental in this approach, as has recently been shown for the analysis of the microbiota of over 1,000 subjects that revealed the presence of networks of specific microbial taxa

Table 1 Potential biomarkers of GI microbiota in health and disease

Disease	Association with disease	Association with health
Irritable bowel syndrome	*R. torque* like species	Bifidobacteria
Inflammatory bowel disease	*R. gnavus*, *R. torque*	*F. prausnitzii*, *A. muciniphila*
Celiac disease	–	Bifidobacteria
Metabolic syndrome-related diseases[a]	Bacteroidetes/Firmicutes ratio	

R = *Ruminococcus*
F = *Faecelibacterium*
A = *Akkermansia*
- indicates that no clear association has been reported
[a] Obesity, T1D, T2D and NAFLD

(Nikkilä and De Vos 2010). What has been observed so far is that the microbial imbalance is manifested in a decrease in protective bacteria and this results in a compromised mucosal barrier. This in turn may result in mucosal inflammation as potential pathogenic taxa are exposed that in a healthy intestine are prevented from interacting. This inflammation may in turn affect the microbial composition, leading to a vicious circle (see Fig. 2). Support for this series of events, the actual order of which may vary, stems from the observation that in the intestinal microbiota of many compromised subjects there is an increased number of bacteria that are likely to induce inflammation whereas bacterial taxa that are associated with anti-inflammatory properties are reduced. However, the studies conducted so far have generated considerable insight into the role of the intestinal microbiota as will be summarized below for major aberrations such as celiac disease, irritable bowel syndrome, inflammatory bowel disease, obesity and diabetes, and non-alcoholic fatty liver disease (Table 1).

2.1 Celiac Disease

Celiac disease (CeD) is a chronic immune-mediated inflammatory disease of the small intestine induced by intolerance to gluten. Active CeD is characterized by mucosal injury with villous atrophy affecting also the nutrient absorption and increased numbers of lymphocytes in the lamina propria (Green and Jabri 2006). CeD occurs in genetically predisposed of all ages with the initial symptoms appearing from infancy (after introduction of gluten-containing food) to old age. Typical symptoms of CeD include malabsorption or even malnutrion in severe cases as well as other gastrointestinal disorders. Moreover, extra-intestinal symptoms such as dermatitis herpetiformis (skin rash) may also occur. Individuals who carry the alleles HLA-DQ2 or HLA-DQ8 have a heightened risk of developing the disease, but only part of them eventually get CeD indicating that yet unknown genetic factors and/or environmental factors are important in the pathogenesis (Trynka et al. 2010).

Commensal microbiota is considered to be an important factor affecting the homeostasis of the gut epithelium and therefore, it has been suggested that alterations in the intestinal microbiota could play a role in the onset of celiac disease. Herein we discuss the possible role of intestinal microbiota in the onset of celiac disease in the light of results from pediatric CeD patients. Several research groups have addressed this question by comparing the microbiota composition of pediatric CeD patients and healthy controls by using both fecal and biopsy samples. While increased bacterial diversity and changes in several bacterial groups in the microbiota of Spanish pediatric CeD patients have been reported in several studies (Nadal et al. 2007; Sanz et al. 2007; Collado et al. 2009; De Palma et al. 2010), recent Scandinavian studies have failed to show major microbiota differences between children with and without CeD (Ou et al. 2009; Kalliomäki et al. 2011). However, duodenal biopsies from CeD patients born during the Swedish CeD epidemic were enriched with rod-shaped bacteria (Ou et al. 2009) and in the Finnish subjects different duodenal expression of Toll-like receptors (TLRs) and their inhibitor was found (Kalliomäki et al. 2011). Thus, these studies also indicated a possible association of microbiota with the disease. Also Italian pediatric CeD patients showed increased bacterial diversity of duodenal biopsies (Schippa et al. 2010).

The found microbiota changes may be either primary and contribute to the pathology of CeD or be a consequence of the disease. It has to be taken into account that CeD profoundly affects the morphology, physiology and immunology of the small intestinal epithelium, which thereby represents a completely different ecological niche for bacteria as compared to the normal healthy mucosa. On the other hand, a recent study showed that carriers of another genetic risk factor of CeD, a risk allele of SH2B3, have stronger activation of the Nucleotide-binding oligomerization domain containing 2 (NOD2) recognition pathway, which is important in bacterial pathogen recognition (Zhernakova et al. 2010). The combination of enhanced response to bacterial ligands and certain microbiota composition may create an immunological environment that can trigger the development of this immune-mediated disease. In this respect viral agents should also be taken into consideration (Plot and Amital 2009). Indirectly, the role of microbiota is supported by the finding that CeD children are more likely to have been born by cesarean section (Decker et al. 2010), which in turn is known to cause altered microbiota colonization process during infancy. The idea of microbes being involved in the etiology is also supported by the association found between infantile infections and the risk of developing celiac disease (Sandberg-Bennich et al. 2002) and thus, further studies are warranted to study the role of microbiota in CeD.

2.2 Irritable Bowel Syndrome

Irritable bowel syndrome (IBS) refers to a common disorder characterized by gastrointestinal (GI) dysfunction. It is not known exactly what causes IBS, but its incidence could be associated with visceral hypersensitivity, aberrant gut motility and autonomous nervous system malfunction. The interactions of these etiological factors make bowel susceptible to some risk and other factors like GI microbiota, diet, infection, hormones, or stress. In general, IBS can occur at any age, but it often begins in adolescence or early adulthood. In addition, studies show that IBS is more commonly developed in women and people who have a family history of this syndrome (Longstreth et al. 2006; Talley 2007).

Generally, IBS is characterized by chronic abdominal pain, discomfort, bloating gas as well as changes in bowel movements. Moreover, IBS patients may have complication of constipation or diarrhea, or switch between both. The above-mentioned symptoms vary from person to person, ranging from mild to severe, but most patients have mild symptoms. Unlike more serious intestinal disease such as IBD, IBS does not cause changes in bowel tissue or increase risk of colorectal cancer, but low-level inflammation has also been observed in IBS subjects (Salonen et al. 2010). According to the stool frequency, form and defaecatory symptoms, described in Rome II criteria (Drossman 2000), IBS can be subdivided into several subtypes. They are diarrhea predominant (IBS-D), constipation predominant (IBS-C) and mixed subtype (IBS-M). Recently, Rome III criteria have also been described (Longstreth et al. 2006).

Although the pathophysiology of IBS is not well determined, GI tract microbiota is suggested to be critical due to their factorial roles in IBS, as described in the following. First, with culture-based methods some species-level differences have been found in earlier studies. To date, with a variety of molecular methods alterations in the GI microbiota in IBS have also been described. For example, significant differences in the GI microbiota of the different IBS-subgroups and healthy controls (HCs) were recently described based on extensive sequencing of percentage of G+C profiled fecal bacterial DNA sample (Kassinen et al. 2007). However, no uniform compositional microbiota alterations have been defined due to the differences in the analytical power and specificity of the study methods used. Salonen et al. compared eight recent cohort studies, in which five studies were done on the same Finnish population. In these studies, the majority of the observed changes in microbial compositions occurred in Firmicutes, which is the largest GI phylum. In addition, genera *Streptococcus*, *Lactobacillus*, *Veillonella*, *Bifidobacterium*, *Clostridium* and families *Lachnospiraceae* and *Ruminococcaceae* have been identified for accounting for the differences (Salonen et al. 2010). Regarding the microbial diversity between each subtype of IBS patients and HCs, the most deviation was found in IBS-D patients, while IBS-C was the least (Rajilić-Stojanović 2007; Lyra et al. 2009).

Generally, the diversity and temporal stability of microbiota is the criteria for defining host-specific microbiota core. As a potential health-beneficial genus, bifidobacteria has been highlighted in several IBS studies. In IBS patients, reduced counts of bifidobacteria have been observed (Kerckhoffs et al. 2009; Malinen et al. 2005). Interestingly, in healthy individuals low counts of bifidobacteria was reported in the subjects suffering from abdominal pain as compared to the ones without pain (Jalanka-Tuovinen et al. 2011).

Second, certain intestinal microbiota alteration seems to be causally related with specific form of IBS. Marshall et al. (2006) found that 36% of human individuals who suffered from *Campylobacter jejuni* and *Escherichia coli* O157:H7 infection, later developed post-infectious IBS (PI-IBS). In addition, a phylotype, having 94% 16S rDNA sequence similarity with *Ruminococcus torques* and found to significantly differ between IBS-diarrhoeas (IBS-D) subjects and HCs (Kassinen et al. 2007), was also found to positively correlate with various bowel symptoms in IBS subjects in a separate study (Malinen et al. 2010). Although the IBS-related microbiota has been studied to some extent in community-level, more statistically significant comparisons need to be implemented between communities in order to find individual biomarker organisms or defined microbiota profiles associated with IBS.

Third, IBS symptoms can be improved by treatments that target the microbiota (antibiotics, probiotics, prebiotics). Recently, Kajander and coworkers demonstrated that a combination of multiple probiotics species alleviate IBS symptoms, and concurrently stabilize the intestinal microbiota (Kajander et al. 2007). It is also noteworthy that the amount of the above-mentioned *Ruminococcus torques*-related phylotype, associated with IBS-D, could be significantly lowered with this probiotics combination (Lyra et al. 2010).

2.3 Inflammatory Bowel Disease

Inflammatory bowel disease (IBD) refers to chronic diseases that cause inflammation of the intestine. IBD contains two groups of diseases: ulcerative colitis (UC) and Crohn's disease (CD). The main difference between UC and CD is the location and nature of the inflammation. In UC, the inflammation usually affects large intestine or colon, compared to terminal ileum and parts of large intestine in CD. Microscopically, UC is primarily affect mucosa, while CD targets the whole intestinal wall. The typical symptoms of IBD are abdominal pain, GI tract bleeding, malnutrition, bloody diarrhea and sometimes, extraintestinal manifestations (liver, lungs, eyes, joints).

GI microbiota and pathogenesis of IBD have a very strong relationship (Cho 2008). The role of GI microbiota in the mechanisms of IBD can be seen in the following aspects.

1. Microbial pathogens make initiation of inflammatory immune response and the disturbed GI microbiota in turn promotes inflammation leading to viscous circle of IBD and microbiota (Fig. 2). In some studies, *Mycobacterium avium spp. paratuberculosis* (MAP) has been proposed as an etiological agent (Sartor 2005), while other studies have not confirmed it (Frank et al. 2007).
2. Microbial dysbiosis in the GI tract microbiota, which leads to reduced proportion of protective species. Normally, microbiota produces butyrate and short chain fatty acid (SCFA), which are fuel for epithelial cells and thus supports

enhanced barrier function. Microbiota alteration has been constituted in several studies (Frank et al. 2007; Nell et al. 2010; Qin et al. 2010). As Frank et al. (2007) reported, both UC and CD samples contained abnormal microbiota, characterized by depletion of commensal bacteria, notably members of the phyla Firmicutes and Bacteroidetes, and relatively increased proportion of Proteobacteria, containing many opportunistic pathogens. Even if microbiota is not a causative agent, it is still likely to contribute to the disease severity.
In human studies, a reduction of a major member of Firmicutes, *Faecalibacterium prausnitzii*, was reported to associate with a higher risk of post-operative recurrence of ileal CD. A lower proportion of *F. prausnitzii* on resected ileal Crohn mucosa also has contribution to endoscopic recurrence at 6 months (Sokol et al. 2008). *F. prausnitzii* exhibits anti-inflammatory effects on cellular and 2,4,6-trinitrobenzenesulphonic acid (TNBS) colitis models, partly due to secreted metabolites able to block NF-κB activation and IL-8 production (Sokol et al. 2008). Hence, all evidence indicates that *F. prausnitzii* could be a candidate of bacteriotheraphy treatment for IBD. In addition, Png et al. (2010) found that mucolytic species *Ruminococcus gnavus* and *R. torque* were increased in normal intestinal epithelium of IBD, while *Akkermansia muciniphila,* another commensal bacterium with a more moderate mucolytic activity, was decreased many folds as compared to healthy individuals (Png et al. 2010).
3. Host genetic predisposition. The known risk genes point to the inappropriate regulation of the mucosal immune system and gut epithelial barrier (Zucchelli et al. 2009; Petnicki-Ocwieja et al. 2009). Nod2 has been found to play a critical role in regulating commensal microbiota in the intestine, which might correlate with the onset of CD (Petnicki-Ocwieja et al. 2009). Another regulator factor, PepT1, is an intestinal transporter, which is supposed to induce inflammation and NF-κB activation upon binding to Nod2. Zucchelli et al. reported its potential role for inflammation and antibacterial responses in IBD (Zucchelli et al. 2009). More recently, two new transcription factors, activating transcription factor 3 and hypoxia-inducible transcription factor 1α, have been demonstrated their potential roles in the initiation and maintenance of IBD (Pierdomenico et al. 2011).

2.4 Obesity

Obesity is characterized by elevated adiposity due to the changes in the body energy metabolism. Obesity is often associated with chronic inflammation and low-level of insulin resistance. Constant insulin resistance will lead to excessive circulating insulin concentrations over time and cause the other related syndrome such as hypertension and hyperlipidemia, and finally type 2 diabetes mellitus. Recently obesity has been reviewed as one of the most prevalent public health problems where its velocity of propagation has been increasing sharply in children.

The high incidence of obesity in children might be related to the high body mass index (BMI) at infant age (Kalliomäki et al. 2008).

Recent breakthrough observations with animal models point out the factors contributing to the development of overweight/obesity in addition to overnutrition, which are systemic low-grade inflammation and intestinal microbiota (Ley et al. 2005; Turnbaugh et al. 2006; Vijay-kumar et al. 2010). Firstly, microbes enable hydrolysis of indigestible polysaccharides to easily absorbable monosaccharides and also induce *de novo* hepatic lipogenesis (Bäckhed et al. 2004; Turnbaugh et al. 2006). Consequently, the glucose is rapidly absorbed and fatty acids are excessively stored. Both processes will increase weight. Second, microbiota in the gut epithelium might suppress a fasting-induced adipose factor (FIAF), which leads to decreased fatty acid metabolism and promote obesity (Bäckhed et al. 2004, 2007). Third, gut microbiota composition has an effect on the intestinal permeability and consequently on the leakage of lipopolysaccharide (LPS) of gut bacteria to the systemic circulation leading to low-grade endotoxemia and inflammation (Cani et al. 2008; Cani and Delzenne 2009). On the other hand, defect function of Toll-like receptor 5 (TLR5) recognizing bacterial flagellins might be also related to the induction of metabolic syndrome and obesity. TLR5 knock-out mice develop symptoms of metabolic syndrome and moreover, the symptoms can be partly transferred to wild-type germ-free mice by fecal transplantation demonstrating the involvement of the GI microbiota (Vijay-Kumar et al. 2010).

Altered microbiota composition has been found between normal and overweight individuals both in animal models and human cohort studies. In obese mice the gut microbiota contains more Firmicutes and less Bacteroidetes than in lean mice (Ley et al. 2005; Turnbaugh et al. 2006; Ley 2010). Cani and colleagues showed in a mouse model that a selective increase of *Bifidobacterium spp.* reduces the occurrence of systemic and hepatic inflammation, possibly via the process of preventing gut/intestinal permeability and bacterial translocation (Cani and Delzenne 2009). Furthermore, fecal transplantation of germ-free wild-type mice with gut microbiota of either lean or obese mice led to increased fat storage, when obese mice acted as fecal donors (Turnbaugh et al. 2006).

Similarly, gut microbiota alterations with reduced Bacteroidetes to Firmicutes ratio have also been linked to obesity in human studies (Ley et al. 2006; Turnbaugh et al. 2009). However, there are some contradictions in the obesity-related gut microbiota changes in human studies. In Turnbaugh et al. (2009), the proportion of Actinobacteria (including bifidobacteria) was increased in obese twins, but bifidobacteria were decreased in infants becoming overweight later in childhood (Kalliomäki et al. 2008). The difference might be due to the different populations (USA vs Finland) used in the studies, and also different characteristics of individuals (age, gender, BMIs). In addition, different techniques involved might also have contribution. For example, Turnbaugh et al. used high-throughput 454 pyrosequencing of V2 and V6 variable region of 16S rRNA, while Kalliomäki et al. used FISH and PCR for profiling gut microbiota composition. Ducan et al. (2008) showed significant changes of gut microbiota composition associated with weight-loss diets. They found reduced levels of *Roseburia* and *Eubacterium*, and increased levels of *Clostridium* spp., but no difference in Bacteroidetes levels in obese people on weight-loss diet. In addition, the gut microbiota profile changes were linked with the reduced production of short chain fatty acids (SCFAs). In Schwiertz et al. (2010) report, more Bacteroidetes were found in overweight ($n = 35$) and obese individuals ($n = 33$) than in lean individuals ($n = 30$), while fecal SCFA amounts were higher in obese. Thus, these studies could not confirm the altered Bacteroidetes to Firmicutes ratio to be related to human obesity, but indicate that SCFA metabolism may be an important factor (Schwiertz et al. 2010).

2.5 *Type 2 Diabetes*

Unlike type 1 diabetes (T1D), the high level of blood glucose in type 2 diabetes (T2D) patients is caused by insulin resistance. Insulin resistance is characterized by the insufficient sensitivity of the body cells that are required to respond to insulin properly. In order to keep the blood sugar level normal, the pancreas has to secrete more insulin. This would lead to excessive levels of insulin and eventually a drop in insulin production due to the dysfunction of pancreas. With the long-term high blood sugar levels, obesity and T2D can affect many major organs and may increase the risk of various health problems including renal failure and coronary artery disease (Daniels 2006).

Insulin resistance in T2D is often associated with obesity (Larsen et al. 2010). However, the cause and effect between obesity and insulin resistance remains controversial. Recent studies have suggested the mechanisms of gut microbiota in the development of obesity-linked insulin resistance, as compositional changes in gut microbiota exist between diabetic and non-diabetic individuals as well (Tsai and Coyle 2009; Diamant et al. 2010). As discussed above, the role of gut microbiota has been proposed in harvesting energy and promoting weight gain by inducing fat storage and low-grade endotoxemia and inflammation (Donath et al. 2008; Diamant et al. 2010).

Several research groups have conducted human studies linking T2D with microbiota composition. Wu et al. (2010) detected low-level *Bacteroides vulgatus* and *Bifidobacterium* genus in microbiota profiles in diabetic group by denaturing gradient gel electrophoresis (DGGE). The significant decrease of *Bifidobacterium* was also observed by real-time PCR. Slightly different results were observed by Larsen et al. (2010). In their studies, the proportions of phylum Firmicutes and class Clostridia were significantly decreased in the diabetic group compared to control group. In contrast, phylum Bacteroidetes and Proteobacteria were somehow enriched in diabetic group. The enrichment of Actinobacteria was not significantly different between groups (Larsen et al. 2010).

2.6 Type 1 Diabetes

Type 1 diabetes (T1D) accounts for around 15% of the total diabetes incidents and typically affects young children up to 14 years old. T1D is characterized by insulin deficiency due to the death of insulin-secreting beta cells of the pancreas. In the case of T1D, the beta cells are usually destroyed inappropriately by the autoantibodies that are produced by the immune system itself. T1D is therefore regarded as an autoimmune disease. However, the reason why such autoimmunity develops is largely unknown.

A number of endogenous and exogenous factors have been studied in association with the development of type 1 diabetes. There exists some evidence for genetic basis of T1D, but the general heritability is rather elusive at the population level. On the other hand, an increasing number of dietary factors, such as the shortage of D vitamin, have been suggested as potential risk factors for T1D (Knip et al. 2010).

Gut microbiota may be involved in the genesis of type 1diabetes. It seems that disturbed microbiota, compromised integrity of the intestinal epithelium and imbalanced intestinal immunity, all play an important role in the development of T1D (Vaarala et al. 2008). Although the effects of gut microbiota on the host physiology remained undetermined, recent evidence in mouse model has suggested its role on the modulation of the host innate immune system and the protective role of "normal" microbiota on the onset of T1D (Wen et al. 2008; Valladares et al. 2010). Two genera of bacteria, *Lactobacillus* and *Bifidobacterium* have been reported as dominate groups negatively correlated with the onset of T1D by comparing Biobreeding diabetes -resistant and -prone rats (Roesch et al. 2009). Recent human cohort studies support the idea that altered gut microbiota may be associated with T1D. Giongo et al. have observed a diminished diversity and stability of gut microbiota in T1D children as compared to age-matched, genotype-matched healthy individuals (Giongo et al. 2011).

2.7 Non-Alcoholic Fatty Liver Disease

Non-alcoholic fatty liver disease (NAFLD) refers to chronic liver disease that is not caused by excessive alcohol uptake. Simple steatosis (fat deposited in liver) and non-alcoholic steatohepatitis (NASH, outright inflammation developed) are the main forms. NAFLD might progress to more severe liver histological changes after NASH, ranging from fibrosis, cirrhosis and even hepatocellular carcinoma (Miele et al. 2009; Wong et al. 2010). The pathogenesis of NAFLD is still under investigation, however, genetic, metabolic, inflammatory and environmental factors have been proposed to have made contributions (Abu-Shanab and Quigley 2010).

Recent research findings point out the correlations between GI microbiota and potential etiological factors in NAFLD (for an extensive review, see Abu-Shanab and Quigley 2010). First, both mouse model and human studies have showed disease-related gut microbiota composition changes and the microbial role in extracting energy from diets, which results in increased caloric intake and ultimately fat deposition in liver and other organs as discussed above in relation to the metabolic diseases, such as obesity and T2D. Both NAFDL and NASH are linked to these diseases.

Second, choline-deficient diets were found to induce a NASH-like syndrome in patients in long-term parenteral nutrition (Buchman et al. 1995). Microbial enzymes are involved in conversion of choline to trimethylamine, which is subsequently cleared from the system (Abu-Shanab and Quigley 2010). Thus, microbiota composition and activity can affect the choline bioavailability.

Third, small intestinal bacterial overgrowth (SIBO) may promote NAFLD. SIBO is characterized by changes in bacterial composition and counts in upper gastrointestinal tract. Usually, it is not related to a single bacterial species, but the bacterial community (Bures et al. 2010). Higher prevalence of SIBO was found in NASH than in healthy controls (Wigg et al. 2001). SIBO has also been associated with increased gut permeability in NAFLD patients (Miele et al. 2009). Further, in SIBO the endogenous production of ethanol by microbiota can be increased and contributes to the disturbance of the intestinal barrier function and facilitates LPS leakage and endotoxemia (Abu-Shanab and Quigley 2010). Also LPS production is likely to be increased due to SIBO. Endotoxemia triggers an inflammatory cascade and finally leads to increased insulin resistance (Abu-Shanab and Quigley 2010). In conclusion, more investigations are warranted related to the SIBO involvement of the NAFLD pathogenesis.

3 Concluding Remarks

It is evident that the human microbiota is highly complex and dynamic. In addition, the technologies that allow us to address their composition, function and interaction with the host are just developing (Zoetendal et al. 2008). However, the great number of studies reported in recent years indicates that the interest in the impact of the intestinal microbiota is increasing rapidly and the field is experiencing a renaissance. While studies relating to antibiotic treatments result in clear cause-effect relations, the comparison of microbiota in healthy and diseased subjects will at most provide correlations or associations. However, these are of great interest as they could contribute to the diagnosis of patients, their differentiation and, ultimately, the development of therapies. Remarkable progress has been obtained with therapies that target the microbiota via the consumption of specific probiotic bacteria. For example, several pioneering studies on probiotics and weight management have shown interesting results. The consumption for 12 weeks of *Lactobacillus gasseri* SBT 2055 in fermented milk, lead to reduced body weight in healthy overweight adults (Kadooka et al. 2010). The same strain SBT2055 was shown to affect both lipid absorption and inflammation (Kadooka et al. 2010). Furthermore, the perinatal maternal application of *Lactobacillus rhamnosus GG* (LGG) may assist in preventing excessive weight gain in infant after birth (Luoto et al. 2010). The latter research group reported also that the combination of dietary counseling and probiotics was more effective than dietary counseling alone, in helping women to maintain control of glucose metabolism (Laitinen et al. 2009). These examples indicate that bacterial therapies may have potential in changing the microbiota and impacting a complex disease such as obesity.

Acknowledgments The work was supported by the Academy of Finland and The Finnish Funding Agency for Technology and Innovation (Tekes).

References

Abu-Shanab A, Quigley EMM (2010) The role of the gut microbiota in nonalcoholic fatty liver disease. Nat Rev Gastroenterol Hepatol 7:691–701

Adlerberth I, Wold AE (2009) Establishment of the gut microbiota in Western infants. Acta Paediatr 98:229–238

Adlerberth I, Lindberg E, Aberg N et al (2006) Reduced enterobacterial and increased staphylococcal colonization of the infantile bowel: an effect of hygienic lifestyle? Pediatr Res 59:96–101

Adlerberth I, Strachan DP, Matricardi PM et al (2007) Gut microbiota and development of atopic eczema in 3 European birth cohorts. J Allergy Clin Immunol 120:343–350

Andersson AF, Lindberg M, Jakobsson H et al (2008) Comparative analysis of human gut microbiota by barcoded pyrosequencing. PLoS One 3:e2836

Bäckhed F, Ding H, Wang T et al (2004) The gut microbiota as an environmental factor that regulates fat storage. Proc Natl Acad Sci USA 101:15718–15723

Bäckhed F, Manchester JK, Semenkovich CF et al (2007) Mechanisms underlying the resistance to diet-induced obesity in germ-free mice. Proc Natl Acad Sci U S A 104:979–984

Bartosch S, Fite A, Macfarlane GT et al (2004) Characterization of bacterial communities in feces from healthy elderly volunteers and hospitalized elderly patients by using real-time PCR and effects of antibiotic treatment on the fecal microbiota. Appl Environ Microbiol 70:3575–3581

Biagi E, Nylund L, Candela M et al (2010) Through ageing, and beyond: gut microbiota and inflammatory status in seniors and centenarians. PLoS One 5:e10667

Buchman AL, Dubin MD, Moukarzel AA et al (1995) Choline deficiency: a cause of hepatic steatosis during parenteral nutrition that can be reversed with intravenous choline supplementation. Hepatol 22:1399–1403

Bures J, Cyrany J, Kohoutova D et al (2010) Small intestinal bacterial overgrowth syndrome. World J Gastroenterol 16:2978–2990

Cani PD, Delzenne NM (2009) The role of the gut microbiota in energy metabolism and metabolic disease. Curr Pharma Des 15:1546–1558

Cani PD, Bibiloni R, Knauf C et al (2008) Changes in gut microbiota control metabolic endotoxemia-induced inflammation in high-fat diet-induced obesity and diabetes in mice. Diabetes 57:1470–1481

Cho JH (2008) The genetics and immunopathogenesis of inflammatory bowel disease. Nat Rev Immunol 8:458–466

Claesson MJ, Cusack S, O'Sullivan O et al (2010) Composition, variability, and temporal stability of the intestinal microbiota of the elderly. Proc Natl Acad Sci U S A. doi:10.1073/pnas.1000097107
Collado MC, Donat E, Ribes-Koninckx C et al (2009) Specific duodenal and faecal bacterial groups associated with paediatric coeliac disease. J Clin Pathol 62:264–269
Collado MC, Isolauri E, Laitinen K et al (2010) Effect of mother's weight on infant's microbiota acquisition, composition, and activity during early infancy: a prospective follow-up study initiated in early pregnancy. Am J Clin Nutr 92:1023–1030
Daniels SR (2006) The consequences of childhood overweight and obesity. Future Child 16:47–67
De La Cochetière MF, Durand T, Lepage P et al (2005) Resilience of the dominant human fecal microbiota upon short-course antibiotic challenge. J Clin Microbiol 43:5588
De Palma G, Nadal I, Medina M et al (2010) Intestinal dysbiosis and reduced immunoglobulin-coated bacteria associated with coeliac disease in children. BMC Microbiol 10:63
Decker E, Engelmann G, Findeisen A et al (2010) Cesarean delivery is associated with celiac disease but not inflammatory bowel disease in children. Pediatrics 125:e1433–e1440
Dethlefsen L, Relman DA (2010) Microbes and health sackler colloquium: incomplete recovery and individualized responses of the human distal gut microbiota to repeated antibiotic perturbation. Proc Natl Acad Sci U S A. doi:10.1073/pnas.1000087107
Dethlefsen L, Huse S, Sogin ML et al (2008) The pervasive effects of an antibiotic on the human gut microbiota, as revealed by deep 16S rRNA sequencing. PLoS Biol 6:e280
Diamant M, Blaak EE, De Vos WM (2010) DO nutrient-gut-microbiota interactions play a role in human obesity, insulin resistance and type 2 diabetes? Obes Rev 12:272–281
DiGiulio DB, Romero R, Amogan HP et al (2008) Microbial prevalence, diversity and abundance in amniotic fluid during preterm labor: a molecular and culture-based investigation. PLoS One 3:e3056
DiGiulio DB, Romero R, Kusanovic JP et al (2010) Prevalence and diversity of microbes in the amniotic fluid, the fetal inflammatory response, and pregnancy outcome in women with preterm pre-labor rupture of membranes. Am J Reprod Immunol. doi:10.1111/j.1600-0897.2010.00830.x
Dominguez-Bello MG, Costello EK, Contreras M et al (2010) Delivery mode shapes the acquisition and structure of the initial microbiota across multiple body habitats in newborns. Proc Natl Acad Sci U S A 107:11971–11975
Donath MY, Schumann DM, Faulenbach M et al (2008) Islet inflammation in Type 2 Diabetes. Diabetes Care 31(Suppl 2):161–164
Drossman DA (2000) The functional GI disorders and the Rome II process. In: Drossman DA et al (eds) The functional gastrointestinal disorders, 2nd edn. Degnon Associates, McLean
Ducan SH, Lobley GE, Holtrop G et al (2008) Human colonic microbiota associated with diet, obesity and weight loss. Int J Obe (Lond) 32:1720–1724
Eckburg PB, Bik EM, Bernstein CN et al (2005) Diversity of the human intestinal microbial flora. Science 308:1635–1638
Fallani M, Young D, Scott J et al (2010) Intestinal microbiota of 6-week-old infants across Europe: geographic influence beyond delivery mode, breast-feeding, and antibiotics. J Pediatr Gastroenterol Nutr 51:77–84
Favier CF, Vaughan EE, De Vos WM et al (2002) Molecular monitoring of succession of bacterial communities in human neonates. Appl Environ Microbiol 68:219–226
Frank DN, Amand AL, Feldman RA et al (2007) Molecualr-phylogenetic characterization of microbial community imbalances in human inflammatory bowel diseases. PNAS 104:13780–13785
Giongo A, Gano KA, Crabb DB et al (2011) Toward defining the autoimmune microbiome for type 1 diabetes. ISME J 5:82–91
Green PH, Jabri B (2006) Celiac disease. Annu Rev Med 57:207–221
Grönlund MM, Lehtonen OP, Eerola E et al (1999) Fecal microflora in healthy infants born by different methods of delivery: permanent changes in intestinal flora after cesarean delivery. Pediatr Gastroenterol Nutr 28:19–25

Grönlund MM, Gueimonde M, Laitinen K et al (2007) Maternal breast-milk and intestinal bifidobacteria guide the compositional development of the Bifidobacterium microbiota in infants at risk of allergic disease. Clin Exp Allergy 37:1764–1772
Hanson LA, Korotkova M, Håversen L et al (2002) Breast-feeding, a complex support system for the offspring. Pediatr Int 44:347–352
Jalanka-Tuovinen J, Salonen A, Nikkilä J et al (2011) Intestinal microbiota in healthy adults: temporal analysis reveals individual and common core and relation to intestinal symptoms. PLoS One 6(7):e23035
Jiménez E, Fernández L, Marín ML et al (2005) Isolation of commensal bacteria from umbilical cord blood of healthy neonates born by cesarean section. Curr Microbiol 51:270–274
Jiménez E, Marín ML, Martín R et al (2008) Is meconium from healthy newborns actually sterile? Res Microbiol 159:187–193
Kadooka Y, Sato M, Imaizumi K et al (2010) Regulation of abdominal adiposity by probiotics (*Lactobacillus gasseri* SBT 2055) in adults with obese tendencies in a randomized controlled trial. Eur J Clin Nutr 64:636–643
Kajander K, Myllyluoma E, Rajilić-Stojanović M et al (2007) Clinical trial: multispecies probiotic supplementation alleviates the symptoms of irritable bowel syndrome and stabilizes intestinal microbiota. Aliment Pharmacol Ther 27:48–57
Kalliomäki M, Collado MC, Salminen S et al (2008) Early differences in fecal microbiota composition in children may predict overweight. Am J Clin Nutr 87:534–538
Kalliomäki M, Satokari R, Lähteenoja H et al (2011) Expression of toll-like receptors and local microbiota in the duodenum of children with celiac disease (Revised version submitted)
Kassinen A, Krogius-Kurikka L, Mäkivuokko H et al (2007) The fecal microbiota of irritable bowel syndrome patients differs significantly from that of healthy subjects. Gastroenterology 133:24–33
Kerckhoffs AP, Samsom M, van der Rest ME et al (2009) Lower bifidobacteria counts in both duodenal mucosa-associated and fecal microbiota in irritable bowel syndrome patients. World J Gastroenterol 15(23):2887–2892
Knip M, Virtanen SM, Åkerblom HK (2010) Infant feeding and the risk of type 1 diabetes. Am J Clin Nutr 91(Suppl):1506S–1513S
Krogius-Kurikka L, Lyra A, Malinen E et al (2009) Microbial community analysis reveals high level phylogenetic alterations in the overall gastrointestinal microbiota of diarrhoea-predominant irritable bowel syndrome sufferers. BMC Gastroenterol. doi:10.1186/1471-230X-9-95
Kuitunen M, Kukkonen K, Juntunen-Backman K et al (2009) Probiotics prevent IgE-associated allergy until age 5 years in cesarean-delivered children but not in the total cohort. J Allergy Clin Immunol 123:335–341
Kunz C, Rudloff S (1993) Biological functions of oligosaccharides in human milk. Acta Paediatr 82:903–912
Laitinen K, Poussa T, Isolauri E et al (2009) Probiotics and dietary counseling contribute to glucose regulation during and after pregnancy: a randomized controlled trial. Br J Nutr 101:1679–1687
Larsen N, Vogensen FK, van den Berg FW et al (2010) Gut microbiota in human adults with type 2 diabetes differs from non-diabetic adults. PLoS One 5:1–10
Lay C, Rigottier-Gois L, Holmstrøm K et al (2005) Colonic microbiota signatures across five northern European countries. Appl Environ Microbiol 71:4153–4155
Ley RE (2010) Obesity and the human microbiome. Curr Opin Gastroenterol 26:5–11
Ley RE, Bäckhed F, Turnbaugh P et al (2005) Obesity alters gut microbial ecology. PNAS 102:11070–11075
Ley RE, Turnbaugh PJ, Klein S et al (2006) Human gut microbes associated with obesity. Nature 444:1022–1023
Longstreth GF, Thompson WG, Chey WD et al (2006) Functional bowel disorders. Gastroenterology 130:1480–1491

Luoto R, Kalliomäki M, Laitinen K et al (2010) The impact of perinatal probiotic intervention on the development of overweight and obesity: follow-up study from birth to 10 years. Int J Obes (Lond) 34:1531–1537
Lyra A, Rinttilä T, Nikkilä J et al (2009) Diarrhea-predominant irritable bowel syndrome distinguishable by 16S rRNA gene phylotype quantification. World J Gastroenterol 15: 5936–5945
Lyra A, Krogius-Kurikka L, Nikkilä J et al (2010) Effect of a multispeicies probiotic supplement on quantity of irritable bowel syndrome-related intestinal microbioal phylotypes. BMC Gastroenterl. doi:10.1186/1471-230X-10-110
Mackie RI, Sghir A, Gaskins HR (1999) Developmental microbial ecology of the neonatal gastrointestinal tract. Am J Clin Nutr 69:1035S–1045S
Malinen E, Rinttilä T, Kajander K et al (2005) Analysis of the fecal microbiota of irritable bowel syndrome patients and healthy controls with real-time PCR. Am J Gastroenterol 100(2): 373–382
Malinen E, Krogius-Kurikka L, Lyra A et al (2010) Association of symptoms with gastrointestinal microbiota in irritable bowel syndrome. World J Gastroenterol 16:4532–4540
Marshall JK, Thabane M, Garg AX et al (2006) Incidence and epidemiology of irritable bowel syndrome after a large waterborne outbreak of bacterial dysentery. Gastroenterology 131:445–450
Martín R, Langa S, Reviriego C et al (2003) Human milk is a source of lactic acid bacteria for the infant gut. J Pediatr 143:754–758
Martín R, Jiménez E, Heilig H et al (2009) Isolation of bifidobacteria from breast milk and assessment of the bifidobacterial population by PCR-denaturing gradient gel electrophoresis and quantitative real-time PCR. Appl Environ Microbiol 75:965–969
Maukonen J, Mättö J, Satokari R et al (2006) PCR DGGE and RT-PCR DGGE show diversity and short-term temporal stability in the Clostridium coccoides-Eubacterium rectale group in the human intestinal microbiota. FEMS Microbiol Ecol 58:517–528
Miele L, Valenza V, Torre GL et al (2009) Increased intestinal permeability and tight junction alterations in nonalcoholic fatty liver disease. Hepatology 49:1877–1887
Mueller S, Saunier K, Hanisch C et al (2006) Differences in fecal microbiota in different European study populations in relation to age, gender, and country: a cross-sectional study. Appl Environ Microbiol 72:1027–1033
Nadal I, Donat E, Ribes-Koninckx C et al (2007) Imbalance in the composition of the duodenal microbiota of children with coeliac disease. J Med Microbiol 56:1669–1674
Nell S, Suerbaum S, Josenhans C (2010) The impact of the microbiota on the pathogenesis of IBD: lessons from mouse infection models. Nature 8:564–574
Nikkilä J, de Vos WM (2010) Advanced approaches to characterize the human intestinal microbiota by computational meta-analysis. J Clin Gastroenterol 44(Suppl 1):S2–S5
Nylund L, Satokari R, Nikkilä J et al Global phylogenetic analysis of intestinal microbiota reveals marked microbial aberrancy in young children developing atopy (in preparation)
Ou G, Hedberg M, Hörstedt P et al (2009) Proximal small intestinal microbiota and identification of rod-shaped bacteria associated with childhood celiac disease. Am J Gastroenterol 104: 3058–3067
Palmer C, Bik EM, DiGiulio DB et al (2007) Development of the human infant intestinal microbiota. PLoS Biol 5:e177
Penders J, Thijs C, Vink C et al (2006) Factors influencing the composition of the intestinal microbiota in early infancy. Pediatrics 118:511–521
Perez PF, Doré J, Leclerc M et al (2007) Bacterial imprinting of the neonatal immune system: lessons from maternal cells? Pediatrics 119:e724–e732
Petnicki-Ocwieja T, Hrncir T, Liu YJ et al (2009) Nod2 is required for the regulation of commensal microbiota in the intestine. PNAS 106:15813–15818
Pierdomenico M, Stronati L, Costanzo M et al (2011) New insights into the pathogenesis of inflammatory bowel disease: transcription factors analysis in bioptic tissues from pediatric patients. JPGN 52:271–278

Plot L, Amital H (2009) Infectious associations of celiac disease. Autoimmun Rev 8:316–319
Png CW, Linden SK, Gilshenan KS et al (2010) Mucolytic bacteria with increased prevalence in IBD mucosa augment in vitro utilization of mucin by other bacteria. Am J Gastroenterol 105:2420–2428
Qin J, Li R, Raes J et al (2010) A human gut microbial gene catalogue established by metagenomic sequencing. Nature 464:59–65
Rajilić-Stojanović M (2007) Diversity of the human gastrointestinal microbiota—novel perspectives from high throughput analyses. PhD thesis, University of Wageningen, The Netherlands
Rajilić-Stojanović M, Heilig HG, Molenaar D et al (2009) Development and application of the human intestinal tract chip, a phylogenetic microarray: analysis of universally conserved phylotypes in the abundant microbiota of young and elderly adults. Environ Microbiol 11:1736–1751
Rinne M, Gueimonde M, Kalliomäki M et al (2005) Similar bifidogenic effects of prebiotic-supplemented partially hydrolyzed infant formula and breastfeeding on infant gut microbiota. FEMS Immunol Med Micr 43:59–65
Roesch LF, Lorca GL, Casella G et al (2009) Culture-independent identification of gut bacteria correlated with the onset of diabetes in a rat model. ISME J 5:536–548
Roger LC, Costabile A, Holland DT et al (2010) Examination of faecal bifidobacterium populations in breast- and formula-fed infants during the first 18 months of life. Microbiology 156:3329–3341
Salonen A, De Vos WM, Palva A (2010) Gastrointestinal microbiota in irritable bowel syndrome: present state and perspectives. Microbiology 156:3205–3215
Sandberg-Bennich S, Dahlquist G, Källén B (2002) Coeliac disease is associated with intrauterine growth and neonatal infections. Acta Paediatr 91:30–33
Sanz Y, Sánchez E, Marzotto M et al (2007) Differences in faecal bacterial communities in coeliac and healthy children as detected by PCR and denaturing gradient gel electrophoresis. FEMS Immunol Med Microbiol 51:562–568
Sartor RB (2005) Does Mycobacterium avium subspeicies paratuberculosis cause Crohn's disease? GUT 54:900–903
Satokari RM, Vaughan EE, Akkermans AD et al (2001a) Polymerase chain reaction and denaturing gradient gel electrophoresis monitoring of fecal bifidobacterium populations in a prebiotic and probiotic feeding trial. Syst Appl Microbiol 24:227–231
Satokari RM, Vaughan EE, Akkermans AD et al (2001b) Bifidobacterial diversity in human feces detected by genus-specific PCR and denaturing gradient gel electrophoresis. Appl Environ Microbiol 67:504–513
Satokari RM, Vaughan EE, Favier CF et al (2002) Diversity of *Bifidobacterium* and *Lactobacillus* spp. in breast-fed and formula-fed infants as assessed by 16S rDNA sequence differences. Micr Ecol Health Dis 14:97–105
Satokari R, Grönroos T, Laitinen K et al (2009) *Bifidobacterium* and *Lactobacillus* DNA in the human placenta. Lett Appl Microbiol 48:8–12
Schell MA, Karmirantzou M, Snel B et al (2002) The genome sequence of *Bifidobacterium longum* reflects its adaptation to the human gastrointestinal tract. Proc Natl Acad Sci U S A 99:14422–14427
Schippa S, Iebba V, Barbato M et al (2010) A distinctive 'microbial signature' in celiac pediatric patients. BMC Microbiol. doi:10.1186/1471-2180-10-175
Schwiertz A, Taras D, Schäfer K et al (2010) Microbiota and SCFA in lean and overweight healthy subjects. Obesity 18:190
Sela DA, Chapman J, Adeuya A et al (2008) The genome sequence of *Bifidobacterium longum* subsp. infantis reveals adaptations for milk utilization within the infant microbiome. Proc Natl Acad Sci U S A 105:18964–18969
Shkoporov AN, Khokhlova EV, Kulagina EV et al (2008) Application of several molecular techniques to study numerically predominant *Bifidobacterium* spp. and Bacteroidales order strains in the feces of healthy children. Biosci Biotechnol Biochem 72:742–748

Sokol H, Pigneur B, Watterlot L et al (2008) *Faecalibacterium prausnitzii* is an anti-flammatory commensal bacterium identified by gut microbiota analysis of Crohn disease patients. PNAS 105:16731–16736
Talley NJ (2007) Functional gastrointestinal disorders: irritable bowel syndrome, dyspepsia, and noncardiac chest pain. In: Goldman L, Ausiello D (eds) Cecil medicine, 23rd edn. Saunders Elsevier, Philadelphia chap 139
Tap J, Mondot S, Levenez F et al (2009) Towards the human intestinal microbiota phylogenetic core. Environ Microbiol 11:2574–2584
Tiihonen K, Tynkkynen S, Ouwehand A et al (2008) The effect of ageing with and without non-steroidal anti-inflammatory drugs on gastrointestinal microbiology and immunology. Br J Nutr 100(1):130–137
Trynka G, Wijmenga C, van Heel DA (2010) A genetic perspective on coeliac disease. Trends Mol Med 16:537–550
Tsai F, Coyle WJ (2009) The microbiome and obesity: is obesity linked to out gut flora? Curr Gastroenterol Rep 11:307–313
Turnbaugh PJ, Ley RE, Mahowald MA et al (2006) An obesity-associated gut microbiome with increased capacity for energy harvest. Nature 444:1027–1031
Turnbaugh PJ, Hamady M, Ystsunenko T et al (2009) A core gut microbiome in obese and lean twins. Nature 457:480–484
Vaarala O, Atkinson MA, Neu J (2008) The "Perfect Storm" for type 1 diabetes- the complex interplay between intestinal microbiota, gut permeability, and mucosal immunity. Diabetes 57:2555–2562
Valladares R, Sankar D, Li N et al (2010) *Lactobacillus johnsonii* N6.2 mitigates the development of type 1 diabetes in BB-DP rats. PLoS One 5:1–9
van der Aa LB, Heymans HS, van Aalderen WM et al (2010) Effect of a new synbiotic mixture on atopic dermatitis in infants: a randomized-controlled trial. Clin Exp Allergy 40:795–804
van Tongeren SP, Slaets JP, Harmsen HJ et al (2005) Fecal microbiota composition and frailty. Appl Environ Microbiol 71:6438–6442
Vijay-kumar M, Aitken JD, Carvalho FA et al (2010) Metabolic syndrome and altered gut microbiota in mice lacking toll-like receptor 5. Science 328:228–231
Wen L, Ley RE, Volchkov PY et al (2008) Innate immunity and intestinal microbiota in the development of type 1 diabetes. Nature 455:1109–1113
Wigg AJ, Roberts-Thomson IC, Dymock RB et al (2001) The role of small intestinal bacterial overgrowth, intestinal permeability, endotoxaemia, and tumour necrosis factor alpha in the pathogenesis of non alcoholic steatohepatitis. Gut 48:206–211
Wong VWS, Wong GLH, Choi PCL et al (2010) Disease progression of non-alcoholic fatty liver disease: aprospective study with paried liver biopsies at 3 years. Gut 59:969–974
Woodmansey EJ, McMurdo ME, Macfarlane GT et al (2004) Comparison of compositions and metabolic activities of fecal microbiotas in young adults and in antibiotic-treated and non-antibiotic-treated elderly subjects. Appl Environ Microbiol 70:6113–6122
Wu XK, Ma CF, Han L et al (2010) Molecular characterisation of the faecal microbiota in patients with type II diabetes. Curr Microbiol 61:69–78
Zhernakova A, Elbers CC, Ferwerda B, Finnish Celiac Disease Study Group, Joosten LA, Saavalainen P, van Heel DA, Catassi C, Netea MG, Wijmenga C (2010) Evolutionary and functional analysis of celiac risk loci reveals SH2B3 as a protective factor against bacterial infection. Am J Hum Genet 86:970–977
Zhou X, Brotman RM, Gajer P et al (2010) Recent advances in understanding the microbiology of the female reproductive tract and the causes of premature birth. Infect Dis Obstet Gynecol 2010:737425 doi:10.1155/2010/737425
Zivkovic AM, German JB, Lebrilla CB et al (2011) Human milk glycobiome and its impact on the infant gastrointestinal microbiota. Proc Natl Acad Sci U S A 108(Suppl 1):4653–4658
Zoetendal EG, Akkermans AD, de Vos WM (1998) Temperature gradient gel electrophoresis analysis of 16S rRNA from human fecal samples reveals stable and host-specific communities of active bacteria. Appl Environ Microbiol 64(10):3854–3859

Zoetendal EG, Rajilic-Stojanovic M, de Vos WM (2008) High-throughput diversity and functionality analysis of the gastrointestinal tract microbiota. Gut 57:1605

Zucchelli M, Torkvist L, Bresso F et al (2009) PepT1 Oligopeptide Transporter (SLC15A1) Gene Polymorphism in Inflammatory Bowel Disease. Inflamm Bowel Dis 15:1562–1568

Zwielehner J, Liszt K, Handschur M et al (2009) Combined PCR-DGGE fingerprinting and quantitative-PCR indicates shifts in fecal population sizes and diversity of Bacteroides, bifidobacteria and Clostridium cluster IV in institutionalized elderly. Exp Gerontol 44: 440–446

Index

Current Topics in Microbiology and Immunology (2013) 358: 347–354
DOI: 10.1007/978-3-642-36560-7

FSC
www.fsc.org
MIX
Papier aus verantwortungsvollen Quellen
Paper from responsible sources
FSC® C105338

If you have any concerns about our products,
you can contact us on
ProductSafety@springernature.com

In case Publisher is established outside the EU,
the EU authorized representative is:
Springer Nature Customer Service Center GmbH
Europaplatz 3, 69115 Heidelberg, Germany

Printed by Libri Plureos GmbH
in Hamburg, Germany